Computational Methods in Applied Sciences

Volume 56

Series Editor

Eugenio Oñate, Universitat Politècnica de Catalunya, Barcelona, Spain

This series publishes monographs and carefully edited books inspired by the thematic conferences of ECCOMAS, the European Committee on Computational Methods in Applied Sciences. As a consequence, these volumes cover the fields of Mathematical and Computational Methods and Modelling and their applications to major areas such as Fluid Dynamics, Structural Mechanics, Semiconductor Modelling, Electromagnetics and CAD/CAM. Multidisciplinary applications of these fields to critical societal and technological problems encountered in sectors like Aerospace, Car and Ship Industry, Electronics, Energy, Finance, Chemistry, Medicine, Biosciences, Environmental sciences are of particular interest. The intent is to exchange information and to promote the transfer between the research community and industry consistent with the development and applications of computational methods in science and technology.

Book proposals are welcome at
Eugenio Oñate
International Center for Numerical Methods in Engineering (CIMNE)
Technical University of Catalunya (UPC)
Edificio C-1, Campus Norte UPC Gran Capitán
s/n08034 Barcelona, Spain
onate@cimne.upc.edu
www.cimne.com
or contact the publisher, Dr. Mayra Castro, mayra.castro@springer.com

Indexed in SCOPUS, Google Scholar and SpringerLink.

More information about this series at https://link.springer.com/bookseries/6899

Martti Lehto · Pekka Neittaanmäki

Editors

Cyber Security

Critical Infrastructure Protection

 Springer

Editors
Martti Lehto ⓘ
Faculty of Information Technology
University of Jyväskylä
Jyväskylä, Finland

Pekka Neittaanmäki ⓘ
Faculty of Information Technology
University of Jyväskylä
Jyväskylä, Finland

ISSN 1871-3033   ISSN 2543-0203   (electronic)
Computational Methods in Applied Sciences
ISBN 978-3-030-91295-6   ISBN 978-3-030-91293-2   (eBook)
https://doi.org/10.1007/978-3-030-91293-2

Foreword

As much of the worldwide economy has moved into cyberspace, protecting and assuring information flows over these networks have become a priority. Most networks today rely on the successive discovery of vulnerabilities and deployment of patches to maintain security. Even after patching, new vulnerabilities are often introduced in successive releases and may even be introduced by the patches themselves. The proposed defensive cyber portfolio is largely focused on changing this paradigm through a variety of methods such as heterogeneity, formal methods proofs, secure code generation, and automation. Exploration of offensive methods is essential to expand and inform defensive work.

Many sources emerge as a significant cyberthreat on every aspect of individuals and states. The size and sophistication of the nation's hacking capabilities have grown markedly over the last few years, and they have already penetrated well-defended networks while seized and destroyed sensitive data. We must anticipate that the cyberthreat may well begin to grow much more rapidly. The first requirement of developing a sound response is understanding the nature of the problem, which is the aim of this volume.

The standard approach for securing (critical) infrastructure over the past 50 years, classified as "walls and gates," has failed. There is no longer any reason to believe that a system of barriers between trusted and untrusted components with policy-mediated pass-throughs will become more successful as the future unfolds. Within the security context, widely used traditional rule-based detection methodologies, including firewalls, signatures/patterns that govern IDS/IPS, and antivirus are irrelevant for the detection of new and sophisticated malware. Malware is masked as legitimate streams and penetrates every state-of-the-art commercial barrier on the market. In the current era of data deluge, protection against cyber-attacks/penetrations becomes more critical and requires sophisticated approaches.

July 2021   Prof. Amir Averbuch
School of Computer Science
Tel Aviv University
Tel Aviv, Israel

Preface

In the cyber world, the most important threat focuses on critical infrastructure (CI). CI encompasses the structures and functions that are vital to society's uninterrupted functioning. It is comprised of physical facilities and structures as well as electronic functions and services.

The modern and efficient countermeasures against cyber-attacks need multidisciplinary scientific computing methods when we focus on the behavior of an actor from tactic, technique, and procedure (TTP) perspectives. A tactic is the broadest-level description of this behavior, while techniques give a more detailed description of behavior in the context of a tactic, and procedures are the narrowest-level, highly detailed description in the context of a technique. Computational science is a great tool to solve cyber security challenges.

In this edited volume, we have chosen contributors that will share their perspectives on cyber security in critical infrastructure from a broad perspective. This volume is focused on critical infrastructure protection. It is comprised of the latest research that researchers and scientists from different countries have discovered. The selected chapters reflect the essential contributions of these researchers and scientists who conducted a detailed analysis of the issues and challenges in cyberspace and provided novel solutions in various aspects. These research results will stimulate further research and studies in the cyber security community.

The content of this volume is organized into three parts. Part I is focused on the digital society. It addresses critical infrastructure and different forms of digitalization, such as strategic focus on cyber security, legal aspects on cyber security, citizens in digital society, and cyber security training. Part II is focused on the critical infrastructure protection in different areas of the critical infrastructure. It investigates the possibility of using new technologies to improve current cyber capability, as well as new challenges brought about by new technologies. Part III is focused on computational methods and applications in cyber environment.

The purpose of this book is to bring together academic researchers from different countries. This book is addressed to researchers, technology experts, and decision-makers in the fields of critical infrastructure protection, ranging from critical infrastructure environment and analysis to some areas like health care, electric power system, maritime, aviation, and built environment. This book also contains a strong societal view such as information influence and ethical concerns. This book presents cyber security solutions from a technology and computational methods perspective. This book is based on invited articles collected from several research programs and papers presented in different cyber security and cyber warfare conferences.

The editors would like to thank Research Assistant Marja-Leena Rantalainen for helping in the technical editing of the book. We would also express our gratitude to Ms. Mythili Settu, and Ms. Mayra Castro from Springer Nature for the project coordination and Prof. Eugenio Oñate, CIMNE Director and Editor of the series *Computational Methods in Applied Sciences* for their fair patience in receiving the material of this volume.

Jyväskylä, Finland   Martti Lehto
July 2021   Pekka Neittaanmäki

Contents

Contributors

Chris Bronk University of Houston, Houston, TX, USA

Andrei Costin University of Jyväskylä, Jyväskylä, Finland

Paula deWitte Texas A&M University, College Station, TX, USA

Virginia A. Greiman Boston University, Boston, USA

Timo Hämäläinen Faculty of Information Technology, University of Jyväskylä, Jyväskylä, Finland

Monika Hanley University of London, London, UK

Kirsi M. Helkala Norwegian Defence University College, Cyber Academy, Lillehammer, Norway

Aki-Mauri Huhtinen Finnish National Defence University, Helsinki, Finland

Aarne Hummelholm Faculty of Information Technology, University of Jyväskylä, Jyväskylä, Finland

Antti Kariluoto Faculty of Information Technology, University of Jyväskylä, Jyväskylä, Finland

Mika Karjalainen Institute of Information Technology, JAMK University of Applied Sciences, Jyväskylä, Finland

Michael Kiperberg Department of Software Engineering, Shamoon College of Engineering, Beer-Sheva, Israel

Mikko Kiviharju Finnish Defence Research Agency, FDRA, Riihimäki, Finland

Tero Kokkonen Institute of Information Technology, JAMK University of Applied Sciences, Jyväskylä, Finland

Pyry Kotilainen Faculty of Information Technology, University of Jyväskylä, Jyväskylä, Finland

Martti Lehto Faculty of Information Technology, University of Jyväskylä, Jyväskylä, Finland

Pekka Neittaanmäki Faculty of Information Technology, University of Jyväskylä, Jyväskylä, Finland

Jouni Pöyhönen Faculty of Information Technology, University of Jyväskylä, Jyväskylä, Finland

Juhani Rauhala University of Jyväskylä, Jyväskylä, Finland

Amit Resh Department of Software Engineering, Shenkar College of Engineering and Design, Ramat Gan, Israel

Carsten F. Rønnfeldt Norwegian Defence University College, Military Academy, Oslo, Norway

Tomi Salmenpää University of Jyväskylä, Jyväskylä, Finland

Miika Sartonen Finnish National Defence University, Helsinki, Finland

Jussi Simola Faculty of Information Technology, University of Jyväskylä, Jyväskylä, Finland

Petteri Simola Finnish Defence Research Agency, Tuusula, Finland

Niko Taari Institute of Information Technology, JAMK University of Applied Sciences, Jyväskylä, Finland

Petri Vähäkainu Faculty of Information Technology, University of Jyväskylä, Jyväskylä, Finland

Richard L. Wilson Philosophy and Religious Studies and Computer and Information Sciences Towson University Towson Md, Senior Research Scholar Hoffberger Center for Professional Ethics University of Baltimore Baltimore Maryland, Towson, USA

Naomi Woods Faculty of Information Technology, University of Jyväskylä, Jyväskylä, Finland

Nezer Zaidenberg School of Computer Sciences, College of Management Academic Studies, Rishon LeZion, Israel;
Faculty of Information Technology, University of Jyväskylä, Jyväskylä, Finland

Mikhail Zolotukhin Faculty of Information Technology, University of Jyväskylä, Jyväskylä, Finland

Part I
Digital Society

Chapter 1
Cyber-Attacks Against Critical Infrastructure

Martti Lehto

Abstract In the cyber world, the most important threat focuses on critical infrastructure (CI). CI encompasses the structures and functions that are vital to society's uninterrupted functioning. It comprises physical facilities and structures as well as electronic functions and services. Critical infrastructure systems comprise a heterogeneous mixture of dynamic, interactive, and non-linear elements. In recent years, attacks against critical infrastructures, critical information infrastructures and the Internet have become ever more frequent, complex and targeted because perpetrators have become more professional. Attackers can inflict damage or disrupt on physical infrastructure by infiltrating the digital systems that control physical processes, damaging specialized equipment and disrupting vital services without a physical attack. Those threats continue to evolve in complexity and sophistication.

Keywords Critical infrastructure · Cyber security · Systems of systems

1.1 Introduction

Most countries have a detailed definition regarding their critical infrastructure, including its importance to society, associated threats, its various parts and sectors, and often the continent by which it is safeguarded. The definitions have normally been published in the context of cyber security strategies. In most countries, this definition has evolved over the years to include an ever-broader range of infrastructures. National definitions differ slightly in the criteria used to define the criticality of an infrastructure. Most countries and institutions use crosscutting criteria, which cover all infrastructures in all sectors.

In the USA there are 16 critical national infrastructure sectors [117]. The United States describes the critical infrastructure as

> the physical and cyber systems and assets that are so vital to the United States that their incapacity or destruction would have a debilitating impact on our physical or economic

M. Lehto (✉)
Faculty of Information Technology, University of Jyväskylä, Jyväskylä, Finland
e-mail: martti.lehto@jyu.fi

© The Author(s), under exclusive license to Springer Nature Switzerland AG 2022
M. Lehto and P. Neittaanmäki (eds.), *Cyber Security*, Computational Methods
in Applied Sciences 56, https://doi.org/10.1007/978-3-030-91293-2_1

security or public health or safety. The nation's critical infrastructure provides the essential services that underpin American society. [40]

In the UK there are 13 Critical national infrastructure sectors. The United Kingdom States describes the critical infrastructure:

National Infrastructure consists of those facilities, systems, sites, information, people, networks, and processes necessary for a country to function and upon which daily life depends. It also includes some functions, sites and organizations which are not critical to the maintenance of essential services, but which need protection due to the potential dangers they could pose to the public in the event of an emergency (civil nuclear and chemicals sites for example). [63]

According to the definition used by Finland's National Emergency Supply Agency, critical infrastructure consists of devices, services, and IT systems that are so vital to the nation that their failure or destruction would degrade national security, the national economy, general health and safety, and the efficient functioning of the central government. Finland has identified seven vital societal functions and eight critical infrastructure areas [82, 83].

According to the EU commission green book

the critical infrastructure includes those physical resources, services, and information technology facilities, networks and infrastructure assets which, if disrupted or destroyed, would have a serious impact on the health, safety, security or economic well-being of citizens or the effective functioning of governments. [43]

There are a certain number of critical infrastructures in the community, the disruption or destruction of which would have significant cross-border impacts. This may include transboundary cross-sector effects resulting from interdependencies between interconnected infrastructures [50]. The goal of the European Programme for Critical Infrastructure Protection (EPCIP) would be to ensure that there are adequate and equal levels of protective security on critical infrastructure, minimal single points of failure and rapid tested recovery arrangements throughout the European Union [44].

In general, critical infrastructure describes the physical and cyber systems and assets that are so vital to the nation that their incapacity or destruction would have a debilitating impact on physical or economic security or public health or safety. So, the nation's critical infrastructure provides the essential services that underpin society. Figure 1.1 illustrates where dependencies and interdependencies exist in a critical infrastructure system, and highlights the existence of dependencies and the inherent and potential complexity of these relationships for infrastructures [103, p. 196].

It is possible to identify three dimensions in safeguarding CI: political, economic, and technical. The political dimension arises from different countries' shared interests in securing their CI systems and the ensuing increased cooperation. The political dimension entails national legislation and national security needs as well as associated international cooperation around these two topics. International cooperation aims to achieve analogous solutions in countries whose needs are comparable. Uniform security legislation and security policies facilitate technical cooperation,

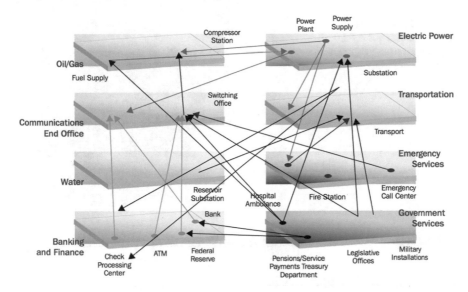

Fig. 1.1 Critical infrastructure network with its interactions

especially when several countries have shared infrastructure. The economic dimension affects all companies and business actors which build, own and administer infrastructure systems and installations, and whose operations are driven by economic interests. The economic dimension also includes a fair apportionment of security costs between the stakeholders. The technical dimension encompasses technological advances, including their utilization, and all practical solutions and measures which states, and businesses incorporate in securing the functioning of their critical infrastructure during possible disruptions [72].

The key aspects of critical national infrastructure issues in cyberspace are the Industrial Control System (ICS), Supervisory Control and Data Acquisition (SCADA) system, Distributed Control System (DCS), and Operational Technology (OT). These systems are key components of infrastructure. Industrial Control System (ICS) is an umbrella term that includes both SCADA and DCS. ICSs are the interfaces where virtual commands generate physical reality in industrial environments. SCADA systems are the software-based elements of those ICSs. ICS and SCADA systems provide real-time, two-way data flow between sensors, workstations, and other networked devices throughout a system. They allow continuous and distributed monitoring and control. DCS is a type of process control system that connects controllers, sensors, operator terminals and actuators. Operational Technology (OT) encompasses the computing systems that manage industrial operations. These systems likewise support both human-to-machine and machine-to-machine interfaces with industrial processes, often to promote efficiency and automation. Figure 1.2 illustrates the environment of ICS, SCADA, DCS and OT [108, 124].

Fig. 1.2 Environment of
ICS, SCADA, DCS and OT
(formulated from Securicon
(2017))

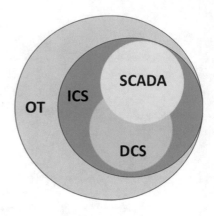

1.2   Cyber Security Threats Against Critical Infrastructure

1.2.1   Motivation of the Attackers

The global community continues to experience an increase in the scale, sophistication, and successful perpetration of cyber-attacks. As the quantity and value of electronic information has increased, so too have the efforts of criminals and other malicious actors who have embraced the Internet as a more anonymous, convenient, and profitable way of carrying out their activities. Of primary concern is the threat of organized cyber-attacks capable of causing debilitating disruptions to a nation's critical infrastructures, functions vital to society, economy, or national security [84].

Threats in cyberspace are difficult to define, as it is hard to identify the source of attacks and the motives that drive them, or even to foresee the course of an attack as it unfolds. The identification of cyber threats is further complicated by the difficulty in defining the boundaries between national, international, public, and private interests. Because threats in cyberspace are global in nature and involve rapid technological developments, the struggle to meet them is ever-changing and increasingly complicated [84].

For this study, a practical threat taxonomy based on the motivation of the attacker has been developed. The threats included in the suggested threat model are all applicable to the critical infrastructure assets presented in the chapter. The presented threat taxonomy mainly covers cyber-security threats; that is, threats applying to ICT, ICS, and SCADA assets.

One of the most common threat models is a six-fold classification based on motivational factors:

1. Cyber Vandalism,
2. Cybercrime,
3. Cyber Espionage,
4. Cyber Terrorism,
5. Cyber Sabotage, and

6.   Cyber Warfare.

With a typology such as these motives can be reduced to their very essence:

1.   Egoism,
2.   Money,
3.   Power,
4.   Paralysis, and
5.   Destruction.

This six-fold model is modified from Dunn Cavelty's structural model [4, 42].

Level 1: Cyber vandalism

Cyber vandalism encompasses cyber anarchy, hacking and hacktivism. Hackers find interfering with computer systems an enjoyable challenge. Hacktivists wish to attack companies for political or ideological motives. It is the act of damaging someone's data from the computer that in a way disrupts the victim's business or image due to editing the data into something invasive, embarrassing, or absurd.

Level 2: Cybercrime

Cyber criminals are interested in making money through fraud or from the sale of valuable information. The Commission of the European Communities defines cybercrime as "criminal acts committed using electronic communications networks and information systems or against such networks and systems" [45].

According to the Commission, cybercrime can be divided into three categories of criminal activities:

1.   Traditional forms of crime committed over electronic communication networks and information systems, such as harassment, threats, or fraud;
2.   The publication of illegal content over electronic media, e.g., child sexual abuse material or incitement to racial hatred;
3.   Crimes unique to electronic networks, e.g., network attacks, denial-of-service attacks, and hacking.

Cybercrime is a crime in which a computer or smart device is the object of a crime and/or is used to commit a crime. A cybercriminal may use a device to access a user's personal information, confidential business information, government information, or disable the device.

Level 3: Cyber espionage

Intelligence services are interested in gaining an economic, military, or political advantage for their companies, organizations or countries. So, cyber espionage can be defined as an action aimed at obtaining secret information (sensitive, proprietary, or classified) from individuals, competitors, groups, governments, and adversaries for the purpose of accruing political, military, or economic gain by employing illicit techniques on the Internet, networks, programs, or computers [87].

Level 4: Cyber terrorism

Cyber terrorism utilizes networks in attacks against critical infrastructure systems and their controls [6]. The purpose of the attacks is to cause damage and raise fear among the public, and to force the political leadership to give into the terrorists' demands. Although cyber terrorist attacks have not yet materialized, an increased level of "know-how" will arguably make them more likely to occur [115].

Level 5: Cyber sabotage

It is an activity in which an attacker (a state actor or a state sponsored group) operating below the threshold of war or executing Military Operations Other Than War (MOOTW). The goals may be to cause instability in the target country, to test one's own offensive cyber-attack capabilities, to prepare for hybrid operations, or to prepare warfare actions. An example is the Stuxnet operation. Stuxnet was a malicious computer worm, which was targeted supervisory control and data acquisition (SCADA) systems and caused substantial damage to the nuclear program of Iran [116, 128].

Level 6: Cyber warfare

No universally accepted definition for cyber warfare exists; it is quite liberally being used to describe the operations of state-actors in cyberspace. It is typically defined as an act of war using internet-enabled technology to perform an attack on a nation's digital infrastructure (civilian or military). Cyber warfare per se, requires a state of war between states, with cyber operations being but a part of other military operations (air, land, naval, space).

During the Russo-Georgian War, a series of cyber-attacks swamped and disabled websites of numerous South Ossetian, Georgian, Russian, and Azerbaijani organizations. The attacks were initiated three weeks before the shooting war began in what is regarded as the first case in history of a coordinated cyberspace domain attack synchronized with major combat actions in the other warfighting domains [71].

Cyber threats can be categorized based on the attacker's skills as follows [1]:

- Unstructured threats consisting of individuals with low or moderate skills who use easily available hacking tools;
- Structured threats by people who know system vulnerabilities and can understand, develop, and exploit codes and scripts (cyber weapons).

1.2.2 Vulnerabilities

According to the Department of Homeland Security the risk environment affecting critical infrastructure is complex and uncertain, so

> Growing interdependencies across critical infrastructure systems, particularly reliance on information and communications technologies, have increased the potential vulnerabilities to cyber threats and potential consequences resulting from the compromise of underlying

systems or networks. In an increasingly interconnected world, where critical infrastructure crosses national borders and global supply chains, the potential impacts increase with these interdependencies and the ability of a diverse set of threats to exploit them. [35]

Implementing ICS/DCS/SCADA-based cyber-physical systems into critical infrastructures brings benefits and also introduces a new set of vulnerabilities and risks to system operators and society.

Threat, vulnerability, and risk form an intertwined entirety in the cyber world. First, there is a valuable physical object, competence or some other immaterial right which needs protection and safeguarding. A threat is a harmful cyber event which may occur. The numeric value of the threat represents its degree of probability. Vulnerability can be defined as an "exploitable weakness or deficiencies in a system, device or its design that allow threat agents/actors to execute commands, access unauthorized data, and/or conduct Distributed Denial of Service (DDoS) attacks" [8]. Vulnerabilities may be the outcome of a weakness in system security procedures, software applications, policies and procedures and regulatory compliance. Vulnerability is the inherent weakness in the system which increases the probability of an occurrence or exacerbates its consequences.

Vulnerabilities can be divided into those that exist in [85].

- People's actions,
- Processes, or
- Technologies.

People like to click all the links.

Very often cybersecurity threats are due to employee errors. The Kaspersky Lab's report says that employee errors accounted for 90% of the data breaches that occurred in the cloud environment. Employees are often victims of social engineering tactics [69]. So, quite often human actors are the weakest link in cybersecurity.

Processes are key to the implementation of an effective cyber security strategy. Processes are crucial in defining how the organization's activities, roles and documentation are used to mitigate the risks to the organization's information. Process vulnerabilities among others lack written security policy, poor regulating policy, lack of security awareness and training, and poor adherence to security, lack of access control and non-existence of disaster/contingency plan.

The main goal of the cybersecurity process is to protect and preserve the confidentiality, integrity, and availability of organizational information assets [69]. But processes are nothing if people do not follow them correctly [52].

Technology solutions protect against cyber risks that may arise from network vulnerabilities but technology itself contains vulnerabilities (HW and SW). So, technological vulnerabilities are security holes in a system.

A software vulnerability is a bug in program coding, configuration, or management. A program can be an algorithm, application, operating system, or browser and control software like communication protocols and devices drives. Hackers use vulnerabilities in software attacks to force systems to give them access to unauthorized data, execute malicious code, obtain remote control, or cause the system to spread infections. The CyLab Sustainable Computing Consortium at Carnegie Mello University estimates that "commercial software has 20–30 code bugs for every 1000 lines of code" [18]. Applied Visions, Inc. estimates that 111 billion lines of new software code containing billions of vulnerabilities are coded every year [18].

A hardware vulnerability is an exploitable weakness in a computer system that enables attack through remote or physical access to system hardware. Hardware vulnerabilities are very difficult to identify. In January 2018, the entire computer industry was put on alert by two new processor vulnerabilities dubbed Meltdown and Spectre that defeated the fundamental OS security boundaries separating kernel and user space memory. The flaws stemmed from a performance feature of modern CPUs known as speculative execution [21].

The hardware vulnerabilities are:

- Semiconductor doping: the process of adding impurities to silicon-based semi-conductors to change or control their electrical properties,
- Manufacturing backdoors for malware or other penetrative purposes including embedded Radio-Frequency Identification (RFID) chips and memory,
- Manufacturing backdoors for bypassing normal authentication systems,
- Eavesdropping by gaining access to protected memory without opening other hardware,
- Hardware modification with invasive procedures, appliances, or jailbroken software,
- Counterfeiting product,
- Hardware side-channel attacks.

Since vulnerabilities can occur anywhere within the network, deploying a single-point solution will expose the system to numerous threats of attack. Solutions that can be integrated and automated into the security framework to provide distributed protection across the network are the best protection against attacks.

General vulnerabilities relate to areas that communally affect all ICT systems (i.e., individual privacy and personal data, and publicly accessible devices). This also includes vulnerabilities in commercially available mainstream IT products and systems. General vulnerabilities are also in wireless and cellular communication. For example, inadequate security protocols, inadequate authentication mechanisms, energy constrain, poor security and unreliable communication [49].

Each critical infrastructure contains specific vulnerabilities in ICS/SCADA systems. The primary causes of ICS and SCADA vulnerabilities fall into three general categories: insecure design, the human element, and configuration issues. An inse-cure design approach failed to consider the contested, interdependent, and complex environment in which these systems would operate. Poorly or negligently configured equipment provide opportunities for attackers to compromise systems that otherwise would have been secure [124].

1.2.3 Attack Vectors

In cybersecurity, an attack vector is a path or means by which an attacker can gain unauthorized access to a computer, network, or information infrastructure to deliver a payload or malicious outcome. Attack vectors allow attackers to exploit system vulnerabilities, install different types of malware and launch cyber-attacks. There are also many different attack vectors that attackers can effectively exploit to gain unauthorized access to IT infrastructure [114].

There is two major types of cyber-attack, non-targeted and targeted attacks. Non-targeted attacks are cyber-attacks that target a wide variety of targets. For example, these attacks include ransomware campaigns and non-targeted malware infections. In un-targeted attacks, attackers indiscriminately target as many devices, services, or users as possible. They do not care about who the victim is as there will be several machines or services with vulnerabilities. To do this, they use techniques that take advantage of the openness of the Internet, which include port scanning, phishing, water holing, ransomware, scanning and other malware infections [64, 80].

Targeted attacks are focused on a specific target and the attack campaign requires resources, such as skill and time. This type of threat is often known as Advanced Persistent Threat (APT). Targeted attacks that have been seen so far are focused on espionage or sabotage without destroying any infrastructure. Advancements in attack techniques show that attacks are evolving and reaching the finesse seen in attacks focusing traditional IT networks. Traditional security measures are not enough to counter these attacks as the adversary has time and skills to bypass them. Having a strong and diverse defense in action makes these attacks more time consuming and increase the probability of detection before the adversary's goal is reached [64, 80].

A cyber-attack on critical infrastructure occurs when a hacker gains access to a computer system that operates equipment in a manufacturing plant, oil pipeline, a refinery, an electric generating plant, or the like and can control the operations of that equipment to damage those assets or other property. Cyber-attacks may aim to cause disruption in the production system, resulting in unanticipated downtime, wasted production efforts, and/or ruined equipment [107].

Cyber threats vary but may include, for example, attacks that [115].

- Manipulate systems or data such as malware that exploits vulnerabilities in computer software and hardware components necessary for the operation of CIs,
- Shut down crucial systems such as DDoS attacks,
- Limit access to crucial systems or information such as through ransomware attacks.

In a targeted attack an organization is targeted because the attacker has a specific interest in the business or has been paid to target a specific organization. The groundwork for the attack could take months so that they can find the best route to deliver their exploit directly to systems (or users). A targeted attack is often more damaging than an un-targeted attack because it has been specifically tailored to attack organizations systems, processes, or personnel, in the office and sometimes at home. Targeted attacks may include spear-phishing, deploying a botnet, subverting the supply chain [64].

In general, attack vectors can be split into passive or active attacks:

- Passive: attempts to gain access or make use of information from the system but does not affect system resources, such as typosquatting, phishing and other social engineering-based attacks;
- Active: attempts to alter a system or affect its operation such as malware, exploiting unpatched vulnerabilities, email spoofing, man-in-the-middle attacks, domain hijacking and ransomware.

The most common attack vectors of cyber-attacks are among others [114]:

- Compromised credentials,
- Weak and stolen credentials,
- Using malicious insiders,
- Missing or poor encryption,
- Misconfiguration,
- Ransomware,
- Phishing and other social engineering-based attacks,
- Exploit unpatched vulnerabilities,
- Brute force attack,
- Spoofing,
- Distributed Denial of Service (DDoS),
- SQL injections,
- Trojans,
- Cross-site scripting (XSS),
- Session hijacking,
- Man-in-the-middle attacks,
- Third and fourth-party vendors.

A typical attack works like this: The cyber attacker starts by establishing a beachhead on the endpoint of the organization that they are aiming to breach. After gaining initial access and establishing persistence, the attacker escalates privileges to gain access to another system that brings them one step closer to their target. From there, the attacker can continue to move laterally until the target is reached, data is stolen, and operations are disrupted—or completely taken over. Cyber operations or attack vectors themselves do not tell about the attacker's motives and goals. For example, hacking and DDos-attacks can be used by all 1–6 level actors [97].

Example 1. Russian government cyber threat actors have been targeting the U.S. critical infrastructure sectors since at least March 2016 in a coordinated campaign of malware attacks, collectively named Dragonfly. The threat actors used a combination of spear-phishing (highly targeted emails with malicious attachments) and watering hole attacks (introducing malware through well-known industry trade publications' websites) to collect user credentials. The threat actors were able to establish footholds in the target networks and conduct network reconnaissance, move laterally, and collect information pertaining to ICSs [20].

Example 2. (Hatman, also known as TRITON and TRISIS). This attack platform targets safety controllers manufactured by a major international ICS provider. Safety controllers play an essential role in ICS environments to ensure the safe and predictable shutdown of operational equipment. Hatman malware was specifically designed to allow changes to the safety controller to introduce new functionality that would likely degrade the safety controller's ability to shut down equipment safely [20].

1.3 Cyber-Attacks Against Critical Infrastructure

Critical infrastructure can be damaged, destroyed or disrupted by deliberate acts of terrorism, natural disasters, negligence, accidents, or computer hacking, criminal activity and malicious behaviour [43]. National critical infrastructure could be targeted by hostile states, cyber criminals, terrorists, or criminals for the purposes of disruption, espionage and/or financial gain. In this chapter, the critical infrastructure taxonomy is based on the US taxonomy, in which the government facilities sector and the commercial facilities sector are combined, and the new sector is governmental institutions sector.

1.3.1 Chemical Sector

The chemical sector manufactures, stores, uses, and transports potentially dangerous chemicals. Chemical sector facilities typically belong in four key functional areas: manufacturing plants, transport systems, warehousing and storage systems, and chemical end users. Most chemical companies have internet-connected devices as part of their process control systems.

For the chemical sector, major cybersecurity issues include impacts to both IT and operational technology (OT) systems and operations due to targeted or opportunistic attacks (e.g., advanced persistent threat, distributed denial of service, or malware and ransomware), disruptions of cloud-based services, or the manipulation of industrial control systems. The sector is vulnerable to the threat of malicious actors physically or remotely manipulating network-based systems designed to control chemical manufacturing processes or process safety systems [20, 36].

Example 3. A notable attack, 'Nitro', occurred in 2011 whereby hackers used a malware called 'PoisonIvy' to steal sensitive data and information from several chemical companies throughout the U.S. [10].

Example 4. In 2017, a petrochemical facility in Saudi Arabia was attacked using Hatman. The sophisticated attack was intended to sabotage the facility's operations such that safety controls would fail, triggering an explosion. Though the attack was unsuccessful in causing an explosion or hazmat release (owing to an error in the code), the incident demonstrated how similar cyber-attacks may be used to cause physical destruction to critical infrastructure [20].

Example 5. Hexion, Momentive, and Norsk Hydro all hit by ransomware cyber-attacks. Those chemical manufacturing companies based in Norway and the US have fallen victim to ransomware attacks, after a program called LockerGoga gained access to systems, encrypted files, and disrupted operations. On 19 March 2019, the global aluminum producer Norsk Hydro was forced to shut down its plants and its worldwide network after a security breach blocked access to files and changed the passwords to user accounts across several of its corporate and production control systems. The malware issued a ransom note stating that files had been encrypted and demanding payments in bitcoin to restore access to data [110].

1.3.2 *Commercial and Government Facilities Sector*

The commercial facilities sector includes a diverse range of sites that draw large crowds of people for shopping, business, entertainment, or lodging. Facilities within this sector operate on the principle of open public access, meaning that the public can congregate and move freely without highly visible security barriers. Most of these facilities are privately owned and operated. Sector operates on the principle of open public access, meaning that the public can move freely throughout these facilities without the deterrent of highly visible security barriers [36, 39].

The government facilities sector includes a wide variety of buildings. Many government facilities are open to the public while others are not. These facilities include general-use office buildings and special-use military installations, embassies, courthouses, national laboratories, and structures that may house critical equipment, systems, networks, and functions [36] The education facilities subsector encompasses early childhood, pre-primary, basic, upper secondary, vocational, and higher education facilities (public, and private).

Cyber intrusions into automated security and supervisory control and data acquisition systems are risks. The increasing reliance on automated security systems and automated building management systems will likely increase vulnerabilities and the likelihood of cyber intrusion, especially in the form of sabotage by current or former insiders with malicious intent. Cyber intrusion into the security systems of government facilities could compromise the protection of facilities, civil servants, and the public and allow for exploitation and attacks with significant consequences [36].

Higher education institutions often collect and store sensitive, personal student data and databases (identity numbers, health, financial, and educational data). Disruptions to institutional data systems could impact the capacity to effectively perform essential business operations and could cause a temporary to long-term school closure. Although a cyber-attack on an education facility would not likely impose cascading effects for the nation, it can have such effects on the campus community through the compromising of personal data, security systems, and research facilities that rely on cyber elements or of emergency management data housed electronically [36].

Example 6. In 2011, two research labs, Pacific Northwest National Laboratory (PNNL) and Thomas Jefferson National Laboratory in Newport News, Virginia were victims of a cyber-attacks. The attacks eventually caused these labs to shut down all internet access and website access for a couple days [53].

1.3.3   Communications Sector

The communications sector is an integral component of the economy, underlying the operations of all businesses, public safety organizations, and government. The communications sector is comprised of telecommunications, internet, postal services, and broadcast. The sector provides services in terrestrial, satellite, and wireless transmission systems. The transmission of these services has become extremely interconnected; satellite, wireless, and wireline providers depend on each other to carry and terminate their traffic, and companies routinely share facilities and technology to ensure interoperability and efficiency. The private sector owns and operates the majority of communications infrastructure [36, 63].

The telecommunication industry has always been an integral part of every aspect whether it is related to the business or individuals, providing a variety of services that connect and communicate with millions of people worldwide. In recent years, the industry has experienced a fundamental transformation with the developments of network technologies. Today's threats are a realization of traditional IP based threats within the all-IP 4G network combined with insecure legacy 2/3G generations. Moving into the 5G era, the threat landscape will increase due to the new services and technologies being introduced [66, 81].

The sector builds and operates complex networks and stores voluminous amounts of sensitive data associated with individuals and corporations. These are among the reasons that make this field more lucrative to malicious actors or hackers. Over the years, the security vulnerabilities of telecom devices have been increased dramatically and now equipping a major space of the threat landscape [81].

Cyber disruptions of communications systems present unique challenges due to global connectivity. The exploitation of vulnerabilities around the world can begin affecting critical communications components in a matter of minutes. A successful cyber-attack on a telecommunications operator could disrupt service for thousands of phone customers, sever Internet service for millions of consumers, cripple businesses, and shut down government operations. Malicious actors may pose many risks, which can impact data, networks, and components, which create financial losses for organizations and severe disruptions in the operations of organizations [36, 88].

CrowdStrike published its 2020 Global Threat Report [24] which shows that the telecommunications and government sectors were the most targeted by the threat groups monitored by the cybersecurity firm. In the case of the telecom sector, many of the attacks were attributed to China-linked hacker groups [79].

DDoS (Distributed Denial of Service) attack is one of the most common types of direct cyber-attacks that can make a machine or network resource unavailable to its intended users by temporarily or indefinitely disrupting services of a host connected to the Internet. The telecom industry experiences more DDoS attacks than any other industry. These attacks can condense network capacity, swell traffic costs, disrupt the availability of service, and even compromise internet access by hitting ISPs [81].

Other threats that exist are the exploitation of vulnerabilities in network and consumer devices, attacks against supply chain, cloud services, IoT environment, and compromising subscribers with social engineering, phishing, or malware. Growing numbers of cyber attackers now combine data sets from different sources, including open sources, to build up detailed pictures of potential targets for blackmail and social engineering purposes. Insiders from cellular service providers are recruited mainly to provide access to data, while staff working for Internet service providers are chosen to support network mapping and man-in-the-middle attacks. Also aging protocols are a significant vulnerability [66, 76].

Example 7. The 2016 Dyn cyber-attack was a series of DDoS attacks on October 21, 2016, targeting systems operated by Domain Name System (DNS) provider Dyn. The attack caused major Internet platforms and services to be unavailable to large swathes of users in Europe and North America. The Mirai worm affected 100,000 UK Post Office broadband customers and 900,000 customers of Deutsche Telecom and was used to mount a DDoS attack, which in turn resulted in outages across Twitter, Spotify, Netflix, Paypal and other services (Wikipedia).

Example 8. In 2018 hackers infected more than 500,000 routers with malware that could cut off internet access and steal login credentials. The hackers have the power to simultaneously kill the devices and take down the internet for vast numbers of people as a result. The hackers have installed a malware known as VPNFilter on all those routers from a range of vendors, including Linksys, MikroTik, Netgear and TP-Link, which had publicly known vulnerabilities. Victims were spread across a total of 54 countries, but most of the targets were based in the Ukraine [62].

1.3.4   Critical Manufacturing Sector

The critical manufacturing sector is crucial to economic prosperity and continuity. This sector has identified the following industries to serve as the core of the sector [36]:

- Primary Metal Manufacturing,
- Machinery Manufacturing,
- Electrical Equipment, Appliance, and Component Manufacturing,
- Transportation Equipment Manufacturing.

Cyberthreats against manufacturing sector focus among others compromised on-site or remote ICS and SCADA systems. Manipulation of these systems can paralyze individual equipment or systems as well as entire production lines. Supply chain systems are vulnerable because of increased reliance on advanced information technology (IT) systems. State-sponsored and other actors could potentially defeat competition and/or obtain competitive secrets through cyber-intrusion [36].

Example 9. SHAMOON virus directed against Saudi Arabian Oil Company (ARAMCO) in 2012. The virus spread throughout the company's network and affected as many as 30,000 computers. Without a way to pay them, gasoline tank trucks seeking refills had to be turned away. Saudi Aramco's ability to supply 10% of the world's oil was suddenly at risk. In addition to affecting ARAMCO, the virus spread and was found on the system of RasGas, a Qatari owned liquefied natural gas company [5, 98, 99].

Example 10. In November 2014, in the hack of a German steel mill, the attackers targeted emails using a 'spear phishing' technique to obtain log-in information, which gave them access to critical production systems at the mill, leading to massive damage [109].

Example 11. In November 2016, hackers destroyed thousands of computers at six Saudi Arabian organizations, including those in the energy, manufacturing, and aviation industries. The attack was aimed at stealing data and planting viruses; it also wiped the computers, so they were unable to reboot. Hackers used a version of a specific type of cyberweapon, which operates like a time bomb [99].

1.3.5 Dams Sector

The dams sector is comprised of assets that include dam projects, hydropower generation facilities, navigation locks, levees, dikes, hurricane barriers, mine tailings, industrial waste impoundments, and other similar water retention and water control facilities. These dams, locks, pumping plants, canals, and levees provide water supply, power generation, navigable waterways, flood protection, and unique environmental stability and enhancements to habitats across the country [36–38].

The increasing use of standardized industrial control systems technology increases the sector's potential vulnerability to direct cyber-attacks and intrusions, which are a constant potential threat across the dam system environment. Opening the flood gates by cyber-attack can cause significant damage, and, if hydropower governors are cyber vulnerable, then generators and turbines could be destroyed in a cyber-attack. In 2016 the US ICS-CERT performed 98 assessments and recorded 94 instances of weak boundary protection of the control system which could facilitate unauthorized access. There were also incidences of unnecessary services, devices, and ports on control systems, as well as weak identification and authentication management [36, 122].

Example 12. In 2013, Iranian attackers were accused of infiltrating a dam in New York and stealing information from the energy company Calpine Corp. The small dam is not built for energy production purposes but to control water levels. The system was directly connected to the Internet and there was no need for the attacker to go past business network nor DMZ. The automation system of controlling the operation of the gate was not active and therefore it is not possible to evaluate the adversary's capabilities regarding the control system because it could not have been operated remotely [2].

Example 13. In 2016, an Iranian nation-state committed a cyber-attack against the United States at the Rye Brook Dam in New York. The hackers accessed industrial control systems within the dam but were fortunately unable to release the water behind the dam due to scheduled maintenance. However, this could have been a disaster waiting to happen with just a few clicks [113]

1.3.6   Defense Industrial Base Sector

The Defense Industrial Base (DIB) is the worldwide industrial complex that enables research and development, as well as design, production, delivery, and maintenance of military weapons systems, subsystems, components, and parts. The DIB sector has become heavily dependent on IT infrastructure, operating within an increasingly information-driven environment. DIB sector IT infrastructure is vulnerable to denial-of-service attacks, data theft and malicious modification of information. These vulnerabilities contribute greatly to the risk in the sector. Foreign entities and non-state actors seek to acquire access to sensitive and classified DIB sector information and technologies by expanding their cyber intelligence/espionage activities [36].

In cyber-attacks, foreign actors are stealing large amounts of sensitive data, trade secrets, and intellectual property (IP) every day from DIB firms. This comes in many forms (e.g., insider threat, phishing). The biggest issue confronting the DIB is how information security is being implemented, i.e., system users not following procedures or system administrators not applying fixes to known vulnerabilities. The DIB relies on commercial-off-the-shelf (COTS) information system products that are often flawed in their design and implementation, thus offering a host of vulnerabilities to those who would exploit them [9, 32].

Cyber-attacks designed to steal IP from the unclassified networks of companies have increased. The small firms are particularly vulnerable because they have challenges to acquire the costly cybersecurity tools (CSTs) and skilled professionals required to adequately protect their networks. In addition, ransomware attacks, possibly by different perpetrators, have also recently increased and have resulted in the destruction of data held on the unclassified networks of small companies and local governments [61].

The Defense Industrial Base is always under constant ransomware attacks. The malicious actors behind these attacks often block access to sensitive government data, intellectual property, and even trade secrets until they get paid. This can potentially harm a government's military capabilities and operations. DIB networks host critical operational assets and data that is crucial to national security. If the systems get breached, national security will be compromised [91].

Example 14. In 2007 Chinese hackers stole technical documents related to the data on the F-35 Joint Strike Fighter, the F-22 Raptor fighter jet, and the MV-22 Osprey. This espionage reduced the adversaries' costs and accelerated their weapon systems development programs, enabled reverse-engineering and countermeasures development, and undermined the U.S. military, technological, and commercial advantage [118].

1.3.7 Emergency Services Sector

The Emergency Services Sector (ESS) is a community of emergency personnel, along with the physical and cyber security resources, providing a wide range of preparedness and recovery services during both day-to-day operations and incident response. ESS is comprised of the following disciplines [36, 89]:

- Law Enforcement: Maintaining law and order and protecting the public from harm.
- Fire and Rescue Services: Prevention and minimizing loss of life and property during incidents resulting from fire, medical emergencies, and other all-hazards events.
- Emergency and Medical Services (EMS): Providing emergency medical assessment and treatment at the scene of an incident, during an infectious disease outbreak, or during transport and delivery of injured or ill-individuals to a treatment facility as part of an organized EMS system.
- Emergency Management: Leading efforts to mitigate, prepare for, respond to, and recover from all types of multijurisdictional incidents. Emergency management increasingly depends on computational and communication systems for coordination, communication, information gathering, training, and planning.
- Public Works: Providing essential emergency functions, such as assessing damage to buildings, roads, and bridges; clearing, removing, and disposing of debris; restoring utility services; and managing emergency traffic.

Through partnerships with public and private sector entities, ESS's mission is to save lives, protect property and the environment, assist communities impacted by disasters (natural or manmade), and aid recovery from emergency situations [34, 36].

Cyber targeting of the ESS will likely increase as systems and networks become more interconnected and the ESS becomes more dependent on information technology for daily operations. For example, cyber disruption of communications systems, computer networks in service vehicles, or GPS during an emergency operation could dramatically disrupt or delay the initial response to an event [36].

Many ESS activities, such as emergency operations communications, database management, biometric activities, telecommunications, and electronic systems (e.g., security systems), are conducted by partners virtually. These activities are vulnerable to cyber-attack. Additionally, the Internet is widely used by the sector to provide information as well as alerts, warnings, and threats relevant to the ESS [33, 75].

The following risks and impacts may occur in the ESS because of cyber-attacks [34]:

- Compromised ESS database causes disruption of mission capability or corruption of critical information,
- Public alerting and warning system disseminates inaccurate information,
- Loss of communications lines results in disrupted communications capabilities,
- Closed-circuit television (CCTV) jamming/blocking results in disrupted surveillance capabilities,
- Loss of communications lines results in disrupted communications capabilities for ESS troops,
- The DDoS attack against services of public safety and emergency services communications networks may lead to a paralysis of the services and loss of life.

Example 15. DDoS attacks are often used as a form of protest. After officer-involved shootings in 2014 in Denver and Albuquerque, divisions of Anonymous launched DDoS attacks to shut down the online service of both police departments [101].

In December 2016, a law enforcement agency near Dallas, was the victim of a ransomware attack when an employee clicked on a link in a phishing email that appeared to be from another law enforcement agency. The agency lost a substantial number of digital files, including video evidence [104].

Example 17. On March 22, 2018, a ransomware attack encrypted data on the City of Atlanta's government servers, affecting various internal and customer applications, including those of the Atlanta Police Department. During the same month, the City of Baltimore, Maryland, had its dispatch system taken offline for more than 17 h due to a cyber-attack [104].

1.3.8 Energy Sector

The energy sector is usually divided into three interrelated segments: electricity, petroleum, and natural gas. Electricity infrastructure is highly automated and controlled by utilities and regional grid operators that rely on sophisticated digital energy management systems. The modern electric grid is dependent upon cyber-physical systems, engineered systems that are built from, and depend upon, the seamless integration of computational algorithms and physical components. Also, oil and natural gas infrastructure is highly automated and controlled by pipeline operators, terminal owners, and natural gas utilities that rely on digital sophisticated energy management systems [36, 48, 73].

The energy infrastructure is arguably among one of the most complex and critical infrastructures since other sectors depend upon it to deliver their essential services. The energy industry is an IP-intensive industry, meaning it holds massive intellectual property. The energy sector consists of all the industries involved in the processes of energy production, distribution, and transmission. ICSs are used for controlling these processes. Energy infrastructures have turned into highly distributed systems, which require proactive protection [36, 48, 73].

Mainly for that reason, it is an attractive target for cyber criminals and cyber espionage. Cyber espionage against the energy sector may be rooted in political and economic motives, which may give the actor access to knowledge that presents a technological advantage, constituting a potential threat to the energy security [90].

Attacks on ICS of the energy sector have become more targeted than in the past. Attackers have become more knowledgeable about how to go after industrial control systems, using attacks customized to exploit ICS. Additionally, threat actors are paying close attention not only to payload, but delivery as well, focusing on ICS trusted relationships.

The power system has evolved into an ICS-enabled industry that increasingly relies on intelligent electronic devices (IEDs) using bidirectional communication to execute operations. Assets may be vulnerable if an infrastructure's industrial control systems are connected to the Internet, either directly or indirectly. For example, control system networks may be connected to the corporate business network, which, in turn, is connected to the Internet. These connections increase the network's vulnerability to direct cyber-attacks that could potentially disrupt movement and increase risk to the sector [36, 73].

Example 18. In 2014 the "Energetic Bear" virus was discovered in over 1,000 energy firms in 84 countries. This virus was used for industrial espionage and because it infected industrial control systems in the facilities, it could have been used to damage those facilities, including wind turbines, strategic gas pipeline pressurization and transfer stations, liquefied natural gas (LNG) port facilities, and electric generation power plants. It has been suggested that state-sponsored attackers wanted to disrupt national scale gas suppliers [107].

Example 19. On 23 December 2015, hackers compromised information systems of three energy distribution companies in Ukraine (Ivano-Frankivsk Oblast) and temporarily disrupted the electricity supply to consumers. This was a multistage, multisite attack that disconnected seven 110 kV and twenty-three 35 kV substations were switched off, and about 225 000 people were without electricity for a period from 1 to 6 h [123].

Example 20. The second attack against the Ukraine was in December 2016 when power cut had amounted to a loss of about one-fifth of Kiev's power consumption. Workstations and SCADA systems, linked to the 330-kilowatt sub-station "North", were compromised. In the latest attack, hackers are thought to have hidden in Ukrenergo's IT network undetected for six months, acquiring privileges to access systems and figure out their workings, before taking methodical steps to take the power offline [102].

Example 21. In August 2017 the Irish electricity transmission system operator EirGrid was a target of a man-in-the-middle attack. The attack first breached Vodafone's Direct Internet Access (DIA) service which was providing Internet access to EirGrid's interconnector site in Wales. Attackers were able to create a Generic Router Encapsulation (GRE) tunnel into the router used by Eirgrid. All traffic through the DIA router were intercepted by the attacker. It was discovered that System Operator for Northern Ireland (SONI) had their data intercepted too. Vodafone and National Cyber Security Centre attribute this attack to state sponsored actor but do not elaborate that estimation further [80].

Example 22. In mid-November 2017, attackers utilized the TRITON sophisticated attack framework to control industrial safety systems at a critical infrastructure facility and accidentally led to a process shutdown. This malware specifically targeted the Triconex Emergency Shut Down (ESD) system. During this sophisticated attack, the attacker utilized many custom intrusion tools to obtain and maintain access to the target's IT and operational technology networks [106].

1.3.9 Financial Services Sector

The financial services sector represents a vital component of a nation's critical infrastructure. Financial institutions provide a broad array of products from the largest institutions to the smallest community banks and credit unions. The finance sector is intricately woven into the daily lives of people around the world and is at the very core of global economies. Financial entities allow citizens and organizations worldwide to manage finances, trade, and to operate in different ways [36, 56].

Banks are 300% more likely to be attacked than the average industry and were the most attacked target in 2019. Those attacks have increased dramatically since COVID-19 [57].

Threat actors have much to benefit from a successful cyber-attack against any financial institution. This threat not only applies to banks, but also to exchanges, asset managers, technology providers, insurers, clearing and settlement houses, as well as supply chains to these institutions. Both state-sponsored and criminal actors have targeted the finance sector to [56]:

- Steal personal data,
- Monitor the financial activities of specific clients,
- Disrupt or tamper with critical operations,
- Steal money.

The attackers use offensive techniques, which is a more sophisticated type of attacks such as; distractive attacks, targeted ransomware attacks, supply chain attacks, and cryptojacking. Differently motivated cyber attackers using computer viruses, Trojan horses, worms, logic bombs, eavesdropping sniffers, and other tools that can destroy, intercept, degrade the integrity of, or deny access to data. Other potential cyberthreats to the sector include confidentiality and identity breaches [36, 56].

Example 23. In 2015 and 2016 a series of cyber-attacks using the SWIFT banking network were reported, resulting in the successful theft of millions of dollars. The attacks were perpetrated by a hacker group known as APT38. If the attribution to North Korea is accurate, it would be the first known incident of a state actor using cyber-attacks to steal funds. The attacks exploited vulnerabilities in the systems of member banks, allowing the attackers to gain control of the banks' legitimate SWIFT credentials. The thieves then used those credentials to send SWIFT funds transfer requests to other banks, which, trusting the messages to be legitimate, then sent the funds to accounts controlled by the attackers [22].

Example 24. The Equifax data breach which occurred between May and July of 2017 at the American credit bureau Equifax. Private records of 147.9 million Americans, along with 15.2 million British citizens and about 19,000 Canadian citizens were compromised in the breach, making it one of the largest cybercrimes related to identity theft. The data breach into Equifax was principally through a third-party software exploit that had been patched, but which Equifax had not updated on their servers. Equifax had been using the open-source Apache Struts as its website framework for systems handling credit disputes from consumers. A key security patch for Apache Struts was released on March 7, 2017 after a security exploit was found and all users of the framework were urged to update immediately. Security experts found an unknown hacking group trying to find websites that had failed to update Struts as early as March 10, 2017, as to find a system to exploit [55].

Example 25. On September 6, 2020, Banco Estado, the only public bank in Chile and one of the three largest in the country, had to shut down its nationwide operations due to a ransomware cyber-attack launched by REvil [16].

Example 26. On October 23, 2020, a software defect led to a disruption to the European Central Bank's main payment system for almost 11 h [16].

1.3.10 Food and Agriculture Sector

Agriculture is essential for modern society and has been involved in the adoption of information technology to manage production, processing, transportation, distribution and retailing of commodities and food products for decades. The food and agriculture sector is composed of farms, restaurants, and food manufacturing, processing, and storage facilities. It is composed of complex production, processing, and delivery systems and encompasses huge amounts of critical assets. The sector has a highly effective and resilient food supply chain, owing to the size, geographic diversity, and competitive nature of the industry [36, 63, 92, 96].

In the heavily mechanized landscape of agriculture, smart technologies and remote administration used in Precision Agriculture (PA) and smart farming creates a new cyber-physical environment. The incorporation of cyber-based technologies and data driven solutions in farm production, food processing, supplier industries, transport of goods, regulatory oversight, marketing sales and communication with consumers creates a paradigm shift. Cloud-based storage of large data sets, use of open-sourced or internet/cloud-based software, and corporate management of proprietary software each increase opportunities for data access by unauthorized users [31, 41].

Agricultural cybersecurity is a rising concern because farming is becoming ever more reliant on computers and Internet access. Like many industries, agriculture is undergoing a digital revolution powered by big data. Computers, robots, sensors, and big data analytics are driving decision making in the search for higher and more sustainable yields. Farms are driving towards a concept of precision agriculture, a data driven methodology for optimizing crop production. Key areas include soil sampling, yield monitors and maps, GPS guidance systems, satellite imaging and automatic section control [23, 94].

Whether it is wired-up off-road equipment and machinery, high-tech food and grain processing, radio frequency ID-tagged livestock, or global-positioning-system tracking, the agriculture sector depends on information systems to sustain and improve operations, competitiveness, and profitability. Agricultural production and operations will increase dependency on software and hardware applications which are vulnerable to cyber-attacks [74].

Potential attacks in various smart agricultural systems are mostly related to cyber-security, data integrity and data loss. Threat scenarios against the food and agriculture sector are [31, 92, 96, pp. 25–27]:

- Leaking of confidential farm data,
- Loss of availability of distribution and storage systems,
- Loss of availability of processing systems,
- Compromised integrity of food assurance systems,
- Farm vehicle collisions with power assets,
- Publishing confidential information that could be damaging from suppliers,
- Rogue data introduction into network,
- Falsification of data to disrupt both crop and livestock,
- The disruption to navigational, positioning and time systems,
- Disruption to communication networks.

In 2018, the U.S. Council of Economic Advisers reported the agricultural sector experienced 11 cyber incidents in 2016. Compared to other sectors the agricultural sector experienced a relatively low number of reported cyber incidents. The total number of incidents were 42,068 [17].

1.3.11 Governmental Institutions Sector

In the European system of accounts, the general government sector is defined as consisting

> of institutional units which are non-market producers whose output is intended for individual and collective consumption and are financed by compulsory payments made by units belonging to other sectors, and institutional units principally engaged in the redistribution of national income and wealth. [47]

The general government sector has subsectors, like central government, state government, local government, and social security funds.

A ministry is a high-level governmental organization headed by a minister and intended to manage a specific sector of public administration. A government or state agency is a permanent or semi-permanent organization in the machinery of government that is responsible for the oversight and administration of specific functions, such as an administration (Wikipedia).

The main functions of general government units are [47]:

- To organize or redirect the flows of money, goods and services or other assets among corporations, among households, and between corporations and households; in the purpose of social justice, increased efficiency or other aims legitimized by the citizens,
- To produce goods and services to satisfy households' needs (e.g., state health care) or to collectively meet the needs of the whole community (e.g., defense, public order and safety).

Many government agencies are tasked with providing new technology and services to citizens as quickly and efficiently as possible for a plethora of functions. Government entities frequently have access to a lot of personally identifiable information and other types of data that would be disastrous if an attacker got their hands on it.

Cyber-attacks against state and local governments have been dramatically increasing. In 2019 in the U.S. alone, there were 140 ransomware attacksan average of 3 per day—targeting public, state, and local government and healthcare providers in the US. This is up 65% from the previous year [97].

A key factor contributing to the rise of attacks on local government agencies is the commoditization of attack techniques. Ransomware is not new, but the hacker community is able to deploy the attacks quickly, easily, and highly successfully. Hackers are using more sophisticated attack methods and are sharing their knowledge readily with others. Ransomware is also growing in popularity because attackers know government agencies are highly likely to pay since their cyber-attack recovery readiness is often low, and because the alternative—denial of government services— is unacceptable [77].

Synack's trust report [112] tells that the government sector is globally the most hardened against cyber-attacks in 2020. It was discovered that governments scored 15% higher respectively than all other industries when it came to preventing attacks and responding to breaches. Government agencies earned the top spot in part due to reducing the time it takes to remediate exploitable vulnerabilities by 73%.

Example 27. Two separate attacks were launched on the U.S. Office of Personnel Management between 2012 and 2015. Hackers stole around 22 million records. The information that was obtained and exfiltrated in the breach included personally identifiable information such as social security numbers, as well as names, dates and places of birth, and addresses. Chinese state-sponsored hackers were accused of the attack (Wikipedia).

Example 28. The 2013 data hack at the Finnish Foreign Ministry was perpetrated by a group of Russian hackers and was part of a wider campaign against targets in nearly fifty countries. Experts have identified that the attack was perpetrated by the Turla group. It is widely believed that the Turla group is the premier Russian hacker organization and it targets ministries, embassies, and militaries in Russia's neighbors. Kaspersky has seen traces in malware and servers, that point to the fact that the authors are Russian speaking, and they seem to have a lot of resources for their cyber espionage operation [126].

Example 29. The Finnish Parliament was the target of a cyber-attack during the autumn of 2020. As a result of the attack, the security of several parliamentary email accounts was compromised, some of which belonged to MPs. The attack was detected by Parliament's internal technical surveillance [127].

Example 30. In September 2020 the Norwegian parliament announced it had been the target of a significant cyber-attack which breached the email accounts of several members and staff of Norway's Labor Party [95].

Example 31. In the fall of 2020, a major cyber-attack by a group backed by a foreign government, penetrated multiple parts of the US Federal Government, leading to a series of data breaches. The hacking group Cozy Bear (APT29) was identified as the cyber attackers. The cyber-attack and data breach were reported to be among the worst cyber-espionage incidents ever suffered by the U.S., due to the sensitivity and high profile of the targets and the long duration (eight to nine months) in which the hackers had access. The attackers exploited software from at least three U.S. firms: Microsoft, SolarWinds, and VMware [59].

1.3.12 Healthcare Sector

The healthcare sector is responsible for protecting and sustaining the citizen's health. Health services are divided into primary health care and specialized medical care. This widespread and diverse sector includes acute care hospitals, ambulatory health-care, national, and local public health systems; disease surveillance; and private sector industries that manufacture, distribute, and sell drugs, biologics, and medical devices [36].

Cyber-attacks against the healthcare sector are especially concerning because these attacks can directly threaten not just the security of systems and information but also the health and safety of patients. Hospitals are especially sensitive to cyber-attacks as any disruption in operations or even disclosure of patient personal information can have far-reaching consequences.

The healthcare sector is one of the most frequently breached industries in the world. This sector has an abundance of sensitive data and Personal Identifiable Information (PII) that can be exploited by hackers within healthcare organizations. According to a CBS report, medical records can sell for up to $1000 each on the dark web, while social security numbers and credit cards sell for $1 and up to $110, respectively [120]. Data breaches cost the health care industry approximately $5.6 billion every year.

High demand for patient information and often-outdated systems are among the nine reasons healthcare is now the biggest target for online attacks [111]:

• Private patient information is worth a lot of money to attackers,
• Medical devices are an easy entry point for attackers,

- Staff need to access data remotely, opening-up more opportunities for attack,
- Workers do not want to disrupt convenient working practices with the introduction of new technology,
- Healthcare staff are not educated in online risks,
- The number of devices used in hospitals makes it hard to stay on top of security,
- Healthcare information needs to be open and shareable,
- In smaller healthcare organizations, cybersecurity is often poorly managed,
- Outdated technology means the healthcare industry is unprepared for attacks.

The threat vectors against healthcare systems are: E-mail phishing attacks, ransomware attacks, man-in-the-middle attacks, loss or theft of medical equipment or data (PII), insider, accidental or intentional data loss, attacks against connected medical devices.

Based on [120] research, the highest risks and the percentage of healthcare organizations affected by each one is:

- Malicious network traffic: 72%,
- Phishing: 56%,
- Vulnerable OS (old version): 48%,
- Man-in-the-middle attack: 16%,
- Malware: 8%.

The US Department of Health and Human Services' (HHS) breach portal contains information about breaches of Protected Health Information (PHI). According to the HHS breach portal, data breaches affected 27 million people in 2019 in U.S. Top 10 breaches by number of individuals affected, currently listed on HHS's breach portal, are given in Table 1.1. [70].

Medical devices include all those devices (hardware and software) used in patient care for diagnosis, treatment, and monitoring. This extends to ancillary support devices that are required for the medical device to function properly and are hosted on a clinical network such as external disk storage, database servers, and gateway or middleware interface devices. Security incidents related to medical devices also have the potential to impact patient safety and do meaningful harm to patients connected to these networked devices [28].

Medical devices have become a cyber-attack target for many reasons. Medical devices are increasingly network connected, using default passwords, missing patches, remote access, and other weaknesses. So, the main vulnerabilities are application vulnerabilities, unpatched software, and configuration vulnerabilities [28].

Example 32. In May 2017 WannaCry malware spread like a worm laterally in the network which targeted computers running the Microsoft Windows operating system by encrypting data and demanding ransom payments in the Bitcoin cryptocurrency. It propagated through EternalBlue exploit. One infected organization was the National Health Service's (created in 1948) hospitals in England and Scotland. Up to 70,000 devices including computers, MRI scanners, blood-storage refrigerators and theatre equipment have been affected. Ambulances diverted and disruption to surgeries. No

Table 1.1 Top 10 breaches by number of individuals affected in US

Name of covered entity	Covered entity type	Individuals affected	Type of breach	Location of breached information	Year
Anthem Inc	Health plan	78,800,000	Hacking/IT incident	Network server	2015
American Medical Collection Agency	Business associate	26,059,725	Hacking/IT incident	Network server	2019
Optum360, LLC	Business associate	11,500,000	Hacking/IT incident	Network server	2019
Premera Blue Cross	Health plan	11,000,000	Hacking/IT incident	Network server	2015
Laboratory Corporation of America	Health plan	10,251,784	Hacking/IT incident	Network server	2019
Excellus Health Plan, Inc	Health plan	10,000,000	Hacking/IT incident	Network server	2015
Community Health Systems Professional Services Corporations	Healthcare provider	6,121,158	Hacking/IT incident	Network server	2014
Science Applications International Corporation	Business associate	4,900,000	Loss	Other	2011
Community Health Systems Professional Services Corporation	Business associate	4,500,000	Theft	Network server	2014
University of California, Los Angeles Health	Healthcare provider	4,500,000	Hacking/IT incident	Network server	2015

patient data was affected and there was no access to prescriptions or medical histories for treatment. 200,000 machines in 150 countries were affected (Wikipedia).

Example 33. A ransomware attack struck a hospital in Düsseldorf on September 10, 2020. The cyber-attack caused network outages that forced the clinic to reroute patients in need of emergency care elsewhere. One 78-year-old woman who required immediate attention for an aneurysm died after being sent to another city. The case is still under investigation [67].

Example 34. In autumn of 2020 the healthcare industry detected cyber-attacks from three nation-state actors targeting seven prominent companies directly involved in researching vaccines and treatments for Covid-19. The targets include leading pharmaceutical companies and vaccine researchers in Canada, France, India, South

Korea, and the United States. The attacks came from Strontium, an actor originating from Russia, and two actors originating from North Korea that are called Zinc and Cerium [15].

1.3.13 Information Technology Sector

The information technology sector is central to the nation's security, economy, and public health and safety as businesses, governments, academia, and private citizens are increasingly dependent upon information technology sector functions. These virtual and distributed functions produce and provide hardware, software, and information technology systems and services, and the Internet [36].

The IT sector is highly concerned about cyber threats, particularly those that degrade the confidentiality, integrity, or availability of the sector's critical functions. Depending on its scale, a cyber-attack could be debilitating to the IT sector's highly interdependent critical infrastructures. The cyber threats include unintentional acts (e.g., the accidental disruption of Internet content services) and intentional acts (e.g., the exploitation of IT supply chain vulnerabilities or the loss of interoperability between systems as the result of an attack) [36].

Cyber-attacks are increasingly targeting the technology sector. Technology became the most attacked industry for the first time, accounting for 25% of all attacks (up from 17%). Over half of attacks aimed at this sector were application-specific (31%) and DoS/DDoS (25%) attacks, as well as an increase in weaponization of IoT attacks [125].

Organizations in the high tech and information technology industry face cyber threats from the following actors [54]:

- Advanced persistent threat (APT) groups seeking to steal economic and technical information to support the development of domestic companies through reducing research and development costs, or otherwise providing a competitive edge.
- Hacktivists and cyber vandals with disruptive motivations may target Internet service providers to gain attention for their cause.

Some parts of the IT and high-tech sector provide an attack path into other sectors since IT products are a key infrastructure component for all kinds of organizations. Cloud storage providers, cloud computing services, developers of cyber security software, or a file-sharing solution provider, are often the targets of cyber-attacks. Partnerships with government or military entities would likely also place companies at risk, as foreign state-sponsored threat actors would probably target such companies [54, 100].

The development of new technologies will likely spur threat activity against the industry. One of the biggest threats for the high-tech sector companies is the loss of intellectual property. Having intellectual property lost or stolen after years of investment can dramatically reduce an organization's competitive advantage. High-tech companies create products that some technically skilled people want to use maliciously. IT tools can be used to implement hacking and cyber intelligence [30].

1.3.14 Nuclear Sector

The nuclear sector covers most aspects of civilian nuclear infrastructure: from the power reactors that provide electricity to citizens, to the medical isotopes used to treat cancer patients. The nuclear sector includes nuclear power plants, research and test reactors, fuel cycle facilities, radioactive waste management, decommissioning reactors, nuclear and radioactive materials used in medical, industrial, and academic settings, and nuclear material transport [36].

The vulnerability of nuclear plants to create a deliberate attack is of concern in nuclear safety and security. Cyber-attacks on the nuclear sector infrastructure and assets by terrorists, extremists, or foreign actors can target the industrial control systems or SCADA. Attacks may pose a significant threat to the sector, allowing malicious actors to manipulate or exploit facility operations. Cyber-attacks on nuclear power plants could have physical effects, especially if the network that runs the machines and software controlling the nuclear reactor are compromised. This can be used to facilitate sabotage, theft of nuclear materials, or in the worst-case scenario a reactor meltdown [27, 36].

Example 35. The Stuxnet worm affected Iran's nuclear development capabilities in 2010. Stuxnet specifically targets Programmable Logic Controllers (PLCs), which allow the automation of electromechanical processes such as those used to control machinery and industrial processes including gas centrifuges for separating nuclear material. Stuxnet caused a malfunction that was invisible to human operators because the SCADA screens in the control room suggested normal operation [51].

Example 36. The South Korean nuclear and hydroelectric company Korea Hydro and Nuclear Power (KHNP) was hacked in December of 2014. Hackers stole and posted online the plans and manuals for two nuclear reactors, as well as the data of 10,000 employees. South Korean authorities traced the IP addresses to Shenyang, a city in north-east China [90].

Example 37. In April 2016 Gundremmingen nuclear power facility had been harboring malware, including remote-access trojans and file-stealing malware, on the computer system that is used to monitor the plant's fuel rods. Fortunately, the computer was not connected to the Internet, and the malware was never able to be activated [58].

Example 38. The Nuclear Power Corporation of India Limited (NPCIL) reported that there was a cyber-attack on the Kudankulam Nuclear Power Plant (KKNPP) in September of 2019. The nuclear power plant's administrative network was breached in the attack but did not cause any critical damage [27].

1.3.15   Transportation Systems Sector

The transportation systems sector is comprised of the road, aviation, rail, and maritime sub-sectors. Most of the transport operates on a commercial basis, with responsibility for resilience delegated to owners and operators. The transportation systems sector moves people and goods through the country and overseas. Transport networks are essential for maintaining the health, safety, security, and social and economic well-being of citizens [36, 49, 63].

The aviation sector is a cornerstone of national and international commerce, trade, and tourism, which means even an isolated incident could spark a crisis of confidence in the entire sector. The aviation sector consists of airports, air traffic control facilities, and air navigation facilities [49, 86].

The maritime sector is a vital part of the global economy, whether it is carrying cargo, passengers, or vehicles. Ships are becoming increasingly complex and dependent on the extensive use of digital and communications technologies throughout their operational life cycle. The maritime transportation system is a geographically and physically complex and diverse system consisting of waterways, ports, and intermodal landside connections [36, 86].

The cyber security risk landscape in transport is currently evolving towards the point that risks that were once considered unlikely began occurring with regularity. Disruption to the transport network has significant impacts on the everyday life of citizens, national defense, security, and the vital functions of the state. One challenge in the transportation sector is that some legacy transportation systems now interface with public applications for ticketing and scheduling and rely on networked devices for routing, positioning, tracking and navigation. This presents multiple potential entry points for hackers [36, 78, 86].

The threat vectors against transportation systems are among others [49]:

- Distributed Denial of Service attacks (DDoS),
- Manipulation of hardware and/or software,
- Malware and viruses,
- Tempering and/or alteration of data including insertion of information,
- Hacking of wireless, connected assets,
- Identity theft,
- Exploitation of software bugs,
- Abuse of authorization,
- Abuse of information leakages,
- E-mail phishing attacks,
- Ransomware attacks,
- Man-in-the-middle attacks,
- Insider, accidental or intentional data loss (data breaches).

Cybersecurity is a growing concern for civil aviation, as organizations increasingly rely on electronic systems for critical parts of their operations, including safety–critical functions. From ransomware attacks to data breaches, the transportation sector is not immune to malicious hackers. A concerted, well-orchestrated attack on any aviation sub system and network could cause a considerable disruption sector-wide [36, 86].

The maritime transport industry is highly exposed to cyber-attacks. Vessels do not need to be attacked directly. An attack can arrive via a company's shore-based information technology systems and very easily penetrate a ship's critical onboard operational technology systems. These systems are used for a variety of purposes, including access control, navigation, traffic monitoring, and information transmission. Although the interconnectivity and utilization of the cyber systems facilitate transport, they can also present opportunities for exploitation, contributing to risk for the maritime systems [36, 86].

There are several reasons for conducting cyber-attacks against the transportation sector. Due to the reliance of trade on the sector, an attack could be used to affect trade in general, or even target a specific commodity and its availability. Due to the interdependence of the various transport infrastructures, there are a variety of targets to impact the trade: Railways or roads could be targeted to prevent goods reaching the ports and disrupting the ports themselves would hinder any import or export. Airports can be targeted to affect tourism, material transportation or business travel. Similarly, disrupting operations could delay military deployments or operations. Cyber-attacks could potentially be seen as the modern version of a Naval blockade. The greatest fear faced by transportation agencies is the potential for accidents, mass chaos, and even injuries or loss of life due to disruptions to critical infrastructure [119].

Example 39. In 2015 an attack on the IT network of the LOT airline of Poland caused at least 10 flights to be grounded. It was one of the first reported cases of hackers causing cancellations. LOT encountered an IT attack that affected the ground operation systems. As a result, LOT was not able to create flight plans and outbound flights from Warsaw are not able to depart. The attack caused journeys from Warsaw to a range of European destinations to be terminated, including trips to Munich, Hamburg, Copenhagen, and Stockholm, amongst others. As many as 1500 passengers were said to have been affected [12].

Example 40. Cyber breach affecting Cosco's operations in the US Port of Long Beach, on 24 July 2018, which affected the giant's daily operations. The company's network broke down, and some electronic communications were not available as a result [105].

Example 41. Petya malware variant infected the IT systems of the world's largest shipping company Maersk with 600 container vessels handling 15% of the world's seaborne trade in June 2017. The breakdown affected all business units at Maersk, including container shipping, port and tugboat operations, oil, and gas production, drilling services, and oil tankers. Maersk reporting up to $300 million in losses [65].

Example 42. In 2016, the light rail system in San Francisco was hacked, halting access to agency emails and the computer system while hackers demanded 100 bitcoin payment to unlock the hacked computer systems, which the department refused to pay. But rather than shut down the network, the attack simply led to machines being turned off and passengers were allowed to grab free rides [13].

Example 43. A team of experts in the US Homeland Security team remotely hacked a Boeing 757. This hack was not conducted in a laboratory, but on a 757 parked at the airport in Atlantic City. The team got the airplane on September 19, 2016 and two days later, an expert was successful in accomplishing a remote, non-cooperative, penetration [25].

1.3.16 Water and Wastewater Systems Sector

The water and wastewater systems sector offers drinking water to citizens which is a prerequisite for protecting public health and all human activity. Properly treated wastewater is vital for preventing disease and protecting the environment. The sector is comprised of public drinking water systems (includes both community and non-community water systems, such as schools, factories, and other commercial or governmental facilities) and wastewater treatment utilities. Water utilities consist of water sources, treatment facilities, pumping stations, storage sites, and extensive distribution, collection, and monitoring systems [36].

Water utilities around the world are vulnerable to attacks because they are usually small and have almost no cybersecurity expertise among staff members [14]. Cyber-attacks on water and wastewater systems sector infrastructure and assets by terrorists, extremists, or foreign actors can target the industrial control systems or SCADA. The cyber-attacks targeting the water sector are complex and sophisticated, and they are often orchestrated by state-sponsored bodies whose objective is to destabilize a country's economy [7].

Our dependency on water—e.g., for consumption, hygiene, agriculture, industry, or energy production—provides an inviting environment for cyber attackers. The effects of a cybersecurity attack on the critical water sector operations could cause devastating harm to public health and safety, threaten national security and result in costly recovery and remediation efforts to address system issues as well as data loss [36, 60].

Example 44. A water department in the state of North Carolina was targeted by a cyber-attack using ransomware in 2018. It began on October 4th when the system was hit with Emotet, an advanced, modular banking Trojan that primarily functions as a downloader or dropper of another banking Trojan. IT staff members were unsuccessful in stopping the ransomware infection from spreading, so the crypto virus spread quickly along the network, encrypting databases and files. The water utility did not pay the ransoms [26].

Example 45. Two cyber-attacks hit Israel's water system in 2020. The first attack hit in April when hackers tried to modify the waters chlorine levels. The first attack hit agricultural water pumps in upper Galilee, while the second one hit water pumps in the central province of Mateh Yehuda. The attempted attacks were unsuccessful [19].

1.4   Critical Infrastructure Protection

Industrial control system cyber-attacks require a lot of knowledge and planning. ICS attacks take a long time to launch because adversaries must know a lot about the systems. Industrial control systems are not just digital; they are also analog and mechanical. Usually, critical infrastructure system has all different kinds of combinations of those things in industrial environment as well as having digital PLCs that are programmed to do things, as well as DCS and SCADA systems [14].

Building a cybersecurity critical infrastructure program takes time, careful planning, and ongoing support from political leadership, state level agencies, and public and private entities overseeing critical infrastructure. The first step is helping key players in government understand the severity, urgency, and potential impacts of different types of cyber threats and the need to take immediate action [29].

The implementation of protective measures aimed at securing critical infrastructure systems requires a considered and holistic approach, as there are many variables involved in establishing and maintaining a balance between security and functionality of service delivery and system availability. A key part of the greater national infrastructure security situational picture is the continued availability of critical infrastructure systems [121].

In the EU's view, strong cyber resilience needs a collective and wide-ranging approach. This calls for more robust and effective structures to promote cybersecurity and to respond to cyber-attacks in the Member States but also in the EU's own institutions, agencies, and bodies. It also requires a more comprehensive, cross-policy approach to building cyber-resilience and strategic autonomy, with a strong Single Market, major Advances in the EU's technological Capability, and far greater numbers of skilled experts. At the heart of this is a broader acceptance that cybersecurity is a common societal challenge, so that multiple layers of government, economy and society should be involved [46].

Managing the risks from significant threat and hazards to physical and cyber critical infrastructure requires an integrated and comprehensive approach across this diverse community. The U.S. National Infrastructure Protection Plan (NIPP) defines the objectives as identify, deter, detect, disrupt, and prepare for threats and hazards to the nation's critical infrastructure, reduce vulnerabilities of critical assets, systems, and networks, mitigate the potential consequences to critical infrastructure of incidents or adverse events that do occur [35].

There are controls that every critical infrastructure organization should consider. These controls are categorized as [68]:

- Basic Controls, such as inventory and control of hardware/software assets, continuous vulnerability management, or controlled use of administrative privileges,
- Foundational Controls, such as email and web browser protections, malware defenses, or secure configuration for network devices like firewalls, routers, and switches,

- Organizational Controls, such as the implementation of a security awareness and training program, incident response and management, penetration tests, and red team exercises.

The U.S. President's National Infrastructure Advisory Council (NIAC) [93] suggests following decisive cyber security actions:

1. Establish separate, secure communications networks specifically designated for the most critical cyber networks, including "dark fiber" networks for critical control system traffic and reserved spectrum for backup communications during emergencies.
2. Identify best-in-class scanning tools and assessment practices, and work with owners and operators of the most critical networks to scan and sanitize their systems on a voluntary basis.
3. Strengthen the capabilities of today's cyber workforce by sponsoring a public– private expert exchange program.
4. Establish a set of limited time, outcome-based market incentives that encourage owners and operators to upgrade cyber infrastructure, invest in state-of-the-art technologies, and meet industry standards or best practices.
5. Establish clear protocols to rapidly declassify cyber threat information and proactively share it with owners and operators of critical infrastructure, whose actions may provide the nation's front line of defense against major cyber-attacks.

1.5 Conclusion

The continuous and rapid digitization at a global level, underpinned by the progress made in smart ICT as well as the integration of IoT and automation, is a trend that deeply transforms many market sectors and has created opportunities in several areas of global economies and societies. So, smart ICT and IoT is the backbone of the fourth industrial revolution and are basic elements of the critical infrastructure [31].

The issue of cyber security in the fourth industrial revolution is currently having and will continue to have a significant impact on the critical infrastructure. As Industrial Control Systems, SCADA, Distributed Control System, Operational Technology, and other process control networks are Internet-connected, they expose crucial services to cyber-attacks. Extremely sophisticated cyber-attacks such as the Duqu and the Stuxnet worm show how effectively critical infrastructure can be attacked.

Cyber-attacks directed against critical infrastructure organizations can be conducted in many forms, which may consist of a single act or a combination of discrete steps threaded together. Such acts may be a Complicated exploitation of coding or the simple use of social engineering to reveal or to gain access to confidential information. Once the targeted system is compromised, perpetrators might implement "back door" Gates or install Stealth code allowing information to be monitored or removed without detection. "Kill switches" can be implemented, which can be activated at a specified time or under a specified set of conditions. System control allows an attacker to paralyze or even destroy a system [3].

The structure and operation of modern highly networked critical infrastructure systems fundamentally depends on networked information systems, some of which have unfortunately been inadequately secured from cyber-attacks. The vulnerabilities also make CI systems highly vulnerable to hybrid warfare tactics of both state and non-state actors. The combined complexities of these networked systems interacting together stands to amplify threats and vulnerabilities that exist in any of the major systems, as well as risk to other dependent systems.

Major attacks on critical infrastructure such as power, gas, and water stations, as well as transportation control systems, have become the new face of warfare. In October 2019, hackers knocked out more than 2000 websites hosted in the nation of Georgia. According to the U.K., the U.S., and Georgia, Russia carried out this attack to destabilize the country as part of its hybrid warfare activities [11].

It will be increasingly difficult to distinguish cyber sabotage operations from hybrid warfare operations. Developments could lead to serious crises around the world. Many security officials have been warned of a "cyber tsunami or 9/11," which could trigger a cyberwar that will unleash reactions that cannot be controlled.

References

1. Abomhara M, Køien GM (2015) Cyber security and the Internet of Things: vulnerabilities, threats, intruders and attacks. J Cyber Secur Mobility 4(1):65–88.https://doi.org/10.13052/jcsm2245-1439.414
2. AP (2015) Iranian hackers infiltrated U.S. power grid, dam computers, reports say. Posted by the Associated Press, CBC/Radio-Canada. https://www.cbc.ca/news/technology/hackers-infrastructure-1.3376342
3. APTA (2014) Cybersecurity considerations for public transit. Report APTA SS-ECS-RP-001-14. American Public Transportation Association, Washington, DC. https://www.apta.com/wp-content/uploads/Standards_Documents/APTA-SS-ECS-RP-001-14-RP.pdf
4. Ashenden D (2011) Cyber security: time for engagement and debate. In: European conference on information warfare and security. Academic Conferences, pp 11–16
5. Ballou T, Allen JA, Francis KK (2016) U.S. energy sector cybersecurity: hands-off approach or effective partnership? J Inf Warfare 15(1):44–59
6. Beggs C (2006) Proposed risk minimization measures for cyber-terrorism and SCADA networks in Australia. In: ECIW 2006—5th European conference on information warfare and security. Academic Conferences
7. Ben Boubaker K (2020) Water infrastructure: when states and cyber-attacks rear their ugly heads. Stormshield. https://www.stormshield.com/news/water-infrastructure-when-states-and-cyber-attacks-rear-their-ugly-heads/
8. Bertino E, Martino LD, Paci F, Squicciarini AC (2010) Web services threats, vulnerabilities, and countermeasures. In: Security for web services and service-oriented architectures. Springer, pp 25–44
9. Biancuzzo MR (2017) Cybersecurity & critical infrastructure. Briefing Papers, issue 17–13, Thomson Reuters

10. Brenner B (2011) Nitro attack: points of interest. CSO. https://www.csoonline.com/article/2134921/nitro-attack--points-of-interest.html
11. Breth J, Douglas C (2020) Cybersecurity needs its place in emergency management now. CPO magazine. https://www.cpomagazine.com/cyber-security/cybersecurity-needs-its-place-in-emergency-management-now/
12. Brewster T (2015) Attack on LOT Polish airline grounds 10 flights. Forbes. https://www.forbes.com/sites/thomasbrewster/2015/06/22/lot-airline-hacked/?sh=3862c062124e
13. Brewster T (2016) Ransomware Crooks demand $70,000 after hacking San Francisco transport system. Forbes. https://www.forbes.com/sites/thomasbrewster/2016/11/28/san-francisco-muni-hacked-ransomware/?sh=ae56b3847061
14. Brumfield C (2020) Attempted cyberattack highlights vulnerability of global water infrastructure. CSO. https://www.csoonline.com/article/3541837/attempted-cyberattack-highlights-vulnerability-of-global-water-infrastructure.html
15. Burt T (2020) Cyberattacks targeting health care must stop. Microsoft. https://blogs.microsoft.com/on-the-issues/2020/11/13/health-care-cyberattacks-covid-19-paris-peace-forum/
16. Carnegie (2021) Timeline of cyber incidents involving financial institutions. Carnegie. https://carnegieendowment.org/specialprojects/protectingfinancialstability/timeline. Retrieved on 24 Jan 2021
17. CEA (2018) The cost of malicious cyber activity to the U.S. economy. Council of economic advisers, White House, Washington, DC
18. Chong J (2013) Why is our cybersecurity so insecure? New Republic, https://newrepublic.com/article/115145/us-cybersecurity-why-software-so-insecure
19. Cimpanu C (2020) Two more cyber attacks hit Israel's water system. ZDNet. https://www.zdnet.com/article/two-more-cyber-attacks-hit-israels-water-system/
20. CISA (2019) Chemical sector landscape. Cybersecurity and infrastructure security agency, U.S. Department of Homeland Security
21. Constantin L (2021) 33 hardware and firmware vulnerabilities: a guide to the threats. CSO. https://www.csoonline.com/article/3410046/hardware-and-firmware-vulnerabilities-a-guide-to-the-threats.html
22. Corkery M (2016) Once again, thieves enter swift financial network and steal. New York times. https://www.nytimes.com/2016/05/13/business/dealbook/swift-global-bank-network-attack.html
23. CRI (2020) Cyber threats to the agriculture sector. Cyber risk international. https://cyberriskinternational.com/2020/04/07/cyber-threats-to-the-agriculture-sector/
24. CrowdStrike (2020) 2020 global threat report. CrowdStrike
25. CSO (2017) Homeland Security team remotely hacked a Boeing 757. CSO, https://www.csoonline.com/article/3236721/homeland-security-team-remotely-hacked-a-boeing-757.html
26. CSO (2018) Ransomware attack hits North Carolina water utility following hurricane. CSO. https://www.csoonline.com/article/3314557/ransomware-attack-hits-north-carolina-water-utility-following-hurricane.html
27. Das D (2019) An Indian nuclear power plant suffered a cyberattack: here's what you need to know. The Washington Post. https://www.washingtonpost.com/politics/2019/11/04/an-indian-nuclear-power-plant-suffered-cyberattack-heres-what-you-need-know/
28. Department of Health (2018). Medical device cyber security—draft guidance and information for consultation, Australia's Therapeutic Goods Administration (TGA), 19.12.2018
29. Deloitte (2017) Cybersecurity for critical infrastructure: growing, high-visibility risks call for strong state leadership. Deloitte. https://www2.deloitte.com/content/dam/Deloitte/us/Documents/public-sector/us-public-sector-cybersecurity-critical-infrastructure.pdf
30. Deloitte (2021) Global cyber executive briefing: high technology. Case studies, Deloitte Development LLC. https://www2.deloitte.com/ba/en/pages/risk/articles/High-Technology-Sector.html. Retrieved on 13 Jan 2021
31. Demestichas K, Peppes N, Alexakis T (2020) Survey on security threats in agricultural IoT and smart farming. Sensors 20(22):6458. https://doi.org/10.3390/s20226458

32. DHS (2010a) Defense industrial base sector-specific plan: an annex to the national infrastructure protection plan. U.S. Department of Homeland Security
33. DHS (2010b) Emergency services sector-specific plan: an annex to the national infrastructure protection plan. U.S. Department of Homeland Security
34. DHS (2012) Emergency services sector cyber risk assessment. U.S. Department of Homeland Security
35. DHS (2013) NIPP 2013: Partnering for critical infrastructure security and resilience. U.S. Department of Homeland Security
36. DHS (2014) Sector risk snapshots. U.S. Department of Homeland Security
37. DHS (2015a) Dams sector-specific plan: an annex to the NIPP 2013. U.S. Department of Homeland Security
38. DHS (2015b) Commercial facilities sector-specific plan: an annex to the NIPP 2013. U.S. Department of Homeland Security
39. DHS (2016) Introduction to the commercial facilities sector-specific agency. https://zahp.org/wp-content/uploads/2018/01/commercial-facilities-ssa-fact-sheet-2016-508.pdf
40. DHS (2020) Critical infrastructure security. U.S. Department of Homeland Security. https://www.dhs.gov/topic/critical-infrastructure-security
41. Duncan SE, Reinhard R, Williams RC, Ramsey F, Thomason W, Lee K, Dudek N, Mostaghimi S, Colbert E, Murch R (2019) Cyberbiosecurity: a new perspective on protecting U.S. food and agricultural system. Front Bioeng Biotechnol 7:63
42. Dunn Cavelty M (2010) The reality and future of cyberwar. Parliamentary Brief. www.parliamentarybrief.com/2010/03/the-reality-and-future-of-cyberwar Can't reach this page!
43. EC (2005) On a European programme for critical infrastructure protection. Green paper, COM (2005) 0576 final, European Commission
44. EC (2006) On a European programme for critical infrastructure protection. Communication from the Commission, COM(2006) 786 final, European Commission
45. EC (2007) Towards a general policy on the fight against cyber crime. Communication from the Commission to the European Parliament, the Council and the Committee of the Regions, COM(2007) 267 final, European Commission
46. EC (2017) Resilience, deterrence and defence: Building strong cybersecurity for the EU. Joint communication to the European Parliament and the Council, JOINT(2017) 450 final, European Commission
47. EC (2020) Glossary: general government sector. European commission. https://ec.europa.eu/eurostat/statistics-explained/index.php/Glossary:General_government_sector. Retrieved on 25 Jan 2021
48. EECSP (2017) Cyber security in the energy sector: Recommendations for the European commission on a European strategic framework and potential future legislative acts for the energy sector. EECSP Report, Energy Expert Cyber Security Platform
49. ENISA (2015) Cyber security and resilience of intelligent public transport: good practices and recommendations. European union agency for network and information security (ENISA)
50. EU (2008) On the identification and designation of European critical infrastructures and the assessment of the need to improve their protection, Council Directive 2008/114/EC. Official J Euro Union L 345:75–82
51. Falliere N, Murchu LO, Chien E (2011) W32.Stuxnet dossier, version 1.4. Wired. https://www.wired.com/images_blogs/threatlevel/2011/02/Symantec-Stuxnet-Update-Feb-2011.pdf
52. FCC (2014) Cyber security planning guide. Federal Communications Commission (FCC)
53. Finkle J (2011) Government facilities targets of cyber attacks. Reuters. https://www.reuters.com/article/us-usa-hackers-idUSTRE7656M020110706
54. FireEye (2016) Cyber threats to the high tech and IT industry. FireEye, Milpitas, CA. https://www.fireeye.com/content/dam/fireeye-www/current-threats/pdfs/ib-high-tech.pdf
55. Fruhlinger J (2020) Equifax data breach FAQ: what happened, who was affected, what was the impact? CSO. https://www.csoonline.com/article/3444488/equifax-data-breach-faq-what-happened-who-was-affected-what-was-the-impact.html
56. F-Secure (2019). Cyber threat landscape for the finance sector. F-Secure.

57. Fuchs J (2020) Why the biggest threat to financial firms is cyber attacks. Avanan. https://www.avanan.com/blog/biggest-threat-financial-firms-cyber-attacks
58. Gallagher S (2016) German nuclear plant's fuel rod system swarming with old malware. Ars Technica. https://arstechnica.com/information-technology/2016/04/german-nuclear-plants-fuel-rod-system-swarming-with-old-malware/
59. Geller E (2020) 'Massively disruptive' cyber crisis engulfs multiple agencies. Politico. https://www.politico.com/news/2020/12/14/massively-disruptive-cyber-crisis-engulfs-multiple-agencies-445376
60. Germano JH (2019) Cybersecurity risk & responsibility in the water sector. American Water Works Association
61. Gonzales D, Harting S, Adgie MK, Brackup J, Polley L, Stanley KD (2020) Unclassified and secure: a defense industrial base cyber protection program for unclassified defense networks. RAND Corporation, Santa Monica, CA
62. Goodin D (2018) Hackers infect 500,000 consumer routers all over the world with malware. Ars Technica. https://arstechnica.com/information-technology/2018/05/hackers-infect-500000-consumer-routers-all-over-the-world-with-malware/
63. GOV.UK (2017) Public summary of sector security and resilience plans. Cabinet Office, London
64. GOV.UK (2019) Common cyber attacks: reducing the impact. National Cyber Security Centre
65. Gronholt-Pedersen J (2017) Maersk says global IT breakdown caused by cyber attack. Reuters. https://www.reuters.com/article/us-cyber-attack-maersk-idUSKBN19I1NO
66. GSMA (2019) Mobile telecommunications security threat landscape. GSM Association, London
67. Hackett R (2020) Ransomware attack on a hospital may be first ever to cause a death. Fortune. https://fortune.com/2020/09/18/ransomware-police-investigating-hospital-cyber-attack-death/
68. Hassanzadeh A, Rasekh A, Galelli S, Aghashahi M, Taormina R, Ostfeld A, Banks MK (2020) A review of cybersecurity incidents in the water sector. J Environ Eng 146(5):03120003
69. Hess E (2019) People, process, and technology: the trifecta of cybersecurity programs. Helical. https://helical-inc.com/blog/people-process-and-technology-the-trifecta-of-cybersecurity-program/
70. HHS (2020) Breach portal: notice to the secretary of HHS breach of unsecured protected health information. U.S. Department of Health & Human Services, Washington, DC. https://ocrportal.hhs.gov/ocr/breach/breach_report.jsf. Retrieved on 13 Jan 2021
71. Hollis D (2011) Cyberwar case study: Georgia 2008. Small wars journal. https://smallwarsjournal.com/jrnl/art/cyberwar-case-study-georgia-2008
72. HVK (2020). Kriittinen infrastruktuuri. Huoltovarmuuskeskus, https://www.huoltovarmuuskeskus.fi/sanasto#k
73. INL (2016) Cyber threat and vulnerability analysis of the U.S. electric sector. Mission Support Center Analysis Report, Idaho National Laboratory
74. ISA (2020) Cybersecurity in the food and agriculture sector. Internet security alliance, Arlington, VA. https://isalliance.org/sectors/agriculture/
75. Jones SR (2017) The impact that a cyber-attack would cause within the emergency services sector. Master's thesis, Utica College, ProQuest LLC, Ann Arbor, MI
76. Kaspersky (2016) Threat intelligence report for the telecommunications industry. Kaspersky Lab
77. Kennedy C (2019) Government networks are under cyber attack: here's how cities, agencies can fight back. Homeland security today. https://www.hstoday.us/subject-matter-areas/infrastructure-security/government-networks-are-under-cyber-attack-heres-how-cities-agencies-can-fight-back/
78. Knott F (2020) The threat of cybercrime for state and local transportation systems. Attila security. https://www.attilasec.com/blog/transportation-systems-cybercrime. Retrieved on 13 Jan 2021

79. Kovacs E (2020) Telecom sector increasingly targeted by Chinese hackers: CrowdStrike. Security week. https://www.securityweek.com/telecom-sector-increasingly-targeted-chinese-hackers-crowdstrike

80. Kovanen T, Nuojua V, Lehto M (2018) Cyber threat landscape in energy sector. In ICCWS 2018: Proceedings of the 13th international conference on cyber warfare and security. Academic Conferences International, pp 353–361

81. Kumar V (2020) Cybersecurity challenges and solutions in the telecom industry. Industry wired. https://industrywired.com/cybersecurity-challenges-and-solutions-in-the-telecom-industry/

82. Kuokkanen N (2020). Kriittisen infrastruktuurin suojaaminen Suomessa. Kandidaatin tutkielma, Jyväskylän yliopisto

83. Laiho M (2020) Toimenpidealoite yhteiskunnan toiminnan kannalta kriittisten alojen työntekijöiden tai sen määrittelyjen perusteiden säätämiseksi. Toimenpidealoite TPA 27/2020 vp, Suomen eduskunta

84. Lehto M (2013) The cyberspace threats and cyber security objectives in the cyber security strategies. Int J Cyber Warfare Terrorism 3(3):1–18

85. Lehto M (2015) Phenomena in the cyber world. In: Lehto M, Neittaanmäki P, (eds) Cyber security: analytics, technology and automation. Springer, Cham, pp 3–29. https://doi.org/10.1007/978-3-319-18302-2_1

86. Lehto M (2020) Cyber security in aviation, maritime and automotive. In: Diez P, Neittaanmäki P, Periaux J, Tuovinen T, Pons-Prats J (eds) Computation and big data for transport. Springer, Cham, pp 19–32. https://doi.org/10.1007/978-3-030-37752-6_2

87. Liaropoulos A (2010) War and ethics in cyberspace: cyber-conflict and just war theory. In: Proceedings of the 9th European conference on information warfare and security (Thessaloniki, 2010), pp 177–182

88. Lobel M (2014) Security risks and responses in an evolving telecommunications industry. PwC (network of member firms of PricewaterhouseCoopers International Limited)

89. Loukas G, Gan D, Vuong T (2013) A review of cyber threats and defence approaches in emergency management. Future Internet 5(2):205–236. https://doi.org/10.3390/fi5020205

90. Macola IG (2020) The five worst cyberattacks against the power industry since 2014. Power technology. https://www.power-technology.com/features/the-five-worst-cyberattacks-against-the-power-industry-since2014/

91. Mallon S (2020) Ransomware and the defense industrial base. SmartData collective. https://www.smartdatacollective.com/ransomware-and-defense-industrial-base/

92. NCC (2019) Cyber security in UK agriculture. NCC group, https://research.nccgroup.com/wp-content/uploads/2020/07/agriculture-whitepaper-final-online.pdf

93. NIAC (2017) Securing cyber assets: addressing urgent cyber threats to critical infrastructure. The President's National Infrastructure Advisory Council (NIAC)

94. Nikander J, Manninen O, Laajalahti M (2020) Requirements for cybersecurity in agricultural communication networks. Comput Electron Agric 179:105776

95. OAGOV (2020) Cyber security threats against global governments increase exponentially. Open access government. https://www.openaccessgovernment.org/cyber-security-threats-global-governments-increasing/96789/

96. Okupa H (2020) Cybersecurity and the future of agri-food industries. Master's thesis, Kansas State University

97. Orr K (2020) Cyber attacks against state and local governments surge. CyberArk Software Ltd. https://www.cyberark.com/resources/blog/cyber-attacks-against-state-and-local-governments-surge

98. Pagliery J (2015) The inside story of the biggest hack in history. CNN business. https://money.cnn.com/2015/08/05/technology/aramco-hack/

99. Pagliery J (2016) Hackers destroy computers at Saudi aviation agency. Cable news network (CNN). https://money.cnn.com/2016/12/01/technology/saudi-arabia-hack-shamoon/

100. Papesh J (2019) When tech is the target: cyber risks for tech companies. AXA XL. https://axaxl.com/fast-fast-forward/articles/when-tech-is-the-target_cyber-risks-for-tech-companies

101. Police1 (2017) 9 cyberattacks that threatened officer safety and obstructed justice. Police1, Lexipol, Frisco, TX. https://www.police1.com/cyber-attack/articles/9-cyberattacks-that-thr eatened-officer-safety-and-obstructed-justice-dCWXReoa54CkcH3y/
102. Polityuk P, Vukmanovic O, Jewkes S (2017) Ukraine's power outage was a cyber attack: Ukrenergo. Reuters. https://www.reuters.com/article/us-ukraine-cyber-attack-energy-idUSKBN1521BA
103. Pye G, Warren M (2011) Analysis and modelling of critical infrastructure systems. In:10th European conference on information warfare and security (ECIW 2011). Academic Conferences, Reading, pp 194–201
104. Quinn C (2018) The emerging cyberthreat: cybersecurity for law enforcement. Police Chief Magazine. https://www.policechiefmagazine.org/the-emerging-cyberthreat-cybersecurity/
105. Safety4Sea (2018) 2018 highlights: Major cyber attacks reported in maritime industry. Safety4Sea. https://safety4sea.com/cm-2018-highlights-major-cyber-attacks-rep orted-in-maritime-industry/
106. SCF (2020) All you need to know about cyber security threats in energy sector. Swiss cyber forum. https://www.swisscyberforum.com/all-you-need-to-know-about-cyber-security-threats-in-energy-sector/
107. Scheuermann JE (2017) Cyber-physical attacks on critical infrastructure: What's keeping your insurer awake at night? K&L Gates. https://www.klgates.com/Cyber-physical-Attacks-on-Critical-Infrastructure--Whats-Keeping-Your-Insurer-Awake-at-Night-01-24-2017
108. Securicon (2019) What's the difference between OT, ICS, SCADA and DCS? Securicon, Alexandria, VA. https://www.securicon.com/whats-the-difference-between-ot-ics-scada-and-dcs/
109. Spiegel (2014). Hacker legten deutschen Hochofen lahm. Spiegel, https://www.spiegel.de/net zwelt/web/bsi-bericht-hacker-legten-deutschen-hochofen-lahm-a-1009191.html
110. Stoye E (2019) Hexion, Momentive and Norsk Hydro all hit by ransomware cyber attacks. Chemistry world. https://www.chemistryworld.com/news/hexion-momentive-and-norsk-hydro-all-hit-by-ransomware-cyber-attacks/3010328.article
111. Swivel (2020) 9 reasons why healthcare is the biggest target for cyberattacks. Swivel secure. https://swivelsecure.com/solutions/healthcare/healthcare-is-the-biggest-tar get-for-cyberattacks/
112. Synack (2020) The 2020 trust report: measuring the value of security amidst uncertainty. Synack
113. Thompson M (2016). Iranian cyber attack on New York dam shows future of war. Time. https://time.com/4270728/iran-cyber-attack-dam-fbi/
114. Tunggal AT (2020) What is an attack vector? Common attack vectors. UpGuard. https://www.upguard.com/blog/attack-vector
115. UN (2018) The protection of critical infrastructures against terrorist attacks: compendium of good practices. United Nations
116. US-Army (1995) Joint doctrine for military operations other than war. Joint Pub 3-07, US Army
117. US-GOV (2001) Uniting and strengthening America by providing appropriate tools required to intercept and obstruct terrorism (USA Patriot Act) act of 2001. Public Law 107–56, U.S. Congress
118. US-GOV (2019) Foreign cyber threats to the United States: hearing before the committee on armed services, United States Senate, one hundred fifteenth congress, first session, Jan 5, 2017. U.S. Government Publishing Office, Washington, DC
119. van Niekerk B (2018) Analysis of cyber-attacks against the transportation sector. In: Cyber security and threats: concepts, methodologies, tools, and applications. IGI Global, pp 1384–1402
120. Wandera (2020) Cybersecurity in the healthcare industry. Wandera. https://www.wandera.com/cybersecurity-healthcare/. Retrieved on 25 Jan 2021
121. Warren M, Pye G, Hutchinson W (2010) Australian critical infrastructure protection: a case of two tales. In: SECAU 2010: proceedings of the 11th australian information warfare and security conference. SECAU Security Research Centre, pp 30–36

122. WaterPower (2019) Hydropower facilities: vulnerability to cyber attacks. Water Power Magazine
123. WEC (2016) The road to resilience: managing cyber risks. World Energy Council
124. Weed AS (2017) US policy response to cyber attack on SCADA systems supporting critical national infrastructure. Air University Press
125. Williams S (2020) Tech industry most attacked sector. IT Brief Australia. https://itbrief.com.au/story/report-tech-industry-most-attacked-sector
126. Yle (2016) Russian group behind 2013 Foreign Ministry hack. Yle. https://yle.fi/uutiset/osasto/news/russian_group_behind_2013_foreign_ministry_hack/8591548
127. Yle (2020) Emails compromised in cyber-attack on Finland's Parliament. Yle. https://yle.fi/uutiset/osasto/news/emails_compromised_in_cyber_attack_on_finlands_parliament/11716393
128. Zetter K (2015) Countdown to zero day: Stuxnet and the launch of the world's first digital weapon. Broadway Books, New York

Chapter 2
Key Elements of On-Line Cyber Security Exercise and Survey of Learning During the On-Line Cyber Security Exercise

Mika Karjalainen, Tero Kokkonen, and Niko Taari

Abstract Cyber security exercises have experienced broad evolution in their rela-tively short lifetime. Cyber security exercises have been changing from individual technical skill based trainings or even competitions to the team based organisational learning experiences where different work-roles are trained and exercised during the cyber security incidents. Nowadays the modern requirements for cyber security exercises are collaboration between different training platforms and on-line remote participation of the learning audience. In the domain of cyber security, the most valuable assets are skills and know-how, so the basic ambition for conducting the cyber security exercises for individuals and for organisations is the learning. In this research, the learning experience during the state-of-the-art on-line remote cyber security exercise is studied. NIST NICE cyber security framework is used as a base for knowledge categories of used questionnaire. The results from the on-line cyber security exercise are analysed with and concluded with future research topics.

Keywords Cyber arena · Cyber security · Cyber security exercise · Learning · On-line training

2.1 Introduction

Modern digitalisation have brought novel threats in the cyber domain. That trans-formation of cyber domain has reflected to the learning requirements of the cyber security exercises. Not only technical evolution is changing the behaviour in the dig-ital ecosystem. Current Corona-virus (Covid-19) pandemic [25] has induced transi-

---

M. Karjalainen (✉) · T. Kokkonen · N. Taari
Institute of Information Technology, JAMK University of Applied Sciences, Jyväskylä, Finland
e-mail: mika.karjalainen@jamk.fi

T. Kokkonen
e-mail: tero.kokkonen@jamk.fi

N. Taari
e-mail: niko.taari@jamk.fi

tions globally. People are working remotely on-line from their homes which brings new considerations from the viewpoint of cyber security. According to the European Union Agency for Cybersecurity (ENISA), "the outbreak of Covid-19 has brought an immense change in the way we conduct our lives" [3]. ENISA have released several articles including guidance for cyber security during the pandemic.

That new norm of remote working raises major requirement for conducting the cyber security exercises: There shall be *remote on-line capability in the cyber security exercise* where learning audience shall be capable of joining exercise remotely from their homes. Supposedly, that raises new technical requirements for the exercise platforms and also for the processes of controlling the exercise. Generally technical platforms for cyber security exercises are described with the term Cyber Range. However, the existing spectrum of cyber ranges is heterogeneous and that term is inconsistent. In the modern cyber security exercises the global complex cyber domain shall be simulated and such holistic platform with modern on-line capabilities shall be described with the term Cyber Arena [9].

Various teams with separate missions and functions are utilized for organising the cyber security exercises. Establishment and assignment of the team is formed according to training objectives, exercise category, personnel and other obtainable resources. Blue Team (BT) is the group of exercise learning audience that are responding to the cyber incidents and defending the valuable assets against cyber threats according to the incident response procedures of the particular organization. Traditionally, BT is modelled according to the real organisation structure and there can be one to several BTs operating in the exercise. Red Team (RT) is the threat actor of the exercise. RT is executing real (or simulated) cyber attacks and intrusions against information technology assets of the BT according to the exercise scenario and guidance of the exercise control team titled White Team (WT). WT is responsible for controlling the exercise and maintain situational awareness of the exercise by observations and collected data. WT is also assessing the learning audience of BTs [1, 13, 17, 22, 26].

The life-cycle of cyber security exercise can be considered as a process with three phases: (i) planning, (ii) implementation/exercise execution and (iii) feedback/post exercise [13, 26]. The Institute of Information Technology of the JAMK University of Applied Sciences has organised cyber security exercises since 2011 for the national security authorities, for the private companies of critical infrastructure and for the university students. Overall, during those years, there have been nearly 2000 individuals as exercise target audiences. This research is conducted for the cyber security students of the bachelor's and master's programs during the academic course of cyber security exercise.

In our earlier publication [10], we studied learning outcome in the cyber security exercise by the questionnaire based on National Initiative for Cyber security Education (NICE) Cyber security Workforce Framework (NICE Framework) [20]. That study was based on on-site exercise and because of the new requirements of on-line exercise capability, the same questionnaire is utilized for the learning audience of the on-line cyber security exercise. That enables analysis for learning outcomes of the on-site exercise similarly as done earlier for the on-line exercise.

Structure of the research is as follows. In Sect. 2.2, the Learning during the cyber security exercises is discussed with the relevant theories and frameworks. After that, in Sects. 2.3 and 2.4, the questionnaire based survey is presented with the analysed results. Finally, conclusions are derived with the found future research topics in Sect. 2.5.

2.2 Pedagogical Framework for Learning in On-Line Cyber Security Exercises

From the theoretical framework perspective, the cyber security exercises consist a multidimensional theory frame. It is a learning element for the adult learner, so the rationale for learning must be understood within an andragogy framework of theory [14]. According to the theory of andragogy, the adult learner is most often self-directed, and is able to apply prior knowledge in to learning new things [18]. In cyber security exercises, the learner's operating environment is a learning environment that conforms to the most authentic operating environment as possible, in which the learner monitors and acts independently, as part of a team. According to authentic learning theory [5], a sufficiently authentic learning environment stimulates learning and enables transfer of the learning to work environment. In order for the learning environment to support the competence requirements of modern digital operating domain, the learning environment should be a comprehensive Cyber Arena as described in paper of Karjalainen and Kokkonen [9], thus being able to express sufficient real-life complexity. When the learning environment is as realistic as possible, it can be stated that authentic learning environment theory also includes the thought of experiential learning theory [15], where in addition to experience, communication between the actors is needed. The importance of communication and interaction cannot be ignored, as learning through collegial reflection deepens the learning and makes it possible to bond with the existing competence. In the study [17], Maennel proposed a learning analytics reference model to be used in the life-cycle of cyber security exercise.

A key element in cyber security exercise is the learner's role as part of a team, so collaborative learning theory must also be applied as a learning theory [21]. Students act as in their given roles, as a member of an imaginary organization's IT infrastructure maintenance team. Thus, the ability of the students to work as a part of a team, and to be able to communicate their own observations and build a collective situational picture of the events in the operating environment, and this way to enable learning, is crucial [16]. The cyber security exercise as a teaching method is best suited for a learner who already has the basic cyber security skills. Based on these existing skills and knowledge's learner can construct the new lessons offered by a realistic simulation environment during the exercise. In a cyber security exercise, the student's entry level should be at the highest level of Miller's pyramid [19].

In present study, we examine learning in an on-line form of cyber security exercise. According to Kersley, on-line implementation can be as much a social event for students as on-site implementation [11]. Critical elements of an on-line exercise course implementation can be considered the planning of the course content the methods how the interaction between the participants in the course is built [24], especially when designing a cyber security exercise course where the interaction between students, lecturers and formed exercise teams is a key, to achieve the learning objectives of the course. A critical factor is the ability of the teaching staff to build the necessary interaction between the teaching environment (Cyber Arena) and the required interaction framework [12]. When building an on-line cyber security exercise, special attention must be paid to the engagement of co-operation between students within and between the teams, which is the core of the exercise and has been found to contribute to the quality of the course [2]. However, in addition to these qualitative elements, special attention must also be paid to the secured implementation of the course. When dealing with genuine cyber threat vectors during the course, special attention should be paid to the isolation of the environments and instructions for handling the data.

2.3   Methods and Data

In our previous research, we have studied the requirements of the cyber security learning environment from the perspective of the functional requirements of the training platform [9]. In this research, we also used the questionnaire from our previous study [10] about learning during cyber security exercises in on-site exercise by using the NIST NICE framework basic question battery [20]. The original research plan was to supplement the 2019 survey sample by collecting new set of answers and by conducting qualitative interviews to deepen the interpretation of the data. In March 2020, just when the course began, the Corona-virus pandemic forced us teaching transferring to an on-line mode. Therefore, the cyber security exercise course [7, 8] was implemented in on-line mode. The students accessed the university's Cyber Arena through VPN tunnel, and thus the entire exercise was planned and implemented in on-line mode.

Virtual collaboration groups and rooms were formed for the exercise by using various collaboration tools. Same collaboration tools were used also earlier during the planning process of the course. The collaboration infrastructure of the on-line cyber security exercise is illustrated in Fig. 2.1. By using the built-in training environment and collaboration tools, the training was conducted in full on-line mode in June 2020. The training was carried out using the planned scenario. The active phase of the exercise consisted of a two-day exercise. The planned original research set-up changed, but the changed research set-up allowed to study the on-line exercise arrangements, as well as the analysis of learning during the on-line exercise.

According to the curriculum, students participate in the planning of the exercise event and contribute to build the IT infrastructure and its cyber security architecture

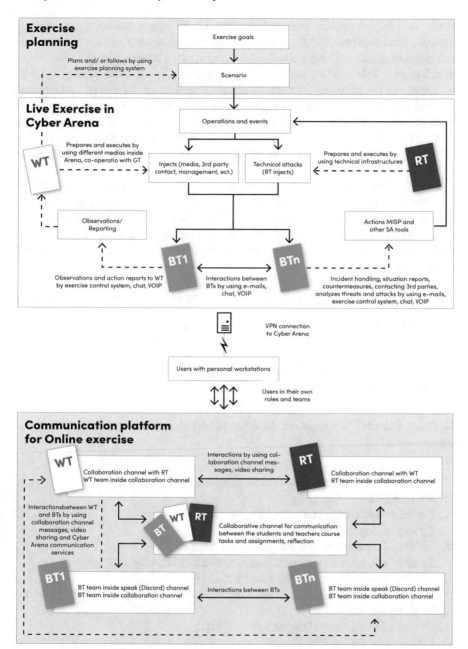

Fig. 2.1 Communication infrastructure of the on-line cyber security exercise

to be used. In previous courses, students have been taught various cyber security controls, their construction, configuration, and management. Environmental vulnerability analysis and auditing methods are also part of the course prerequisites and has been taught to them. When students from both Bachelor and Master degree levels participate in the exercise, the roles of the participants are divided so that Masters level students have more responsibility for organizational entities, architectural level functionalities, and in the event management process for tier 2 and 3 level analysis and investigation. Correspondingly, Bachelor level students are mostly responsible for monitoring security controls, and troubleshooting in the event management process for tier 1 and possibly tier 2 level tasks.

In a research sample that was conducted in 2019, answering to the questionnaire was voluntary for the students. However, we found that among the respondents the disappearance of the respondents was significant. We changed the requirements of the curriculum so that answering of the questionnaire was a mandatory for students to complete the course. Thus, the 2020 sample includes all 33 students who participated in the course.

Similarly as in our earlier study, for evaluating the learning of the topic, five questions were selected to addressing the knowledge level before and after the exercise:

1. (Topic) was/were present in the exercise [Yes/No]
2. (Topic) was/were something I personally encountered during the exercise [Yes/No]
3. My knowledge of (topic) increased during the exercise [Yes/No]
4. Level of knowledge before the exercise [1–10]
5. Level of knowledge after the exercise [1–10]

Similarly, from NICE framework 44 relevant knowledge topics were chosen to the questionnaire:

1. Cyber threats and vulnerabilities
2. Organization's enterprise information security and architecture
3. Resiliency and redundancy
4. Host/network access control mechanisms
5. Cybersecurity and privacy principles
6. Vulnerability information dissemination sources
7. Incident categories, incident responses, and timelines for responses
8. Incident response and handling methodologies
9. Insider Threat investigations, reporting, investigative tools and laws/regulations
10. Hacking methodologies
11. Common attack vectors on the network layer
12. Different classes of attacks
13. Cyber attackers
14. Confidentiality, integrity, and availability requirements and principles
15. Intrusion Detection System (IDS)/Intrusion Prevention System(IPS) tools and applications
16. Network traffic analysis (tools, methodologies, processes)

17. Attack methods and techniques (DDoS, brute force, spoofing,etc.)
18. Common computer/network infections (virus, Trojan, etc.) and methods of infection (ports, attachments, etc.)
19. Malware
20. Security implications of software configurations
21. Computer networking concepts and protocols, and network security methodologies
22. Laws, regulations, policies and ethics as they relate to cybersecurity and privacy
23. Risk management processes (e.g., methods for assessing and mitigating risk)
24. Cybersecurity and privacy principles
25. Specific operational impacts of cybersecurity lapses
26. Authentication, authorization, and access control methods
27. Application vulnerabilities
28. Communication methods, principles, and concepts that support the network infrastructure
29. Business continuity and disaster recovery continuity
30. Local and Wide Area Network connections
31. Intrusion detection methodologies and techniques for detecting host or network -based intrusions
32. Information technology security principles and methods (e.g., firewalls, demilitarized zones, encryption)
33. Knowledge of system and application security threats and vulnerabilities
34. Network traffic analysis methods
35. Server and client operating systems
36. Enterprise information technology architecture
37. Knowledge of organizational information technology (IT) user security policies (e.g., account creation, password rules, access control)
38. System administration, network, and operating system hardening techniques
39. Risk/threat assessment
40. Knowledge of countermeasures for identified security risks. Knowledge in determining how a security system should work (including its resilience and dependability capabilities) and how changes in conditions, operations, or the environment will affect these outcomes
41. Packet-level analysis using appropriate tools (e.g., Wireshark, tcpdump)
42. Hacking methodologies
43. Network protocols such as TCP/IP, Dynamic Host Configuration, Domain Name System (DNS), and directory services
44. Methods and techniques used to detect various exploitation activities

In addition to the learning survey presented above, we conducted an interview to the course lecturers to find out about the key elements of planning and construction of the on-line exercise environments, as well as possible suggestions for improvement that would have arisen from the lecturers. We also wanted to ask for the lecturers experiences and views in relation to the learning outcomes measured from students during the on-line exercise. The interviews were conducted in the fall of 2020 as face-to-face on-line interviews by video conference system due to Covid-19 situation.

The data consists of three semi-structured interviews. The third author of this paper, a member of the course teaching staff, was excluded from the interviews in order to be able to rule out his preconceived notions about the material. The interviews were started with background questions about the role and tasks of the lecturer during the course. Lecturers were then asked to describe the communication environment built for the on-line exercise, what learning environments were used in the course, and how they were used. Informants were also asked to describe in particular how and on what platforms the students and lecturers communicated and what kind of visibility the teaching staff had to the mentioned forums. Lecturers were also asked to qualitatively evaluate how successful the exercise arrangements were in terms of the technical arrangements of the learning environment, and the success of the students' learning. The duration of the interviews ranged from 22 to 42 min. The interviews were recorded and transcribed. The interviews were conducted by the first author.

The data analysis started by differentiating from the data the sections for descriptions of the training arrangements as well as the sections dealing with the students' learning and suggestions for improvement of on-line arrangements. In the analysis, we used conventional method of qualitative analysis, where data is structured, categorized, and merged in higher-level themes [6]. In the analysis, we combined the approach of inductive and abductive analysis [4] in an effort to understand the specific requirements of cyber security exercise in relation to the more general theories of on-line pedagogy presented above.

2.4   Results

The survey data was comprehensively examined. The calculations of average, median and standard deviation was analysed for each knowledge before and after the on-line exercise. In addition to analysing the statistical significance, p-values for each knowledge were calculated and analysed with the null hypothesis of *no learning during cyber security exercises*. Calculated p-values for individual knowledges can be found from Table 2.1. Commonly referred p-values are $p < 0.05$ as statistically significant and $p < 0.001$ as statistically highly significant [23]. If we use those commonly used p-values as the basis of the analysis, we can see that the learning in almost all of the knowledge areas was statistically highly significant and in all except one knowledge it was statistically significant. That specific knowledge where the learning was not statistically significant is the *Packet-level analysis using appropriate tools*, which can be explained that only limited amount of students (one blue team) used packet-level capture and analysis software tools. For the rest of the students there were no appropriate deep packet-level analysis executed during the hectic exercise event.

For visualising the trends of learning during the on-line and on-site exercises, the box plot figures were produced. Figure 2.2 illustrates the box plot statistics containing the interquartile ranges (IQR) of answers in different years. The red (left) box plot shows the level of the knowledge before the exercise while the blue (right) box plot

Table 2.1 P-values calculated for each knowledge of the survey

Knowledge	P-value		Knowledge	P-value
1	0.000000015		23	0.001277018
2	0.000079831		24	0.001534458
3	0.001048385		25	0.000421841
4	0.001472250		26	0.002135130
5	0.001254890		27	0.000067056
6	0.000016276		28	0.001534458
7	0.000008653		29	0.011447886
8	0.000176480		30	0.006124649
9	0.000803151		31	0.008672679
10	0.000377142		32	0.000107062
11	0.000004378		33	0.000435434
12	0.000003624		34	0.000501850
13	0.002590328		35	0.000226935
14	0.036821182		36	0.006124649
15	0.004182652		37	0.001762796
16	0.000057269		38	0.004789459
17	0.000012334		39	0.000939544
18	0.000028078		40	0.001534458
19	0.005926692		41	0.152397681
20	0.000698501		42	0.000004218
21	0.001230732		43	0.000008632
22	0.031610717		44	0.001393172

shows level of the knowledge after the exercise. The median line of the answers is drawn inside of the box plot. Outlier answers are presented as bullets outside of the box plots.

It can be easily seen from the numerical data that there is increasing of the knowledge during the exercise covering all the selected knowledge areas. In advance, there was significant amount of learning during the on-line exercise even if on-line exercise is more complex for the learning audience and it suffers lack of face-to-face communication. That is the most prominent observation based on the numerical estimations.

Because number of samples was limited (number of students participating the course), there was also qualitative analysis done. As qualitative analysis, the lecturers of the course were interviewed. The aim of the interview was to map the essential structures and functionalities of the on-line exercise. The interview also sought to find out the lecturers views and experiences of students learning during the on-line exercise. All of the lecturers have several years of experience from this particular exercise course. The informants saw their own role more as an adviser role who

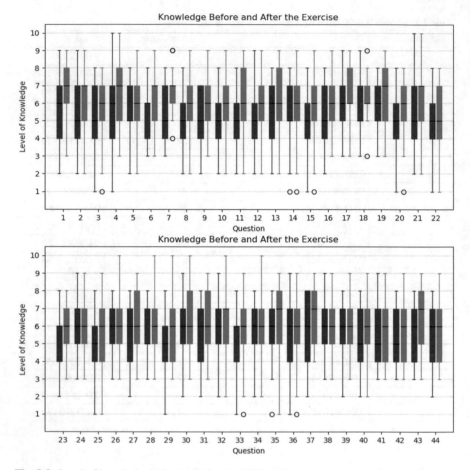

Fig. 2.2 Level of knowledge before and after the 2020 on-line exercise

ensures that the set learning goals will be achieved and supports the students by answering the questions from them. Lecturers had no previous experience of cyber security exercises as an on-line exercise, so the implementation design was experimental in nature and the introduction of best practices arising from other on-line activities:

> The role is to guide, how such an exercise is built from the BT/WT/RT team perspectives. That is, what needs to be considered, how threat activity should be built so that other participants can learn. Role is more of a mentor role, less of a lecturer, for the student there is a big need for self-study and we support the issues that arise for students.

The first contact of the course had been implemented by using traditional classroom teaching, after that first contact lessons the situation quickly changed, so the lecturers need to start planning the on-line execution of the course. Due to the urgency of the schedule, the decision was made to implement the necessary communication platforms and channels by using a communication system familiar to students and lecturers. So they made a decision to use the Microsoft Teams[1] collaboration platform, as the schedule did not allow the construction of a custom made platform inside the Cyber Arena. The GitLab[2] environment was used as the platform for the distribution and saving of the general material of the course. Students logged in to the Cyber Arena from their own workstations using the VPN tunnel to the VMware Cloud Director.[3] An additional chat service was built inside the Cyber Arena, which thus also made it possible to communicate also within the training environment:

> A significant issue was to make the decisions on which (communication) system to install all the information and whether to build systems related to, for example, study or exercise guidance within the Cyber Arena environment, or whether to use out of the Cyber Arena systems/communication channels, such as Teams. At this point, students and lecturers came to the conclusion that Microsoft Teams would be strongly used for its various channels and screen-sharing technology for communicating and sharing information. Students' internal meetings and information sharing was held at Teams and a lot of documentation was shared also at Teams. In the Cyber Arena environment, the actual technical systems, defended targets and attack computers were then used. The planning and conducting of the exercise was carried out with such out-of-game solutions, another option would have been to build everything inside the training system. Here, the students felt that this was (using Teams) a more natural way for them, to use a tool that they also normally use in other studies and communication in everyday life.

In addition to this, BT used the Discord[4] system for intra-team communication, which was mainly used as a voice channel. WT / RT used the Rocket.Chat[5] service for their own mutual and internal communication. The implemented platform for communication and information sharing is shown in Fig. 2.1.

Lecturers had access to all communication channels created and they were able to monitor events and discussions in existing forums without interfering with the content of the discussions. There was also a mode of operation how students or the team were able to invite the lecturer to the channel when they experienced problems or ambiguities during the exercise. Lecturers found the procedure even more effective than in a traditional on-site exercise where they visit classrooms used by students. The transition between the facilities was quick and when problems arose, it was easy for the lecturer to join the conversation when the students invited them to the channel. All lecturers stated that the technical arrangements for the on-line exercise went smoothly. The implemented environment was able to be seamlessly integrated into the exercise, even though it was differentiated from the Cyber Arena. Lecturers felt that the on-line exercise was an encouraging, good experience and gave insight and reassurance that the on-line cyber security exercise also enabled students to learn:

> This was an interesting experience and it is especially interesting to see that students learned so well in this on-line exercise. It shows that, at least for myself, I had doubts about how

---

[1] https://www.microsoft.com/teams/.

[2] https://about.gitlab.com/.

[3] https://www.vmware.com/products/cloud-director.html.

[4] https://discord.com/.

[5] https://rocket.chat/.

well an individual is able to learn when there are no elements of live/on-site exercise around them. At least this survey shows that learning during an exercise has been even better than an on-site exercise, whether it's because of a lower starting level or a careful assessment of one's own skills and then it feels like this went well, so I can't explain that. This was an encouraging, good experience and gave me the reassurance that this exercise should continue to be facilitated in this way. Through further development, this will certainly be a good way to organize an on-line exercise.

As further development needs, the lecturers identified the need to build the entire communication platform inside the Cyber Arena. This was pointed because there is a risk that the training content becoming entangled with other content in the Teams platform. When the Teams platform is also used for other study or work, other possible communication on the Teams platform interferes with the focus on the exercise. Lecturers also found it difficult to monitor an individual student's performance during the exercise. As a result, the evaluation of the exercise was simplified and the numerical evaluation was abandoned. The development of evaluation and the analysis tool used for it, was also seen as a future development task. The situation awareness of students activities during the exercise should also be improved by bringing new situational awareness tools to the exercise. In the future, the communication channels used by the students will also be defined more precisely by the lecturers. Some lecturers also expressed the need to simplify the cyber environment modelled in the Cyber Arena, as in on-line mode students are more passive to ask and thus some threat activities may be left without any actions.

As an overall result, combined from quantitative and qualitative result, it can be said that significant learning takes place during the on-line cyber security exercise. Cyber security exercises are extremely effective for gaining the understanding of the complex cyber incidents and the unexpected behaviour and dependences of the cyber incidents. Infirmity of the on-line exercise versus on-site exercise is the lack of face-to-face communication which reduces analysis of scenarios. However it can be solved with the accomplished technical implementation of the communication infrastructure supporting the requirements of the on-line exercise event.

2.5   Conclusion

The present study examined learning in on-line exercise with a questionnaire built from the NIST NICE framework. The result of the study confirms the analysis of the data collected in the previous study [10], whereby the cyber security exercise serves as an excellent teaching platform and as an tool for teaching cyber security contents with a versatile focus. The on-line exercise also showed that the on-line exercise achieves the set learning objectives well. The self-assessment carried out by the students, where they assess their own level of competence before and after the cyber security exercise, shows statistically significant learning in 43 from 44 content areas of the questionnaire. The result also correlates well with the previous sample, which helps to eliminate the result uncertainties raised by the loss of the respondents in the previous sample.

The qualitative part of the study retrieved information on the aspects of organizing an on-line cyber security exercise, which allows the exercise to be organized in such a way that the exercise can reflect the needed collaborative learning elements, between the individuals and the teams co-operations, problem solving and learning. As a core result, the requirement to build an adequate collaboration platform was emerged. In the exercise under review, the collaboration platform was built on a so-called out of the game style, i.e. outside the actual Cyber Arena. The arrangement was successful in a technical sense, but the lecturers of the course also highlighted areas for development. As things to be developed, the lecturers saw the construction of an collaboration platform inside the Cyber Arena. This avoids security risks and reduces the disadvantages of concentration that may arise from a more general collaboration forum in relation to other activities. Lecturers were positively surprised by the measured learning outcomes, which contributed to strengthening their perception of the future transition of the course to a permanent on-line format. The lecturers found the individual assessment of students difficult, as the rapidly constructed collaboration platform did not allow for sufficiently detailed monitoring of the actions taken by the individual student. For the assessment of the individual, it was felt that better visibility into the student's performance was needed.

The results indicates the difference between students level of knowledge between those who have been in on-line course and those who have been in on-site course. However, the data collected do not provide enough information to analyse the reasons for the difference between the levels of learning outcomes. Future research should seek to analyse the reasons that explain the differences in levels of knowledge between on-line and on-site teaching methods.

Acknowledgements The authors would like to thank Ms. Tuula Kotikoski for proofreading the manuscript and Ms. Heli Sutinen for finalizing Fig. 2.1. This research is funded by the Regional Council of Central Finland/Council of Tampere Region and European Regional Development Fund as part of the Health Care Cyber Range (HCCR) project of JAMK University of Applied Sciences Institute of Information Technology.

References

1. Almroth J, Gustafsson T, (2020) CRATE exercise control: a cyber defense exercise management and support tool. In: IEEE European symposium on security and privacy workshops (EuroS&PW), pp 37–45. IEEE, New York. https://doi.org/10.1109/EuroSPW51379.2020.00014
2. Chen PSD, Gonyea R, Kuh G (2008) Learning at a distance: engaged or not? Innov: J Online Educ 4(3)
3. ENISA (2020) Covid19. The EU Agency for Cybersecurity (ENISA). https://www.enisa.europa.eu/topics/wfh-covid19

4. Graneheim UH, Lindgren BM, Lundman B (2017) Methodological challenges in qualitative content analysis: a discussion paper. Nurse Educ Today 56:29–34. https://doi.org/10.1016/j.nedt.2017.06.002
5. Herrington J, Oliver R (2000) An instructional design framework for authentic learning environments. Educ Technol Res Dev 48(3):23–48. https://doi.org/10.1007/BF02319856
6. Hsieh HF, Shannon SE (2005) Three approaches to qualitative content analysis. Qual Health Res 15(9):1277–1288. https://doi.org/10.1177/1049732305276687
7. JAMK (2020) Cyber security exercise, 5 cr (YTCP0400). JAMK University of Applied Sciences. https://opetussuunnitelmat.peppi.jamk.fi/fi/YTC2020SS/course_unit/YTCP0400. Accessed 30 Nov 2020
8. JAMK (2020) Kyberturvallisuusharjoitus, 5 op (TTC7530). JAMK University of Applied Sciences. https://opetussuunnitelmat.peppi.jamk.fi/fi/TTV2020SS/course_unit/TTC7530. Accessed 30 Nov 2020
9. Karjalainen M, Kokkonen T (2020) Comprehensive cyber arena; the next generation cyber range. In: 2020 IEEE European symposium on security and privacy workshops (EuroS&PW). IEEE, New York, pp 11–16. https://doi.org/10.1109/EuroSPW51379.2020.00011
10. Karjalainen M, Puuska S, Kokkonen T (2020) Measuring learning in a cyber security exercise. In: Proceedings of the 12th International conference on education technology and computers, ICETC 2020. ACM (to be published)
11. Kearsley G (2000) Online education: learning and teaching in cyberspace. Wadsworth Thomson Learning
12. Kearsley G (2008) Tips for training online instructors. Unpublished article
13. Kick J (2014) Cyber exercise playbook. Tech. Rep. MP140714, MITRE Corporation, Wiesbaden. https://www.mitre.org/sites/default/files/publications/pr_14-3929-cyber-exercise-playbook.pdf
14. Knowles MS (1995) Designs for adult learning: practical resources, exercises, and course outlines from the father of adult learning. American Society for Training and Development (ASTD)
15. Kolb DA, Boyatzis RE, Mainemelis C (2001) Experiential learning theory: previous research and new directions. In: Perspectives on thinking, learning, and cognitive styles. Lawrence Erlbaum Associates Publishers, pp 227–247
16. Laal M (2013) Collaborative learning; elements. Proc Soc Behav Sci 83:814–818
17. Maennel K (2020) Learning analytics perspective: evidencing learning from digital datasets in cybersecurity exercises. In: 2020 IEEE European symposium on security and privacy workshops (EuroS&PW). IEEE, New York, pp 27–36. https://doi.org/10.1109/EuroSPW51379.2020.00013
18. Merriam SB, Bierema LL (2013) Adult learning: linking theory and practice. Wiley
19. Miller GE (1990) The assessment of clinical skills/competence/performance. Acad Med 65(9):S63–S67
20. Newhouse W, Keith S, Scribner B, Witte G (2017) National initiative for cybersecurity education (NICE) cybersecurity workforce framework. NIST Special Publication 800-181, National Institute of Standards and Technology. https://doi.org/10.6028/nist.sp.800-181
21. Panitz T (1999) Collaborative versus cooperative learning: a comparison of the two concepts which will help us understand the underlying nature of interactive learning. ED 448 443, ERIC Institute of Education Sciences. https://files.eric.ed.gov/fulltext/ED448443.pdf
22. Seker E, Ozbenli HH (2018) The concept of cyber defence exercises (CDX): planning, execution, evaluation. In: 2018 International conference on cyber security and protection of digital services (Cyber Security). IEEE, New York, pp 1–9. https://doi.org/10.1109/CyberSecPODS.2018.8560673
23. StatsDirect (2020) P values. StatsDirect. https://www.statsdirect.com/help/basics/p_values.htm. Accessed 30 Nov 2020
24. Swan K (2001) Virtual interaction: design factors affecting student satisfaction and perceived learning in asynchronous online courses. Distance Educ 22(2):306–331

25. WHO (2020) Coronavirus disease (COVID-19) pandemic. World Healt Organization (WHO). https://www.who.int/emergencies/diseases/novel-coronavirus-2019. Accessed 5 Nov 2020
26. Wilhelmson N, Svensson T (2014) Handbook for planning, running and evaluating information technology and cyber security exercises. Swedish National Defence College

Chapter 3
Cyber Law and Regulation

Virginia A. Greiman

Abstract The rise of cyber attackers and concerns about cyber espionage, cyber-crime, and cyber warfare have focused the attention of governments, private industry and policy makers in recent years creating a greater need for laws and regulation in the cyber domain. It is no longer sufficient to understand the laws of your own country when it comes to our networked global society. Courts and parliaments around the world must update the law as technology continues to evolve and as artificial intelligence (AI), robotic process automation (RPA), the Internet of Things (IoT), Cloud computing and autonomous machines continue to advance and impact the security of national governments, businesses, and individual privacy. Since it would be an impossible task to cover all laws and regulations governing cyberspace, this chapter focuses on an overview of cyber regulation and the cyber regulators from a distributive, transnational, regional, and national perspective.

Keywords Cyber law · Governance · Regulations · Cyber operations

3.1 Introduction

The principles and over-arching structure of global cyber law is drawn from decades of law making through international organizations including the UN Charter and NATO's North Atlantic Treaty, court decisions, customary law, statutes, and regulations, and more recently risk strategies, policies, and standards. The development of cyber law as an international body of law is complex and raises challenges for cyber-related contract law, tort law governing property rights and personal injury, privacy law, the prosecution of cyber-crimes, law enforcement, civil procedure, and intellectual property protection.

The Internet has had a profound influence on our everyday lives and recent cyber intrusions causing serious social and economic harm to millions of people around the globe, highlights the reality that cyberspace is not a private or protected place. Thus,

V. A. Greiman (✉)
Boston University, Boston, USA
e-mail: ggreiman@bu.edu

in order to protect our basic human rights including the right to privacy, freedom of expression, to be treated fairly, and to security of our property and reputation we need to care about how regulation can be used to provide better protection in the cyber domain. A heightened concern and awareness of cybersecurity issues have led to a number of regulatory responses and driven the development and implementation of national cybersecurity strategies by all global states. In addition to governments and regulators, cybersecurity issues are also receiving greater attention from other stakeholders. For instance, there have been industry-wide private sector initiatives on the cyber front.

Cyber security is just beginning to evolve as a legal duty worldwide. Though much harm can be inflicted on a person through ransomware, cyberbullying, identity theft and other crimes it has been difficult to bring these cases before the courts often due to ambiguity in the laws, failure of attribution, or lack of sufficient evidence. The cases decided worldwide demonstrate that cyber or Internet-related conduct is subject to jurisdiction by foreign governments and plaintiffs, and that defendants can be expected to be subject to litigation anywhere in the world. Cyber law is further complicated by the stability of global connectedness and the potential for diminution of global boundaries.

The concern about global stability raises the question of who is in charge of cyber space and who is accountable when things go wrong. To understand the regulation of cyber security, we need to first explore the governance of cyberspace. In this next section we explore the four dimensions of cyber governance—distributive, transnational, nationalism and regional through the characteristics and frameworks that encompass each dimension.

3.2 Governance of the Internet and Cyberspace

In current practice, cyberspace includes, but is not coextensive with, the Internet [28]. The U.S. government has defined cyberspace as "the interdependent network of information technology infrastructures," which "includes the Internet, telecommunications networks, computer systems, and embedded processors and controllers in critical industries" [65]. The Oxford English Dictionary defines cyberspace similarly as "[t]he space of virtual reality; the notional environment within which electronic communication (esp. via the internet) occurs" [47]. The Joint Chiefs of Staff of the United States Department of Defense define cyberspace as one of five interdependent domains, the remaining four being land, air, maritime, and space [13].

While there is a common understanding of the Internet, there is not yet a shared view of Internet governance. The term 'Internet governance' has been defined in various ways to reflect differing cultures and political, legal, and economic interests. In the United States Internet governance as it is commonly known, has been defined differently within the federal government, the private sector, and military operations to protect the respective interests of each sector. The key challenge is to satisfy vital national security interests, while achieving the critical interrelated objectives of peace, security, and economic and technological advancement. The rise of the Internet centered in the United States has been described by scholars as "a disruptive event in the system of international relations formed around communication and information policy" [41, p. 55].

Table 3.1 illustrates characteristics and organizational frameworks of the common governance models for cyberspace. As listed in Table 3.1, these governance models vary widely and derive from various cyber powers that interact and overlap and encompass (1) distributive, (2) transnational, (3) regional, and (4) national governance models. Table 3.1 has identified the scale and scope of cyber "governance" models, the characteristics of these models and examples of organizational frameworks.

Table 3.1 Cyber governance, characteristics and framework [31]

Cyber governance	Characteristics	Organization/frameworks
Distributive	Self-regulating: code, architecture and markets, decentralized	Open-source communities/WIPO/Linux/software code, social norms
Globalism/transnationalism	Treaties, norms, cooperative problem-solving arrangements, multi-stakeholderism, soft power, networks, interlocking governance systems, multilateralism, bilateralism	ICANN, Internet Society (ISOC), World Wide Web Consortium (W3C), NATO Cyber Centre of Excellence, Cybercrime Conventions, United Nations Charter, OSCE, Council of Europe, The Organization for Economic Cooperation and Development (OECD), WTO, INTERPOL
Regionalism	Geographic contiguity, democratic, co-regulation, shared values and cultures, interaction, common political institutions, military interaction, norms pertaining to conflict resolution	Regional agreements such as: The Shanghai Cooperation Agreement, The Arab Convention on Combating IT Offenses, The African Union Convention on Cyber Security, The Organization of American States Inter-American Cybersecurity Strategy, The Agreement on Cooperation Among the States Members of the Commonwealth of Independent States, and the European Council Convention on Cybercrime
Nationalism	Geographically bounded, single government, state-centric laws, and frameworks, state control, top-down decision making, historic cultural and linguistic cohesiveness	National and international cyber security strategies, privacy policies, laws and regulations, cybercrime law enforcement, standards (NIST) (ISO) frameworks, information sharing

3.2.1 Distributed Governance

Distributed governance exists in many forms on the Internet and has drawn its inspiration from the theory and practice of the open governance movement [63]. Distributive governance facilitates cooperation between actors and organizations, moving away from a top-down, bureaucratic system in which a single authority sets agendas, enabling more flexibility, fluidity and creativity in decision making. John Perry Barlow, an Internet visionary, co-founder of the Electronic Frontier Foundation (EFF) and a former fellow at Harvard's Berkman Center for the Internet and Society, became the voice of the Utopian view of cyberspace in 1996 when he issued a declaration of independence for cyberspace [1]. Barlow contended that legitimacy comes from the consent of Internet users around the globe, and not from government authority and control. Harvard Professor Lawrence Lessig supports this structure but contends that the regulator of cyberspace is Code–the software and hardware that make cyberspace as it is [37, p. 5]. This Code, or architecture, sets limits or constraints on how cyberspace is experienced. For example, Code can provide access to knowledge, limit your privacy, or it can be used to censor speech (p. 7). The architect of the Internet can shape the path of the law, but the architecture can also be disrupted and controlled by the interests of powerful nations and the conflicts within and between them [28].

3.2.2 Transnational Governance

In recent years, a growing body of transnational governance is represented by the evolution of multistakeholderism. The concept of multistakeholderism and the flexible Internet governance vision it embraces has evolved from its original structure as nations have expressed concern over a U.S. Centric governance structure. For example, the transfer of some of the Internet Corporation for Assigned Names and Numbers (ICANN's) technical functions by the U.S. Department of Commerce to the "multistakeholder community" in October 2016 represented the final phase of a plan to privatize the coordination and management of the Domain Name Severs (DNS). The Council of Europe (CoE) in its Internet Governance Strategy (2016–2019) is committed to multistakeholder governance with leading actors in the field of Internet governance, including relevant international organizations, the private sector, and civil society [8].

"Generally, non-governmental stakeholders agree that multistakeholder collaboration and cooperation are the best means to develop effective cybersecurity policies that respect the fundamentally global, open and interoperable nature of the Internet" ([45], p. 18). As described in Finland's cyber security strategy, "[i]nternational cooperation relies on the existing international law, international treaties and respect for human rights also in the cyber environment" [23]. On 16 December 2020, the European Commission and the High Representative of the Union for Foreign Affairs and Security Policy presented a new EU Cybersecurity Strategy. The Strategy covers the security of essential services such as hospitals, energy grids and railways and ever-increasing number of connected objects in our homes, offices and factories, building collective capabilities to respond to major cyberattacks and working with partners around the world to ensure international security and stability in cyberspace. Among the aims of the strategy is to advance a global and open cyberspace [15].

Global transnational governance has become critical to national security and has emerged in various forms. The significance of global transnational governance is evidenced through the enactment of multinational and regional agreements and treaties that can impact cyberspace in a profound way. For example, the role of NATO, as described by Klimburg [34] in the National Cyber Security Framework Manual, is designed to be a political-military alliance, with its interests coalescing around counter-crime, intelligence and counter-intelligence, critical infrastructure protection and national crisis management, and diplomacy and internet governance among its 30 member nations.

With the increase in espionage and surveillance related activities, the obligations of States have become a central concern in the cyber context. The customary law of state responsibility for violations of international law, and with it the law of attribution, is drawn primarily from the long-term work of the International Law Commission (ILC) and its Draft Articles on Responsibility of States for Internationally Wrongful Acts [32]. Although not a treaty, the ILC rules were commended to the member states by the UN General Assembly in 2001 and have become the authoritative guidepost for public international cyber law. The Rules have been cited repeatedly by courts, tribunals, and other bodies. Drawing from the ILC rules, the threshold point in the cyber context is that "a State bears international responsibility for a cyber-related act that is attributable to the State and that constitutes a breach of an international legal obligation" [55, p. 84].

3.2.3 Regional Governance

Regionalism has been identified as the growth of societal integration within a given region, including the undirected processes of social and economic interaction among nation-states [20]. For example, Regional Trade Agreements (RTAs) administered by the World Trade Organization often have broad geopolitical, developmental, macroeconomic, social, and environmental goals, going well beyond trade policy [19].

Regionalism is of increasing significance in cyber strategies, particularly as it relates to national security. One of the most advanced areas in regionalization as it relates to cyber security is the enforcement of international criminal law. There is a wide range of cyber threats, including war, espionage, sabotage, and disruption, [52] and international law is ambiguous about their status as a crime, an act of war, or act of espionage. To address these concerns, groups of states have developed interstate agreements to grant jurisdiction over international cybercrimes to various international tribunals. Some scholars have argued that regionalization of international law enforcement would offer various benefits and drawbacks that are in inherent tension and would offer a means of balancing the benefits and dangers of both supranational and national enforcement [4, p. 730].

As highlighted in Table 3.1, presently, there are six major agreements governing cybercrime representing every region of the world, with the largest regional group represented by the Council of Europe Convention on Cybercrime. The Convention on Cybercrime, popularly known as the Budapest Convention, has served as an important governance tool on many levels including through its mandatory provisions on the enactment of legislation to combat cybercrime, extradition, mutual assistance, and law enforcement [7]. However, there are many that believe the Convention on Cybercrime is stalemated by opposition from countries that use cybercrime as a political tool and by new powers who object to signing a treaty that they did not negotiate. Regionalized organizations must develop ways to penalize those who do not support and assist in cybercrime prevention, deterrence, and prosecution. As of 2020, 64 countries are a member of the Convention. Though not currently a member, Brazil has been invited to accede by the Council of Europe [9].

3.2.4 National System Governance

If one were to look at cyberspace as a single system, national governance by far dominates the existing law and regulation of cyberspace. From the British vote for Brexit, and America's move towards restraints on immigration and free trade, the move toward nationalism is causing major shifts in political power throughout the world. In part, the uprising is a clash of values between the United States advocating for an open cyberspace, while a few other nations including China and Russia push for greater control and regulation of data within their borders. The role of nation-states in Internet governance has been the subject of much debate by the international community. National governments continue to focus on the development of risk strategies, policies, and standards to protect critical infrastructure. National and international cyber security strategies aim to foster better relationships between the public and private sector and among nation-states.

National cyber security encompasses many areas for regulation. A few of the most important are discussed in this chapter including national criminal laws, data privacy regulations, breach notification statutes, reasonable cyber security practices, standards and risk management, and net neutrality regulation.

Recently, there has been a focus on national cyber security law that provides for stricter government controls over 'critical information infrastructure.' For example, in 2017 China's government enacted a broad new cybersecurity law aimed at tightening and centralizing state control over information flows and technology equipment, raising concerns among foreign companies operating in the country [5].

In 2016 Russia passed a new law, which increases government access to online content [53]. The Yarovaya law, as it is known, has been condemned by opposition, human rights activists, and even Edward Snowden, the fugitive NSA contractor, who was granted asylum in 2013 and permanent residency in October 2020 by Russia. But the legislation requires Russian citizens to inform the government whenever they believe they have "reliable" information on a possible terror attack, uprisings, and a slew of other crimes [53]. Art. 16 (Protection of Information) of the Information Law requires the protection of information through various measures, including preventing unauthorised access, hacking, cyber-attacks, and other protections of information. Art. 17 of the Information Law provides for civil remedies and criminal penalties for violation of the law.

In the United States, high level government officials with national security and law enforcement agencies have argued that expanded surveillance powers are needed, especially because of recent threats to America's democratic process. The U.S. Amendments to Rule 41 of the Federal Criminal Rules of Procedure have expanded the hacking powers of law enforcement agencies in the United States giving U.S. judges the ability to issue a warrant for a person or property within or outside their district.

3.3 Cyber Operations

3.3.1 Cyber Warfare

The research on national security strategies, national "cyber security" strategies and military or national defense cyber strategies has shown that there is not a common understanding among nations of an armed cyber-attack [30]. Cyber-attack, cyber warfare, cyber-crimes, and cyber espionage have different and conflicting meanings. Cyberwarfare does not merely change the "weaponry" of modern wars, but it represents a radical shift in the nature of the "wartime battlefield" requiring a reframing of our understanding of conventional kinetic warfare and the new realm of cyber warfare. To develop a framework for cyber warfare, there must first be an agreement among nations as to what constitutes a cyber-attack. This requires agreement on when a cyber-attack has occurred and the extent of damage that must occur to be characterized as a cyber-attack. The Tallinn Manual 2.0 has not discussed the prohibition of intervention below the armed conflict threshold thus, leaving open the question of acceptable responses to cyber-attacks. In accordance with the United Nations Charter Article 51, official doctrine reserves the right to respond to a cyberattack by any means that are felt to be necessary and proportional [55].

Late in 2019, Maj. Gen. Thomas Murphy, Director of the Protecting Critical Technology Task Force (PCTTF), a special Pentagon Task Force established to protect industrial security, affirmed that foreign nations steal billions per year in technology and intellectual property from the United States [38]. The need remains acute today to better protect the information security of defense contractors. A different kind of attack, directed against infrastructure and industrial operations, seeks not to steal data so much as to corrupt it, or host systems, and damage or deny the utility, or safety of physical assets operated through computer instruction or network connection. A well-known example of such a "cyber-physical" attack is the NotPetya malware attack in 2017 that crippled Maersk shipping and impaired the operations of the Merck pharmaceutical giant, which are among many consequences to companies in Europe and elsewhere [29].

3.3.2 Cyber Espionage

The definition of a cyber-attack varies widely. For example, the United States definition of a cyber-attack does not include espionage as the U.S. has a separate Espionage Act (Espionage Act of 1917), while Germany makes no distinction between cyberattack and probe or espionage [26, pp. 14–15]. However, espionage's permissibility under international law remains largely unsettled; no global regulation exists for this important state activity [50, pp. 360—361]. The contradiction of espionage is evident as states deem their own espionage activities legitimate and essential for national security, while aggressively pursuing criminal actions against foreign espionage activity. "The law of espionage is, therefore, unique in that it consists of a norm (territorial integrity), the violation of which may be punished by offended states, however, states have persistently violated the norm" [56, p. 218]. No international convention prohibits the practice of espionage "because all states have an interest in conducting such activity"[56, p. 223].

Although it is unclear under international law whether states, in general, have a lawful right to spy on other states, the disallowance of certain activities within espionage is clearer [50, pp. 360—361]. The treatment of those involved in spying activities as well as the use of torture to extract information has been held unlawful by Courts in many nations [24]. In 2013, fifteen countries, including the United States and China, agreed that international law, in particular, the United Nations Charter applies in cyberspace and explicitly highlighted the need to elaborate confidencebuilding measures and norms, rules, or principles of responsible behavior of States [61].

3.3.3 Cyber Crime

While 154 countries (79%) have enacted cybercrime legislation, the pattern varies by region. Europe has the highest adoption rate (93%) and Asia and the Pacific the lowest (55%) [62]. The evolving cybercrime landscape and resulting skills gaps are a significant challenge for law enforcement agencies and prosecutors, especially for cross-border enforcement.

The UN Office on Drugs and Crime (UNODC) has identified 14 acts that generally constitute cybercrime offences in most nation States which have the relevant law organized into three broad categories:

1. acts against the confidentiality, integrity and availability of computer data or systems, illegal access, interception or acquisition of computer data, breach of privacy or data protection measures;
2. computer-related acts for personal or financial gain, identity offences, intellectual property offences, and solicitation or grooming of children;
3. computer contents-related acts, hate speech, pornography and acts in support of terrorism offences.

The list is not exhaustive, and some states may criminalize certain acts which are not crimes in other states [61].

3.4 Computer Crime Law

3.4.1 Computer Crime Law in the United States

The 1986 U.S. Computer Fraud and Abuse Act (CFAA) remains the most relevant, if dated, applicable law shaping the U.S. cybersecurity landscape. In particular, the CFAA, as amended in 2008, criminalizes "unauthorized access" to a computer or "unauthorized transmission" of things like malware (malicious software), DDoS attacks, identity theft, and electronic theft as well as damaging a protected computer or network, obtaining and trafficking private information, and affecting the use of a computer (such as by using a computer to form a botnet). Violators may face up to 20 years in prison, restitution, criminal forfeiture, and/or a fine. In addition, in certain circumstances, the CFAA allows the victims of computer crimes to bring private civil actions against violators for compensatory damages and injunctive or other equitable relief.

In addition to a comprehensive federal statute, in 1978, Arizona and Florida became the first state legislatures to enact computer crime statutes. Today, according to the National Conference of State Legislators, all 50 states in the United States have some form of computer specific criminal legislation [43]. Though it is beyond the scope of this paper to cover the laws of all 50 States, a brief comparative analysis highlights the challenges of harmonizing these laws and developing a uniform approach to computer crime.

While some state laws are duplicative of federal laws; other laws address computer specific problems including (1) online harassment, (2) spam, (3) spyware, (4) data privacy, and (5) cyberbullying [12]. Some state laws also directly address other specific types of computer crime, such as spyware, phishing, denial of service attacks, and ransomware [43]. Some state statutes also punish actions that cross borders and liability can be imposed in all jurisdictions where the wrongful act occurs. For example, Florida's Computer-Related Criminal statute chapter 815, provides that a "person who causes, by any means, the access to a computer, computer system, computer network, or electronic device in one jurisdiction from another jurisdiction is deemed to have personally accessed the computer, computer system, computer network, or electronic device in both jurisdictions." In recent years' statutes have been enacted to address offenders who have targeted minors on the Internet for advertising and other illicit purposes (Cal. Bus. & Prof. Code; Del. Code sec. 1204).

3.4.2   Computer Criminal Law of Nations

The development of the U.K. Computer Misuse Act (CMA) in 1990 in many ways mirrors the development of the CFAA in 1984. Both were enacted before the World Wide Web was developed; both regulate the concept of "unauthorized access"; and both provide expansive protection to covered computer systems [10]. However, while the CFAA relies upon broad definitions of "protected computer", the CMA provides no explicit definition of "computer", allowing this definitional ambiguity to be resolved by the courts. Canada passed the relevant provisions of its Criminal Code in 1985 shortly after the CFAA was introduced in the U.S. Congress. Other G7 members—including Germany and the United Kingdom–passed relevant laws in the 1990s, while France, Italy, and Japan did not regulate cyber conduct until the 2000s. In New South Wales, Australia, unauthorised access to computer systems is criminalised by both state and federal legislation, namely, the Crimes Act 1900 (NSW) ("the Crimes Act") and the Commonwealth Criminal Code ("the Code"). Most commonly, persons suspected of engaging in cybercrime are charged pursuant to the Code, given its universal application in all states and territories in Australia.

Since their original passage many of these laws have been updated, amended, or supported with additional laws to reflect the changing technology and the needs of the cyber environment. For example, the UK's recent Serious Crime Act (2015) amends the Computer Misuse Act 1990 to ensure sentences for attacks on computer systems fully reflect the damage they cause. The Act included a new Section (3ZA) creating a new category of unauthorized acts if the person does any unauthorized act in relation to a computer including causing, or creating risk of, serious damage to human welfare in any place; damage to the environment of any place; damage to the economy of any country; or damage to the national security of any country. As defined in the Act, a person causes damage to human welfare only if it causes—loss to human life; human illness or injury; disruption of a supply of money, food, water, energy or fuel; disruption of a system of communication; disruption of facilities for transport; or disruption of services relating to health. A person guilty of an offence, unless the act causes damage to human welfare, is liable, on conviction on indictment, to imprisonment for a term not exceeding 14 years, or to a fine, or to both. However, where an offense is committed as a result of an act causing or creating a significant risk of serious damage to human welfare under this subsection, or serious damage to national security, then the person guilty of the offence is liable, on conviction on indictment, to imprisonment for life, or to a fine, or to both.

Under the Criminal Law of the People's Republic of China, cybercrimes are mainly provided in Chapter VI, Section 1 "Crimes of Disturbing Public Order." Articles 285, 286, and 287 are the three major articles that directly relate to cybercrimes. The punishments for violating these Articles include imprisonment, detention, and fines. For example, the offender may be sentenced to up to seven years' imprisonment for illegally obtaining data from a computer information system in serious cases [6].

In Singapore stiff penalties can be assessed for hacking. In one case, Lim Siong Khee v Public Prosecutor [2001] 1 SLR(R) 631, the accused hacked the victim's email account by answering correctly the hint question to successfully retrieve passwords and to gain unauthorised access. He was sentenced to 12 months' imprisonment. In October 2016, the ICO in England issued a then-record £400,000 fine to telecoms company TalkTalk for security failings that allowed a cyber-attacker to access customer data. The ICO investigation found that the attack took advantage of a technical weakness in TalkTalk's systems which could have been prevented if the company had taken "basic steps" to protect customer data [27].

The focus on the need for stiffer criminal penalties is also recognized at the transnational level. In 2013, the European Parliament and the Council issued EU Directive (2013) on attacks against information systems, replacing Council Framework Decision 2005/222/JHA. The Directive establishes minimum rules concerning the definition of criminal offences and sanctions in the area of attacks against information systems. It also aims to facilitate the prevention of such offences and to improve cooperation between judicial and other competent authorities.

3.5 Regulations in Cyber Space

3.5.1 National and Transnational Data Privacy Regulations

As more and more social and economic activities take place online, the importance of privacy and data protection is increasingly recognized [64]. Of equal concern is the collection, use and sharing of personal information to third parties without notice or consent of consumers. Remarkably, 132 out of 194 countries have put in place legislation to secure the protection of data and privacy. Africa and Asia show a similar level of adoption with 55% of countries having adopted such legislations from which 23 are least developed countries [62].

The concept of Privacy by Design (PbD) has proven to be both popular and malleable [48]. "'Privacy by Design' has become a populist term in the privacy community, but it means different things to different people" [40]. For Microsoft, Privacy by Design means an inherent respect for privacy, backed by mature and comprehensive privacy policies and protections [40]. Google, Twitter, and Mozilla have all drawn praise for implementing PbD-style policies, including most notably default encryption.

In the EU, the European Union's General Data Protection Regulation (GDPR), a comprehensive privacy and data security law, was first proposed in 2012 and formally adopted in late 2015. Article 23 of this law requires that controllers of the data of EU citizens "implement mechanisms for ensuring that, by default, only those personal data are processed which are necessary for each specific purpose of the processing and are especially not collected or retained beyond the minimum necessary for those purposes" [25].

Similarly, the U.S. Federal Trade Commission definition of privacy by design can be seen as quite comparable (methodologically and even substantially) to what is in the EU law in all its dimensions... and is clearly formulated with a view to the practical implementation of the principle. Moreover, the National Institute of Standards and Technology (NIST), an agency of the U.S. Department of Commerce, issued an internal report on privacy engineering and risk management in federal systems that includes a privacy risk model and a methodology to implement privacy requirements when engineering systems are processing personal data [3].

Another important piece of privacy legislation, the US/EU Privacy Shield Agreement [49] has been challenged by other countries in the European Court of Justice to ensure that data transferred by companies across the Atlantic would be afforded the same level of protection as in Europe. This makes it extremely costly for U.S. businesses to protect the data of citizens transferred from other countries around the world. In July 2020, the European Court of Justice (CJEU) struck down the Privacy Shield that secured unrestricted EU-US data flow on the grounds that personal data transferred to and stored in the U.S. could not be guaranteed an adequate level of data protection as that under the GDPR. As a result of that decision, the EU-U.S. Privacy Shield Framework is no longer a valid mechanism to comply with EU data protection requirements when transferring personal data from the European Union to the United States. However, that decision does not relieve participants in the EU-U.S. Privacy Shield of their obligations under the EU-U.S. Privacy Shield Framework. Australia has gone even further under its Information Privacy Act of 2014 as amended in 2017 to require mandatory disclosure of data breaches by all companies that control data of Australian nationals.

Joining the global trend originating in Europe with the General Data Protection Regulation (GDPR), Brazil recently enacted its own omnibus law governing the use of personal data, the Lei Geral de Proteção de Dados (LGPD), or General Law for the Protection of Privacy. Similar to the EU's GDPR and California's Consumer Privacy Act (CCPA), LGPD is intended to regulate the processing of personal data. The LGPD applies to any natural person or legal entity, including the government, that processes the personal data of the people of Brazil, even if the entity processing the data is based outside of Brazil.

In the United States privacy is mainly enforced at the state level. Many states have laws or regulations that impose cybersecurity, data protection, or notification requirements on covered organisations. For example, in 2017, New York's Department of Financial Services implemented its Cybersecurity Regulation (23 NYCRR 500), which requires banks, insurance companies, and other covered entities to establish and maintain a cybersecurity programme designed to protect consumers and ensure the safety and soundness of New York State's financial services industry. Similarly, in 2018 the state of Colorado enacted privacy and cybersecurity legislation that will require covered entities to implement and maintain "reasonable procedures" around the protection and maintenance of confidential information.

One of the more comprehensive regulations is the Massachusetts cybersecurity regulation (Reg 201 CMR 17), which requires companies to maintain cybersecurity programs that, at a minimum and to the extent technically feasible, should have:

1. secure authentication protocols;
2. secure access control measures;
3. encryption of all transmitted records containing personal information;
4. reasonable monitoring of systems for unauthorized use;
5. encryption of all information on laptops;
6. up to date firewall protection and security patches;
7. malware protection and virus definitions, and
8. education and training of employees on all aspects of cyber systems.

Thera are two special regulatory privacy provisions at the federal level in the United States. The Gramm–Leach–Bliley Act (GLBA), also known as the Financial Services Modernization Act of 1999, requires financial institutions to explain their information-sharing practices to their customers and to protect their customer's private information. The second law, the Health Insurance Portability and Accountability Act of 1996 (HIPAA) required the Secretary of the U.S. Department of Health and Human Services (HHS) to establish regulations for the privacy and security of identifiable health information known as protected health information (PHI). PHI entails "individually identifiable health information" such as an individual's past, present, or future physical or mental health or conditions. The HIPAA Breach Notification Rule, 45 CFR §§ 164.400–414, requires HIPAA-covered entities and business associates to provide notifications in the event of certain incidents impacting PHI.

3.5.2 Breach Notification Statutes

Even though many countries including lesser developed countries have laws that mandate data breach notification (e.g., Philippines, Data Privacy Act of 2012; Qatar, Law No. (13) of 2016 Concerning Personal Data Protection; and Indonesia, Regulation No. 82 of 2012 regarding Provisions of Electronic Systems and Transactions and its implementing regulation, Regulation No. 20 of 2016 regarding the Protection of Personal Data in an Electronic System), data breach notifications are not mandatory in most countries and/or are mandatory for the private sector and not the public sector in other countries, or only for certain sectors in society (e.g., Angola and Serbia). In Argentina, while data breach notification is not required, agencies are required to keep records of data breaches in the event they occur in case they are requested during an investigation or audit [62].

While there is no general federal law on data breach notifications in the U.S. other than a few sector specific statutes, all 50 U.S. states and four territories have now passed breach notification statutes with varying requirements [42]. Typically, breach notification statutes require notification be sent to individuals whose electronic personal Information, as defined therein, was acquired in an incident, though some states require notification based on access to such information alone. Personal identification can include social security numbers and driver license numbers, however, increasingly, states are also including in the definition of personal Information, health, and biometric information, as well as usernames and passwords that provide access to an online account.

In 2017, the SEC announced the creation of a Cyber Unit within its Enforcement Division to specifically focus on cyber-related misconduct. In February 2018, the SEC also released an interpretive statement and guidance on public company cyber-security disclosures, which expanded previous guidance and stressed that regulated companies have an affirmative obligation to disclose material cybersecurity risks and incidents and that an ongoing incident investigation does not allow a company to avoid making such disclosures [57]. And, in April 2018, the SEC charged Altaba Inc. (formerly Yahoo! Inc.) with failing to disclose a material cybersecurity breach. Altaba agreed to pay a $35 million civil penalty to settle the SEC's charges [58].

3.5.3   Regulation of Cyber Security: The Reasonableness Standard

Many regulators expect regulated companies to have implemented "reasonable" security measures, taking into account factors such as the sensitivity of the data protected. In light of the proliferation of standards, many companies rely on omnibus cybersecurity frameworks like the NIST Cybersecurity Framework [44], which recommends that companies take steps to identify and assess material foreseeable risks (including with vendors), design and implement policies and controls to protect the organisation in light of those risks, monitor for and detect anomalies and realised risks, respond promptly and adequately to incidents and then recover from any incident.

Ukraine is one among many countries that has adopted new laws on cyber security aimed at reducing the likelihood of a cyber-attack and speeding up the recovery process after an attack. Following the practices of the leading EU countries and the U.S. on the issues of cyber protection in cyberspace, the Law "On Basic Principles of Cyber Security of Ukraine" was adopted by Ukraine on October 5, 2017. According to Article 1 of the Law the concept of cyber protection is defined as a set of organizational, legal, engineering, and technical measures, as well as measures of cryptographic and technical protection of information. The law is aimed at preventing cyber incidents, detecting, and protecting against cyber-attacks, eliminating their consequences, and restoring the sustainability and reliability of functioning systems of communication technological systems.

In the United States, in the absence of a federal law, the question of whether a company or organization has exercised "reasonable" cyber security practices has been the purview of the Federal Trade Commission. As an example, in August 2013, the U.S. Federal Trade Commission brought an administrative action against LabMD, Inc., a small, little-known medical testing company in Atlanta, Georgia, alleging violations of the Act in connection with alleged security breaches in 2008 and 2012. The Commission's legal argument against LabMD was their failure to employ "reasonable and appropriate measures" to protect consumers' information constituted an "unfair" act or practice under Section 5 of the FTC Act [35]. This decision was overturned by the Eleventh Circuit Court of Appeals holding that the LabMD consent order was void for lack of specificity [36].

3.5.4 Standards for Cybersecurity Regulation

In the absence of legislation governing reasonable cyber security practices, standards play a key role in improving cyber defense and cyber security across different geographical regions and communities. The number of standards development organizations and the number of published information security standards have increased in recent years, creating significant challenges. Nations are using standards to meet a variety of objectives, in some cases imposing standards that are competing and contradictory, or excessively restrictive and not interoperable. Other standards favor companies that are already dominant in their field. The European Union, with the support of the European Union Agency for Cyber Security (ENISA), has started to include standards in its strategies and policies, but much remains to be done that requires the involvement of public and private sector actors working in tandem [51].

Since 2009, ENISA has been identifying and elaborating on the work performed by standardization bodies (such as ISO, ETSI, ITU, CEN, CENELEC) relevant to its areas of work. One of the first deliverables summarized and presented findings covering the importance of correctly defining resilience in the context of standardization, the identification and presentation of the major activities undertaken by Standards Developing Organizations (SDOs) in security, and identification of key areas where further work is necessary [17].

The European Commission in December 2020 issued a new standard setting law-the Digital Services Act (DSA) for comment. This Act builds on the rules of the E-Commerce Directive, and addresses the issues emerging around online intermediaries. Member States have regulated these services differently, creating barriers for smaller companies looking to expand and scale up across the EU and resulting in different levels of protection for European citizens [14]. Whereas the GDPR harmonized and raised data protection standards, the DSA aims to establish a comprehensive framework for how digital services operate in Europe to address among other factors illegal content and societal harm [2].

The move towards bottom-up regulatory frameworks is best evidenced in the National Institute for Standards and Technology (NIST) cybersecurity framework [44], which aims to improve private sector cybersecurity through voluntary standards. The framework integrates industry standards and best practices to help organizations manage their cybersecurity risks. It provides a common language that allows staff at all levels within an organization—and at all points in a supply chain—to develop a shared understanding of their cybersecurity risks. NIST worked with private sector and government experts to create the framework. The effort went so well that Congress ratified it as a NIST responsibility in the Cybersecurity Enhancement Act of 2014.

Companies from around the world have embraced the use of the Framework, including JP Morgan Chase, Microsoft, Boeing, Intel, Bank of England, Nippon Telegraph and Telephone Corporation, and the Ontario Energy Board. Similarly, the Federal Financial Institutions Examination Council (FFIEC) provides a "repeatable and measurable process for financial institutions to measure their cybersecurity preparedness over time." Like the NIST RMF, the FFIEC CAT offers core principles and goals but relies on the company's own risk-management assessment and strategies [22].

3.5.5   Net Neutrality Regulation

In recent years, net neutrality has been an important legal issue in cyber security. Net neutrality is the core principle that Internet service providers and governments should treat all data on the Internet equally and not engage in price discrimination or arbitrage. In the United States, the Federal Communications Commission (FCC) has jurisdiction over interstate and foreign communications by wire and radio. Neutrality is threatened when broadband providers block or slow down Internet traffic based upon the customer's wealth or financial condition. Net neutrality concerns an Internet users' access and expression in addition to regulatory issues. The FCC has adopted three basic principles to promote net neutrality: (1) transparency which requires disclosure of network management practices, (2) no blocking which prohibits the blocking of lawful content, and (3) no unreasonable discrimination in transmitting lawful network traffic [21].

The European Union approved strong rules in 2015, requiring companies that provide internet access to handle all traffic equally, leaving flexibility to restrict traffic when network equipment was operating at its maximum capacity [16]. EU rules also allow traffic restrictions to protect network security and handle emergency situations. In 2016, European Union electronic communications regulators detailed potential problems in agreements between telecommunications companies and content providers. And they explained that quality of service could vary, but no specific applications should be discriminated against. In 2017, they highlighted the importance of Europe's emphasis on proactively monitoring compliance with net neutrality rules, rather than waiting for violations to happen before reacting. This gives European residents much stronger consumer protection than exists in the U.S. On June 4, 2012, the Netherlands became the first country in Europe and the second in the world, after Chile, to enact a network neutrality law. The importance of net neutrality rules by all countries cannot be overstated, yet it remains to be seen how this issue is prioritized in a nation's cyber security regulatory regime.

3.6   Summary

In summary, the cyberspace ecosystem is global and complex, and regulating it is a challenge. There is no single set of rules, and no single definition of what governance structure serves the best interests of all participants. The emerging area of cyber human rights including privacy and freedom of expression must be integrated into cyber governance structures and continually evaluated for effectiveness and sustainability and alignment with global norms of behavior. In defending cyberspace, governance models should be evaluated for performance, adaptable to the changing needs of its ecosystem, and transformative to address not only the evolving technology, but also the evolving practices that are developed through the infusion of a broader perspective of ideas from multiple stakeholders engaged in the advancement of a sustainable and legitimate cyberspace.

The tension between nation-states and globalization will continue as international actors and non-governmental organizations press for inclusion in governance and multistakeholder organizations like ICANN. As we have seen, many nations are concerned about common issues such as critical infrastructure protection, fighting cybercrime, and promoting an open Internet, but even those advocates and nations that favor a state-centric approach to cybersecurity have noted the important roles that the international community and international law play in enhancing cybersecurity [59]. Nation-states can be strengthened through regionalism and globalization, however, the concern for national security can lead to the decline of globalization as evidenced by increasing national focus on safeguarding the nation, the building of public private partnerships, and increased protectionism in national cyber security strategies and legal frameworks.

Looking to the future, all countries need to address these challenges in the formulation of its national cyber security strategies. This requires negotiation of agreements with the private sector on cyber security and Internet governance, addressing the need for sanctions at the global level against countries that have not complied with international law on cyber security, and balancing the needs of national security, while protecting the privacy and civil liberty interests among the world's users of cyberspace.

References

1. Barlow JP (1996) A declaration of the independence of cyberspace. Electronic Frontier Foundation, Davos. https://www.eff.org/cyberspace-independence
2. Blankertz A, Jaursch J (2020) How the EU plans to rewrite the rules for the Internet. Brookings Tech Stream. https://www.brookings.edu/techstream/how-the-eu-plans-to-rewrite-the-rules-for-the-internet/
3. Brooks S, Garcia M, Lefkovitz N, Lightman S, Nadeau E (2017) An introduction to privacy engineering and risk management in federal systems. NISTIR 8062, National Institute for Standards and Technology (NIST). https://nvlpubs.nist.gov/nistpubs/ir/2017/NIST.IR.8062.pdf
4. Burke-White WW (2003) Regionalization of international criminal law enforcement: a preliminary exploration. Faculty Scholarship at Penn Law 959, https://scholarship.law.upenn.edu/faculty_scholarship/959
5. Chin J, Dou E (2016) China's new cybersecurity law rattles foreign tech firms. The Wall Street Journal
6. China (1997) Criminal law of the People's Republic of China. https://www.ilo.org/dyn/natlex/docs/ELECTRONIC/5375/108071/F-78796243/CHN5375%20Eng3.pdf
7. CoE (2001) Convention on cybercrime. ETS No. 185, Council of Europe (CoE). http://www.coe.int/en/web/conventions/full-list/-/conventions/treaty/185
8. CoE (2016) Internet governance strategy 2016—2019: Democracy, human rights and the rule of law in the digital world. Council of Europe (CoE). https://rm.coe.int/16806aafa9
9. CoE (2019) Budapest Convention: Brazil invited to accede. T-CY News, Council of Europe (CoE). https://www.coe.int/en/web/cybercrime/-/budapest-convention-brazil-invited-to-accede
10. Craig AN, Shackelford SJ, Hiller JS (2015) Proactive cybersecurity: a comparative industry and regulatory analysis. Amer Bus Law J 52(4):721–787
11. de Bossey C (2005) Report of the Working Group on Internet Governance. Working Group on Internet Governance (WGIG). http://www.wgig.org/docs/WGIGREPORT.pdf
12. Decker C (2008) Cyber crime 2.0: an argument to update the United States criminal code to reflect the changing nature of cyber crime. Southern Calif Law Rev 81(5):959—1016
13. DoD (2013) Cyberspace operations. Joint Publication 3–12(R), Department of Defense (DoD). https://fas.org/irp/doddir/dod/jp3_12r.pdf
14. EC (2020a) Digital services act—questions and answers. European Commission (EC), Brussels, https://ec.europa.eu/commission/presscorner/detail/en/QANDA_20_2348
15. EC (2020b) The EU's cybersecurity strategy for the digital decade. European Commission (EC). https://ec.europa.eu/digital-single-market/en/news/eus-cybersecurity-strategy-digital-decade
16. EC (2020c) Open Internet. European Commission (EC). https://ec.europa.eu/digital-single-market/en/policies/open-internet.
17. ENISA (2011) Resilience metrics and measurements: technical report. European Union Agency for Cybersecurity (ENISA). https://www.enisa.europa.eu/publications/metrics-tech-report
18. ENISA (2018) ENISA's PETs maturity assessment repository: populating the platform. European Union Agency for Network and Information Security (ENISA)
19. Fawcett L (1995) Regionalism in historical perspective. In: Fawcett L, Hurrell A (eds), Regionalism in world politics: regional organization and international order. Oxford University Press, Oxford
20. Fawcett L, Hurrell A (eds) (1996) Regionalization in world politics: regional organization and international order. Oxford University Press, Oxford
21. FCC (2018) Restoring internet freedom. Federal Communications Commission (FCC). https://www.fcc.gov/restoring-internet-freedom (Public access to this page has been disabled by the content owner.)
22. FFIEC (2017) Cybersecurity assessment tool. Federal Financial Institutions Examination Council (FFIEC). https://www.ffiec.gov/cyberassessmenttool.htm

23. Finland (2019) Finland's cyber security strategy 2019. Security Committee, Helsinki
24. Forcese C (2011) Spies without borders: international law and intelligence collection. J Nat'l Secur Law Policy 5:179–210
25. GDPR (2016) General data protection regulation GDPR. European Union (EU). https://gdpr-info.eu/
26. Germany (2011) Cyber security strategy for Germany. Federal Ministry of the Interior
27. Glick B (2016) TalkTalk hit by record £400,000 fine over data breach. Computer Weekly. https://www.computerweekly.com/news/450400451/TalkTalk-hit-by-record-400000-fine-over-data-breach
28. Goldsmith J, Wu T (2006) Who controls the internet? Oxford University Press, Illusions of a borderless world
29. Greenberg A (2018) The untold story of NotPetya, the most devastating cyberattack in history. Wired. https://www.wired.com/story/notpetya-cyberattack-ukraine-russia-code-cra shed-the-world/
30. Greiman VA (2015) Cybersecurity and global governance. J Inf Warfare 14(4):1–14
31. Greiman VA (2019) The winds of change in world politics and the impact on cyber stability. Int J Cyber Warfare Terrorism 9(4):27–43. https://doi.org/10.4018/IJCWT.2019100102
32. ILC (2001) Responsibility of states for internationally wrongful acts. In: Report of the International Law Commission on the work of its fifty-third session. International Law Commission (ILC), United Nations, pp 31—208
33. IoT (2020) H.R.1668—IoT cybersecurity improvement act of 2020. 116th Congress (2019–2020), U.S. House or Senate
34. Klimburg A (ed) (2012) National cyber security framework manual. NATO CCD COE Publication, Tallinn, NATO
35. LabMD (2013). Complaint counsel's opposition to respondent LabMD, Inc.'s motion to stay proceedings. Docket No. 9357, U.S. Federal Trade Commission
36. LabMD (2018) On petition for review of an order of the Federal Trade Commission (FTC Docket No. 9357). No. 16-16270 (11th Cir.), U.S. Court of Appeals
37. Lessig L (2006) Code: and other laws of cyberspace, Version 2.0. 2nd revised ed, Basic Books
38. Lopez T (2019) Task force curbs technology theft to keep joint force strong. DoD News, U.S. Dept of Defense. https://www.defense.gov/Explore/News/Article/Article/2027555/task-force-curbs-technology-theft-to-keep-joint-force-strong/
39. Maurer T (2011) Cyber norm emergence at the United Nations: an analysis of the UN's activities regarding cyber-security. Discussion Paper 2011-11, Belfer Center for Science and International Affairs, Harvard Kennedy School, Cambridge, MA
40. Meisner J (2010) Privacy by design at Microsoft. Official Microsoft Blog. https://blogs.micros oft.com/on-the-issues/2010/11/30/privacy-by-design-at-microsoft/
41. Mueller M (2010) Networks and states: the global politics of internet governance. MIT Press, Cambridge, MA
42. NCSL (2019) Security breach legislation. National Conference of State Legislatures (NCSL). https://www.ncsl.org/research/telecommunications-and-information-technology/sec urity-breach-notification-laws.aspx
43. NCSL (2020) Computer crime statutes. National Conference of State Legislatures (NCSL). https://www.ncsl.org/research/telecommunications-and-information-techno logy/computer-hacking-and-unauthorized-access-laws.aspx
44. NIST (2014) Framework for improving critical infrastructure cybersecurity. Version 1.0, National Institute of Standards and Technology (NIST). https://www.nist.gov/cyberframework/framework
45. OECD (2012) Cybersecurity policy making at a turning point: analyzing a new generation of national cybersecurity strategies for the internet economy; and Non-governmental perspectives on a new generation of national cybersecurity strategies: Contributions from BIAC, CSISAC and ITAC. Organization for Economic Cooperation and Development (OECD), Paris
46. OECD (2021) Recommendation of the council on digital security of critical activities. OECD/LEGAL/0456, Organization for Economic Cooperation and Development (OECD)

47. OED (2017) Cyberspace, n. Oxford English Dictionary (OED)
48. Pardau SL, Edwards B (2017) The FTC, the unfairness doctrine, and privacy by design: new legal frontiers in cybersecurity. J Bus Tech Law 12(2):227—276. https://digitalcommons.law.umaryland.edu/jbtl/vol12/iss2/5
49. Privacy (2020). Privacy shield program overview. Privacy Shield Framework. https://www.privacyshield.gov/Program-Overview
50. Pun D (2017) Rethinking espionage in the modern era. Chicago J Int Law 18(1):353–391
51. Purser S (2014) Standards for cyber security. In: Hathaway ME (ed) Best practices in computer network defense: incident detection and response. IOS Press, Amsterdam, pp 97–106
52. Rid T (2013) Cyber war will not take place. Oxford University Press, Oxford
53. Russia (2016) Yarovaya Law. Russian federal bills 374-FZ and 375-FZ.
54. Rustad ML (2020) Global internet law, 3rd edn. West Academic Publishing, St. Paul, MN
55. Schmitt MN (ed) (2017) Tallinn manual 2.0 on the international law applicable to cyber operations, 2nd edn. Cambridge University Press, Cambridge
56. Scott RD (1999) Territorially intrusive intelligence collection and international law. Air Force Law Rev 46:217–226
57. SEC (2018a) Commission statement and guidance on public company cybersecurity disclosures. Release Nos. 33-10459 and 34-82746, U.S. Securities and Exchange Commission (SEC). https://www.sec.gov/rules/interp/2018/33-10459.pdf
58. SEC (2018b) Altaba, formerly known as Yahoo!, charged with failing to disclose massive cybersecurity breach; Agrees to pay $35 million. Press Release, U.S. Securities and Exchange Commission (SEC)
59. Shackelford SJ, Kastelic A (2015) Toward a state-centric cyber peace? Analyzing the role of national cybersecurity strategies in enhancing global cybersecurity. New York Univ J Legislation Public Policy 18:895–984
60. UNIDIR (2017) The United Nations, cyberspace and international peace and security: Responding to complexity in the 21st century. United Nations Institute for Disarmament Research (UNIDIR). https://www.unidir.org/files/publications/pdfs/the-united-nations-cyberspace-and-international-peace-and-security-en-691.pdf
61. UNODC (2013) Comprehensive study on cybercrime. United Nations Office on Drugs and Crime (UNODC), Vienna. https://www.unodc.org/documents/organized-crime/UNODC_CCPCJ_EG.4_2013/CYBERCRIME_STUDY_210213.pdf
62. UNODC (2020) Data breach notification laws. United Nations Office on Drugs and Crime (UNODC). https://www.unodc.org/e4j/en/cybercrime/module-10/key-issues/data-breach-notification-laws.html
63. Verhulst SG, Noveck BS, Raines J, Declercq A (2014) Innovations in global governance: toward a distributed internet governance ecosystem. Paper Series, No. 5, Centre for International Governance Innovation and Chatham House
64. WEF (2014) Rethinking personal data: a new lens for strengthening trust. World Economic Forum (WEF). http://www3.weforum.org/docs/WEF_RethinkingPersonalData_ANewLens_Report_2014.pdf
65. WH (2009) Cyberspace policy review: assuring a trusted and resilient information and communications infrastructure. The White House, Washington, D.C.
66. WSIS (2003) Declaration of principles. Adopted in World Summit on the Information Society (WSIS), Geneva, 10–12 Dec 2003, International Telecommunication Union (ITU). http://www.itu.int/net/wsis/docs/geneva/official/dop.html
67. WSIS (2005) Tunis agenda for the Information Society. Adopted in World Summit on the Information Society (WSIS), Tunis, 16–18 Nov 2005, International Telecommunication Union (ITU). https://www.itu.int/net/wsis/docs2/tunis/off/6rev1.html

Chapter 4
Understanding and Gaining Human Resilience Against Negative Effects of Digitalization

Kirsi M. Helkala and Carsten F. Rønnfeldt

Abstract Digitalization of society has, like everything else, both positive and negative sides. Influencing people via cyber domain is one of the aspects digitalization has brought. Consequences of the influence vary depending on who is behind an information operation and what the goals of the operation are, even though the methods might still be the same. In this chapter, we discuss influencing an individual who is a part of an information system; what influence is based on, what kinds of influences there are and how we can resist being influenced. While we focus on military matters, the findings also apply for civilians in society at large.

Keyword Digitalization · Cybersecurity · Human Resilience

4.1 Introduction

Digitalization as the "conversion of text, pictures, or sound into a digital form that can be processed by a computer" by Lexico [1] has brought information and services closer to the users. When living in highly digitalized societies, such as the Scandinavian, it is hard to come up with examples of services, occupations and hobbies that do not benefit from digitalized forms of information. With mobile phones, which almost everyone in highly digitalized societies has, access to all information and services is literally within a reach of hand.

Digitalization as everything else has both positive and negative sides. For instance, by connecting people in shared information systems it has made services more equally available for everyone and arguably created more equality among different user groups. However, the same forms of connectivity also enables cyberattacks. Recent examples of attacks that have received media attention in Nordic countries include the lockdown of heating systems in housing blocks in Finland in 2016 [2], the data breach

K. M. Helkala (✉)
Norwegian Defence University College, Cyber Academy, Lillehammer, Norway
e-mail: khelkala@mil.no

C. F. Rønnfeldt
Norwegian Defence University College, Military Academy, Oslo, Norway

that compromised all health records of the South-East Regional Health Authority in Norway in 2018 [3], and the ransomware attack that affected global production lines of the Norwegian organization Hydro in 2019 [4]. These attacks were all based on vulnerabilities in IoT devices. Connectivity might also create domino effects among services that are dependent on each other and cause even more significant consequences for the society.

On a user level, the consequences of cyber-physical attacks towards services vary from a mild irritation when an online service is not available to life threatening when, for example, water and power supplies are damaged for a longer period of time. There is very little a person can do to stop a cyber-physical attack on a service he is a user of. The users can mostly mitigate consequences that they are facing by being proactive and keeping back-up resources available.

However, cyber-physical attacks are not the only concern in a digital society. As mentioned, digitalization has made information widely accessible, which is of course positive. The negative side is that it also enables malicious information actions, which in a military setting are the 'visible' part of information operations. The latter also includes analysis of information environment, planning process, synchronization, integration and coordination of information actions [5]. The effects of such information actions should be in line with the overall mission's aims and objectives. However, information operations are not only delivered by armed forces as shown later in Sect. 4.3, but others also use similar techniques to affect people's minds, values, and attitudes [6].

In this chapter we look at the factors that predispose us as individuals to be victims of information operations. Focusing on military related studies we also discuss how education can be used to make people more resilient against information operations and the negative effects of digital society in general. The argument builds on our presentation at the International Society of Military Sciences 2020 conference under the CC BY 4.0 license [7].

4.2 Human—Part of Information System

Any individual is part of information systems [8]. As individuals, we collect, analyze, store, edit and share information. Information is not only lingual, oral or written. We collect information from our surroundings via our sensory system and its five senses—seeing, hearing, feeling, tasting and touching [9]. Moreover, every piece of information that we sense will affect us, whether we acknowledge it or not [10].

The business sector exploits these features of our sensory system in an effort to make us buy items and services. Sensory marketing [11, 12] is based on the perception capabilities of our senses, and the impact that different signals will have on us.

Vision, for instance, has the most significant role in the perception [11, 12]. In sensory marketing colors, color brightness, designs and patterns of products and lighting conditions in stores have four functions: draw attention, impress us, give information out easily and satisfy customers' expectations [11]. The intended and most likely outcome is the customers purchasing more products when all four functions have a positive impact on them.

Sounds create feelings and emotions and have an impact on mood and cognition [11]. The impact varies depending on the rhythm, ambience of melody and decibel level. For example, rhythmical music has been found to energize people [11], moderate level noise to enhance performance on creative tasks [13] and high-level noise to impair creativity [13].

It is interesting to note that in the physical world, smell influence our feelings more so than do sight and sound [11, 12]. Taste is closely associated with smell, and together they create the flavour sensation and recall memories [11]. However, associations are individual; the impact varies from person to person. These features of our sensory system are well known to most of us both in physical world but also in cyberspace. However, smell, taste and touch are still far behind from audio-visual influencing in cyberspace.

The Internet is an environment [14] that expands cognitive memory [15] as memories are stored on social platforms [16] meaning that we have almost instant access to other memories and experiences online. Memories and stories stored with text, figures and films lay the foundation for emotional contagion [17]. Emotional contagion means transferring emotional states from one person to another [18] and is common in-person interaction. Emotional contagion also occurs online without in-person interaction as shown in a study on Facebook posts and their comments [17].

4.3   Types of Influence

Sensory marketing and emotion arousal techniques aim to influence target groups by conveying information in a manner expedient to the sender [19, 20]. Also here "knowledge of the target groups is essential as people tend to accept agreeable information that pleases them and stories are sold by playing with audiences' desires, prejudices, and allegiances" [21]. This section discusses different types of influence that an individual is exposed to daily. Figure 4.1 gives an overview.

4.3.1   Information Operations

Information operations are part of military operations, as already noted. However, the concept can be widened to include all actions where information is used to influence behavior of other people. The same tools are often used to influence a person even if the actor behind them might have different goals. All examples, except addiction and technostress in Fig. 4.1, can be appreciated as examples of information operations that individuals frequently experience. However, the lines between the categories shown in Fig. 4.1 are blurred as discussed below.

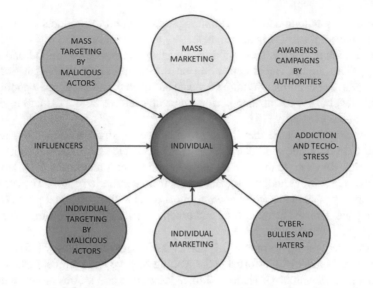

Fig. 4.1 Example of factors that have an effect on an individual on a daily basis

Targeting Faceless Masses with Specific Goals

As part of the masses, i.e., society, individuals may be indirectly targeted. Mass marketing, awareness campaigns by authorities and mass targeting by malicious actors highlighted in Fig. 4.1 are examples of generic information widely spread to large populations. Electronic marketing, digital marketing, and social media marketing are all names for mass marketing on electronic platforms online. Demographics of the populations are used to select the potential target groups. In e-marketing cases the advertisement campaigns are pre-planned, and the goal is well defined: to sell products.

However, the same techniques can be used for other ends, for example, information security awareness campaigns and the current hand hygiene campaign related to Covid-19 run by state authorities. In these cases, the expectation is for the society to gain a positive impact as correct information is spread to population. Phishing emails—an infamous tool in the cyber domain—also belongs to this category.

Phishing mails have several malicious goals. They try to convince people to reveal access to credentials, give confidential data, infect their computers with malware or open links to infected sites [8, 22]. The goal can be to use credentials to cash in right away, to sell them further, or to use them in further cyberattacks on a physical or a logical layer of the cyber domain. Phishing emails often contain audio-visual media content (sensory marketing) that catches a person's attention and arouses emotions (excitement for winning something, annoyance to act in a hurry) that contribute to action.

Targeting Individuals with Specific Goals

Individuals are more directly targeted in customized information operations. While general e-marketing profiles customers by the broad demographic groups (age, gender, race etc.) they belong to, digitalized information operations go further and profile individuals by their specific behavior. Psychographics, collections of people's activities, interests and opinions, are used to understand people's cognitive factors that drive their behaviors. Psychographics are subsets of demographic groups and therefore they are more suited for individual targeting [23].

In the same vein, but more specifically tailored than phishing emails are spear phishing messages. They appear to be legitimate and arriving from an employer, a colleague, a friend or another legitimate correspondence [8, 22]. Catphishing specially plays with our emotions. Catphishing is a form of spear phishing on dating sites and peer-group chatting rooms (i.e. widow groups), where visually attractive profiles and stories that are emotionally arousing are used to hook the victims [24–26].

Yet another example of a more individualized targeting is the use of professional trolls in digital communication platforms, where discussions are enabled. Trolls are known to use suitable emotional content to arouse public emotions and to harden attitudes by continually starting arguments, criticizing others or complaining at the cost of others [27]. Studies have shown that we share information more willingly while being emotionally aroused (for example, when feeling anxiety or amusement) than when we are not [28]. Loss of self-awareness and anonymity can be side factors that together with external emotional influence make an ordinary actor who considers himself a normal law-abiding citizen aroused (manipulated) enough to work as a troll [27, 29].

Targeting a Specific Person

Cyberbullying and hate speech emails are examples of more personalized, direct attacks against a person. Cyberbullying has different forms including flaming, harassment, cyberstalking, denigration, masquerading, outing and exclusion [30, 31]. Hate speech includes similar aspect as cyberbullying but is more extreme. Whereas the purpose of cyberbullying in most cases is to humiliate and harm a person with no considerations to physical acts, the purpose of hate speech is to promote physical violence against a person.

Influencers

An influencer is a person or thing that influences someone or something [32]. Such have always existed, and all previously discussed cases involve a person or a group whose goal it is to make others behave in some specific way. Yet, in today's digital world, the term influencer generally refers to a person who has a public and a very popular social media account, for example, in YouTube, Twitter, Facebook, Snapchat, TikTok etc. Such influence is often based on expertise in some area, but can also be based on influencer's knowledge, position, or authority.

Due to the high number of followers, influencers often make their living by advertising products and endorsing brands [33], but endorsements can also be a part of an organization's strategic communication [34]. In general, influencers' opinions can have a large impact on public opinions. Endorsement of a candidate in an election is often given as an example, but influencers have given their support to more general topics, such as to the act of voting in itself [35] and to public health campaigns [36]. Economics is not a bystander either. Kylie Jenner's Tweet that was said to cause Snapchat's 1.3 billion dollar drop in value [37] can be given as an example. There are several economic studies investigating the effect tweets have on stock market. Results of a recent study [38] show that factors like sentiment and emotions derived from tweets predict stock market movements.

Also, hitherto unknown people can become influencers in today's digitally connected world. There are plenty of people who have used the cyber domain in order to gain large publicity for a cause. Two examples are the whistle-blowers Manning and Snowden who leaked information via social media platforms with significant impact on public debate and possibly also on US national security. Another example is photojournalist Nilüfer Demir generally credited for changing the European discourse on the Syrian refugee crisis in 2015 with her photo of the dead boat-refugee toddler Alan Kurdi [39]. Also, Rahaf Mohammed al-Qunun comes to mind: The Saudi woman, who used twitter messages to gain asylum in Canada [40].

While we collect and are affected by all the information that surrounds us, we also influence others by our activities in a host of information networks in the physical and cyber domain. This is particularly so in the latter with its large connectivity. The term influencers, however, is generally restricted to actors who intentionally exploit these features of the cyber domain.

4.3.2   Addiction and Technostress

Last, but not least, we address the actual digitalization itself as it can also have an impact on a person. Technostress and Internet addiction are some of the negative sides of digitalization.

Internet addiction [41] is "characterized by excessive or poorly controlled preoccupations, urges or behaviors regarding computer use and internet access that lead to impairment or distress." Similarly, to other addictions, Internet addiction has a negative effect on academic achievements, work, health, economic situation and relationships [42]. Smartphone addiction is close to Internet addiction, as smart phones are devices for using the Internet and online services. Studies regarding Internet and smartphone addiction have shown that the failure in self-regulation is the factor causing the addiction [43, 44].

Digital society and fast developing technology affect us with their speed and demands and this is called technostress. According to [45] technostress is "a condition of constant high cognitive demand and physiological arousal. The condition is observable in people who, over time, have experienced reduced possibility of understanding, and gaining overview and control over information and workplace processes. The condition ensues from interaction with technology that lacks in usability, and (or) inapt organizational demands and conditions for its use".

The causes of technostress are many and include demands of a high working pace, frequent interruptions, multitasking (techno-overload), highly complex digital technology that a worker feels unable to master (techno-complexity), constant feeling of uncertainty due to the fast changing technology (techno-uncertainty), fear of losing a job because either digital technology or more qualified person will replace him (techno-insecurity), and blurred lines between work and leisure time (techno-invasion) [45, 46]. Also, frustration due to technical errors and low usability (techno-unreliability) as well as irritation due to unpredictability of machine behavior (stress in human–machine interaction) increase technostress [46]. In addition, technological workplace surveillance can augment stress [46].

4.4 Gaining Individual Resistance

In view of the foregoing, we shall now consider how individuals can become more resilient against information operations and the negative effects of digital society. In all cases discussed earlier, our emotions are under influence. Since emotionally aroused and stressful situations may spur us to act in ways contrary to our own interest, we should pay attention to personalized and emotional content in digital media and carefully consider our options of response before acting. Critical thinking and self-regulation are therefore essential to act properly, as also recommended in [47].

At an individual level, the key to resist the effect of information operations is to be aware of their existence. Moreover, we need to learn how to calm ourselves and to gain a better picture of the situation before we decide how to act. We can also enhance our ability to explore response options.

Gaining situational awareness and understanding of "the bigger picture" are normal military activities and always part of military operations, cyber operations included. However, they can also be applied in everyday life, and more importantly, their usage can be trained. Such training is an important part of military education, also in the Norwegian Defence Cyber Academy, NDCA, where cadets have one or two exercises per school year. Last years the cyber component in form of code and human factor has increasingly been added to the exercise's scenarios at NDCA, and the cadets' performance in the exercises has been studied. They have been participants in most of the studies discussed in the following sections.

4.4.1 Handling Stress

Almost every one of us has felt stress or has been in a stressful situation ranging from a lost phone or giving a speech in the front of an audience, to being in an accident or in a fight. Using coping strategies is a way to reduce stress and increase the ability to act. NDCA has studied its cadets' usage of the coping strategies that were part of their military education. During one of the semesters, cadets were self-reporting their use of coping strategies in different military and classroom contexts [48]. The results indicated that "having control" was the main factor in order to perform well both in a military and classroom context. Cadets found the same coping strategies and for the same reason useful in the cyber domain as well [49].

Gaining control begins from understanding the situation one is faced with. In the above-mentioned study of technostress [46], the usage of coping strategies and self-regulation are also mentioned as tools to reduce it. User education, technical support and planned implementation of technologies reduced stress caused by techno-complexity, -uncertainty, -unreliability and stress in human–machine interaction. Coping strategies and self-regulation, however, are constantly needed to overcome techno-overload and techno-invasion stress.

The stress factor is also discussed in [50] presenting roundtable discussions on resilience for military readiness and preparedness from five domains in a point counter point format: physiological versus psychological resiliency, differences of sex, contributions of aerobic and strength training, thermal tolerance, and the role of nature versus nurture. The authors concluded that interconnectedness of those five domains calls for interdisciplinary approach to build resilience. They argued for the importance of both psychological and physiological resilience. Physiologically resilient soldiers were found to be capable of taking good decisions in stressful situations. Psychological resilience, on the other hand, has an impact on both physical and cognitive performance and health. The adaptive resilience is based on both physiological and psychological resilience, and even though resilience is considered an individual trait, it can be enhanced by training in reality-based scenarios [50]. Though the roundtable discussions primarily focus on war in the physical domain, some points are valid also for the cyber domain. This includes their characterization of modern military operations as conducted in volatile, uncertain, complex and ambiguous environments which for the soldiers involved cause physical exertion, cognitive overload, sleep restriction and caloric deprivation. This is not necessarily unfamiliar to civilians in digital societies, which are also complex and contain uncertainty. Even if the physical exertion or caloric deprivation are not the first to occur for civilian users of the cyber domain, cognitive overload and sleep restriction can be easily associated with it.

4.4.2 Critical Thinking, Self-Regulation and Educational Approach

In the studies [51–53] cyber operator performance in a military setting was studied. The studies were carried out in NDCA and specifically cadets participating in cyber defence exercises in their last year were used as a test group.

This exercise is run in a cyber range, a closed virtual system, where different information systems can be built, operated, used, attacked and defended. The exercise is based on real life scenarios and from the human point of view we have used, for example, malicious mass targeting and malicious individual targeting as a way to enter the system. We have influenced the cadets by media coverage and human intelligence about demonstrations and political disturbance and put cadets to handle the communication with the public. An important issue is that the cadets are responsible for how their group behaves and works. Each group has a mentor giving guidance related to the actual cyber defence, but the cadets are responsible for solving both internal and exercise-related issues.

Self-governance and cognitive competences were found to be important both in supporting cyber operator performance and in dealing with the cyber power capacity. These cognitive skill sets are applicable in both offensive and defensive cyber activities. In addition, management of cyber power needs skills such as unstructured problem solving, critical thinking, learning, and reasoning [54] and building a mental model helps to enhance situational awareness [52]. These skills are valuable not only for cyber operators but also for civilians who may also be targets of or—in our digital society—collaterally influenced by information operations.

The context of an incident is always an important factor. In order to gain good situational awareness, the knowledge about surroundings, not just the actual incident, needs to be included. Paul et al. [55] include cultural and society dependent factors when describing critical consumers of information. According to [55] critical consumer of information is a person who knows that within every given society or culture, the dominant viewpoints are given a privileged place in media. Therefore, critical consumers seek other aspects of the information using intellectual standards that are independent on cultural or ideological views. Being a critical consumer of information is exercising critical thinking in a larger, informative context.

Critical consumers as well as producers of information ask questions such as "what", "how" and "why", which all belong to a critical thinking process. Critical thinking is, for example, to consider the different viewpoint of each case and to analyze concepts, theories and explanations [55]. It is a cognitive process of clarifying issues and exploring implications and consequences. Moreover, it is transferring concepts to new contexts. Paul and Elder [55] divide critical thinking into two forms: general competencies that apply to all thinking within all domains, and specific thinking within particular domains. In order to learn about content, a person needs to learn a systematic way of asking the right questions.

Learning how to ask the right questions leads to self-regulation. According to Baumeister, Heatherton and Tice [56], self-regulation is one's own ability to control one's own thoughts, emotions and actions. Based on Center on Alcoholism, Substance Abuse, and Addictions (CASAA) Self-Regulation Questionnaire (SRQ) [57] (which was also used when the NCDA-cadets' self-regulation was measured), behavioral self-regulation skills include the ability to: Receive relevant information; Evaluate the information and comparing it to norms; Trigger change; Search for options; Formulate a plan; Implement the plan; and Assess the plan's effectiveness. According to this view, efforts to improve self-regulation should focus on each of these seven components at a time. Cleary and Zimmerman [58] presented the Self-Regulation Empowerment Program (SREP) that empowers adolescent students' learning and is based on coaching and cyclical feedback loops. This suggests that military educations accustomed to use mentoring and continued feedback, are well suited to develop this kind of self-regulation.

To cultivate critical thinking and self-regulation NCDA has adopted what we call "slow education". This is an educational approach that gives space and time for thinking processes, for exchanging and reusing knowledge across contexts and for encouraging individual ideas. During military exercises, cadets are exposed to working in multi domain environments. We use mentoring to guide cadets to understand and to use different problem-solving strategies. When appropriate we also encourage cadets to learn from their own mistakes. In a study about cognitive performance in military cyber defence scenarios [53] we found that such slow education may help to support a cyber operator's self-governance in cyberspace operations and in the utilization of cyber power.

4.5   Discussion and Conclusion

In this chapter we have sought to clarify how an individual as an asset in an information system can be influenced by information actions via cyber domains. Information in this context is presented in the form of text, colors, figures, sounds, lights, music, films etc., and we have focused on the emotions that information triggers. From an attribution perspective, literately anyone can run an information operation and execute information actions: me, you, a cyberbully, a criminal, a terrorist, a corporation, a state. Anyone who can use digital means to post a piece of information influences others in digital reality.

Therefore, we as individuals should always be aware of a fundamental human feature: We are constantly formed by the information around us, whether we want it or not, both in the physical world and in cyberspace. In other words, we are always influenced by as well as influencing others. Yet, we can limit the extent to which this affects us.

We can develop our individual resilience by gaining situational awareness of who we as individuals are digitally connected to and which information systems, we are part of. We can make critical mental models on how connected we actually are and how our behavior can affect these systems. Critical thinking, self-regulation, adaptability and controlling our own situation enhance our resilience and are therefore factors that both prevent us from being easy victims of information actions and help us to mitigate the consequences. Moreover, these skills prevent addiction and technostress.

Although the studies and exercises discussed here are military related, it is equally relevant for civilian educational institutions to look into the concept of using real life training sessions that mirrors information systems used in special sectors. We have introduced educational approaches that can support students' own thinking process and reflection skills to develop critical thinking, self-regulation and adaptability to gain control. These approaches are useful beyond higher level cyber education. They may, for instance, be introduced in primary school using physical objects and tangible cause-consequence examples. Gradually teachers can help their pupils transfer these skills to the cyber context. Such endeavors can in different ways help future generations to cope with the negative effects of our increasingly digitalized society and to develop resilience against information operation.

References

1. Digitalization. Lexico. https://www.lexico.com/en/definition/digitalization
2. Janita (2016) DDoS attack halts heating in Finland amidst winter. Metropolitan.fi https://met ropolitan.fi/entry/ddos-attack-halts-heating-in-finland-amidst-winter. Accessed 26 June 2020
3. Dataangrepet mot Helse Sør-Øst. Norsk Rikskringkasting (NRK). https://www.nrk.no/nyheter/ dataangrepet-mot-helse-sor-ost-1.13873606. Accessed 26 June 2020
4. Cyber-attack on hydro. Hydro (2019) https://www.hydro.com/en/media/on-the-agenda/cyber-attack/. Accessed 26 June 2020
5. NATO military policy for information operations (2018) Draft MC 0422/6 (unclassi- fied), NATO. https://shape.nato.int/resources/3/images/2018/upcoming%20events/MC%20D raft_Info%20Ops.pdf. Accessed 09 July 2020
6. Lehto M, Limnéll J (2020) Digitalisaatiokehityksen vaikutuksia yhteiskunnassa ja asevoimissa 2020-luvulla. Sotilasaikakauslehti 95(2):9–16
7. Helkala KM, Rønnfeldt CF (2020) Human resilience against negative effects of digitalization. Submission of abstract to international society of military sciences. Helsinki. http://urn.fi/URN: NBN:fi-fe2020110589357
8. Whitman ME, Mattord HJ (2012) Principles of information security. 4th edn. Cengage Learning
9. Lindsay PH, Norman DA (1977) Human information processing: an introduction to psychology, 2nd edn. Academic Press
10. Halász V, Cunnington R (2012) Unconscious effects of action on perception. Brain Sci 2(2):130–146
11. Erenkol AD, Merve A (2015) Sensory marketing. J Adm Sci Policy Stud 3(1):1–26
12. Rathee R, Rajain P (2017) Sensory marketing—investigating the use of five senses. Int J Res Financ Mark 7(5):124–133
13. Mehta R, Zhu RJ, Cheema A (2012) Is noise always bad? Exploring the effects of ambient noise on creative cognition. J Consum Res 39(4):784–799
14. Musetti A, Corsano P (2018) The internet is not a tool: reappraising the model for internet- addiction disorder based on the constraints and opportunities of the digital environment. Front Psychol 9:558
15. Clowes RW (2015) Thinking in the cloud: the cognitive incorporation of cloud-based technology. Philos Technol 28(2):261–296
16. Clowes RW (2013) The cognitive integration of E-memory. Rev Philos Psychol 4:107–133
17. Kramer ADI, Guillory JE, Hancock JT (2014) Experimental evidence of massive-scale emotional contagion through social networks. Proc Natl Acad Sci USA 111(24):8788–8790
18. Hatfield E, Cacioppo JT, Rapson RL (1993) Emotional contagion. Curr Dir Psychol Sci 2(3):96–100
19. Garcia-Retamero R, Cokely ET (2011) Effective communication of risks to young adults: using message framing and visual aids to increase condom use and STD screening. J Exp Psychol Appl 17(3):270–287

20. Otieno C, Spada H, Liebler K, Ludemann T, Deil U, Renkl A (2014) Informing about climate change and invasive species: how the presentation of information affects perception of risk, emotions, and learning. Environ Educ Res 20(5):612–638. https://doi.org/10.1080/13504622. 2013.833589
21. Lazer DMJ, Baum MA, Benkler Y, Berinsky AJ, Greenhill KM, Menczer F, Metzger MJ, Nyhan B, Pennycook G, Rothschild D, Schudson M, Sloman SA, Sunstein CR, Thorson EA, Watts DJ, Zittrain JL (2018) The science of fake news. Science 359(6380):1094–1096
22. What is social engineering? Kaspersky, 2020. https://usa.kaspersky.com/resource-center/def initions/social-engineering
23. What is psychographics? Understanding the tech that threatens elections. CBINSIGHTS, 2020. https://www.cbinsights.com/research/what-is-psychographics. Accessed 07 July 2020
24. Romance scams. RomanceScams.org. https://www.romancescams.org/. Accessed 20 March 2019
25. Social engineering scams on social media. Norton, https://us.norton.com/internetsecurity-onl ine-scams-social-engineering-scams-on-social-media.html. Accessed 20 March 2019
26. The romance scam. The Antisocial engineering, 2018. https://theantisocialengineer.com/2018/ 09/11/the-romance-scam/. Accessed 20 March 2019
27. Binns A (2012) Don't feed the trolls! managing troublemakers in magazines online communi- ties. J Pract 6(4):547–562
28. Berger J (2011) Arousal increases social transmission of information. Psychol Sci 22(7):891– 893
29. Cheng J, Bernstein M, Danescu-Niculescu-Mizil C, Leskovec J (2017) Anyone can become a troll: causes of trolling behavior in online discussions. In: CSCW '17: proceedings of the 2017 ACM conference on computer supported cooperative work and social computing. ACM, pp 1217–1230
30. Cowie H (2013) Cyberbullying and its impact on young people's emotional health and well- being. Psychiatrist 37(5):167–170
31. Perren S, Dooley J, Shaw T, Cross D (2010) Bullying in school and cyberspace: associations with depressive symptoms in Swiss and Australian adolescents. Child Adolesc Psychiatry Ment Health 4:28
32. Influencer. Oxford Learner's Dictionaries. https://www.oxfordlearnersdictionaries.com/defini tion/english/influencer?q=Influencer
33. Lou C, Yuan S (2019) Influencer marketing: how message value and credibility affect consumer trust of branded content on social media. J Interact Advert 19(1):58–73. https://doi.org/10.1080/ 15252019.2018.1533501
34. Sng K, Au TY, Pang A (2019) Social media influencers as a crisis risk in strategic commu- nication: impact of indiscretions on professional endorsements. Int J Strateg Commun 13(4):301–320. https://doi.org/10.1080/1553118X.2019.1618305
35. Internett-kjendis skal få unge til å stemme (2017) Aftenposten. https://www.aftenposten.no/ norge/i/QVvPV/internett-kjendis-skal-faa-unge-til-aa-stemme. Accessed 09 July 2020
36. Kostygina G, Tran H, Binns S, Szczypka G, Emery S, Vallone D, Hair E (2020) Boosting health campaign reach and engagement through use of social media influencers and memes. Soc Media + Soc 6(2):1–12
37. Hills MC (2018) Snapchat's $1.3 billion drop in value is linked to a Kardashian. Forbes. https://www.forbes.com/sites/meganhills1/2018/02/23/snapchat-stock-value/?sh=306 947a6e457. Accessed 08 July 2020
38. Steyn DHW, Greyling T, Rossouw S, Mwamba JM (2020) Sentiment, emotions and stock market predictability in developed and emerging markets. GLO discussion paper 502, global labor organization (GLO). Essen. http://hdl.handle.net/10419/215436
39. Gunter J (2015) Alan Kurdi: why one picture cut through. BBC News. https://www.bbc.com/ news/world-europe-34150419. Accessed 02 Dec 2020
40. Chen H, Lin MM (2019) Rahaf al-Qunun: unpicking the tweets that may have saved her life. BBC News. https://www.bbc.com/news/world-asia-46819199. Accessed 02 Dec 2020

41. Shaw M, Black DW (2008) Internet addiction: definition, assessment, epidemiology and clinical management. CNS Drugs 22(5):353–365
42. Young KS (1998) Internet addiction: The emergence of a new clinical disorder. Cyberpsychol Behav 1(3):237–244
43. van Deursen AJ, Bolle CL, Hegner SM, Kommers PA (2015) Modeling habitual and addictive smartphone behavior: the role of smartphone usage types, emotional intelligence, social stress, self-regulation, age, and gender. Comput Hum Behav 45:411–420
44. Jiang Q, Leung L (2012) Effects of individual differences, awareness-knowledge, and acceptance of internet addiction as a health risk on willingness to change internet habits. Soc Sci Comput Rev 30(2):170–183
45. Sellberg C, Susi T (2014) Technostress in the office: a distributed cognition perspective on human-technology interaction. Cogn Technol Work 16:187–201
46. Dragano N, Lunau T (2020) Technostress at work and mental health: concepts and research results. Curr Opin Psychiatry 33(4):407–413
47. Bakir V, McStay A (2018) Fake news and the economy of emotions. Digit J 6(2):154–175. https://doi.org/10.1080/21670811.2017.1345645
48. Helkala K, Knox B, Jøsok Ø (2015) How the application of coping strategies can empower learning. In: 2015 IEEE frontiers in education conference (FIE). IEEE, pp 1–8
49. Helkala K, Knox B, Jøsok Ø, Lugo R, Sütterlin S (2016) How coping strategies influence cyber task performance in the hybrid space. In: HCI international 2016—posters' extended abstracts (Toronto, ON, 2016). Communications in computer and information science, vol 617. Springer, pp 192–196
50. Nindl BC, Billing DC, Drain JR, Beckner ME, Greeves J, Groeller H, Teien HK, Marcora S, Moffitt A, Reilly T, Taylor NAS, Young AJ, Friedl KE (2018) Perspectives on resilience for military readiness and preparedness: report of an international military physiology roundtable. J Sci Med Sport 21(11):1116–1124
51. Jøsok Ø, Lugo RG, Knox BJ, Sütterlin S, Helkala K (2019) Self-regulation and cognitive agility in cyber operations. Front Psychol 10:875
52. Knox BJ, Jøsok Ø, Helkala K, Khooshabeh P, Ødegaard T, Lugo RG, Sütterlin S (2018) Socio-technical communication: the hybrid space and the OLB model for science-based cyber education. Mil Psychol 30(4):350–359. https://doi.org/10.1080/08995605.2018.1478546
53. Knox BJ, Lugo RG, Helkala KM, Sütterlin S (2019) Slow education and cognitive agility: improving military cyber cadet cognitive performance for better governance of cyberpower. Int J Cyber Warfare Terrorism 9(1):48–66
54. Knox BJ (2018) The effect of cyber power on institutional development in Norway. Front Psychol 9:717
55. Paul R, Elder L (2005) A guide for educators to critical thinking competency standards: standards, principles, performance indicators, and outcomes with a critical thinking master rubric. Foundation for critical thinking
56. Baumeister R, Heatherton T, Tice D (1995) Losing control: how and why people fail at self-regulation. Academic Press
57. Self-Regulation Questionnaire (SRQ). Center on Alcoholism, Substance Abuse, and Addictions (CASAA). https://casaa.unm.edu/inst/SelfRegulation%20Questionnaire%20(SRQ).pdf
58. Cleary TJ, Zimmerman BJ (2004) Self-regulation empowerment program: a school-based program to enhance self-regulated and self-motivated cycles of student learning. Psychol Sch 41(5):537–550

Chapter 5
Users' Psychopathologies: Impact on Cybercrime Vulnerabilities and Cybersecurity Behavior

Naomi Woods

Abstract The internet and digital technologies have become an integral part of people's daily lives. The online world provides many benefits to billions of users globally. However, it also brings risks too, because it is easy for criminals to reach their victims and exploit their online behavior. Nevertheless, users often perform risky security behaviors for convenience and usability, and because of their inadequate security awareness. With around 25% of the world's population experiencing mental and/or neurological disorders, it is important to understand how users' psychopathologies manifest themselves in the context of cybersecurity. This chapter has reviewed the symptoms of several mental disorders while considering the online benefits and risks, and these symptoms have been applied to evaluate users' vulnerability to cybercrimes and cybersecurity practices. The findings reveal how the complexity of each mental disorder influences users' online engagement and susceptibility to cybercrimes, and how uniquely, to varying degrees, they affect different cybersecurity behaviors.

Keyword Cybersecurity behavior · Psychopathology · Mental disorder · Cybercrime · Online benefits · Online risks · User psychology · Cyberpsychology

5.1 Introduction

Over the last few decades, digitalization has changed our society. Today, technology is integral to nearly every part of our lives [1]. At home, from the moment we wake, the lights we put on, and the electricity we use to make coffee—it all comes from or is managed by systems that are digitalized. Most of us use our phones and smart devices throughout our day. Digital systems can also be found in nearly all working environments—we do not even have to work in an office to use a computer. From farming to manufacturing and many other industries interacting with machinery, all have digitalized systems [2]. We also communicate mainly online and have so many

N. Woods (✉)
Faculty of Information Technology, University of Jyväskylä, Jyväskylä, Finland
e-mail: naomi.woods@jyu.fi

platforms and services available. We bank, we shop, our entertainment, all online. And as technology develops, more and more services and devices will become smart, including many infrastructures. Furthermore, our personal identity is online too. Therefore, with so many digital services, systems, and personal and organizational information online, cybersecurity and securing these assets can be considered just as important as the assets themselves [3, 4].

There are regular reports in the news or on social media referring to some sort of cyberattack [5]. Even though cybersecurity has developed immensely and is constantly evolving, cybercrime is growing, and criminals are always one step ahead. Many complex protocols have been developed by cybersecurity professionals and, when employed properly, provide a strong defense [6]. If a user, even with the strongest protocols, creates a password from their pet's name or modifies the password for other accounts using predictable patterns, such insecure security behavior will ultimately undermine all security efforts [7, 8].

To attempt to improve users' security practices, users (within organizations and at home) are provided with digital security guidance, made aware of threats, and have technological restrictions imposed [9]. Nonetheless, many users will circumvent security policies and safety protocols for the sake of convenience and usability [10, 11]. This can have catastrophic repercussions with millions of euros and dollars are lost globally due to security breaches, not to mention the threat to public safety when information and records are stolen if breaches succeed [12].

There are several reasons that users are thought to adopt unsafe security practices, including a lack of awareness of security threats, a lack of awareness of the consequences of their actions, and also a lack of knowledge of how security actually works. Many users do not realize how severe a security breach could be or how it could impact their lives and the lives around them [13–15]. Many are not even aware of how vulnerable they are to a potential threat, or to the damage it can cause [16]. However, when one thinks about what creates that awareness, understanding of behavioral consequences, and knowledge, and how a user considers it, processes it, stores it, and acts on it, each user will have many individual factors that affect each part of the process. Through understanding the individual differences of users, allows cybersecurity professionals to understand why the user is the "weakest link in security" [8, 17, 18]. Therefore, it is increasingly acknowledged that cybersecurity professionals not only need to have backgrounds in computer science and engineering, they also need to have backgrounds in the human, behavioral and cognitive sciences to comprehend the user and their security behavior [6, 17].

Due to the immense usage of technology, the digital user can be potentially anyone nowadays, including children, the elderly, people with disabilities, and people with mental disorders. There are over 4.3 billion internet users [19]. However, in 2001 the World Health Organization [20] estimated that one in every four people (25%) has a mental or neurological disorder, at around 450 Million people globally. This number has gone up over the years, with growing populations [20]. This means that about 25% of digital users (just under 1.1 Billion) are living with some form of mental disorder. Therefore, the way these users interact with digital technology and the online world, and especially the way they interact with cybersecurity needs to be taken into consideration.

Over the years, research towards studying the usage of technology has been conducted by computer scientists, mathematicians, and engineers. It has only been in more recent times that researchers with a human sciences background have started to examine how users and societies interact with digital technologies [21]. This has led to the emergence of cyberpsychology as an applied psychological discipline [22]. One area of research within this discipline consists of the examination of psychopathology and abnormal behaviors, and how interacting with the internet and digital technology can affect users' mental and psychological states. One example includes the emergence of the internet, social media, internet abuse and online gaming addictions [23–25]. However, there is still very little research examining how users' psychopathology affects their interaction online, and next to no research considering how users' psychopathology affects the way they interact with cybersecurity. To address this gap, this chapter will discuss common psychopathologies and online interaction. It will consider the characteristics of specific mental disorders, and what online benefits, risks, and security issues bring to these users. It will examine online benefits and risks specific to each disorder, including vulnerability to cybercrimes, and consider the impact psychopathologies have on the users' interaction with cybersecurity in an attempt to secure their digital lives.

5.2 Psychopathology and Abnormal Psychology

There are many areas of psychology examining different aspects of human beings, including social, behavioral, cognitive, physiological, neuropsychological, etc. Clinical psychology is a research-based practice, concerned with examining, diagnosing, and providing care for individuals exhibiting a broad spectrum of abnormal behaviors and psychopathologies [26, 27].

Abnormal psychology studies unusual or atypical behaviors, thoughts, and emotions that are symptomatic of mental, behavioral, or neurological disorders. One criterion is that they are severe enough to negatively impede upon an individual's life and negatively impact (or to the extent to) how the individual interacts with society [24, 27]. Clinical psychologists employ different approaches to understand and treat abnormal behaviors. There are four main schools: psychodynamic, humanistic, behavioral and cognitive-behavioral, as well as systemic/family. These approaches examine and treat the individual, often observing the behavior, considering the biological bases, and the internal thought processes indicative of mental disorders [27, 28]. Many clinicians will combine different approaches to provide successful treatment of different disorders. However, before treatment of abnormal behaviors and psychopathologies can be determined, diagnosis is first needed [29].

5.2.1   Classifying Psychopathology

Diagnosis is essential to the practice of psychology and psychiatry. It can take many years for mental health practitioners and clinicians to learn and become experts in formulating a diagnosis. Diagnostic manuals provide guidance, descriptions of symptoms, and criteria for diagnosing mental disorders. These manuals provide practitioners with a standardized language and an understanding to increase the reliability of diagnoses while reducing susceptibility to practitioner bias [30]. The two main diagnostic manuals for categorizing abnormal behavior, are the International Statistical Classification of Diseases and Related Health Problems (ICD) [31], and the Diagnostic and Statistical Manual of Mental Disorders (DSM) [29]. The ICD is a broader standardized tool, encompassing all medical conditions. The ICD includes a chapter dedicated to "Mental and behavioral disorders" [31]. Both manuals categorize symptoms of abnormal behavior into disorders through the individual meeting specific criteria. While the ICD is thought to be used more frequently in Europe, the DSM is more frequently used in the US. However, both are employed globally. For the purposes of this chapter, we will use the DSM to look at psychopathologies and their impact on the cybersecurity context.

The DSM was first published by the American Psychiatric Association (APA) in 1952 and has seen many revisions over the years as clinical psychology has evolved with the times. The most recent edition, the DSM-5 was published in 2013 and is globally used by mental health practitioners, and researchers. Developed from the scientific advances in brain imaging techniques, neuroscience, and genetics, the manual was reorganized around psycho-physiological relationships rather than just common symptoms. This has led to the classification of 19 main classes with over 100 specific disorders. The most recent version of the DSM has also been modified to be more "harmonized" with ICD-11 (the most recent version) [29]. Although both manuals, (DSM and ICD) have proven to be useful, through the years they have also been considered controversial due to cultural, political bias, societal norms that influenced what is classified as a disorder or not. Further controversy has always arisen regarding the labeling of individuals. Many people believe that once one has been given a diagnosis, one is stuck with that diagnosis or "label". Being labeled as one thing or another often provides others a reason to act in a specific way based on the label rather than the individual [24, 28, 30]. However, the counterargument highlights the need for diagnosis, as, without it, the correct treatment cannot be provided [27].

5.2.2   Mental Disorder

DSM-5 includes an overall definition of a mental disorder as:

> A mental disorder is a syndrome characterized by clinically significant disturbance in an individual's cognition, emotion regulation, or behavior that reflects a dysfunction in the psychological, biological, or developmental processes underlying mental functioning. Mental disorders are usually associated with significant distress or disability in social, occupational, or other important activities. An expectable or culturally approved response to a common stressor or loss, such as the death of a loved one, is not a mental disorder. Socially deviant behavior (e.g., political, religious, or sexual) and conflicts that are primarily between the individual and society are not mental disorders unless the deviance or conflict results from a dysfunction in the individual, as described above. ([29], p. 20)

This definition does not include all the features of every specific mental disorder. However, each mental disorder is characterized and described under each class within the manual. Each disorder can be diverse with a wide range of characteristics. These diverse characteristics and the severity of these characteristics can affect the way in which the user interacts with the internet and digital technology.

5.3   Online Benefits, Risks, and Security Behavior

In the days before mass digitalization and the internet, individuals with mental disorders could find themselves isolated and lonely. As the only form of interaction with society was oftentimes face-to-face encounters, and societal participation proved difficult for many. Nowadays, although engaging with the online world brings many benefits and opportunities, it can also bring risks, which will differ for each disorder. Along with these risks, there are potential vulnerabilities to cybercrimes, and vulnerabilities in the users' cybersecurity behaviors they adopt to protect themselves.

5.3.1   Benefits of Online Interaction

Engaging with digital technologies and interacting online can provide many benefits for users who have psychopathologies. The opportunities for being more integrated into society, contributing to society, and being supported by society have considerably grown with this easily accessible medium. There are increased opportunities for occupation, education, communication, development, entertainment, shopping, creativity, participation and civic engagement, social interaction, and connectedness [32–34]. Previous research has identified the many benefits of using ICT and the internet for users, including for those users with intellectual disabilities [35]. These benefits have been organized into themes such as social utility, accessing information, personal identity, and occupational and enjoyment. These themes were based on the users and gratifications framework [36, 37].

Table 5.1 Themes of the benefits of internet and digital technology usage with examples

Social and communication	• Communicating with other on social media, and other platforms • Developing and maintaining friendships and romantic relationships • Unity with family and friends • Engaging with society
Cognitive	• Develop social learning skills • Learning about themselves, others, and other things • Expressing emotions, and attitudes
Occupational	• Employment • Education
Independence	• Online shopping • Online banking • Entertainment
Supportive	• Support groups and online discussion boards • (Mental) health information • Online therapies

Within this chapter, we will examine the benefits of internet and ICT usage for individuals with different mental disorders, and will organize the benefits into themes based on clinical impairments (DSM-5) such as, *social and communication, cognitive, occupational, and independence*. The additional theme of *supportive* is included due to the supportive benefits the internet and digital technology provide to those with mental disorders (detailed in Table 5.1). Several of these themes can overlap each other, due to their related nature. For example, attending an educational institution online not only has occupational benefits but has cognitive benefits through learning as well as social and communicative benefits through communication with peers.

The supportive theme has been appended, as for many individuals with mental disorders, the services provided by interacting with the internet and the digital world have brought about many benefits to support those who would have found it otherwise difficult to find or receive help [38]. For instance, interacting with online message or discussion boards has been seen to reduce stress [39]. Furthermore, many individuals who engage with online support groups receive plentiful benefits, through being able to express themselves and their emotions, enhancing their knowledge, and maintaining family and friends' relationships [40, 41].

5.3.2 Risks of Online Interaction

There are many benefits of engaging with the online world. But, with these benefits also come several risks. There are two types of risks that online engagement can bring to individuals with mental disorders:

1. The risk of exacerbating the symptoms of the disorder,

2.   The risk of becoming a victim of cybercrime.

Any user is vulnerable to cybercrime. However, do the symptoms of a mental disorder compound the user's vulnerability? To understand this, we will briefly discuss to which cybercrimes users are vulnerable.

There are many definitions of cybercrime, however, within this chapter we will use Nurse's definition and taxonomy of cybercrimes against individuals (2019), as it has been derived from a "comprehensive and systematic review" of research, practice, and real-world cases. Reference [42] defines "any crime (traditional or new) that can be conducted or enabled through, or using, digital technologies". His taxonomy of cybercrimes against an individual includes five main types:

1.   Social engineering and trickery,
2.   Online harassment,
3.   Identity-related crimes,
4.   Hacking,
5.   Denial of services and information.

Social engineering and trickery include phishing and catfishing. Through phishing, for instance, criminals aim to steal the user's confidential information such as authentication credentials (username and password), and/or online banking details [13]. The crime can occur when an individual overly shares personal information about themselves and others, and/or is willing to trust the sender and give them what they request in the communication, e.g., financial help. Criminals will often use techniques to get what they want, through claiming their issue is important and urgent, and presenting themselves in an official capacity, such as an organization, or someone of trust, like a potential partner. They often manipulate the victim in highly stressful situations and preying on anxieties where decision-making is not optimal [43].

Online harassment can occur when an individual has been too trusting with their personal information, and identity online. Through revealing their information, it can be utilized to direct anger and hate towards them. Examples of these types of crime include cyberbullying, stalking, and trolling [44]. The criminal uses the anonymity of the internet to act as they please harassing and manipulating the victim [45].

Identity-related crimes refer mainly to identity theft. It occurs due to the enormous amount of personal information that is available online. It can also be contributed to by a victim sharing personal information online, which is then extorted and used by the criminal in an anonymous environment to gain what they want. The criminal can also use the victim's information to engage in further criminal activity [42].

Hacking can include compromising digital information and computing systems, targeting the cybersecurity principles of confidentiality and integrity [11]. Hacking can lead to the exposure of confidential information and/or modification and deletion of information. Hacking can occur through malware, such as spyware, and through account hacking by means of exploiting insecure password management behaviors (e.g., reusing passwords). Criminals exploit users' awareness (or lack of) regarding their security and privacy and their poor security behaviors, such as creating weak passwords, and reusing and modifying them for several accounts, due to convenience and memory burdening [46, 47].

Denial-of-Services (DoS) and information refers to when criminals will, for instance, bombard organizations with website traffic which can lead to genuine users being unable to access services or information [11]. Another type of crime can include ransomware, where the criminal will apply malware to encrypt an individual's (or organization's) information then request payment for the release of the information. Criminals will exploit the individuals' anxieties and manipulate them into getting what they want, as quickly as possible. These forms of crime target the cybersecurity principle of availability [42].

5.3.3   Cyber Security Behaviors

As mentioned in the previous section, there are many risks to being online and interacting with digital technologies. Cybercriminals attack individuals and organizations, and targets the basic principles of cybersecurity (CIA: confidentiality, integrity, and availability) [17]. There are technical protocols in place to attempt to prevent attacks and decrease the user's vulnerability, such as virus scanning software. However, the user still needs to undertake cybersecurity activities/secure behaviors to ensure they (as a vulnerable factor) are not the weakest link in security.

There are many security behaviors an individual can perform including [48, 49]:

- Securing their devices, systems, and services with good password management,
- Being careful with their privacy (i.e., not oversharing),
- Archiving and backing-up information,
- Virus scanning,
- Updating applications and software,
- Installing and updating security patches,
- Avoiding and not opening suspicious emails and websites,
- Not plugging in suspicious USB drives, etc.

We will discuss some of the main behaviors within this section.

Good or secure password management starts with creating a strong password. Many users believe if the password is strong it is difficult to remember, but this is not necessarily the case. Passwords can be incredibly memorable as long as the individual has contributed the effort to make it so. Individuals can use memory techniques such as mnemonics to help them create strong passwords and to remember them as well [46, 47, 50]. However, creating a secure password is not all about making it strong and meeting the minimum password policy requirements [51, 52]. A strong (long and character-complex) password can be broken if the individual uses the password for many accounts, where the security-levels of some systems are not as strong. Or, the individual uses predictable patterns, such as "Cappuccino1!" and "Cappuccino2!", modifying their passwords for several accounts, that hackers can exploit [53]. Other insecure password behaviors include writing passwords down in an insecure (or not encrypted) document or post-it note, and sharing passwords. Many users adopt these behaviors for two main reasons: convenience, and memory burden. Users are not willing to expend the time and effort into creating strong passwords, and when there are so many accounts, many feel overwhelmed by how many passwords they would be required to create, learn and recall. Therefore, they adopt insecure password behaviors and undermine the authentication mechanism [14].

Oversharing and disclosing personal information and carelessness with privacy can result in the individual being vulnerable to most cybercrimes. It can lead to exposing details of not just the individual but others around them, and organizational information. For instance, in one case, people were using fitness trackers to track their exercise patterns and unintentionally revealed details of a military base as they posted their results to an online application [54]. Through trusting and sharing personal information online, this can unintentionally expose details that allow criminals leverage to manipulate the individual into doing what they want them to do.

Proactive security behaviors include, virus scanning, updating applications, and software, installing and updating patches that require very little effort, as the technology does the work for the individual [55]. All the individual is required to do is regularly remember to start the action (often when prompted) and give the system time to undertake the process. Some users do not undertake these activities due to the inconvenient time when they are required, or because updates may change the application to a less user-friendly version, or because they are unaware of the risks of not taking these actions [56]. Individuals who exhibit risk-taking behaviors have been found to be less likely to undertake these proactive security practices [57].

Backing-up data and information: losing data is common among individuals and organizations, however, using backup solutions can prevent the threat of losing data [58]. Backup solutions include cloud solutions, and external hard drives, and USB sticks. If data is not backed-up it can be impossible to recover if an incident occurs. Regularly backing-up data can ease the burden of data loss in the event of accidental deletion, intentional deletion through malware, hard drive crash, power failure, or a natural disaster [59]. As seen with updating virus software, applications, and patches, backing-up data does not require much effort, however, users contemplate the perceived convenience over perceived threat [58].

Not opening suspicious emails, clicking on suspicious links, or visiting suspicious websites: users can be encouraged to do these actions through social engineering by the medium of phishing emails. A prime example can include, when users receive a scam email and are invited to visit fraudulent websites. The criminal creates a website that is designed to look like an official legitimate website. The user is asked to enter their authentication credentials for a service, they may already use. The criminals can then access the victim's account, potentially other accounts (if their passwords are reused), and use the details for other criminal activity [13, 42, 53].

Using personal USB drives within organizations, and using suspicious USB drives can bring a whole host of security threats to an organization or to an individual. Many users do not have as strict cybersecurity practices at home, with insecure networks. The user will plug-in their USB drive at home, picking up e.g., viruses, then bring it to work and spread the viruses [18]. Another issue comes from users being given

USB drives for free (sometimes as a marketing promotion), or just finding them. Criminals will leave USB drives for individuals to find in the hope they will just use them—and they do. These drives are often full of malicious software to allow the criminals to gain access remotely to secure systems [11, 60].

For many years, it has been acknowledged that psychology plays a role in the adoption of security and protective practices by the users. However, very few studies have considered the role that psychopathology plays in users' safety behaviors and interaction with cybersecurity.

5.4   Understanding Users' Mental Disorders

This section describes some of the major classes of mental disorders with reference to DSM-5 [29]. Eight of the 19 major classes are reviewed. They were chosen due to their prevalence in society and relevance to online interaction. Substance-related and addictive disorders will not be discussed in this chapter. This is because the recognition of internet-based disorders (named currently as internet gaming disorder) represents excessive internet abuse [24, 29] and is classified under this major class. Owing to the growing phenomena of internet-based disorders, a separate chapter would be required to fully review them.

Therefore, using the characteristics of eight classes of mental disorder, the benefits and risk to online interaction will be considered. These characteristics will be applied to the context of cyber security to reflect upon the vulnerabilities to cybercrimes that individuals with these disorders may face and how their disorder traits influence their cybersecurity behavior. Finally, a summary of mental disorders is presented.

5.4.1   Neurodevelopmental Disorders

Neurodevelopmental disorders are a group of disorders that occur during the developmental stage. They are characterized by developmental impairments that vary from specific limitations in learning and in controlling attention to more general impairments in intelligence and social skills. These disorders usually manifest in the early stages of development, usually before a child goes to school. With this early onset, these disorders with a variety of impairments often cause limitations to personal, social, or academic performance. Neurodevelopmental disorders class includes intellectual disabilities, communication disorders, autism spectrum disorder, attention-deficit/hyperactivity disorder (ADHD), and specific learning disorder [29].

Intellectual disabilities are characterized by significantly impaired intelligence, deficits in general mental abilities, and impairments in everyday adaptive behavior. Communication disorders are characterized by difficulties in speech, language, and communication. They have not always been considered as mental disorders.

However, the impairments can cause distress and limit functioning in life, and therefore are now considered as disorders. Autism spectrum disorder is described as a disorder with symptoms that are persistent deficits in shared or common social communication. It is accompanied by nonverbal communicative behaviors used for social interaction, impairments in developing, managing, and understanding relationships, and restricted, repetitive patterns of behavior, interests, or activities. ADHD, on the other hand, is categorized by inattention, and/or excessive activity and impulsivity which can interfere with development and functioning [61], which is not appropriate for the individual's age. Specific learning disorder is characterized by learning difficulties and difficulties in applying academic skills accurately or as quickly as others of the same age. It is more commonly known as the reading disorders such as dyslexia and dyscalculia.

All these disorders can be generally categorized to describe them. However, they will vary in numerous impairments and in their severity. For instance, some individuals may be able to interact and function in society, however, others are affected enough by symptoms such as inability to critically think, lack of foresight, or a delayed response to all information processing; this makes interaction incredibly difficult in society (McHale, 2010). Additionally, more than one neurodevelopmental disorder will frequently occur. For instance, individuals with autism spectrum disorder will often have intellectual disabilities, and many individuals with ADHD will also have a specific learning disorder [29, 30, 62].

Through interacting with the online world, this can afford many benefits but also brings about risks to individuals with neurodevelopmental disorders. The online world provides an environment where individuals can engage and develop relationships, learn and develop their social skills, and learn about themselves and others. It allows many to communicate with many others, providing a platform that can encourage self-expression more easily, often due to anonymity or being physically distanced [63]. In addition to the social and cognitive benefits of engaging with the internet and digital technology, it can also provide occupational and supportive benefits such as employment and education, online health (and mental health) information, online therapies, and support groups. These can be particularly useful as social engagement can be challenging for many individuals. Furthermore, digital technology can also allow many to become more independent, accessing entertainment, services, online shopping, and banking, and provide assistive technologies if required [35, 64–66].

However, regardless of the benefits the internet and digital technology afford users there are many risks for users with neurodevelopmental disorders. Individuals with these disorders can lead to maladaptive use of the internet [23], due to issues regarding critical thinking and judgment [66]. Individuals with neurodevelopmental disorders can also be vulnerable to cybercrime, dependent on the disorder and severity of the disorder. These may include being cyberbullied and harassed [35, 67], and being vulnerable to scammers, social engineering, privacy risks, and account hacking through errors in judgment and/or reading, or misunderstandings.

Individuals with neurodevelopment disorders may have significant challenges with performing cybersecurity behaviors. When creating strong passwords, individuals require intelligence, attention, and learning skills. Those with intellectual disabilities, specific learning disabilities, and ADHD may find it particularly difficult to create and learn strong passwords, and therefore, may adopt insecure password behaviors such as reusing passwords. Those with autism spectrum disorder may also find difficulties coping with different password strength requirement policies due to inflexible thinking patterns, which could lead to frustration and stress. Disclosing and sharing information and personal details, as well as opening suspicious emails and websites, may bring about cybercrimes upon individuals without critical thinking, awareness, and judgment. These characteristics can be present in individuals with intellectual disabilities and autism spectrum disorder.

Virus scanning, installing and updating applications, software, and security patches, and backing-up data often require minimal action. However, those who lack judgment and awareness, may ignore any cues to perform these actions. Individuals with specific learning disorder (i.e., dyslexia) often make errors when reading text. Therefore, they may also make errors when reading scamming emails, leading them to click on suspicious links and disclose personal information as a result. These errors increase their vulnerability to crimes such as hacking and identity theft. They may also fear and experience anxiety updating applications and software due to the worry of making such reading errors.

5.4.2   Schizophrenia Spectrum and Other Psychotic Disorders

Schizophrenia spectrum and other psychotic disorders include schizophrenia, other psychotic disorders, and schizotypal (personality) disorder. They affect more than 20 million people [68] and are characterized by positive and negative symptoms.

Positive symptoms include abnormalities in one or more of the following areas: delusions, hallucinations, disorganized thinking (speech), disorganized or abnormal motor behavior. Delusions are fixed beliefs that the individual has and is not willing to change even when faced with contrary evidence. They can include such beliefs:

- the individual will be harmed in some way by anyone or anything;
- the individual has exceptional abilities or is famous;
- another individual has negative or positive emotions towards them, such as love;
- concern about a major event occurring;
- preoccupation with their health.

They are often considered weird or bizarre as they are obviously not conceivable in most people's lives. Hallucinations are experiences perceived by the individual without any external stimulus. They can include all senses, however, auditory hallucinations are the most common, such as hearing voices. Disorganized thought is often represented through speech, where the individual will switch or jump from topic to topic with loose associations during one conversation. Disorganized motor behavior can manifest ranging from child-like behavior to unpredictable agitation to catatonia. Negative symptoms on the other hand, are represented by the absence of emotions and behaviors, such as becoming withdrawn, unmotivated, or unresponsive [29]. Individuals with these disorders will experience phases of positive and negative symptoms, of which these phases can last for varying periods of time.

These disorders are considered to be severe, with the individual becoming disassociated from reality and potentially dangerous. In situations when they become a danger or risk to themselves or to others, this will lead to periods of hospitalization [69]. The disorders vary in occurrences dependent on factors such as gender, and economic status [70]. Although they have a genetic component, it is often the environmental factor that triggers the disorder or episode, such as stress, drug-taking, and alcohol abuse.

Due to the severity of the schizophrenia spectrum and other psychotic disorders, it is difficult for many to interact with the online world in a positive manner. However, there are some benefits through engaging with the internet and digital technologies such as social and communication, supportive, and possibly independence (dependent on severity). For those, especially while experiencing the negative symptoms (if they are not in a catatonic state), may be able to communicate with others and interact with society over social media and chatrooms. They may be able to also use services such as online shopping, and entertainment, providing some sort of level of independence. Supportive services can also be accessed including support groups, health information, and therapies.

The risks of online interaction arise when considering the positive symptoms of these disorders (if they are able to interact at all). When an individual experiences positive symptoms, they are evaluated for hospitalization based on whether they are considered a danger or risk to themselves or others. These positive symptoms can be transferred to the online world if they are able to interact with technology. Therefore, engaging with the internet can pose a risk to themselves, and they can pose a risk to others through the medium of the online environment. Examples of how interactions with the online world can become a risk to the individual can involve finding information (real or fake) that could support delusional thoughts, such as conspiracy theories [28]. Owning digital technology such as a smartphone, or smartwatch can lead to paranoia and delusions of being monitored, and privacy being invaded. Furthermore, through the increased stimuli that engagement with technology and the internet provide, this can also increase the likelihood of hallucinations, where the individual may believe that, for example, a blogger may be talking about them, or directly to them. In addition, due to impaired and disordered thoughts and delusions, individuals with schizophrenia spectrum and other psychotic disorders are more likely to be vulnerable to cybercrimes, including bullying and harassment, social engineering and scamming, identity theft, and hacking. On the flip side, these disorders can present themselves through maladaptive behavior, and exhibiting intense and unpredictable emotions including, anxiety, aggression, and may even lead to violence—becoming a danger to others [29]. If the individual is able to interact with technology and the online world, they may also become a risk to others. This can involve, creating fake news online, posting offensive or unacceptable information generally directed, or directed at a specific person, participating in online harassment and even cyberstalking, and inciting hate crimes in others, all as a result of delusions.

Engaging with cybersecurity may prove difficult for those individuals with schizophrenia spectrum and other psychotic disorders. With positive symptoms, if the individual is able to interact online, creating and learning strong passwords may not be of concern, if the individual has other goals in mind. Disorganized thoughts may prevent the individual from engaging with any type of password management, which may prevent them from gaining access to online services and accounts. On the other hand, not being able to remember passwords while experiencing positive symptoms may influence them to create weak passwords, or write passwords down, while in a more lucid state. As a result of disorganized thoughts and decreased judgment, these individuals may divulge personal information about themselves and others online, be more willing to plug-in devices, and more easily coerced into clicking on suspicious links and visit suspicious websites, which increases the risk of becoming a victim of cybercrime. While experiencing negative symptoms, individuals may avoid or forget to update software, applications, and patches, again leaving them and their technology potentially open to attacks.

5.4.3 Bipolar and Related Disorders

Bipolar and related disorders bridge the main classes of schizophrenia spectrum and other psychotic disorders and depressive disorders, due to the presence of symptoms from both classes. They include a range of disorders associated with bipolar, such as bipolar I disorder, bipolar II disorder, and bipolar induced by substances, medication, or medical conditions [29].

Bipolar I disorder refers to what many would know as classic manic-depressive disorder, whereas bipolar II is a milder version. The disorders are characterized by swings of extreme emotional highs, manic symptoms, and psychosis to major depressive episodes. During manic phases, an individual can experience excitable moods, endless energy, and increased appetites. Often the individual will feel as they do not need sleep and may binge on food, alcohol, drugs, and sex, and undertake risky behaviors. During the depressive stage, individuals will experience periods of major depression, feelings of hopelessness and guilt, suicidal thoughts, and increased sleep [30]. The different episodes can last weeks and have a significant impact on the individual's life, as they find it hard to manage the extreme moods.

The different stages of bipolar disorders may lead to different ways in which an individual will interact with the online world. The benefits of online interaction can provide social engagement and communication, support and health information, and independence during both manic and depressive stages [71]. During the depressive stage, social interaction and communication can help create a support network for those individuals that may have found it hard to interact without access to the internet.

Online therapy, support groups, can also aid individuals during manic stages (ibid.). Furthermore, the availability of services such as online shopping can provide levels of independence to users as well.

Problems occur, mainly through the manic stages when access to the internet can become detrimental, as with schizophrenia spectrum and other psychotic disorders, the maladaptive behaviors are transferred to the online world. Individuals experiencing a manic stage often exhibit risk-taking behaviors, not fully aware of the danger they may be putting themselves in. Risky online behaviors can include, excessive online shopping, online gambling, and engaging in cyber sexual behavior (e.g., sexting, engaging with strangers online, and visiting suspicious sexual websites) that could result in dangerous situations [28]. As the manic individual often finds it hard to sleep, the 24/7 availability of the online world provides the perfect environment for several cyber risks to manifest, especially when the individual lacks self-control. Although most risks from online interaction can be seen during the manic stage, there are some during the depressive stage too. Many might feel more acutely isolated when they view the social world online and are not engaged with it. Moreover, if the individual is cyberbullied or becomes a victim of other cybercrime, this may fuel their feelings of desperation leading to suicidal thoughts.

The different stages of bipolar disorders can affect how the individual will perform cybersecurity behaviors to protect themselves. During a manic stage, the individual may not be concerned with creating and learning a strong password when their goal is to gain access to, for instance, an online gambling website. They may also have issues with recalling passwords due to their manic unstructured thoughts. Moreover, the depressive stage may affect the motivation to engage in secure password behavior and result in forgetting passwords. These issues may lead to choosing convenience over security with regards to password management. Individuals experiencing a manic stage may also be more willing to share private information and be more easily manipulated to reveal further information, and agree to criminals' demands. These symptoms may also result in clicking on dubious email links and websites, and using suspicious USB devices. With risk-taking behaviors as a common trait in these individuals, users are less likely to perform proactive security behaviors, such as virus software updates [57], and the likelihood of circumventing security policies and protocols are more likely, increasing the risk of being a victim of cybercrimes.

5.4.4 Depressive Disorders

Depressive disorders are one of the most common types of mood disorders [31]. 264 million people are affected by depression, with more women being affected than men [68]. There are many depressive disorders including the most commonly known, major depressive disorder (MDD). The most common symptoms are the presence of sadness, emptiness, irritable mood, a loss of pleasurable feelings, and interests, physical slowness, social withdrawnness, loss of appetite and weight gain, and many more. This can be accompanied by somatic symptoms and cognitive impairments. Somatic symptoms include physical aches and pains, whereas cognitive impairments can include, lack of concentration, and learning and memory impairments. The episodes of symptoms can vary in intensity and duration, and can significantly affect the individual's capacity to function impeding on social and occupational functioning [29].

With depressive disorders, there are many benefits from engaging with digital technology and the internet. Many individuals often withdraw into themselves and avoid contact with others. Therefore, the online world provides them with a platform to interact in social activities and communication. However, they may still not choose to engage in these activities as their symptoms are too severe. Nevertheless, they are able to gain support through online mental health information, participate in support groups and receive therapy, if they are able [72]. Due to the symptoms of depression, individuals may still not participate with supportive online interactions for reasons such as, not believing it will help [73, 74]. They are also able to continue to be independent, as when individuals feel unable to, for example, leave the house, they are able to continue to shop, bank, and even undertake work and education online—maintaining their independence.

The risks of individuals with depressive disorders being online are synonymous with those with the depressive stages of bipolar disorders. Viewing the world through the lens of technology can increase the feeling of isolation, which can exacerbate their symptoms further. Similarly, if the individual becomes the target of online harassment, this can also worsen the symptoms [75], and potentially lead to suicide. Furthermore, through their cognitive abilities slowing, this can result in mistakes that can increase the likelihood of becoming a victim of other cybercrimes, such as hacking and identity theft.

Individuals with depressive disorders are often distracted and consumed by their symptoms, accompanied by the slowing of cognitive processes [30]. This could lead many to not consider cybersecurity too much and becoming passive towards any cues of criminal activity or attack. Moreover, due to memory impairments, the individual may find it difficult to remember passwords, and therefore adopting insecure password behaviors (reusing, writing passwords down, etc.) to continue to allow access to systems, devices, and services. Slowness in cognitive processing may lead to impaired decision-making, resulting in individuals clicking on suspicious links, and visiting dubious websites, and poorly evaluating the security of their actions [76]. Furthermore, the memory impairments could also lead to passive inactivity regarding software, application, and security patch updates, all resulting in increased vulnerability.

5.4.5 Anxiety Disorders

Anxiety disorders are the most prevalent of mental disorders, with over 284 million people affected globally [68]. As with depressive disorders, most people experience at some time in their life, a bit of anxiety. However, the difference between "a bit" and a disorder is determined by the severity—is it severe enough to disrupt a person's everyday life and functioning? There are a variety of anxiety disorders, for instance, generalized anxiety disorder (GAD), separation anxiety disorder, social anxiety disorder, panic disorder, substance/medication-induced anxiety disorder, and more. They all share characteristics of excessive fear and anxiety, and related behavioral disturbances. The DSM-5 defines fear as "the emotional response to a real or perceived imminent threat", whereas it defines anxiety as "anticipation of future threat" ([29], p. 189). Although they are similar, there are some distinctive differences, for example, fear often activates the fight or flight autonomic mechanisms, with thoughts of immediate danger, and strategies and resulting escape behaviors. Anxiety often leads to alertness, preparing for a future threat, resulting in muscle tension, cautiousness, and even avoidance.

Many individuals experience panic attacks with their anxiety disorder as a specific response to fear. However, panic attacks are not only experienced by those who have an anxiety disorder, they are also present with other mental disorders. Like with many mental disorders, there are genetic factors that are often triggered by an environmental factor such as stress.

Generalized Anxiety Disorder (GAD) is one of the most common anxiety disorders and is characterized by frequent, persistent excessive, and uncontrollable worry and anxiety that is often irrational or more excessive than the object of worry warrants [30]. Many individuals also experience feelings of panic, paranoia, and can suffer from paralysis, stimming, stuttering, their mind going blank, and adopt avoidance behaviors. Individuals may also experience bouts of depression [29]. Many people who have never experienced an anxiety disorder may ask, "but why do you feel anxious, what triggered it?" Yet, sometimes the initial trigger may have occurred several years ago, and since then the anxiety persistently occurs with no event or reason and reoccurs with no warning. In many cases, the anxiety and worry are often severe enough that they will impede on the individual's social and occupational functioning.

The benefits that an individual with Generalized Anxiety Disorder (GAD) will gain through interacting with digital technology and the online environment are plenty. Where individuals with GAD would possibly withdraw from society due to their symptoms of social anxiety, the online environment provides a means to engage with society and communicate with people through social media platforms, etc. It allows these individuals to express themselves while still being able to socially distance [77]. Whereas, leaving the house for shopping or going to work/education, or attending a therapy session could prove difficult for many, accessing these services online improves their quality of life [72]. However, being able to access life via the online world instead of the offline world can enable the individual to avoid the actual issue. Furthermore, even though the individual can interact online, the symptoms do not disappear. This means their thoughts and behavior can be transferred online. Many may feel anxious about being misunderstood [77] through the lack of personal or e.g., visual interaction. Many may interpret conversations from others differently to their intended purpose, which could exacerbate their symptoms. They may feel overwhelmed by the amount of communication, avoid important communication as well as worry about not replying immediately. Some individuals may develop a phobia towards technology and feel paranoid towards technology monitoring and invading their privacy. Some individuals could feel paranoid or worried that they are missing out, and feel depressed viewing society continue without them partic-ipating [78]. Symptoms can also be triggered or increased through others gaining too much access to the individual when all they want to do is withdraw and recover but feel pressured to interact. Individuals with GAD are also vulnerable to several cybercrimes, such as social engineering [79] and DoS ransomware attacks, where the criminal can manipulate them preying upon their insecurities and anxieties. They can also be severely affected by cyberbullying and harassment, which can intensify their symptoms [75].

Cybersecurity behavior can be affected by anxiety disorder characteristics in many ways, dependent on the type and severity of symptoms. If the individual is more anxious, they could perhaps be more cautious with their security and privacy? This may be the case for some, however, through the worry and fear of cybercrimes and attacks, and becoming a victim, it can become counterproductive. For example, indi-viduals who are more anxious about remembering their passwords, often adopt inse-cure password behaviors such as password reuse to compensate for their perceived memory capabilities [80]. Some individuals who suffer with anxiety, and worry about their cybersecurity, may install several antivirus software programs ("just in case"), which can often end up working against each other, causing the individual's computer to be vulnerable to attacks. In addition, to the lack of security, having these programs operate at the same time, slows the computer and can overload it when updates are required. This further leads to potential loss of data. What is more, individuals with anxiety disorders such as GAD can often be consumed and preoccupied (as with depressive disorders) with their suffering and with very little thought for cybersecu-rity, leading to errors in judgment such as clicking on suspicious links in emails. Indi-viduals can also become easily overwhelmed, by too much contact with others, too much information to process, and complex information, which leads many to ignore communications and adopt avoidance behaviors. Through becoming overwhelmed and resulting in avoidance, the individual may not recognize or acknowledge warn-ings or cues that their cybersecurity is being threatened (e.g., security warnings or messages to update patches), and may become inactive in protecting themselves and their systems.

5.4.6   Obsessive–Compulsive and Related Disorders

The class of obsessive–compulsive and related disorders include several disorders such as obsessive–compulsive disorder (OCD), body dysmorphic disorder, hoarding disorder, trichotillomania (hair-pulling disorder), excoriation (skin-picking) disorder, and substance/medication-induced obsessive–compulsive and related disorder. Obsessive–compulsive disorders are more common than many realize,   experi-ence by about 2% of the global population, and are considered one of the top causes of disability [20]. Obsessive–compulsive disorders have common characteristics including obsessions and/or compulsions. DSM-5 [29] defines obsessions as "recurrent and persistent thoughts, urges, or images that are experienced as intrusive and unwanted". Whereas, they define compulsions as, "repetitive behaviors or mental acts that an individual feels driven to perform in response to an obsession or according to rules that must be applied rigidly" ([29], p. 235). Other obsessive–compulsive disorders can also show the presence of symptoms of preoccupations and repetitive behaviors or mental acts in response to the preoccupations. Some of the obsessive–compulsive disorders have characteristics with recurrent body-focused repetitive behaviors, such as hair pulling, and skin picking, with repeated attempts to decrease or stop the behaviors [29].

With OCD, the obsessions and compulsions can vary in each individual. However, there are many common themes to the obsession-compulsions such as cleanliness, forbidden thoughts (e.g., sexual, violence, etc.), symmetry, harm and death. The intense anxiety that accompanies the obsession, with the repetitive thoughts can be extremely distressing and often lead to compulsive repetitive actions [30]. The thoughts can be rational, for example, not washing your hands after being in public, could potentially result in catching coronavirus, leading possibly to death. However, sometimes the thoughts are irrational, and even though the individual may understand they are irrational, the fear of the outcome may be too overwhelming to deal with. For example, worrying thoughts of not turn the light on and off five times will result in losing your job. The individual may know they will not lose their job but will do the repetitive behavior anyway, just in case. In really extreme circumstances, the behavior can lead to washing hands until they bleed, acting out dangerous behaviors, such as walking in front of cars, or on the other hand, protective behaviors such as saving data in many forms of back-ups, and never throwing anything away. OCD, like most psychological disorders, is genetically based but requires an environmental trigger. Triggers can include, abuse experienced as a child, or a traumatic event such as abuse, bullying, a death, or loss of some kind.

Interacting with the online world can bring benefits to individuals who have OCD. Like with anxiety disorders, individuals can communicate and socially interact with others and society in general through social media, and platforms. They can get access to services, for instance, news outlets, online shops and banks, education, and work. Many can also express themselves and their thoughts anonymously through the medium of the internet, where they may feel more comfortable. In this way, they can gain access to support through groups and therapy [81].

Nevertheless, the online world poses risks as well as benefits. As with anxiety disorders, the ability to enact many life functions through the medium of the internet can result in individuals ignoring the issues they have and not seeking help or treatment. Using the internet and digital technology can also fuel these disorders, through filter bubbles and personalized internet searches presenting individuals with what they want to see, even if it is fake news—giving them an unbalanced and potentially dangerously biased view [82]. If an individual with OCD is concerned about their health, the effects of personalized search results can fuel their worry. Fake news, privacy invasion, and worries around technology monitoring can also fuel an individual's disorder. Individuals with OCD can become victims of cybercrimes. Cyberbullying, hacking, and social engineering. With the compulsion being the motivation behind an action, this allows cybercriminals to manipulate individuals, preying upon their anxieties, and their obsession in completing their behavior.

When experiencing symptoms of OCD, many feel compelled regardless of the result or cost to enact a behavior to appease the obsessive thoughts. Two themes that can be common amongst those with OCD are hygiene/cleanliness and protection as a response to the threat of potential harm [29]. These themes could potentially lead to more proactive security behaviors [43]. However, the behavior can eventually become more extreme, turning counterproductive. For example, the individual may be less trusting and hypervalent, being careful to not open suspicious emails or click on unknown links, but, if a thought arises in their mind that they have to do these actions, they will probably do them. Another example can include, zealous but rational protective thoughts can result in cybersecurity issues, as with anxiety disorders—overcompensation leading to, e.g., antivirus software working against each other. Moreover, irrational thoughts, such as, *my friend might die if I do not uninstall the antivirus software* will also lead to vulnerability. All cybersecurity behaviors could be potentially affected in the same way, as with anxiety disorders it will depend on the severity and type of issues that the individual has, which will determine how they interact with cybersecurity. Individuals with OCD could potentially have problems with managing their passwords. Issues such as choice of password could potentially be a problem, as the choice may come from preference from an obsessive thought, and not protection, e.g., *will choosing 1 or 2 result in death* (or other outcomes)? The individual may become fixated with a specific password and reuse it for many accounts. However, when a password policy requires other criteria, this may lead to frustration, anger, and anxiety. These issues lie with the motivation behind the behavior—even if the theme is protection, the behavior is driven by a compulsion rather than protection itself. Therefore, individuals with OCD can leave themselves open and vulnerable to attack from all types of cybercrime, and manipulation from criminals.

5.4.7 Neurocognitive Disorders

There are several neurocognitive disorders (NCDs), however, they can be divided into three subtypes, delirium, major neurocognitive disorders, and mild neurocognitive disorders. Delirium refers to a notable decrease in awareness and attention. Major and mild NCDs include disorders that are caused by diseases such as, Alzheimer's, Parkinson's, Huntington's disease, HIV, and from traumatic brain injury, etc. Neurocognitive disorders are symptomatic of changes in the brain structure, function, or chemistry [30]. Although many mental disorders include cognitive impairments (e.g., memory impairments in depression), NCDs are disorders where cognitive impairments are the main characteristic. Furthermore, they are not present since birth or since early life, and therefore, represent a decline from "normal" functioning. NCDs can include cognitive impairments such as complex/divided attention (e.g., paying attention within an environment with many stimuli, for instance holding a conversation and the TV playing at the same time), planning or decision making, illogical thinking, learning, and recalling, using language, and social cognition (e.g., knowing how to behave in different settings). When cognitive decline occurs, it may not only impact upon the individual's life, socially, occupationally and with everyday functioning, it can also be frightening, and frustrating [29]. Furthermore, the individual often requires levels of care from family, friends, and/or professionals due to the inability to undertake everyday activities, and/or be trusted to be able to care for themselves.

Digital technology and internet access can provide many benefits for individuals with NCDs in terms of social and communication, cognitive, occupational, supportive, and independence benefits, depending on their levels of severity [83, 84]. Individuals can establish and maintain social relationships online using social media. They can also find information online when they have forgotten particular information. They can interact with work and education if they are able. Digital technology and the internet are most useful when considering support and independence [85]. Through being more in control of their lives, through access to online services, including banking and shopping, can empower an individual with NCDs, whereas previously, an appointed carer would do these things on their behalf [83]. If the individuals are at the stage of their disorder where a carer is needed, the individual will obviously not be able to engage with the online world fully and therefore would not receive all the benefits of being online. However, digital technology and the internet can support the carer in their role as a caregiver [86]. Nevertheless, individuals with NCDs can gain access to telecare or telehealth [87], providing care, therapies, and support groups remotely. Participation and exchanging issues around symptoms with those who are in similar circumstances can lead to various positive outcomes [88]. Furthermore, with developments in technology, assistive technologies have allowed many to live independently in their homes. Specifically, assistive technology, an umbrella term refers to any device or system that allows an individual to perform a task they would otherwise be unable to do or increases the ease and safety with which the task can be performed [89]. Personal assistants in the form of mobile devices, aid those individuals that have memory impairments, with their contacts, to-do lists, and schedules [90]. Furthermore, cognitive agents are AI that resolves issues that individuals may have when interacting with technology, such as reformatting screens and enhancing relevant information when the individual has attentional impairments.

Digital and internet interaction pose many benefits for users with NCDs, however, there can be issues, if the individual has no close family or friends who are internet users. This can present the problem of getting the help and guidance they need to interact with digital technology and in the online world. Until recent times, security risks were often mitigated through avoidance—an assigned person (family member or carer) would interact online with accounts and services on the individual's behalf. However, nowadays with assistive technologies, more individuals with NCDs are able to go online, and therefore, be at a higher risk of exploitation and vulnerability. Individuals with these disorders are potentially vulnerable to cybercrimes, such as social engineering, identity theft, and risks to privacy, because of attentional and memory impairments, and confusion over knowing who and what to trust, and illogical thinking.

Individuals with NCDs could find password authentication particularly difficult to manage, as the cognitive impairments related to attention and memory would make it incredibly difficult to create, learn and recall strong passwords. This could potentially lead many to adopt insecure password behaviors, and susceptible to hacking. Oversharing personal information, opening suspicious emails, and clicking on dubious links and websites may also become problematic through individuals not knowing whom to trust or possibly not knowing with whom they are talking or interacting. This will leave them especially vulnerable to social engineering and scamming. With regards to proactive security behaviors such as virus scanning, software updating, etc., the individual with NCD may be confused by the security messages and cues to take action and forget to do so. All of these may result in exposure to attacks.

5.4.8 Personality Disorders

Personality disorders have a general definition, and criteria that need to be met before applying one of the 10 specific personality types. DSM-5 defines a personality disorder as "an enduring pattern of inner experience and behavior that deviates markedly from the expectations of the individual's culture, is pervasive and inflexible, has an onset in adolescence or early adulthood, is stable over time, and leads to distress or impairment" ([29], p. 645). Maladaptive traits and patterns of behavior, cognition, and inner experience are present across many contexts within an individual's life. The variety of traits has led to defining differing disorders. Although they may vary in some respects, they are similar in others.

These similarities have resulted in disorders being grouped into three clusters:

Cluster A: Paranoid, schizoid, and schizotypal personality disorders. Individuals with these disorders will seem eccentric or odd to others.

Cluster B: Antisocial, borderline, histrionic, and narcissistic personality disorder. Individuals with these disorders can appear erratic, dramatic, or emotional.

Cluster C: Avoidant, dependent, and obsessive-compulsive personality disorders. Individuals with either of these personality disorders may seem as fearful or anxious.

Details of each personality disorder are represented in Table 5.2. Dependent on which type of personality disorder an individual has will depend on how they are affected and the impact it has upon their life. For instance, an individual with paranoid personality disorder may find it difficult to have or keep an occupation or function within society due to symptoms of distrust and suspiciousness of others.

Table 5.2 DSM-5 [29] brief description of each personality disorder

Cluster A	Paranoid personality disorder	Pattern of distrust and suspiciousness such that others' motives are interpreted as malevolent
	Schizoid personality disorder	Pattern of detachment from social relationships and a restricted range of emotional expression
	Schizotypal personality disorder	Pattern of acute discomfort in close relationships, cognitive or perceptual distortions, and eccentricities of behavior
Cluster B	Antisocial personality disorder	Pattern of disregard for, and violation of, the rights of others
	Borderline personality disorder	Pattern of instability in interpersonal relationships, self-image, and affects, and marked impulsivity
	Histrionic personality disorder	Pattern of excessive emotionality and attention seeking
	Narcissistic personality disorder	Pattern of grandiosity, need for admiration, and lack of empathy
Cluster C	Avoidant personality disorder	Pattern of social inhibition, feelings of inadequacy, and hypersensitivity to negative evaluation
	Dependent personality disorder	Pattern of submissive and clinging behavior related to an excessive need to be taken care of
	Obsessive–compulsive personality disorder	Pattern of preoccupation with orderliness, perfectionism, and control

Many individuals who have personality disorders may be unaware of even having the disorder and can be unaware of the way their thoughts and behaviors affect their own lives or the lives around them. This can be apparent with obsessive–compulsive personality disorder when compared with obsessive–compulsive disorder. With OCD the individual has recurrent thoughts that cause anxiety until the compulsions or behavior are untaken to relieve the obsessional thoughts. With obsessive–compulsive personality disorder, the individual engages with obsessive behaviors but is not necessarily anxious about their thoughts, only frustrated if they cannot complete the behavior. Many individuals with personality disorders are unaware that their thoughts and behaviors are not considered "normal" by others. Many find that others around them do not respond to what they say or how they behave in a manner they expect. However, they do not understand why, but do not necessarily question their own abnormal or maladaptive behavior.

Users with personality disorders (regardless of the type) will benefit to varying degrees from interacting with digital technologies and the online world. This includes social and communication—individuals will be able to communicate through social media, engage with society, and can interact with friends and family. Cognitive benefits can be represented through individuals being able to express their emotions to others, and potentially anonymously if they felt more comfortable. Individuals whose disorder impedes upon their daily functioning may benefit occupationally through interacting with employment and education online. So too, with online shopping and banking, etc., allowing those to live independently. Finally, the online world can also provide support for those with personality disorders, offering health information, support groups, and online therapy.

The type of personality disorder and the severity of the disorder will affect the risks that online engagement will pose to the individual, and so too, the risks the individual will pose to the online world. This is also the case for the interaction with cybersecurity, as the individual may undertake risky security behaviors, and therefore, be vulnerable to attack. Although at the same time, they may undertake behaviors that could violate the cybersecurity of others. These will be discussed per personality disorder type.

Paranoid Personality Disorder

Individuals with this type of disorder could potentially be less likely to become a victim of cybercrime. Due to the suspicious nature and distrust of others, they are more prone to being hypervigilant towards their environment [30, 43], and less likely to overshare personal information increasing the chances of social engineering, hacking identity theft, or harassment. However, many become hostile and aggressive in response to their paranoid thoughts and find conspiracies everywhere to support their paranoia. This could result in the individual becoming a cybercriminal by posting or messaging offensive information towards society in general or towards a specific person, by cyberstalking, by creating fake news, or re-sharing fake news. If they have the skills, they could hack accounts and systems seeking out information to confirm their paranoid beliefs. Regarding their cybersecurity behavior, individuals with paranoid personality disorder may adopt more secure password behaviors due to their paranoia and hypervigilance. They may also perform proactive security behaviors such as virus software updates, back-up files and data, and updating applications.

Schizoid Personality Disorder

Individuals with this disorder could possibly be less likely to be a victim of some cybercrimes such as social engineering and harassment. This could because they are often solitary, do not have many (if at all) close relationships, and find it hard to form meaningful relationships [29]. They tend to not experience strong emotions and therefore, do not have the need to express them. This means that individuals with this type of personality disorder may be less present (or not present at all) on social media sites and would not be sharing their opinions or views online, which decreases the likelihood of becoming a victim of those specific cybercrimes. These individuals if they were to experience cyberbullying, may also be less distressed by this type of crime due to their indifference to criticism. They may although, still be vulnerable to other cybercrimes, for example, identity theft, hacking, and denial of services, especially if they adopt poor security practices. However, these individuals are emotionally restricted and are not risk-takers by nature, and therefore, they may perform proactive security behaviors, be less likely to click on suspicious links, back-up their information [57], and have good password management through the lack of emotive decision-making.

Schizotypal Personality Disorder

Individuals with this type of personality disorder could potentially be vulnerable to cybercrimes such as cyberbullying, harassment, and social engineering due to their peculiar behavior, odd speech (or use of language) and thinking, as well as unusual perceptual experiences [30]. These individuals may feel that they have special powers, like telepathy, or have magical control over others. These symptoms are considered odd by many or eccentric, but they are not psychotic [29]. They may, however, not be as vulnerable to these crimes as some, as even though they exhibit these symptoms, they also find relationships and social interaction difficult and anxiety-inducing. What this means is that they may be less present on social media platforms and have reduced online engagement in social activities, therefore potentially being less likely to be open to these risks. Individuals with this disorder can also be suspicious and paranoid, which may reduce the likeliness of sharing personal information online, opening emails from unfamiliar senders, clicking on links and websites, using dubious USB drives, and more likely to perform proactive cybersecurity activities to protect themselves. However, due to a combination of paranoia, preoccupation with paranormal phenomena, and magical thinking, the internet may provide information such as fake news that will fuel their symptoms.

Antisocial Personality Disorder

This type of disorder is what many would know and refer to as psychopathy or sociopathy [30]. The individual could potentially be a victim of cybercrimes, but it is more likely they will be the perpetrator. This is because deceitfulness and manipulation are central characteristics of the disorder, with poor social conformity and impulsivity. They have little sense of responsibility and can engage in criminal activity, with a lack of remorse [29]. Individuals with this personality disorder may engage in social engineering as they can utilize their manipulative traits, repeatedly lying, using aliases, and conning others to gain personal profit or pleasure. Individuals with this type of disorder are often aggressive and have violent tendencies that could result in cyberbully, online harassment and cyberstalking [91–93]. These individuals have been reported to engage in rule violations, with criminal activity, such as stealing and pursuing illegal occupations. Therefore, these behaviors can transition to the online world, through hacking, identity theft, creating viruses and ransomware, and even becoming an insider threat to an organization [94]. The individual may have no remorse for others' wishes, rights or feelings, and be indifferent to the harm they have caused. They can also believe that the victim "deserved it", blaming them for being stupid or helpless, and showing little or no empathy for their own criminal actions. With regards to cybersecurity, they may, on the one hand, be more aware of cybercrimes because of their own actions, and undertake proactive cybersecurity behaviors. They could also be less receptive to the trickery of other social engineers, distrusting suspicious emails and not clicking on any links. However, due to the lack of empathy, impulsiveness, extreme irresponsibility, and disregard for their own and others' safety [29], they may adopt risky security behaviors, such as insecure password management practices, and not engage with proactive security practices, such as updating virus software, applications, or patches nor backing up their data, thus increasing their chances of becoming a victim of cybercrimes [57].

Borderline Personality Disorder

Individuals with this type of disorder could potentially be vulnerable to social engineering as through an intense fear of abandonment, they could be vulnerable to criminals manipulating them for their own gain. However, they may enact cybercrimes themselves, as because of their intense fear of abandonment [29], they can become inappropriately angry and could express that anger through cyberbullying and harassment. They could also be vulnerable to other cybercrimes such as hacking and identity theft due to their lack of control, and impulsive behavior [61], as they could overshare information, and not engage in cybersecurity behaviors.

Histrionic Personality Disorder

Individuals with histrionic personality disorder exhibit attention-seeking behaviors and are especially concerned with their appearance to gain attention [30]. Owing to their need to be the center of attention, these individuals can be overly trusting, open and flirtatious, which could more easily result in vulnerabilities to cybercrimes, such as, social engineering, especially catfishing, and hacking. In contrast, they too can be manipulative, as well as vain and demanding, and therefore, when they do not receive the attention they require, they can turn aggressive, resulting in online harassment and bullying behaviors. Additionally, individuals with this disorder may believe their relationships to be closer and possibly more profound than they actually are [29], which could also result in online harassment, and cyberstalking. What is more, through their need for social engagement and positive reinforcement from people, when they receive any criticism (particularly with regards to their appearance), this could cause distress. Therefore, with the excessive availability of online complements through selfie posting, the increased probability of criticism is also increased, leading to states of depression and possibly suicidal thoughts and behavior (ibid.). With regards to cybersecurity behavior, individuals with this type of disorder may excessively share personal information and information about those who are around them, due to the individual being more trusting and open [43]. This could result in hacking and identity theft. They may not be overly concerned with protecting themselves and their information, as they have other more pressing concerns of gaining attention and compliments. Therefore, creating strong passwords may seem an inconvenience when needing to access an online account, and adopting insecure password behaviors such as reusing passwords will seem more appealing. As with password management, updating virus software, patches, and applications, or backing-up information may also not take precedence over their other goals. However, if they are aware of the inconvenience and access issues experienced due to a security incident, they may be more motivated to have good security hygiene to ensure they continue to interact with the online world.

Narcissistic Personality Disorder

Due to being exhibitionistic and the need for admiration, individuals with this disorder could be at risk of social engineering, hacking, ransomware, and cyberbullying by means of overly sharing information, opinions, and excessive posting. They could also be potentially manipulated by criminals if they believed that information showing them in a less positive light would be revealed. These individuals are often oblivious to the harm they cause to others through their hurtful comments and remarks [29] which could manifest in cyberbullying and harassment online, and could engage with cyberstalking [92, 93]. They, on the flip side, have an extremely fragile self-esteem, which means they are very sensitive to comments and criticism themselves [29]. Therefore, they will excessively lash out in aggression, rage, as well as counterattack and/or become socially withdrawn with a depressive mood. Online interaction with access to positive and negative comments and communication and exacerbate these symptoms. Cybersecurity behavior could be reflected by their awareness of threats—if they are aware of the threats that could undermine their grandiose online persona, they may be more motivated to protect it. Under the belief that they are special and deserve the best [30], they may purchase the "best" antivirus software, and take pride in their proactive cybersecurity behaviors (while believing others are "less" than themselves if they do not take the same proactive stance). With regards to password security behavior, as with other cybersecurity behavior, it is based on their awareness—if they believe they are creating the "best" password regardless of its strength they will continue to feel secure. However, if they learn that their password behavior is not as secure as they thought, this could lead them to feel unnecessarily inadequate, exacerbating their symptoms.

Avoidant Personality Disorder

Individuals with this type of disorder may be less likely in some ways to be vulnerable to cybercrimes due to the hesitancy in engaging with new activities and social activities. Many individuals with this disorder are anxious about how they are evaluated socially [29]. Although the internet can provide the benefit of anonymous social engagement [77], the disorder could be severe enough that the anonymity would not influence the anxiety levels enough. This also means that if the individual were to engage with online relationships and activities, any negative comments or communication could potentially worsen their condition. These individuals are also less likely to take risks through fear of embarrassment, and they are more likely to appraise "normal" situations as potentially dangerous, needing their lives to be more secure [29]. The overestimation of dangerous contexts may result in many being hypervigilant to online risks and cybercrime and lead many to adopt proactive cybersecurity behaviors and be less likely to share personal information [57]. However, as with anxiety disorders, this can go in the other direction. The fear could drive the individual to install many antivirus software programs, contacting the effectiveness of their purpose. Furthermore, their feelings of adequacy may result in them feeling help-less towards protecting themselves online, and adopt avoidance behaviors, ignoring security warnings.

Dependent Personality Disorder

Individuals with this type of personality disorder will be vulnerable to cybercrimes, for example, hacking, online harassment, social engineering. Through the excessive need for others for emotional support and decision-making, and due to the extreme fear of losing approval this allows the individual to be easily manipulated by crimi-nals. Other online risks are seen by worsening their condition by means of interacting with the online world, as criticism and disapproval support their beliefs of worth-lessness [30]. Individuals with this personality disorder may open suspicious emails and follow the requests within, and use USB drives given to them by criminals in the attempt to being accommodating. Individuals may feel inadequate to protect them-selves against any cybersecurity risks of which they are aware. They may additionally depend on others to enact proactive security behaviors and even choose passwords for them. However, their overall feelings and beliefs of inadequacy will drive these indi-viduals to feel helpless in the face of a security threat and influence their motivation to protect themselves.

Obsessive–Compulsive Personality Disorder

Individuals with this personality disorder may be at risk to all cybercrimes, but not necessarily more than individuals without the disorder. Individuals with this disorder are often orderly, have a preoccupation with detail, and have a need to control their environment, affecting their interpersonal and social functioning [30]. They may however, easily find themselves cyberbullying others online if they do not correspond or agree with their rigid thoughts and beliefs. They are followers of rules, procedures and exhibit compliance behavior. These characteristics could result in good cybersecurity practices reducing the chances of becoming a victim of cybercrime. However, if the individual is not fully aware of proper security practices and policies, they may find it difficult to adapt their existing behavior. Furthermore, due to their perfectionism and high standards, to become a victim of cybercrime would potentially have devastating effects on their condition as they are mercilessly self-critical about their own mistakes [29].

5.4.9   Summary of Mental Disorders

In Table 5.3, eight of the 19 classes of mental disorders are summarized together with the benefits and risks of online interaction, and the cyber security behaviors. Each class has a short description of the disorder. The benefits summarize the beneficial themes of internet and digital technology usage and provide examples that are present for each mental disorder class. The risks of online engagement list the present risks for each disorder, and clarify whether the user would likely be the victim and/or the

Table 5.3 Summary of mental disorders with brief description, online benefits, online risks, and cybersecurity behaviors

Mental disorder	Brief description	Benefits of online interaction: themes and examples	Risks of online engagement: victim or perpetrator and why	Cyber security behaviors: adoption of secure or insecure behaviors
Neurodevelopmental disorders inc: intellectual disabilities, autism spectrum disorder, ADHD, specific learning disorder	Developmental impairments: vary from specific limitations in learning and in controlling attention to more general impairments in intelligence and social skills	*Social and communication*: develop and maintain friendships *Cognitive*: develop social skills *Occupational*: education *Independence*: online gaming *Supportive*: health information	*Social engineering*: victim *Online harassment*: victim *Identity-related crimes*: victim *Hacking*: victim *DoS*: victim	*Password management*: secure /insecure *Oversharing info*: secure /insecure *Proactive security behaviors*: secure /insecure *Backing-up info*: secure /insecure *Suspicious emails, links, websites, USBs*: secure /insecure
Schizophrenia spectrum and psychotic disorders	Positive symptoms: delusions, hallucinations, disorganized thinking (speech), disorganized or abnormal motor behavior Negative symptoms: absence of emotions and behaviors	*Social and communication*: communicate with others *Cognitive*: empress emotions and attitudes *Independence*: online shopping *Supportive*: support groups	*Social engineering*: victim /perpetrator *Online harassment*: victim /perpetrator *Identity-related crimes*: victim *Hacking*: victim *DoS*: victim	*Password management*: insecure *Oversharing info*: insecure *Proactive security behaviors*: insecure *Backing-up info*: insecure *Suspicious emails, links, websites, USBs*: insecure

(continued)

Table 5.3 (continued)

Mental disorder	Brief description	Benefits of online interaction: themes and examples	Risks of online engagement: victim or perpetrator and why	Cyber security behaviors: adoption of secure or insecure behaviors
Bipolar disorders	Manic phase: excitable moods, endless energy, risky behaviors (binge on food, alcohol, drugs, and sex) Depressive phase: hopelessness and guilt, suicidal thoughts, and increased sleep	*Social and communication*: develop and maintain friendships *Cognitive*: emotional expression *Occupational*: work *Independence*: online banking *Supportive*: online therapy and support networks	*Social engineering*: victim *Online harassment*: victim *Identity-related crimes*: victim *Hacking*: victim *DoS*: victim	*Password management*: insecure *Oversharing info*: insecure *Proactive security behaviors*: insecure *Backing-up info*: insecure *Suspicious emails, links, websites, USBs*: insecure
Depressive disorders inc: major depressive disorder (MDD)	Feeling of sadness, emptiness, irritable mood. Loss of pleasurable feelings, physical slowness, social withdrawmness, loss of appetite and weight gain	*Cognitive*: emotional expression *Occupational*: work *Independence*: online shopping *Supportive*: support groups. online therapy	*Social engineering*: victim *Online harassment*: victim *Identity-related crimes*: victim *Hacking*: victim *DoS*: victim	*Password management*: insecure *Oversharing info*: secure /insecure *Proactive security behaviors*: insecure *Backing-up info*: insecure *Suspicious emails, links, websites, USBs*: insecure

(continued)

Table 5.3 (continued)

Mental disorder	Brief description	Benefits of online interaction: themes and examples	Risks of online engagement: victim or perpetrator and why	Cyber security behaviors: adoption of secure or insecure behaviors
Anxiety disorders inc: generalized anxiety disorder (GAD)	Excessive fear and anxiety. Feelings of worry, panic, paranoia, mind going blank, adopt avoidance behaviors, depression	*Social and communication*: communicate with others, develop and maintain friendships *Cognitive*: emotional expression *Occupational*: work, education *Independence*: online shopping and banking *Supportive*: health information, online therapies	*Social engineering*: victim *Online harassment*: victim *Identity-related crimes*: victim *Hacking*: victim *DoS*: victim	*Password management*: insecure *Oversharing info*: secure /insecure *Proactive security behaviors*: insecure *Backing-up info*: insecure *Suspicious emails, links, websites, USBs*: secure /insecure
Obsessive–compulsive disorders	Obsessions and/or compulsions, preoccupation with repetitive thoughts and behaviors. Themes: cleanliness, forbidden thoughts (e.g., sexual, violence, etc.), symmetry, harm and death	*Social and communication*: engaging with society *Cognitive*: expressing emotions and attitudes *Occupational*: work/education *Independence*: online shopping *Supportive*: health information, support groups	*Social engineering*: victim *Online harassment*: victim *Identity-related crimes*: victim *Hacking*: victim *DoS*: victim	*Password management*: secure /insecure *Oversharing info*: secure /insecure *Proactive security behaviors*: secure /insecure *Backing-up info*: secure /insecure *Suspicious emails, links, websites, USBs*: secure /insecure

(continued)

Table 5.3 (continued)

Mental disorder	Brief description	Benefits of online interaction: themes and examples	Risks of online engagement: victim or perpetrator and why	Cyber security behaviors: adoption of secure or insecure behaviors
Neurocognitive disorders: inc: dementia	Cognitive impairments: complex/divided attention, illogical thinking, learning, and recalling, using language, and social cognition	*Social and communication*: communicate with family and friends *Cognitive*: expressing emotions *Independence*: assistive technologies* *Supportive*: for carers* * in more severe cases	*Social engineering*: victim *Online harassment*: victim *Identity-related crimes*: victim *Hacking*: victim *DoS*: victim	*Password management*: insecure *Oversharing info*: insecure *Proactive security behaviors*: insecure *Backing-up info*: insecure *Suspicious emails, links, websites, USBs*: insecure
Personality disorders				
Paranoid personality disorder	Pattern of distrust and suspiciousness such that others' motives are interpreted as malevolent	*Social and communication*: develop and maintain friendships, engaging with society *Cognitive*: express attitudes *Occupational*: work /education *Independence*: online banking, shopping *Supportive*: health information, support groups, therapy	*Online harassment*: perpetrator	*Password management*: insecure *Oversharing info*: secure *Proactive security behaviors*: secure *Backing-up info*: secure *Suspicious emails, links, websites, USBs*: secure
Schizoid personality disorder	Pattern of detachment from social relationships and a restricted range of emotional expression	*Cognitive*: learning *Occupational*: work /education *Independence*: online gaming *Supportive*: health information	*Identity-related crimes*: victim *Hacking*: victim *DoS*: victim	*Password management*: secure *Oversharing info*: secure *Proactive security behaviors*: secure *Backing-up info*: secure *Suspicious emails, links, websites, USBs*: secure

(continued)

Table 5.3 (continued)

Mental disorder	Brief description	Benefits of online interaction: themes and examples	Risks of online engagement: victim or perpetrator and why	Cyber security behaviors: adoption of secure or insecure behaviors
Schizotypal personality disorder	Pattern of acute discomfort in close relationships, cognitive or perceptual distortions, and eccentricities of behavior	*Cognitive*: express emotions and attitudes *Occupational*: work /education *Independence*: online gaming *Supportive*: health information, support groups, therapy	*Online harassment*: victim *Hacking*: victim *DoS*: victim	*Password management*: secure *Oversharing info*: secure *Proactive security behaviors*: secure *Backing-up info*: secure *Suspicious emails, links, websites, USBs*: secure
Antisocial personality disorder	Pattern of disregard for, and violation of, the rights of others	*Social and communication*: communicate with others *Cognitive*: express attitudes *Independence*: online banking, shopping *Supportive*: health information, support groups	*Social engineering*: victim /perpetrator *Online harassment*: victim /perpetrator *Identity-related crimes*: victim /perpetrator *Hacking*: victim /perpetrator *DoS*: victim /perpetrator ara>	*Password management*: secure /insecure *Oversharing info*: secure /insecure *Proactive security behaviors*: secure /insecure *Backing-up info*: secure /insecure *Suspicious emails, links, websites, USBs*: secure /insecure

(continued)

Table 5.3 (continued)

Mental disorder	Brief description	Benefits of online interaction: themes and examples	Risks of online engagement: victim or perpetrator and why	Cyber security behaviors: adoption of secure or insecure behaviors
Borderline personality disorder	Pattern of instability in interpersonal relationships, self-image, and affects, and marked impulsivity	*Social and communication:* develop and maintain relationships *Cognitive:* express emotions and attitudes *Occupational:* work /education *Independence:* online banking, shopping *Supportive:* health information, support groups, therapy	*Social engineering:* victim *Online harassment:* victim /perpetrator *Identity-related crimes:* victim *Hacking:* victim *DoS:* victim	*Password management:* insecure *Oversharing info:* insecure *Proactive security behaviors:* insecure *Backing-up info:* insecure *Suspicious emails, links, websites, USBs:* insecure
Histrionic personality disorder	Pattern of excessive emotionality and attention seeking	*Social and communication:* develop and maintain friendships, engaging with society *Cognitive:* express emotions and attitudes *Occupational:* work /education *Independence:* online shopping *Supportive:* health information, support groups, therapy	*Social engineering:* victim *Online harassment:* victim /perpetrator *Identity-related crimes:* victim *Hacking:* victim /perpetrator *DoS:* victim	*Password management:* insecure *Oversharing info:* insecure *Proactive security behaviors:* insecure *Backing-up info:* insecure *Suspicious emails, links, websites, USBs:* insecure

(continued)

Table 5.3 (continued)

Mental disorder	Brief description	Benefits of online interaction: themes and examples	Risks of online engagement: victim or perpetrator and why	Cyber security behaviors: adoption of secure or insecure behaviors
Narcissistic personality disorder	Pattern of grandiosity, need for admiration, and lack of empathy	*Social and communication:* develop and maintain friendships, engaging with society *Cognitive:* express emotions and attitudes *Occupational:* work /education *Independence:* online shopping *Supportive:* health information, support groups, therapy	*Social engineering:* victim *Online harassment:* victim /perpetrator *Identity-related crimes:* victim /perpetrator *Hacking:* victim /perpetrator *DoS:* victim	*Password management:* secure /insecure *Oversharing info:* secure /insecure *Proactive security behaviors:* secure /insecure *Backing-up info:* secure /insecure *Suspicious emails, links, websites, USBs:* secure /insecure
Avoidant personality disorder	Pattern of social inhibition, feelings of inadequacy, and hypersensitivity to negative evaluation	*Cognitive:* learn *Occupational:* work /education *Independence:* online banking, shopping *Supportive:* health information, support groups, therapy	*Online harassment:* victim *Hacking:* victim *DoS:* victim	*Password management:* secure /insecure *Oversharing info:* secure /insecure *Proactive security behaviors:* secure /insecure *Backing-up info:* secure /insecure *Suspicious emails, links, websites, USBs:* secure /insecure

(continued)

Table 5.3 (continued)

Mental disorder	Brief description	Benefits of online interaction: themes and examples	Risks of online engagement: victim or perpetrator and why	Cyber security behaviors: adoption of secure or insecure behaviors
Dependent personality disorder	Pattern of submissive and clinging behavior related to an excessive need to be taken care of	*Social and communication:* develop and maintain friendships, engaging with society *Cognitive:* express emotions and attitudes *Occupational:* work /education *Supportive:* health information, therapy	*Social engineering:* victim *Online harassment:* victim *Identity-related crimes:* victim *Hacking:* victim *DoS:* victim	*Password management:* insecure *Oversharing info:* insecure *Proactive security behaviors:* insecure *Backing-up info:* insecure *Suspicious emails, links, websites, USBs:* insecure
Obsessive–compulsive personality disorder	Pattern of preoccupation with orderliness, perfectionism, and control	*Social and communication:* develop and maintain friendships, engaging with society *Cognitive:* express emotions and attitudes *Occupational:* work /education *Independence:* online banking, shopping *Supportive:* health information, support groups,	*Social engineering:* victim *Online harassment:* victim /perpetrator *Identity-related crimes:* victim *Hacking:* victim *DoS:* victim	*Password management:* secure /insecure *Oversharing info:* secure *Proactive security behaviors:* secure *Backing-up info:* secure *Suspicious emails, links, websites, USBs:* secure

preparator of these risks. The cyber security behaviors refer to whether the user with the specific mental disorder would likely adopt insecure and/or secure behaviors.

5.5 Conclusion

Out of the 4.3 billion online users [19], around 25% will have mental or neurological disorders [20], and therefore, these disorders need to be considered when reviewing users' cybersecurity behaviors. This chapter has provided an overview of how users' psychopathologies could impact cybersecurity behavior and online interaction from the perspective of different mental disorders. It has brought to light the complexities of how mental disorder characteristics can impact a users' experience while engaging with the online world, including gaining support, being at risk to cybercriminals, and participating in cybercrimes themselves.

An examination of psychopathologies has revealed that of each major class of mental disorder, there are numerous disorders under each category, of which have numerous symptoms. These include social, cognitive, and behavioral impairments, that impede upon the individual's daily functioning and independence. The evidence suggests that whereas engaging with the online world and digital technologies can support the clinical impairments of these disorders through a variety of benefits; users are also at risk of exacerbating their symptoms. Conversely, their symptoms can also in return, exacerbate the risks of becoming victims of cybercrimes, through users' generalized online behavior and their specific cybersecurity behavior. There are several mental disorders whose symptoms, such as disorganized thoughts, lack of judgment, overly trusting, and easily manipulated, could increase a users' likelihood of becoming a victim to various cybercrimes. However, there are many disorders where the symptoms can result in users becoming the perpetrators of cybercrimes. On one hand, several disorders present characteristics that could potentially increase users' cybersecurity, through exhibiting secure/protective behaviors, hypervigilance, and distrust. However, these behaviors, because they are driven by a psychopathology, could in fact have the opposite effect, increasing the chance of becoming a cybercrime victim.

Previous research has only recently begun to examine psychopathology and abnormal behaviors with respect to interacting with the internet and digital technology (e.g., [8, 20, 28, 36, 66]). With regards to cybersecurity, previous research has studied how cybercrimes affects the psychology of users (e.g., [5]). There have also been several research examining how cognitive attributes affect security compliance and awareness (e.g., [15, 76]) and examining personality traits such as, openness, narcissism, impulsiveness, and trust in relation to susceptibility to cybercrimes (e.g., [42, 43, 48, 57]). Nonetheless, there is very little research examining how users' psychopathology affects their interaction online, and next no research examining specific mental disorders and vulnerability to cybercrime [35, 66, 79], and cybersecurity behaviors.

This chapter has begun to address this gap by reviewing a variety of mental disorders and applying their characteristics to the cybersecurity context. Although examining personality traits and cognitive factors can provide some insight into how individuals with mental disorders may perform (or not) cyber secure behaviors, examining specific isolated symptoms does not give an accurate overall representation of how the complexity of these interacting symptoms affects a user's cybersecurity behavior. Furthermore, traits and symptoms when they are "non-clinical" are again, are very different. The criteria of mental disorders are determined by their severity and longevity of symptoms [29]. Therefore, observing the effects of, for example, feeling anxious is very different from observing the effects of the unending, paralyzing pain of worry and paranoia. So too, examining a user who is a "bit particular", is very different from examining a user who experiences the excruciating need to perform an action (even when it is acknowledged as abnormal) because of repetitive intrusive thoughts. However, these symptoms are experienced by a substantial number of users and therefore, need to be considered when examining users' vulnerability to cybercrimes and the protective actions they perform to secure themselves in the online world.

Acknowledgements I would like to thank Assoc. Professor Rebekah Rousi and Dr. Juuli Lumivalo for all their encouragement and feedback in writing this chapter. I would like to thank Miika Luhtala for his patience, technical expertise and fixing my laptop (a lot) while writing this chapter. I am also grateful to Janne Kohvakka for his constant support.

References

1. Legner C, Eymann T, Hess T, Matt C, Böhmann T, Drews P, Mädche A, Urbach N, Ahlemann F (2017) Digitalization: opportunity and challenge for the business and information systems engineering community. Bus Inf Syst Eng 59(4):301–308
2. Li Y, Dai J, Cui L (2020) The impact of digital technologies on economic and environmental performance in the context of industry 40 a moderated mediation model. Int J Prod Econ 229:107777
3. Nye JS (2011) Nuclear lessons for cyber security? Strateg Stud Q 5(4):18–38
4. Von Solms R, Van Niekerk J (2013) From information security to cyber security. Comput Secur 38:97–102
5. Bada M, Nurse JRC (2020) The social and psychological impact of cyberattacks. In: Benson V, McAlaney J (eds) Emerging cyber threats and cognitive vulnerabilities. Academic Press, pp 73–92
6. Patterson W, Winston-Proctor CE (2019) Behavioral cybersecurity: applications of personality psychology and computer science. CRC Press, Boca Raton, FL
7. Bonneau J, Just M, Matthews G (2010) What's in a name? Evaluating statistical attacks on personal knowledge questions. In: Sion R (ed) Financial cryptography and data security: 14th international conference, FC 2010, revised selected papers. Springer, Berlin, pp 98–113
8. Schneier B (2015) Secrets and lies: digital security in a networked world, 15th edn. Wiley, Indianapolis, IN
9. Herath T, Rao HR (2009) Encouraging information security behaviors in organizations: role of penalties, pressures and perceived effectiveness. Decis Support Syst 47(2):154–165
10. Adams A, Sasse MA (1999) Users are not the enemy. Commun ACM 42(12):40–46

11. Anderson R (2020) Security engineering: a guide to building dependable distributed systems, 3rd edn. Wiley, Indianapolis
12. Ponemon Institute (2018) 2018 Cost of a data breach study: global overview. Ponemon Institute LLC
13. Arachchilage NAG, Love S (2014) Security awareness of computer users: a phishing threat avoidance perspective. Comput Hum Behav 38:304–312
14. Grawemeyer B, Johnson H (2011) Using and managing multiple passwords: a week to a view. Interact Comput 23(3):256–267
15. Humaidi N, Balakrishnan V (2015) Leadership styles and information security compliance behavior: the mediator effect of information security awareness. Int J Inf Educ Technol 5(4):311–318
16. Renaud K, Weir GRS (2016) Cybersecurity and the unbearability of uncertainty. In: 2016 cybersecurity and cyberforensics conference (CCC). IEEE, pp 137–143
17. ENISA (2018) Cybersecurity culture guidelines: behavioural aspects of cybersecurity. European union agency for network and information security (ENISA). https://www.google.co.uk/url?sa=t&rct=j&q=&esrc=s&source=web&cd=&ved=2ahUKEwjLrLu84aDvAhXto4sKHT0cCW0QFjAAegQIARAD&url=https%3A%2F%2Fwww.enisa.europa.eu%2Fpublications%2Fcybersecurity-culture-guidelines-behavioural-aspects-of-cybersecurity%2Fat_download%2FfullReport&usg=AOvVaw0R_7y4E2KXvtl0iV3jh8iQ
18. Moallem A (ed) (2018) Human-computer interaction and cybersecurity handbook. CRC Press, Boca Raton, FL
19. World Stats (2019) Usage and population statistics. Internet World Stats. http://www.internetworldstats.com/stats.htm. Accessed 1 May 2019
20. WHO (2001) International classification of functioning, disability and health. World Health Organization. https://www.who.int/classifications/icf/en/
21. Whitty MT, Young G (2017) Cyberpsychology: the study of individuals, society and digital technologies. Wiley, Indianapolis, IN
22. Connolly I, Palmer M, Barton H, Kirwan G (eds) (2016) An introduction to cyberpsychology, 1st edn. Routledge, New York
23. Carli V, Durkee T, Wasserman D, Hadlaczky G, Despalins R, Kramarz E, Wasserman C, Sarchiapone M, Hoven CW, Brunner R, Kaess M (2013) The association between pathological internet use and comorbid psychopathology: a systematic review. Psychopathology 46(1):1–13
24. Flood C (2016) Abnormal cyberpsychology and cybertherapy. In: Connolly I, Palmer M, Barton H, Kirwan G (eds) An introduction to cyberpsychology, 1st edn. Routledge, New York, pp 153–164
25. Morahan-Martin J (2007) Internet use and abuse and psychological problems. In: Joinson AN, McKenna KYA, Postmes T, Reips U-D (eds) Oxford handbook of internet psychology. Oxford University Press, Oxford
26. Barlow DH, Durand VM (2012) Abnormal psychology: an integrative approach. Wadsworth, Cengage Learning, Belmont, CA
27. Kring AM, Davison GC, Johnson SL, Neale JM (2007) Abnormal psychology. 10th edn. Wiley
28. Norman KL (2017) Cyberpsychology: an introduction to human-computer interaction, 2nd edn. Cambridge University Press, Cambridge
29. APA (2013) Diagnostic and statistical manual of mental disorders (DSM-5®), 5th edn. American Psychiatric Publishing
30. Black DW, Grant JE (2014) DSM-5® guidebook: the essential companion to the diagnostic and statistical manual of mental disorders. American Psychiatric Publishing, London
31. WHO (2019) Mental disorders. World Health Organization. https://www.who.int/news-room/fact-sheets/detail/mental-disorders
32. Bannon S, McGlynn T, McKenzie K, Quayle E (2015) The internet and young people with additional support needs (ASN): risk and safety. Comput Hum Behav 53:495–503
33. Chadwick DD, Wesson C, Fullwood C (2013) Internet access by people with intellectual disabilities: inequalities and opportunities. Future Internet 5(3):376–397

34. Livingstone S, Haddon L (2009) EU kids online. Zeitschrift für Psychologie/J Psychol 217(4):236–239
35. Chadwick DD, Chapman M, Caton S (2019) Digital inclusion for people with an intellectual disability. In: Attrill-Smith A, Fullwood C, Keep M, Kuss DJ (eds) The Oxford handbook of cyberpsychology. Oxford University Press, Oxford, pp 261–284
36. Katz E (1974) Utilization of mass communication by the individual. In: Blumler JG, Katz E (eds) The uses of mass communications: current perspectives on gratifications research. Sage Publications, pp 19–32
37. Ruggiero TE (2000) Uses and gratifications theory in the 21st century. Mass Commun Soc 3(1):3–37
38. Parikh SV, Huniewicz P (2015) E-health: an overview of the uses of the Internet, social media, apps, and websites for mood disorders. Curr Opin Psychiatry 28(1):13–17
39. Wright K (2000) Computer-mediated social support, older adults, and coping. J Commun 50(3):100–118
40. Barak A, Boniel-Nissim M, Suler J (2008) Fostering empowerment in online support groups. Comput Hum Behav 24(5):1867–1883
41. Coulson N, Smedley R (2015) A focus on use of online support. In: Attrill A (ed) Cyberpsychology. Oxford University Press, Oxford, pp 197–213
42. Nurse JRC (2019) Cybercrime and you: how criminals attack and the human factors that they seek to exploit. In: Attrill-Smith A, Fullwood C, Keep M, Kuss DJ (eds) The oxford handbook of cyberpsychology. Oxford University Press, Oxford, pp 663–690
43. Moody GD, Galletta DF, Dunn BK (2017) Which phish get caught? an exploratory study of individuals' susceptibility to phishing. Eur J Inf Syst 26(6):564–584
44. Jones LM, Mitchell KJ, Finkelhor D (2013) Online harassment in context: trends from three youth internet safety surveys (2000, 2005, 2010). Psychol Violence 3(1):53–69
45. Barton H (2016) The dark side of the internet. In: Connolly I, Palmer M, Barton H, Kirwan G (eds) An introduction to cyberpsychology, 1st edn. Routledge, New York, pp 58–70
46. Woods N, Siponen M (2018) Too many passwords? How understanding our memory can increase password memorability. Int J Hum Comput Stud 111:36–48
47. Woods N, Siponen M (2019) Improving password memorability, while not inconveniencing the user. Int J Hum Comput Stud 128:61–71
48. Shropshire J, Warkentin M, Sharma S (2015) Personality, attitudes, and intentions: predicting initial adoption of information security behavior. Comput Secur 49:177–191
49. Whitman ME (2003) Enemy at the gate: threats to information security. Commun ACM 46(8):91–95
50. Woods N (2019) The light side of passwords: turning motivation from the extrinsic to the intrinsic. In: Proceedings of the 14th Pre-ICIS workshop on information security and privacy at ICIS 2019
51. Campbell J, Ma W, Kleeman D (2011) Impact of restrictive composition policy on user password choices. Behav Inf Technol 30(3):379–388
52. Shay R, Komanduri S, Durity AL, Huh P, Mazurek ML, Segreti SM, Ur B, Bauer L, Christin N, Cranor LF (2016) Designing password policies for strength and usability. ACM Trans Inf Syst Secur 18(4):1–34
53. Das A, Bonneau J, Caesar M, Borisov N, Wang X (2014). The tangled web of password reuse. In: NDSS '14. Internet Society, pp 23–26
54. Hern A (2018) Strava suggests military users 'opt out' of heatmap as row deepens. The Guardian. https://www.theguardian.com/technology/2018/jan/29/strava-secret-army-base-locations-heatmap-public-users-military-ban
55. Furnell SM, Bryant P, Phippen AD (2007) Assessing the security perceptions of personal Internet users. Comput Secur 26(5):410–417
56. Sasse MA, Smith M, Herley C, Lipford H, Vaniea K (2016) Debunking security-usability tradeoff myths. IEEE Secur Priv 14(5):33–39
57. Egelman S, Peer E (2015). Scaling the security wall: developing a security behavior intentions scale (SeBIS). In: CHI'15: proceedings of the 33rd annual ACM conference on human factors in computing systems. pp 2873–2882

58. Menard P, Gatlin R, Warkentin M (2014) Threat protection and convenience: antecedents of cloud-based data backup. J Comput Inf Syst 55(1):83–91
59. Crossler RE (2010) Protection motivation theory: understanding determinants to backing up personal data. In: 2010 43rd Hawaii international conference on system sciences. IEEE, pp 1–10
60. Tischer M, Durumeric Z, Foster S, Duan S, Mori A, Bursztein E, Bailey M (2016). Users really do plug in USB drives they find. In: 2016 IEEE symposium on security and privacy (SP). IEEE, pp 306–319
61. Coutlee CG, Politzer CS, Hoyle RH, Huettel SA (2014) An abbreviated impulsiveness scale constructed through confirmatory factor analysis of the Barratt impulsiveness scale version 11. Arch Sci Psychol 2(1):1–12
62. Mayes SD, Calhoun SL, Crowell EW (2000) Learning disabilities and ADHD: overlapping spectrum disorders. J Learn Disabil 33(5):417–424
63. Tynes BM (2007) Role taking in online "classrooms": what adolescents are learning about race and ethnicity. Dev Psychol 43(6):1312
64. Chadwick DD, Fullwood C (2018) An online life like any other: identity, self-determination, and social networking among adults with intellectual disabilities. Cyberpsychol Behav Soc Netw 21(1):56–64
65. Chadwick DD, Quinn S, Fullwood C (2016) Perceptions of the risks and benefits of internet access and use by people with intellectual disabilities. Br J Learn Disabil 45(1):21–31
66. Good B, Fang L (2015) Promoting smart and safe internet use among children with neurodevelopmental disorders and their parents. Clin Soc Work J 43(2):179–188
67. Kowalski RM, Fedina C (2011) Cyber bullying in ADHD and asperger syndrome populations. Res Autism Spectrum Disord 5(3):1201–1208
68. GBD 2017 Collaborators (2018) Global, regional, and national incidence, prevalence, and years lived with disability for 354 diseases and injuries for 195 countries and territories, 1990–2017: a systematic analysis for the global burden of disease study 2017. Lancet 392:1789–1858
69. Becker T, Kilian R (2006) Psychiatric services for people with severe mental illness across western Europe: what can be generalized from current knowledge about differences in provision, costs and outcomes of mental health care? Acta Psychiatr Scand 113(Suppl. 429):9–16
70. McGrath J, Saha S, Chant D, Welham J (2008) Schizophrenia: a concise overview of incidence, prevalence, and mortality. Epidemiol Rev 30(1):67–76
71. Conell J, Bauer R, Glenn T, Alda M, Ardau R, Baune BT, Berk M, Bersudsky Y, Bilderbeck A, Bocchetta A, Bossini L et al (2016) Online information seeking by patients with bipolar disorder: results from an international multisite survey. Int J Bipolar Disord 4(1):1–14
72. Sunderland M, Wong N, Hilvert-Bruce Z, Andrews G (2012) Investigating trajectories of change in psychological distress amongst patients with depression and generalised anxiety disorder treated with internet cognitive behavioural therapy. Behav Res Ther 50(6):374–380
73. Breuer L, Barker C (2015) Online support groups for depression: benefits and barriers. SAGE Open 5(2):1–8
74. Nimrod G (2013) Online depression communities: members' interests and perceived benefits. Health Commun 28(5):425–434
75. Kowalski RM, Giumetti GW, Schroeder AN, Lattanner MR (2014) Bullying in the digital age: a critical review and meta-analysis of cyberbullying research among youth. Psychol Bull 140(4):1073–1137
76. Donalds C, Osei-Bryson KM (2020) Cybersecurity compliance behavior: exploring the influences of individual decision style and other antecedents. Int J Inf Manag 51:102056
77. Erwin BA, Turk CL, Heimberg RG, Fresco DM, Hantula DA (2004) The internet: home to a severe population of individuals with social anxiety disorder? J Anxiety Disord 18(5):629–646
78. Wegmann E, Oberst U, Stodt B, Brand M (2017) Online-specific fear of missing out and Internet-use expectancies contribute to symptoms of Internet-communication disorder. Addict Behav Rep 5:33–42

79. Welk AK, Hong KW, Zielinska OA, Tembe R, Murphy-Hill E, Mayhorn CB (2015) Will the "phisher-men" reel you in? Assessing individual differences in a phishing detection task. Int J Cyber Behav, Psychol Learn 5(4):1–17
80. Woods N (2016) Improving the security of multiple passwords through a greater understanding of the human memory. Dissertation, University of Jyväskylä
81. James TL, Lowry PB, Wallace L, Warkentin M (2017) The effect of belongingness on obsessive-compulsive disorder in the use of online social networks. J Manag Inf Syst 34(2):560–596
82. Holone H (2016) The filter bubble and its effect on online personal health information. Croat Med J 57(3):298–301
83. Astell AJ, Bouranis N, Hoey J, Lindauer A, Mihailidis A, Nugent C, Robillard JM (2019) Technology and dementia: the future is now. Dement Geriatr Cogn Disord 47(3):131–139
84. Clare L, Rowlands JM, Quin R (2008) Collective strength: the impact of developing a shared social identity in early-stage dementia. Dementia 7(1):9–30
85. LaMonica HM, English A, Hickie IB, Ip J, Ireland C, West S, Shaw T, Mowszowski L, Glozier N, Duffy S, Gibson AA, Naismith SL (2017) Examining internet and eHealth practices and preferences: Survey study of Australian older adults with subjective memory complaints, mild cognitive impairment, or dementia. J Med Internet Res 19(10):e358
86. Boots LMM, de Vugt ME, van Knippenberg RJM, Kempen GIJM, Verhey FRJ (2014) A systematic review of Internet-based supportive interventions for caregivers of patients with dementia. Int J Geriatr Psychiatry 29(4):331–344
87. Berridge C, Furseth PI, Cuthbertson R, Demello S (2014) Technology-based innovation for independent living: policy and innovation in the United Kingdom, Scandinavia, and the United States. J Aging Soc Policy 26(3):213–228
88. Asbury T, Hall S (2013) Facebook as a mechanism for social support and mental health wellness. Psi Chi J Psychol Res 18(3):124–129
89. WHO (2018) Assistive technology. World Health Organization. https://www.who.int/news-room/fact-sheets/detail/assistive-technology
90. Lopresti EF, Mihailidis A, Kirsch N (2004) Assistive technology for cognitive rehabilitation: state of the art. Neuropsychol Rehabil 14(1–2):5–39
91. Bogolyubova O, Panicheva P, Tikhonov R, Ivanov V, Ledovaya Y (2018) Dark personalities on facebook: harmful online behaviors and language. Comput Hum Behav 78:151–159
92. Moor L, Anderson JR (2019) A systematic literature review of the relationship between dark personality traits and antisocial online behaviours. Pers Individ Differ 144:40–55
93. Smoker M, March E (2017) Predicting perpetration of intimate partner cyberstalking: gender and the dark tetrad. Comput Hum Behav 72:390–396
94. King ZM, Henshel DS, Flora L, Cains MG, Hoffman B, Sample C (2018) Characterizing and measuring maliciousness for cybersecurity risk assessment. Front Psychol 9:39

Chapter 6
Process Ontology Approach to Military Influence Operations

Miika Sartonen, Aki-Mauri Huhtinen, Monika Hanley, and Petteri Simola

Abstract Technology has transformed our information environment into a global village with endless opportunities but also malevolent actors with power to influence the networked world. Militaries are not granted an exception from the new requirements of adaptability and renewal. As always, new technology gives the best rewards to those who use it first. Thus, in a sense, militaries compete with each other for the use of weapons of influence in the global struggle for dominance. Fast-paced changes in communication technology have led to different levels of adaptation by militaries and also different interpretations of the rules of combat regarding the use of influence in a military context. This creates an asymmetry in the selection of target groups, influencing them, and, in turn, protecting them. To help militaries position themselves into the networked ensemble of modern influence operations, the authors suggest a process ontology approach to military influence operations. Finding key interconnected elements of modern influence efforts will help streamline military sense-making processes.

Keywords Influence · Information · Process · Rhizome

M. Sartonen (✉) · A.-M. Huhtinen
Finnish National Defence University, Helsinki, Finland

M. Hanley
University of London, London, UK
e-mail: mh322@student.london.ac.uk

P. Simola
Finnish Defence Research Agency, Tuusula, Finland

© The Author(s), under exclusive license to Springer Nature Switzerland AG 2022   135
M. Lehto and P. Neittaanmäki (eds.), *Cyber Security*, Computational Methods
in Applied Sciences 56, https://doi.org/10.1007/978-3-030-91293-2_6

6.1   Introduction

> ... but suppose we were (as we might be) an influence, and idea, a thing intangible, invulnerable, without front of back, drifting about like a gas?
>
> We might be a vapour, blowing were we listed.
>
> Our kingdoms lay in each man's mind; and as we wanted nothing material to live on, so we might offer nothing material to the killing.
>
> (T.E. Lawrence, Seven pillars of wisdom: A triumph, 1935).

Military influence practices are known by a multitude of names, but the common denominator is the act of influencing single individuals or groups, often referred to as target audiences. Equally important are the defensive processes of countering hostile military deception or disinformation. The militaries are not, however, the only actors, nor do they have information sovereignty in the modern information environment. In addition, in today's security environment militaries often operate with other governmental agencies in order to wield or defend against comprehensive operations, including different means of influence power.

For a military commander, the experience of navigating in the modern information environment with multiple governmental and other partners can be exhausting, especially if the processes, corresponding authorities and legal jurisdiction are not clear. This can lead to exceeding one's authority, but also to indecision. Especially when countering foreign hostile influence operations, it would be practical to have common ground, a model or framework, on which to base discussion of operations.

In this article, the authors suggest a process ontology approach that addresses the elemental factors of information age military influence. The goal is to point out that a process organisational approach can provide new insights into the challenges of the evolving information environment, thus providing a fresh look at the organisations of the "information actors" both inside and outside the armed forces, allowing the reflection of current organisations and responsibilities vis-a-vis the identified entities.

The authors use the word influence, as 'the power or capacity of causing an effect in indirect or intangible ways' [28] in a military context to convey the idea of how militaries are affected by the effects of human cognition. In the article, we use the term *military influence operations* to describe the various flows of influence in the military context. Other related terms are used when in context with the source.

6.2   Modern Influence Environment

Technological development is a fact that we have no possibility of avoiding, or getting rid of, in the online digital solutions of our daily life routines. Today, information is not a mere tool but rather a collection of environmental forces that increasingly affect who we are, how we socialise in our daily life, and how we use this concept as part of our understanding of reality and virtuality:

As connected individuals, part of a modern society, we generate data with everything we do. - It is virtually impossible to get an overview of the data we use and generate—data are everywhere. - As with any new technology, data analytics create opportunities for both use and abuse. [42, 5]

The modern battlefield is becoming increasingly digitised, and thus often approached via technological solutions. One does not need, however, highly advanced cyber capabilities in order to challenge even the most technologically advanced militaries. Data that is openly available can be used to gather intelligence or geolocate military units, and social media can be used for influence purposes without sophisticated systems [6]. By the same token, the overwhelming capability of modern technology can sometimes fail at arbitary tasks, as "online discussions about black-and-white chess pieces are confusing artificial intelligence algorithms trained to detect racism and other hate speech" [10].

The more new information there is, the more there is also new disinformation on social media:

Strikingly, the report found that Instagram is $10\times$ cheaper to manipulate than Facebook, TikTok has virtually no self-regulatory defences, and it remains easy to manipulate US senators' accounts, even during an election period. [13]

Thus, it seems we have to tolerate disinformation more than before, and are becoming resilient to the information boomerang effect:

However, other platforms exhibited a continued inability to combat manipulation. Of the 337,768 fake engagements purchased, more than 98 per cent remained online and active after four weeks, and even discounting fake views, more than 80 per cent of the 14,566 other fake engagements delivered remained active after a month. [5, 3]

Step by step, players experience and react to some of the most common techniques in the production of COVID-19 misinformation, including hyper-emotional language, fake experts and conspiracy theories [15].

The coronavirus disease (COVID-19) has spawned chaos and disinformation around the world, especially in social media. Additionally, all kind of conspiracy theories have spread exponentially [39]. The rhizomatic spread of COVID-19 [21, 27] has become not only one of the deadliest and most uncontrolled pandemics in modern history, but also a dominating flashpoint in the global contest for information among nations, with competing narratives reflective of competing political systems[1]:

Like COVID-19 itself, information can mutate and evolve; it experiences "superspreading events," where influencers use digital and traditional media to amplify rumors that are soon "caught" by unsuspecting members of the public. Just like a pandemic, an infodemic can be mitigated by practicing a digital form of hygiene – employing skepticism, verifying sources and the like. And conversely, an infodemic can spread out of control when influencers and the public make no attempts to contain the rumors around them. [4]

When we look at the political papers and their suggestions concerning information development, we can find almost the same conclusions and recommendations in

---

[1] See https://www.atlanticcouncil.org/event/covid-narrative-arms-race/

them, namely, that, due to information technology, our environment has become more insecure, complex, constantly changing and online dependent. The questions about security, in general, become more difficult to answer, as "to be secure in one's being is paradoxical in the sense that to be is to survive while always becoming otherwise" [1]. What this means is that we cannot build our security anymore as fully-controlled circles, as with traffic lights systems:

> The unsettling logic of one's security of existing creates a possibility for actors to performatively take a leap of faith. [1, 292]

The insecurity of our being means that our being is always becoming otherwise. The key question is how we manage the limits of reflexivity, namely the fear of not knowing enough, as our knowledge is always engaged with the unknown.

We are networked and constantly connected to the digital domain, but cannot control our simultaneous connectedness to the real (material) and digital (virtual) world domains. Thus, we need new a kind of understanding and modelling to make sense of our reality, which can be achieved by artificial intelligence and machine learning. The capabilities are steadily getting more effective:

> A few days ago, Google released a new huge paper that proposes a new method to significantly boost the number of parameters while maintaining the number of Floating-point operations per second (the ML computational cost standard metric). [18]

Soon enough we will have to be ready to answer the question: Under which circumstances should we accept living machines as equal to human beings? The blurring of the distinction between human, machine and nature is obvious in modern life [32]. In the end, there are no clear structures or blocks anymore, instead there are blurring and blending dimensions:

> One can extract a single brick from a Lego-house without too much hassle and generally without influencing the structural integrity of the whole. When considering a watercolour painting, things are different. After the watercolours blend together, as they are supposed to do, there are only vague borders between the colours, so that there is no exact or simple way to separate one colour from the other. This interconnection between the various colours is enhanced by the fact that the colours continue to blend further as time passes (as long as there is enough water). I hope that the analogy of the watercolour painting sheds light on the way temporal beings are fundamentally constituted by the various ways they interact through time, so that it is impossible to separate one being from the rest without oversimplification. [34, 35]

6.3 Psychology of Modern Influence Practices

Military influence practices have many names, and are often cited as psychological operations, psychological influence operations or psychological warfare. The word psychology is used for a reason. These practices draw their success from our psychological and sociological traits and attributes, turning them into vulnerabilities. In modern influence curriculum, few names are often mentioned.

The first one is Robert Cialdini, whose research led to the well-known six principles of persuasion [8]:

1. Reciprocity,
2. Scarcity,
3. Authority,
4. Consistency,
5. Liking,
6. Consensus.

The second and third one are Nobel laureate, Daniel Kahneman and Adam Tversky whose trail-blazing work in understanding human decision-making has provided significant understanding of how our everyday cognitive heuristic can predispose us to erroneous thinking [20, 41]. Cialdini´s principles of persuasion, as well as -theories of cognitive heuristics, provide a general understanding of why we are vulnerable to influence attempts and fake news, as well as why countermeasures based on rational argumentation or direct countering of these fake news often fail.

Fake news is one scourge of current information era. One explanation for our inability to discern truth from falsehood is partisanship and politically-motivated reasoning. However, extensive literature review by Pennycook and Rand [31] challenges this common narrative. According to their review, a lack of careful reasoning and relevant knowledge, and the use of familiarity and other heuristics, are the main reasons why we fall for fake news [31]. Pennycook and Rand identify several cognitive heuristics that predispose us to believe fake news. These heuristics are:

• Familiarity (repetition),
• Source (does it come from a source I think is credible),
• Emotionality (rather than cold fact).

Emotional content also often includes social context, which we are tuned to recognise above all else [38]. Among these, there are several other aspects and factors to take into account when building an understanding of our vulnerabilities.

Endsley [12] also summarises a few cognitive factors and heuristics often relevant, such as attention, anchoring, and confirmation bias. Fake news is often constructed in such a way as to draw our attention (click bait). First, information creates an "anchor" and can have a significant effect on what information is later attended to or believed. Fake news is easy and faster to create compared to more in-depth real news, and thus anchor our decision-making. Confirmation bias makes us vulnerable to seek and recognise only information that supports our previous assumptions, thus possibly reinforcing our previous erroneous assumptions.

Social media of the past decade and the increased use of the internet with more sophisticated methods to collect user data (big data) have provided the opportunity to personalise influence attempts. It is now possible to extend target groups and focus groups from age, gender, and ethnicity to personality, motivations and emotions. One such way is to collect user data and, based on that data, create personality profiles from users and thus use that information to focus influence attempts. Previous research has shown that different personality traits predispose us to different principles of influence [29, 46]. These personality profiles drawn from social media data are far from perfect, accuracy ranging from 0.3 to 0.4 [3]. Nevertheless, computer-based estimations already exceed the human ability to predict personality from social media content [16].

Even with limited accuracy, these new ways already provided more accurate ways of influencing. For example in marketing compared to unpersonalised advertisements, tailored personality based advertisements can increase click rates up to 40% and purchases up to 50% [26]. These user behaviour-based information collection methods are not limited to social media. It has been shown that the way we use our mobile phones can also be used as a profiling method. For example, data from our mobile phone's acceleration sensor can be used to predict our behaviour and personality [14].

Our cognitive heuristics that serve us well in everyday life have been turned against us in modern influence practices. With these well-known cognitive factors, modern possibilities to gather user-specific information produces more specific and more personalised attack vectors. Challenges lies in both our ability to counter and recognise these influence attempts as well as our attempts to educate (or inoculate) our citizens against these means. Unfortunately, this is not an easy task as many of these cognitive heuristics work below our conscious mind.

6.4   Influence in Warfare

Many of the doctrinal solutions for Western militaries stem from the Unites States. After the Cold War, the US military approach concerning matters of influence has mostly been communicated by doctrines addressing concepts such as command and control warfare or information operations. In addition to more technical fields, such as cyber and electronic warfare, military influence has been addressed by psychological operations, and partly by military deception and operations security. A clear line has typically been drawn between the aforementioned fields and (military) public affairs. Concerning the political guidance and influence, public diplomacy and, more specifically, strategic communications, affect, but are not directly a task of, the US armed forces [2, 45–57].

Although being able to provide social legitimacy, this type of systematic, doctrine-based approach may not be the best way of addressing the complexities of modern warfare and providing lasting solutions [30, 36]. A more recent approach, the Joint Concept for Operating in the Information Environment [19] presents a way to 'build information into operational art to design operations that deliberately leverage information and the informational aspects of military activities to achieve enduring strategic outcomes'. This can be seen as an attempt to form a more comprehensive understanding of the information environment and its effects on combat operations.

The Russian approach to influence operations is more holistic, encompassing the continuous information struggle between nations. The main goal of the information struggle is to 'ensure the protection of the national interests in the information-psychological sphere and, ultimately information-psychological security of the state'. At the more practical level, there is a difference between the 'standard information war' of a military operation and the 'strategic information war', asymmetrical attacks on different layers of society that can attack the same person via different spheres of cognition, making the identification of an attack difficult. Although there is currently no identifiable Russian model of information-psychological operations, it is worth noting that they will most likely the target individual and mass consciousness levels both in the conflict zone and beyond [33].

In the contemporary Russian military thinking there exists a concept of a new type of permanent war, with no distinction between military and peaceful means. During an actual conflict, if confrontation seems unavoidable, Russia may try to gain initiative by striking first, and achieving information and situational superiority in the initial period of war. Concerning Russian military thinking, however, it should be acknowledged that trying to understand Russian military thinking with Western concepts, such as "hybrid warfare", runs the risk of mirror-thinking that hides how military thought is actually integrated into operational design in the Russian military [40, 12-1–12-5].

A Chinese perspective points out that, in a world where all things are interdependent, the difference between what is and what is not the battlefield is merely relative, as possible future war will be 'fought and won in a war beyond the battlefield'. This new type of warfare can use violence of be non-violent, and it 'can be a confrontation between professional soldiers, or one between newly emerging forces consisting primarily of ordinary people or experts... Any sphere can become a battlefield, and any force can be used under combat conditions.' [24, 153–184].

From the aforementioned examples we can conclude, that information is being interwoven into modern military thought in a variety of ways that are not immediately comparable. In addition, warfare cannot be effectively discussed without the element of society. What, then, is the role of armed forces in a conflict that targets society as a whole? The distinction between military and non-military from the perspective of influence operations is (and most probably has always been) blurred in practise. To begin, all the soldiers are also citizens, and thus influence attempts directed at militaries will also affect the soldiers as citizens and vice versa. This can clearly be seen in the case of COVID-19, with the "infodemic" affecting citizens regardless of their organisational backgrounds. Secondly, influence operations targeted at military forces will typically also have third parties as unintended receivers of information, especially in the digital information environment. These incidents can have unpredictable consequences, which the armed forces should be prepared to deal with.

Comprehensive influence operations, that more or less resemble military operations and target societies as a whole, are not a new phenomenon. The concept of ontological security can be used to understand the large-scale implications these type of operations have on society, such as perverting the information landscape in order to influence policy, erode society's coherence and foster anxiety [7]. These types of influence operations do not necessarily target militaries as their first option, but militaries will be affected as part of the larger society. Using the aforementioned metaphor, it can be said that the all-pervasiveness of the digital information environment has made the distinction between civilian and military even more of a watercolour-type, although many organisations still seem to employ a Lego-brick approach to this complex phenomenon. In the spirit of the quote from T.E. Lawrence, it can be said that militaries typically are well-prepared to identify and counter *operations*, but less well equipped to counter *influence*.

6.5   Military Influence Operations

How do militaries fare in the modern influence arena, and what does it look like from a military point of view? According to Larson et al. [23, xiv–vv], experiences of the operations in Iraq and Afghanistan show that success in the modern information battlefield depends on how commanders interpret the modern battlespace and how successfully they use their capabilities to influence the outcome of desired end states. The commander's interest and involvement in the integration of information operations with other capabilities is also a factor. Thus, successful commanders have a clear understanding of the factors that affect the information battlespace, the resources to influence these factors, and they strive for a clear end state. Following too many factors, or using resources to respond to daily crises without the end state in mind will not yield successful results.

Legal factors play an important part in the commander's decision-making process, but also add uncertainty if the commander's role as an authority is not clear. Concerning the complex situations militaries have to face in today's information environment, many situations can be unambiguous, paving the way for different interpretations of the legal case. If jurisdictional aspects are not effectively solved, commanders will be uncertain in making decisions, as there will be overlapping laws and their interpretations, and the laws of conflict will not cover all the situations of modern warfare [17].

One of the descriptive features of the Information Age environment is the speed of communications, and thus, of the changes that constantly take place, shaping both the target audiences and the effectiveness of messages. As the speed of communication continues to increase, the process of globalisation intensifies, as well [47, 30–31]. In a military context, this presents challenges as commanders contend with the speed of discourse, and developing strategic foresight to address the constantly developing information environment. Elements of military influence operations can be challenging to identify, especially in a rapidly changing environment. The targets of the operations are rarely a single entity, but can be made up of individuals, groups, networks, leadership, and the general public [22]. As such, leaders must understand this quick nature of information transfer, the rapid psychological impact that these operations may have on people, and to respond rapidly.

Another key challenge is the amount of data available in the information environment. As a human, our ability to process massive and often unstructured data is limited at best. Thus, processing and sorting available data in a form that is under-standable for us is done by programmes and algorithms, which are often referred to as artificial intelligence (AI). For future influence practises, AI will play an increasingly significant part. In the commercial world social media advertisers have long been automatically targeted to match our online behaviour and preferences (likes, pages we visit and so on), with an increased understanding of our online behaviour (such as how our personality is portrayed on social media) future development and usage of AI already going beyond this. Recent advances in natural language processing have increased our capability to analyse and automatically interpret social media posts and profile different personalities, emotions and valence. At the same time deepfake AI algorithms are already taking influencing to next level. We have already seen these rather convincing and disturbing deep fake videos of our leaders [35]. These videos are getting better and are relatively easy make with even rather limited information [48]. As Sample points out in a Guardian article:

> The more insidious impact of deepfakes, along with other synthetic media and fake news, is to create a zero-trust society, where people cannot, or no longer bother to, distinguish truth from falsehood. And when trust is eroded, it is easier to raise doubts about specific events [35].

As mentioned previously, the target audience of information influence operations can be many and, can frequently change based on temporal advantage [36]. The rhizome concept works well to replace the previous notions of a linear process, in that, during the process, elements may be constantly changing, sometimes in a circular fashion in a cycle of revision and refining [9]. It has been suggested that, as opposed to just knowing what impacts the target audience and in what ways, it is more valuable to understand the end process and work backwards to identify the target audience and the effects of the rhizome. No ideas exist in a vacuum nor are they deployed as such. Once a certain narrative or influence operation is released, it comes into contact with already existing narratives or ideas and can latch on to some, change others, and quickly become something much greater than its original start. If we agree that it is not a linear process and instead think of it as an entity in constant development, acting in parallel with other processes, we can better understand the nature of military influence operations.

6.6   Process Ontology Approach

There are many ways in which militaries, as part of society, approach the complexity of the modern information battlespace and its effect on the sense-making and influ-ence processes of both individuals and organisations. The plethora of concepts and definitions make it difficult to reach a common ground regarding whose authority or responsibility it is to respond to multi-faceted hybrid-type operations, especially if the effects are hidden by the complexity of attack vectors. In addition, outside the active influence and counter-influence operations are all the other global streams of information, and their effects on the success of both friendly and hostile operations. These effects cannot be ignored, rather they should be taken advantage of whenever possible, and can be conceptualised as influence management activities [37]. How, then, can militaries make sense of the influence operations' requirements in a constantly changing technological information environment?

As a solution, the authors suggest a process ontology approach. Finding the essential elements and flows of information that affect militaries would enable military planners to have a more educated understanding of the requirements of modern and future military influence operations. What are the requirements for such an approach? Among the maze of modern aspects of the information battlespace, clarity is the first requirement. In addition, the model should acknowledge the global reach of modern communications, and the effect of fast-paced timing. The characteristics of the modern, rhizomatic environment should also be acknowledged, allowing for elements to emerge and change by the effects of the environmental factors. The framework should also be "platform and channel agnostic" in order to be useful in the constantly changing digital information environment.

An example of an ontology in the military sphere related to military influence operations is a Core Ontology for Situation Awareness, developed for [25]. Another related core ontology is ONTO-CIF, developed to improve military intelligence analysis [11]. An example of a more specific ontology is the Military Information Ontology developed as a foundation for semantic architecture models for unmanned aircraft systems [45]. The aforementioned examples highlight what on ontological approach can do, providing a logically consistent model that allows for comprehensive understanding on the subject.

As an example of a military process, the U.S. [44] defines the core task of psychological operations as follows:

- Develop,
- Design,
- Produce,
- Distribute,
- Disseminate,
- Evaluate.

These activities are targeted vis-a-vis an adversary, and do not directly (apart from the contemporary integration with information operations) consider the military role as a governmental agency countering hybrid operations that target the whole of society. Target audience analysis is an example of a single process by which the targets, messages and channels of the influence operation are identified and analysed [43, 5–1]. The authors' idea is to look for similar descriptive doctrinal processes, and provide an ontological model of the elements that constitute a military influence (and counterinfluence) operation.

6.7 Conclusion

As modern, connected individuals, we generate data with everything we do. This data can be analysed and used both by those we give permission to and by those we do not. The cognitive heuristics that serve us well in everyday life can be turned against us by modern influencers in the form of more specific and more personalised attack vectors. As with any new technology, militaries are adapting to the modern influence battlespace with varying approaches and capabilities.

Military influence operations are a difficult subject to approach, partly due to the nature of the modern rhizomatic information environment, as well as, in part, to the variety of the mandates given to different security authorities. The variety in the ways militaries approach the rules of combat regarding the use of influence creates asymmetry in the practises of selection, influence and protection of target audiences.

A process ontological approach provides a way to look at influence operations from an alternative viewpoint, apart from the pre-set of presumptions and restrictions originating from organisational traditions. The attempt is to look at the process itself, and find out the requirements it sets on the militaries' capabilities. Using a conceptual model will aid in clarifying the military sense-making processes and provide a starting point for further discussion on the subject.

The authors are currently in the process of creating a process ontological approach to contemporary military influence operations. This article provides an insight why such an ontology should be created, and what could be gained by it.

References

1. Arfi B (2020) Security qua existential surviving (while becoming otherwise) through performative leaps of faith. Int Theory 12(2):291–305. https://doi.org/10.1017/S1752971920000123
2. Armistead L (2010) Information operations matters: best practises. Potomac Books, Dulles, VA
3. Azucar D, Marengo D, Settanni M (2018) Predicting the big 5 personality traits from digital footprints on social media: a meta-analysis. Pers Individ Differ 124:150–159
4. Bandeira L, Aleksejeva N, Knight T, Le Roux J (2021) Weaponized: how rumors about Covid-19's origins led to a narrative arms race. Atlantic Council, Washington, DC
5. Bay S, Dek A, Dek I, Fredheim R (2020) Social media manipulation 2020: how social media companies are failing to combat inauthentic behaviour online. NATO StratCom COE
6. Bolt N, Haiden L, Hajduk J, Lange-Ionatamišvili E (2020) Clarifying digital terms. NATO StratCom COE, Riga
7. Bolton D (2021) Targeting ontological security: Information warfare in the modern age. Polit Psychol 42(1):127–142. https://doi.org/10.1111/pops.12691
8. Cialdini RB (2006) Influence: the psychology of persuasion. Harper Business
9. Coyne R (2008) The net effect. Design, the rhizome, and complex philosophy. Futures 40(6):552–561
10. Cuthbertson A (2021) AI mistakes 'black and white' chess chat for racism. Independent. https://www.independent.co.uk/life-style/gadgets-and-tech/ai-chess-racism-youtube-agadmator-b1804160.html

11. Dragos V (2013) Developing a core ontology to improve military intelligence analysis. Int J Knowl-Based Intel Eng Syst 17(1):29–36
12. Endsley MR (2018) Combating information attacks in the age of the internet: new challenges for cognitive engineering. Hum Factors 60(8):1081–1094
13. Fredheim R, Van Sant K (2020) Robotrolling 2020. Issue 4, NATO StratCom COE. Available: https://www.stratcomcoe.org/robotrolling-20204
14. Gao N, Shao W, Salim FD (2019) Predicting personality traits from physical activity intensity. Computer 52(7):47–56
15. GCS (2021) GCS International joins the fight against health misinformation worldwide. Government communication service. https://gcs.civilservice.gov.uk/news/gcs-international-joins-the-fight-against-health-misinformation-worldwide/
16. Hinds J, Joinson A (2019) Human and computer personality prediction from digital footprints. Curr Dir Psychol Sci 28(2):204–211
17. Hollis DB (2009) New tools, new rules: international law and information operations. In: David GJ, McKeldin TR (eds) Ideas as weapons: influence and perception in modern warfare. Potomac Books, Dulles, VA, pp 59–72
18. Ibrahim M (2021) Google switch transformers: scaling to trillion parameter models with constant computational costs. Towards data science. https://towardsdatascience.com/goo gle-switch-transformers-scaling-to-trillion-parameter-models-with-constant-computational-costs-806fd145923d
19. JCOIE (2018) Joint concept for operating in the information environment (JCOIE). U.S. Joint chiefs of staff. Available: https://www.jcs.mil/Portals/36/Documents/Doctrine/concepts/joint concepts_jcoie.pdf
20. Kahneman D, Tversky A (1979) Prospect theory: an analysis of decision under risk. Econometrica 47(2):263–291
21. King MA (2020) Humans and nonhumans becoming political: Memphis women's march assemblages. Dissertation, University of Memphis
22. Larson EV, Darilek RE, Gibran D, Nichiporuk B, Richardson A, Schwartz LH, Thurston CQ (2009) Foundations of effective influence operations: a framework for enhancing army capabilities. RAND Corporation, Santa Monica, CA
23. Larson EV, Darilek RE, Kaye DD, Morgan FE, Nichiporuk B, Dunham-Scott D, Thurston CQ, Leuschner KJ (2009) Understanding commander's information needs for influence operations. RAND Corporation, Santa Monica, CA
24. Liang Q, Xiangsui W (2015) Unrestricted warfare, Reprint. Echo Point Books and Media, Brattleboro, VT
25. Matheus CJ, Kokar MM, Baclawski K (2003) A core ontology for situation awareness. In: Proceedings of the sixth international conference of information fusion. IEEE, pp 545–552. https://doi.org/10.1109/ICIF.2003.177494
26. Matz SC, Kosinski M, Nave G, Stillwell DJ (2017) Psychological targeting as an effective approach to digital mass persuasion. Proc Natl Acad Sci 114(48):12714–12719
27. McClure BE (2020) Internet memes and digital public discourse. Dissertation, University of Georgia, Athens
28. Merriam (2021) Influence. Merriam-webster dictionary. https://www.merriam-webster.com/dictionary/influence. Accessed 23 Feb 2021
29. Oyibo K, Orji R, Vassileva J (2017) Investigation of the influence of personality traits on Cialdini's persuasive strategies. In: Orji R, Reisinger M, Busch M, Dijkstra A, Kaptein M, Mattheiss E (eds) Proceedings of the second international workshop on personalization in persuasive technology. CEUR Workshop Proceedings vol 1833. pp 8–20
30. Paparone C (2017) How we fight: a critical exploration of US military doctrine. Organization 24(4):516–533
31. Pennycook G, Rand D (2020) The psychology of fake news. PsyArXiv Preprints. https://doi.org/10.31234/osf.io/ar96c
32. Peters MA, Jandric P (2019) Posthumanism, open ontologies and bio-digital becoming: response to Luciano Floridi's Onlife Manifesto. Educ Philos Theory 51(10):971–980

33. Pynnöniemi KP (2019). Information-psychological warfare in Russian security strategy. In: Kanet R (ed) Routledge handbook of Russian security policy. Routledge, London, pp 214–226. Available: http://hdl.handle.net/10138/308903
34. Röck T (2019) Time for ontology? The role of ontological time in anticipation. Axiomathes 29:33–47. https://doi.org/10.1007/s10516-017-9362-2
35. Sample I (2020) What are deepfakes—and how can you spot them? The Guardian. https://www.theguardian.com/technology/2020/jan/13/what-are-deepfakes-and-how-can-you-spot-them
36. Sartonen M, Huhtinen A-M, Lehto M (2016) Rhizomatic target audiences of the cyber domain. J Inf Warfare 15(4):1–13
37. Sartonen M, Huhtinen A-M, Simola P, Takamaa KT, Kivimäki V-P (2020) A framework for the weapons of influence. Int J Cyber Warfare Terrorism 10(1):34–49
38. Schilbach L, Eickhoff SB, Rotarska-Jagiela A, Fink GR, Vogeley K (2008) Minds at rest? Social cognition as the default mode of cognizing and its putative relationship to the "default system" of the brain. Conscious Cogn 17(2):457–467
39. Sederholm T, Huhtinen A-M, Jääskeläinen P (2021) Coronavirus as a rhizome: the pandemic of disinformation. Int J Cyber Warfare Terrorism 11(2):43–55
40. Thomas TL (2019) Russian military thought: concepts and elements. MITRE product MP190451V1, Mitre corporation. Available: https://www.mitre.org/publications/technical-papers/russian-military-thought-concepts-and-elements
41. Tversky A, Kahneman D (1981) The framing of decisions and the psychology of choice. Science 211(4481):453–458. https://doi.org/10.1126/science.7455683
42. Twetman H, Bergmanis-Korats G (2020). Data brokers and security. Risks and vulnerabilities related to commercially available data. NATO StratCom COE
43. U.S. Army (2003) Psychological operations tactics, techniques and procedures. Field manual 3 May 301. Available: https://www.fas.org/irp/doddir/army/fm3-05-301.pdf
44. U.S. Army (2005) Psychological operations. Field manual 3 May 30, Department of the Army, Washington, DC. Available: https://fas.org/irp/doddir/army/fm3-05-30.pdf
45. Valente A, Holmes D, Alvidrez FC (2005) Using a military information ontology to build semantic architecture models for airspace systems. In: 2005 IEEE aerospace conference. IEEE, pp 1–7. https://doi.org/10.1109/AERO.2005.1559635
46. Wall HJ, Campbell CC, Kaye LK, Levy A, Bhullar N (2019) Personality profiles and persuasion: an exploratory study investigating the role of the big-5, type D personality and the Dark Triad on susceptibility to persuasion. Pers Individ Differ 139:69–76
47. Wenzlhuemer R (2013) Connecting the nineteenth-century world: the telegraph and globalization. Cambridge University Press
48. Zakharov E, Shysheya A, Burkov E, Lempitsky V (2019) Few-shot adversarial learning of realistic neural talking head models. In: 2019 IEEE/CVF international conference on computer vision (ICCV). IEEE, pp 9458–9467

Part II
Critical Infrastructure Protection

Chapter 7
Future Smart Societies' Infrastructures and Services in the Cyber Environments

Aarne Hummelholm

Abstract We hear every day about new services and applications that citizens can use for different purposes and needs around the world in real time every day. Digital innovations in the software and applications of information systems used by the information society have contributed to this rapid development. At the same time, the rapid technological development of new terminals as well as new smart devices and sensors has made it possible to develop and offer completely new services to citizens. In the digitalising intelligent society of the future, there is also a need for various services that can be used by service developers, administrators, responsible organizations, and stakeholders to maintain and develop the infrastructures of the intelligent society and related services. Although the development of the digital world is a positive thing, there is another side to the story. There are many security threats in this kind of environments. The latest challenges for the operating environment are heterogeneous tele-communication networks where new devices and systems are seamlessly interconnected. These systems are now coming into smart homes, smart buildings and into marketplaces, into smart hospitals, ships terminals, airports, train stations, open areas, cars, trains and various control and energy systems. This paper describes a society where all important functions interact with each other through digital networks. The paper provides a technical outlook and enterprise architecture level pictures of smart cities infrastructures for the developers and to the designers.

Keywords Smart cities · Networks · Societies · Wireless · Architectures

7.1 Introduction

The society around us, with all its structures, is currently undergoing major changes. These changes affect the daily lives of all citizens in many ways. This is one reason for dividing smart city environments into different segments with their services that we use every day in our future societies and smart cities. Another reason for this is

A. Hummelholm (✉)
Faculty of Information Technology, University of Jyväskylä, Jyväskylä, Finland
e-mail: aarne.hummelholm@jyu.fi

© The Author(s), under exclusive license to Springer Nature Switzerland AG 2022
M. Lehto and P. Neittaanmäki (eds.), *Cyber Security*, Computational Methods in Applied Sciences 56, https://doi.org/10.1007/978-3-030-91293-2_7

that responsible service operators need to be able to analyze and verify the security and other characteristics of the services end-to-end in a smart society. So, there must be a trusted environment in which these services are used.

The purpose of this study is to describe the architectures of future smart cities and to assess and analyze the services, cyber threats and risks that will affect both the future society and smart cities and the daily lives of people. In addition, the study aims to perform risk and threat analyses. In order to identify the information threats to society and determine their impact on our daily lives, we must first describe our future operating environment and identify the key elements and structures that affect its functioning.

This study uses a completely new approach that divides the structures of society into six segments (see Fig. 7.1):

1.   Basic infrastructure,
2.   Energy,
3.   Mobility,
4.   Buildings and homes,
5.   Public services,
6.   Communications and IoT services.

The approach differs, for example, from the way the International Electrotechnical Commission (IEC) presents the structures and services of smart cities [2].

The IEC divides the future urban environment into five segments:

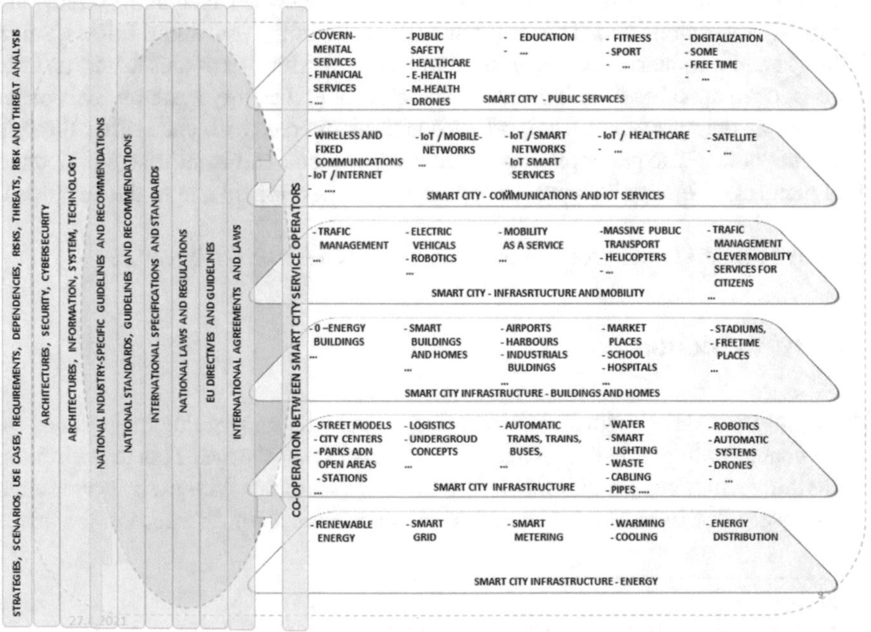

Fig. 7.1 Functional environments in the future smart cities [1]

1. Energy,
2. Mobility,
3. Buildings and Homes,
4. Public Services, and
5. Water.

Smart Cities Council [3] describes another method for building a smart city, but its approach is also different from that used in this paper, because it does not provide the necessary principles for its overall architecture.

One way of dividing is the Beecham Research's Sector Map, in which society is divided into nine distinct entities with different functional backgrounds [4]. In this paper, these segments are grouped in a new way that is more appropriate to the architecture work. Smart cities and smart societies are divided into six segments to get a clear picture of the architecture and services of smart cities. This study applies the Enterprise Architecture Framework method [5] and uses the Quality Function Deployment (QFD) model [6] to define the dependencies of different services and functions.

The society of future will produce a huge amount of information in digital form. Every citizen receives the information they need in real time, regardless of time and place [7]. Changes in social structures take place very quickly, for example, via social media. They also affect the implementations and operating models, structures, and people's daily lives and work environments. The solutions currently in use are likely to be outdated in a few years. The current powerful digitalization trend is increasing the range of services offered and making them easier to use. These developments also have a strong impact on the service chains of the services provided, including subcontracting chains, hardware solutions, service providers, and operating models for each part of the service chain. As a result of the development of intelligent societies described above, people and systems produce a huge amount of information that needs to be processed and stored. The services provided by smart social connections need to be integrated to work together so that we can get services in real time when we need them, no matter where we are. This means that all services are interconnected through integrations, our future smart societies, and smart cities.

However, technical solutions for new service environments are not yet all in line with international standards, such as IoT devices and sensors. They also do not have secure interfaces with telecommunications and service networks. Services based on technically outdated solutions and services using new technology are in use at the same time. Future information and communication systems must be designed to adapt to this challenging environment where security and cyber threats are ubiquitous. We need to create a new type of Secure Digital Building (SDB) for citizens, because we must also consider cyber security issues in our living environment so that we can use the services we are offered safely every day (see Fig. 7.2). The SDB building is divided into different segments so that we can separate those services from each other into different security zones or security levels. Figures 7.1 and 7.2 show well how the systems of future operating environments and, in some cases, the systems of

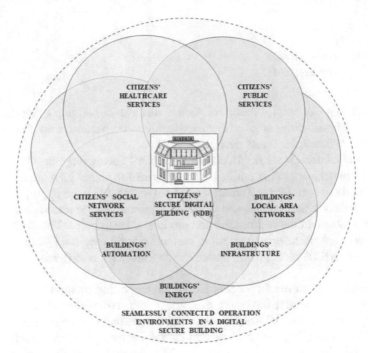

Fig. 7.2 Secure digital building (SDB) for citizen

the modern smart urban environment already in use, as well as the home and living environment systems of citizens are interconnected.

This forms a national and global entity in which each system affects everything indirectly or immediately (Fig. 7.3). When making information security and cyber security solutions for public services, it is no longer enough to look at systems only in their own silos, but we also need to look at the larger whole, the dependencies involved, the resulting risks and threats, and their implications for solutions made.

For example, in terms of information systems and data storage, our systems have long been seamlessly interconnected in our smart societies around the world. Hackers and cyber attackers can easily take advantage of these ubiquitous interconnected telecommunications and information system environments in the same way that citizens use them to manage their day-to-day affairs. Attackers are constantly looking for vulnerabilities and protection deficiencies in the systems so that they can exploit them immediately to achieve their own goals. A good example of an attacker's potential are events in the United States, where network attackers gained access to the source code of Windows OS operating system software and thus management systems [9, 10]. Because operating system updates are made globally, an attack can contaminate smart devices everywhere. When data centers are used around the world to store data, these types of attacks can cause quite unpredictable problems for systems.

The variables affecting the future society that should be considered in development plans of an intelligent society are shown in Fig. 7.4. In Finland, such variables have

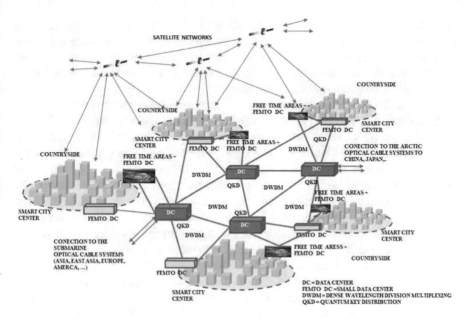

Fig. 7.3 Interconnected world [8]

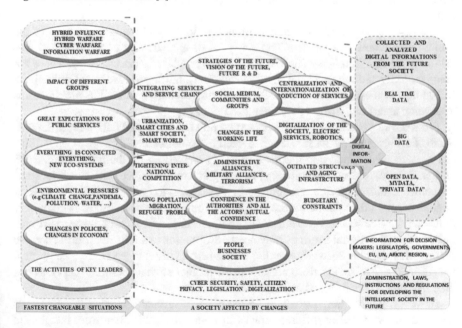

Fig. 7.4 Variables affecting the future society [1]

been discussed with experts from the Ministry of Finance and the University of Lapland, which studies the Arctic region. But these things change and complement over time, so such a broad examination would also require even more extensive research into the factors of change affecting society and their implications for the structures and services of our modern society.

When drafting and enacting laws, network, and service designers as well as authorities must consider the variables that will affect the future society. The systems used in our societies produce a massive amount of societal data. Examples of these data types are 'Big Data', 'Open Data', 'My Data' and 'Private Data'. We get real time information on digitalisation, citizens, aging infrastructure, an aging population, global competition (globalization), urbanization, pollution, environmental pressures, the political climate, interest groups, refugee issues, climate change, etc. The information society constantly generates a lot of information that is difficult to get to the right place if the tools and methods needed to monitor the development of society are not available or if information does not flow smoothly enough between different actors such as users, decision makers, producers, and politicians. If we operate in only one silo, the consequences could be catastrophic for the intelligent society of the future, for example, when looking at cyber-attacks against the functions of society. Critical infrastructures can be disrupted or rendered inoperable in many situations, either intentionally or unintentionally, and due to disruptions caused by various risk factors.

The research topic of this paper is limited to looking at the key elements of the future society and smart cities. The activities of a smart city can be divided into different service segments. These segments are regulated in many ways. There are many national and international recommendations, laws, standards, and guidelines. In addition to these, we need to have visions and scenarios as well as architectures to develop smart city infrastructures and services [1].

From the strategies and scenarios, we can find use cases and derive requirements for the development of the services provided by a smart city. Each set of operations has its own service and communication needs depending on the user group. Such groups include, for example, citizens, planning and maintenance staff, financial staff, telecom operators, service operator staff, virtual service providers and operators, administrators, other public authorities, etc. Each user group operates horizontally in its own service sector. For a smart city to function properly and be able to provide citizens with the digital services they need, information systems in different service sectors must be able to work together both horizontally and vertically. They must also be able to exchange information with each other so that smart city services can be implemented flexibly and efficiently.

However, information systems used by different service sectors or smaller units within them are often at different stages in their life cycle. Therefore, integration between them may not be possible for technical reasons. In this case, data models, operating systems, management systems, and application interfaces do not work well enough together and their coordination and information exchange through integration environments is difficult. Platform solutions can be different, supplied by different vendors, and have their own de facto standard that is not compatible with devices and systems from other vendors. This may lead to the introduction of certain types of open ICT system platforms that guarantee the functionality of some systems until upgrades and are cost-effective until the end of the platform's life cycle, when hardware or software needs to be renewed.

Security solutions for systems and services can also be different and even partially inadequate. This leads to practical challenges. For example, it may be impossible to connect different systems together and thus make different services compatible. When we look at the operating environment of a future smart city, we also need to consider underground structures along with upward structures. The cities of the future can be built into multi-storey structures, including skyscrapers, underground facilities and services, street-level transport services, transport arrangements, drone-based services, taxi arrangements by air and other new types of transport services to facilitate the mobility of people, etc. In the future, all ICT services in a smart city must be automatically adaptive and dynamic services using the new 4G, 5G, 6G, and 7G wireless virtualized mobile network technologies [11, 12].

In this multi-storey structure of smart cities, we must also consider the potential for cyber-attackers and hackers to attack our systems. In this highly complex, ubiquitous, and multidimensional environment, there are many systems, information systems, and physical systems that do not necessarily include security mechanisms to protect them from cyber attackers. These multilayer interconnected telecommunication networks are multidimensional because we use more and more wireless technologies and even the same frequency bands in the same space and thus often form a MESH network there.

7.2 Smart City Communication Infrastructures

Smart city ICT solutions are increasingly using a variety of wireless technology solutions that are highly vulnerable to cyber-attacks and allow attackers to attack the critical infrastructure and services of a smart society. Information security is one of the key issues that must be considered in the design of technical solutions from the very beginning so that information security solutions and security structures can be made considering issues and dependencies shown in Figs. 7.5 and 7.6. We need to have a clear picture of connected and integrated systems. We can then better look at dependencies, assess risks and cyber threats, identify risks and threats as well as residual risks and threats based on real-world situations.

Smart cities also have a lot of underground facilities, such as underground shops in shopping centers, leisure facilities, sports facilities, concert facilities, etc. These facilities form a network of underground facilities in the city with all the necessary services. The underground facilities of smart cities and their functions must also be designed so that there are no space or devices to prepare and carry out cyber-attacks against the systems and services of the city. Underground spaces can be grouped according to the services provided into separate segments, for example on a separate floor or corridor, as follows:

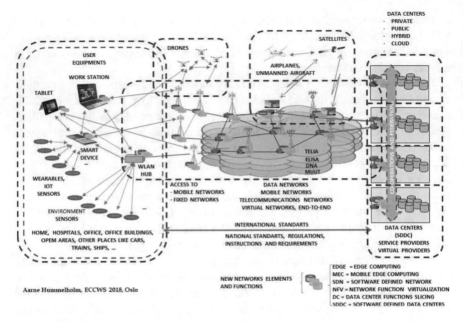

Aarne Hummelholm, ECCWS 2018, Oslo

Fig. 7.5 Network architecture [1]

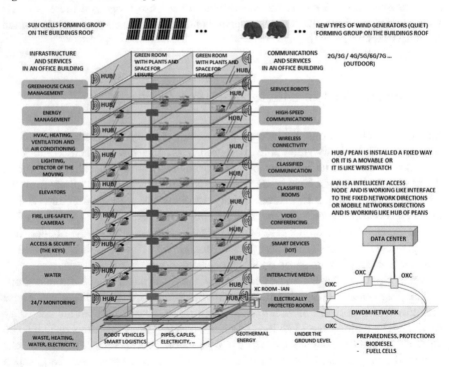

Fig. 7.6 Office building and its infrastructure and communication systems [13]

- Underground stores in shopping malls,
- Underground facilities for sports and leisure,
- Concert, theater, and leisure facilities,
- Heating, water, and electricity pipes,
- Waste management and transportation,
- Robotics logistics / intelligent logistics,
- Automatic robotic metro and train systems,
- Other public transport systems in smart city centers,
- Air conditioning systems,
- Systems required for the use of renewable energy,
- Exit corridors,
- Service corridors.

In order to integrate operating environments and implement smart city operations flexibly and efficiently, we need different types of communication solutions between different buildings, interiors, street environments, and underground spaces to provide services to citizens in these places. Figure 7.5 illustrates how in the smart cities of future, telecommunication systems and services will operate in a virtualized operating environment, where the resources of telecommunication networks and data centers are shared among network and service users by arranging different service operators to operate either together or separately. These virtualized systems operate on wireless or wired networks. Every smart city service and infrastructure segment needs telecommunication network connections to the necessary services that come either for their own use or for citizens, organizational cooperation, development work, management, etc.

When looking at future infrastructures such as buildings, homes, hospitals, leisure facilities, etc., we need to ensure telecommunication arrangements in many ways (Fig. 7.6), including energy-efficient structural solutions and zero-energy buildings [14]. In these buildings, we face challenges in wireless communication arrangements because wall and window solutions in buildings may attenuate outgoing and incoming wireless technology signals too much. Renewable energy, such as that produced by solar cells or wind generators, can be used in buildings. These buildings have many different types of services that operate on the same LANs and use the same network resources. Inside the building, we see different services on the same LAN and wireless networks. The smart devices of all citizens also use the same online resources. There are also green rooms with flowers and small trees for people.

In these virtualized communication environments of the future, we also need to look at security and cybersecurity issues. There are also single-family or multi-family detached houses on the edge of smart cities (Fig. 7.7), where citizens living in need modern telecommunications services to use smart city services flexibly and reliably. Renewable energy solutions should also be made for new building from the outset so that we can take care of the production and use of renewable energy in our own environment, as required by the European Commission's Energy Efficiency Directive [14]. For example, suburban single-family houses would have solar panels and new

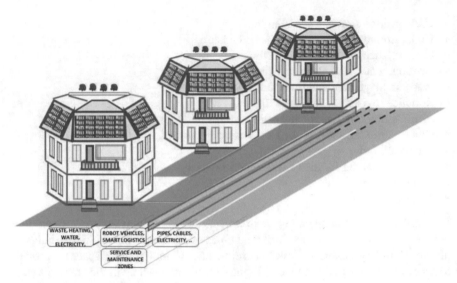

Fig. 7.7 Detached houses using renewable energy solutions in the suburban area

types of quiet wind generators on the roof, as well as a hybrid system of air source heat pumps and geothermal heat.

When we live in the smart cities and smart societies of the future and use public, communication, and mobility services, we get a lot of information about the basic infrastructure: buildings, homes, leisure places, and parks (Fig. 7.8). In such a communication environment, all systems are seamlessly connected to each other.

To ensure privacy, security, and cybersecurity [15, 16] in that whole to interconnected environment of a smart city, we face huge challenges in making information

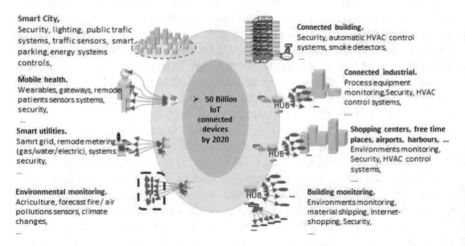

Fig. 7.8 IoT devices, sensors, and actuators for buildings and operating environments

systems work together, as they are often at different stages of their life cycle and may not be compatible with the operating systems and software used. Security solutions may not be adequate or up to date. Furthermore, the services provided to citizens differ in terms of service availability, usability, and confidentiality, and often also in terms of critical services due to time delays. In addition, service requirements vary depending on the service segments and the services available there.

This means that future environments will require a wide variety of security zones, gateways, service interfaces, and integration to make services to work as flexibly as desired. Therefore, it is necessary to manage services, users, user access, management services, shared services, infrastructure services, critical services, and service groups in virtualized networks environments (Fig. 7.9).

We can also divide the services into different groups depending on user groups and needs, for example as follows (Fig. 7.10):

- Mobile broadband,
- Healthcare,
- First aid and rescue,
- Authorities,
- Internet of Things,
- Physical infrastructure.

Each group can then be divided internally into different VPN groups and different encryption keys can be used to distinguish user groups.

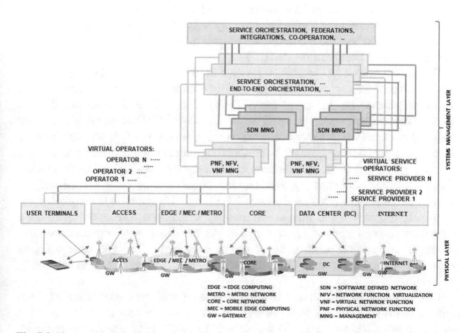

Fig. 7.9 Communications' characterisation for various layers of infrastructure of communications and service providers [17, p. 35; 11, 12]

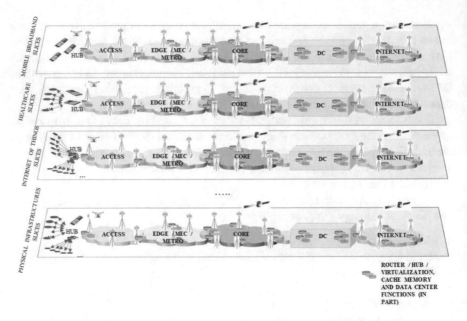

Fig. 7.10 Communication systems slicing [17, p. 37; 11, 12]

There are hundreds of systems to virtualise within virtualized network environments. Along with virtualizations, we need to reduce the power consumption of data centers and telecommunications networks by using the latest technology such as Network Function Virtualization (NFV), Software-Defined Networking (SDN) and Software Defined Data Centres (SDDC).

The energy consumption of ICT systems is growing exponentially despite the virtualization of data center platforms. Profiling the use of equipment and systems is necessary based on the actual use of the systems and services so that they do not consume energy unnecessarily. When performing virtualization calculations and analyses, we need to consider our daily working hours, because we usually only use our terminals during working hours and then shut down our systems. Often, the situation in data centers is different: servers and storage systems are running all the time and maybe at full capacity. When we talk about the increase in service-specific capacity per user generated by the new 5G, 6G and 7G wireless technologies, we completely forget how this increase in capacity is achieved. The use of basic radio technology in base stations is not reported. As the capacity of wireless networks increases, base station needs to be placed closer together and cell sizes decrease (Elders, Shannon's theorem). This means that energy consumption is increasing even more.

This will lead to an increase in energy demand and the construction of fiber-optic connections between base stations and data center access points to meet the need for transmission capacity. Thus, we need to calculate amount of energy used, considering, among other things, the number of base stations, the number of devices

Table 7.1 Basic cells for mobile and wireless networks used in smart city environments

Wireless cell type	Cell radius (m)	Installation location
Femtocell	10–15	Indoors, inside cars
Picocell	100–250	Indoors
Microcell	500–2500	Different smart city areas
Makrocell	Cell radius larger than in a microcell	Outdoors, the output power is typically several of watts
Wi-Fi	50–100	Indoors and outdoors

in the cell, the capacity used, and the amount of energy consumptions (bits/km) per virtualized and sliced networks. From the energy consumption of a service, we can calculate the amount of CO_2 caused by its use. This means the carbon footprint of the data flow from smart devices to data centers and back. When we know the energy consumption of services and the resulting CO_2 emissions, we can compare the carbon footprint of the services we use [18].

Mobile networks in smart cities use many different types of cellular structures and solutions and the same frequency bands simultaneously. The coverage of the cells depends on the location, the number of users of mobile network services, whether the base station is in an office buildings or shopping center or as a separate building, leisure services, whether we use LANs together with mobile networks, etc. (see Table 7.1). Mobile network architectures are also changing rapidly. There are several variations in the cellular structures of mobile networks.

When looking at service provider communications at different levels, we need to consider cyber security issues comprehensively at each level and in each service providers' systems so that in such interconnected environments, cyber attackers cannot attack service provider management systems and databases.

7.3 Cyber Threats and Risks in the Future Network Environments of Smart Cities

Citizens use the services provided by a smart society, regardless of time and place, when they need them or when it is necessary to check things or make an appointment for a community service, such as health services. When citizens use the same smart devices in all the services, their devices may not be protected. They are commercially available devices without security and monitoring mechanisms. These types of devices are problematic from a security perspective because they are devices that can be easily taken over by cyber attackers and used in their attacks on systems used by citizens, such as health care systems or other systems around the world (Fig. 7.11).

At the same time Multiple sensors (IoT sensors, home monitoring sensors) and other condition monitoring devices can be connected to citizen devices in home

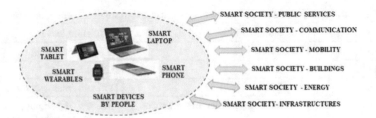

Fig. 7.11 Smart terminals used by citizens and their connections to various service groups and services via the access network

environments. As is well known, these devices and actuators have many vulnerabilities. Their security is not good enough or even non-existent, making them a favourable target for cyber attackers who exploit this vulnerability in their attacks on social systems. The connections of citizens' smart devices in home environments are shown in Fig. 7.12.

Standards used in wireless IoT networks include Bluetooth, 6LoWPAN, Wi-Fi, Wig, Zigbee, RFID, and NFC. Companies are developing products that meet these wireless standards. The idea is to develop an IoT product that overcomes the following key challenges:

- Interoperability of wireless standards,
- Safety assurance,
- Prevention of disturbances and faults.

Because wireless devices have been designed and developed based on various wire- less standards, as outlined above, the biggest challenge is the interoperability of these devices over an IoT network. Another challenge is the interference caused by these devices to each other when they operate in either the same or nearby frequency

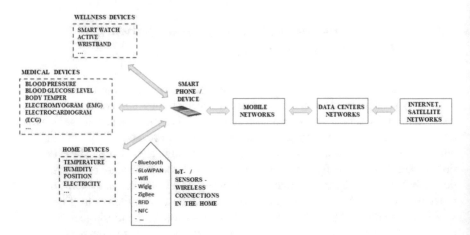

Fig. 7.12 Citizens' smart device connections in home environments

bands. The radiated power is also a critical factor to consider [19]. Table 7.2 shows the security systems in the different operating environments.

The abbreviations in Table 7.2 are.

- WBAN: Wireless Body Area Network (~1–2 m),
- WPAN: Wireless Personal Area Network (~10–100 m),
- WLAN: = Wireless Local Area Network (~50–100 m, with certain solutions and in a specific situation even longer),
- LTE, 3G, 4G, and 5G: mobile network technologies.

In the future, the services will be used mainly via wireless networks. People move from one place to another, for example in cars, trains, trams, or ships, while using the services offered to them. As a result of such use, attackers have increasing access to systems and services. An attacker could exploit vulnerabilities in terminal devices or wireless parts of network because they could not always be delivered in a secure network environment. Once an attacker gets inside the system, he, or she can perform operations he/she wants without anyone noticing or having time to initiate counter-measures. Security levels must also be considered in physical structures such as office buildings, homes, shopping centers, theaters, terminals, recreational area buildings, and telecommunications and information technology systems. With a smart device, the user can connect to any service in society without knowing in which operating environment the service is produced and from which environment the service is provided (Fig. 7.13). Welfare-related services add to challenges, as they are the most critical services in our society. People also use them at home, in the city, in shops and supermarkets, on trains or other means of transportation. The sensors and IoT devices connected to users' smart devices form a MESH network with devices from other users that use the same wireless LAN frequencies because they are freely available. In such situations, interference occurs between the frequencies used. In addition, these technologies contain many security and cybersecurity vulnerabilities.

One truly critical function in such an operating environment is the transfer of data from patients' own smart and welfare devices to hospital information systems. It is quite easy for hackers and cyber attackers to attack systems, for example, for financial gain or to create chaos in society.

These increasing cyber security challenges will have a significant impact on the design and development of systems, devices, and services in the future. For example, there has been a demand for hardening of intelligent devices for patients and nursing staff, as well as information systems used in healthcare. The question therefore arises as to whether this hardening is sufficient in such an all-connected world [20]. If patient devices are standard smartphones with applications purchased directly from the operator's or device manufacturer's stores, they usually do not have the protection or hardening required in healthcare environments [21]. Therefore, information security and cybersecurity related to user data may not be at an adequate level. On top of all that, our smart devices are connected to communications systems and services worldwide. Figure 7.14 illustrates how complex the information flows of the information systems of our intelligent society become when we use our smart devices as

Table 7.2 Security systems in citizen IoT sensors and home environment devices

Operating environment	Devices (from Fig. 7.12)	Types of communication system	Control on the access side	Control in the data center	IoT/sensor device vulnerabilities	Vulnerabilities in access	Probabilities of attack
Medical	Wirelessly connected devices (sensors and IoT devices)	BAN, PAN, WAN LTE, 3G, 4G, 5G	ID code, security VPN, or none	Firewall	Often non-secure solution	Vulnerable wireless connection	Very high
Wellness	Wirelessly connected devices (sensors and IoT devices)	BAN, PAN, WAN	ID code, security VPN, or none	Firewall	Often non-secure solution	Vulnerable wireless connection	Very high
Peoples's home	Wirelessly connected devices (sensors and IoT devices)	BAN, WAN	ID code, security VPN, or none	Local or remote (firewall)	Often non-secure solution	Vulnerable wireless connection	Very high

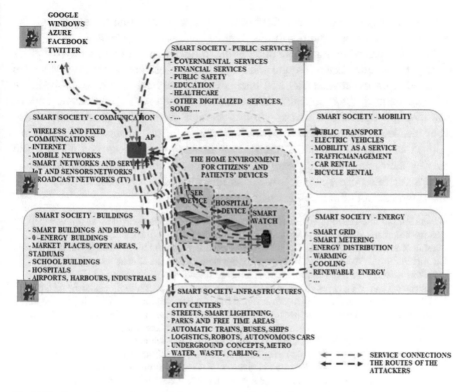

Fig. 7.13 Smart city services and infrastructures, citizen devices, and attacker routes

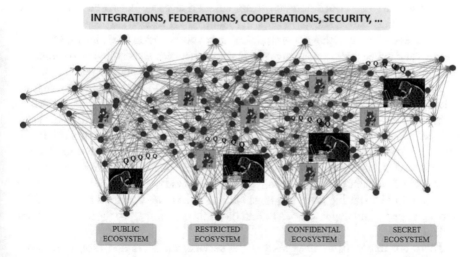

Fig. 7.14 Active nodes in our ecosystems and information flows in the service and social media groups of our smart society [17, p. 41]

active nodes in our ecosystems. Different levels of information travel through the same communication networks and form quite complex ecosystems.

As described above, that mean that our smart devices allow us to access all kinds of services around the world quite easily, but we do not know if these services are protected from cyber attackers and hackers. Such services will also be available in the virtualized communication environments of the future. This means a lot of security challenges more. Just as we use services around the world, hackers and cyber attackers can use them to try to find vulnerabilities in services, data forms, applications, and information systems. It is also easy for attackers to deceive and cover up their tracks by using extensive telecommunications networks and satellite systems for this purpose, as the world is interconnected everywhere [22]. Stefan Tanase writes about the Satellite Turla network:

> To attack satellite-based Internet connections, both the legitimate users of these links as well as the attackers' own satellite dishes point to the specific satellite that is broadcasting the traffic. The attackers abuse the fact that the packets are unencrypted. Once an IP address that is routed through the satellite's downstream link is identified, the attackers start listening for packets coming from the Internet to this specific IP. When such a packet is identified, for instance a TCP/IP SYN packet, they identify the source and spoof a reply packet (e.g., SYN ACK) back to the source using a conventional Internet line. At the same time, the legitimate user of the link just ignores the packet as it goes to an otherwise unopened port, for instance, port 80 or 10,080. There is an important observation to make here: normally, if a packet hits a closed port, an RST or FIN packet will be sent back to the source to indicate that there is nothing expecting the packet. However, for slow links, firewalls are recommended and used to simply DROP packets to closed ports. This creates an opportunity for abuse. [23] Stefan Tanase writes that cyber-attackers attacks through satellite networks to our systems. [23]

When researchers use social media to communicate with each other around the world, they use both intercontinental submarine cable networks and satellite networks. The situation shown in Figs. 7.13 and 7.14 is a reality. There are different groups in social media, such as communications, collaboration, and multimedia [24]. Hackers and cyber-attackers are aware that researchers use social media to communicate. With their information network analysis tools, they can analyze network traffic on a social media network, obtain user-related information, and obtain relevant and valuable information about messages sent by researchers for their own purposes.

Social networks offer a lot of opportunities for hackers and cyber-attackers, because often these networks and the services they implement are not sufficiently protected from the actions of hackers and cyber-attackers. To design and implement the information and cybersecurity architectures of the intelligent society of the future and to define and analyze dependencies, risks, and threats, we need an understanding of large entities, starting with top-level service needs. We can then make descriptions of systems and dependencies to find the multilevel threats to the systems of our society.

There are many different types of ICT service providers and mobile network operators in virtualized ICT environments. The systems used must be integrated, interconnected, and collaborative with each other in order for the systems work well in such a complex integrated environment. This means a lot of collaboration between different service providers. In addition, we need to do a lot of enterprise architecture work, make strategies, outline future scenarios, find use cases, try to find dependencies, and then analyze the risks and threats to the systems and services used. Figure 7.15 shows one hospital operating environment that is truly complex. From it we can get an idea of the challenges related to the functionality of services as well as security and cyber security issues.

In the healthcare environment, we need Intelligent Information Management Systems (IIMS) that allows us to track the data flows to find out if someone is attacking the systems or if someone has accessed data for which they are not authorized. The hospital environment with its services is the most critical environment, so it must be protected as well as possible. Artificial intelligence, machine learning, and deep learning should be used in this environment to find vulnerabilities in systems or to detect attacks on systems. They should also be used for real-time monitoring and in the analysis of at different gateway (GW) points to analyze network threats and security in management systems. If we use IIMS in the services of a smart city, then we can manage the city's infrastructures more efficiently.

In 2017, Gillian Mohney wrote that hospitals remain key targets as ransomware attacks are expected to increase [25]. Since then, ransomware attacks on hospitals and healthcare systems have increased quite fast. Cyber-attacks on medical devices could endanger patient safety [26]. Hackers are increasingly targeting financially motivated cyber-attacks on healthcare organizations to gain access to patient data [27]. One of the latest hacks in Finland was the case of Vastaamo, where a lot of patient data was leaked to the Internet. There is constant news around the world that cyber attackers have attacked hospital systems and been paid millions of euros in ransom money.

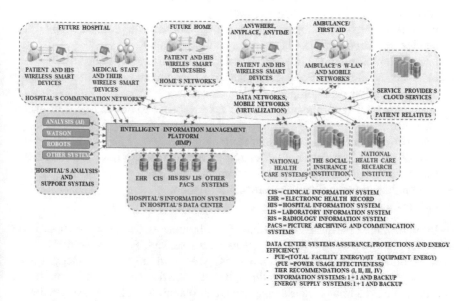

Fig. 7.15 E-health and M-health hospital operating environment [8, p. 643]

7.4   Cyber Threat Risks and the QFD Model

The current strong digitalisation trend in our society increases the range of services offered. It also makes them easier to use. These developments also have a strong impact on the service chains of the services provided, service providers, hardware solutions, and operating models in each part of the service chain.

Services form end-to-end service chains. The service chain begins at the point where the analog data is transferred digitally in the IoT devices and sensors. A huge amount of information comes from such sources, which is incredibly challenging work for cyber and security analysts. This digital signal is sent from the IoT device to the user's smart device using wireless technology. The area of wireless technology is quite vulnerable because there are not good enough encryption systems between IoT sensors and user smart devices. For this reason, we need to look at the service chain immediately after the signal is generated, because at that point, the IoT device produces, for example, critical electronic health data.

In general, cyber security threats and attack vectors, for example in health networks (wireless/fixed), include the following [8, p. 645; 28, p. 5]:

- Monitoring and eavesdropping on patients' vital signs,
- Threats to information during transmission,
- Threat routing in networks,
- Location threats and monitoring of activities,
- Distributed Denial of Service (DDoS) threats,
- Interfering with or blocking radio communications from IoT devices and sensors,
- Exploiting vulnerabilities to access health services,
- Attack hospital health information systems,
- Interfering with or blocking wireless communications throughout the hospital and interfering with the day-to-day operation of the hospital,
- Interruption of treatment or service (including the possibility of death),
- Malware and phishing: Sophisticated malware and phishing programs that install malicious scripts on a computer or steal credentials,
- Scamming staff with phishing email or fake websites to obtain logins or install malware,
- Unintentional or intentional insider threat,
- Loss of patient data, especially Electronic Protected Health Information (ePHI),
- Data breach, data filtering, and loss of assets,
- Blackmail and using filtered sensitive information,
- Intellectual Property (IP) theft.

Until now, the Enterprise Architecture (EA) framework has been used to provide understanding of our information systems in our society. For stakeholder's enterprise architecture visualizes, measures, analyses, and improves strategy, goals, changes, projects, and innovations through data analytics and real-time data-based concept plans, roadmaps, charts, diagrams, and planning books on a single digital collaboration platform [5].

Dragon1-open is one example of an enterprise architecture development platform that we can use to create the smart city architecture of the future (Fig. 7.16). Dragon1-open integrates processes, applications, tools, and analytics. The use of Dragon1-open leads to prioritization, implementation, and management of the digital transformation, IoT, blockchain, artificial intelligence, machine learning, micro-services, cybersecurity, mobile, cloud computing, automatization, data lakes, robotization [5].

The AI chatbot in Dragon1-open enterprise architecture platform imports, improves, and reuses data from Excel to increase and improve the user experience, customer engagement, supply chains, and digital ecosystem, serving as a decision support system [5]. The proposed Dragon1 EA-type tools offer one opportunity to develop the modern intelligent urban architectures of the future based on artificial intelligence, machine learning, and in-depth learning, because the amount of data that will come from these systems in the future is truly enormous. Smart city platforms and systems are a huge entity with all their connections, and we need new tools to identify all possible attack vectors on these systems so that we can protect our systems from cyber-attacks.

We can no longer do this work with our current development tools. We also need a framework to find out what types of information environments we live in and what systems we use. The QFD model is one possibility to use such analyses (Fig. 7.17). With these tools, we can look at, for example, the evolution of architectures and

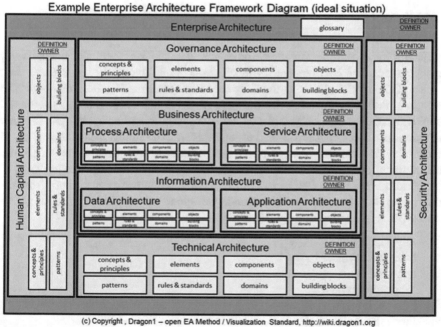

Fig. 7.16 Example of the enterprise architecture (EA) framework: dragon1 [5]

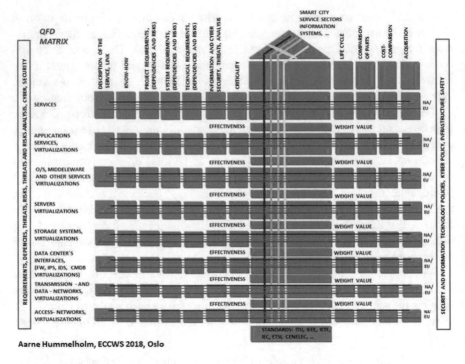

Aarne Hummelholm, ECCWS 2018, Oslo

Fig. 7.17 QFD model [6, 1]

technologies, quantum encryption, and other end-to-end service chains in this overly complex technical environment involving wireless and fixed data networks.

The QDF model allows us to easily see the dependencies and anomalies of the information systems in our environment, the dependencies between the systems and applications, their efficiency, and weight in our systems, and how important they are to operations and about levels of risks and threat. QFD model makes it possible to obtain the information we need in different situations about our future smart city environment and see the whole environment simultaneously from the top level to the bottom level. When analyzing communication and information systems from the bottom up, we normally use the seven-layer Open Systems Interconnection (OSI) model (used in communication standards):

1. Physical layer,
2. Data link layer,
3. Network layer,
4. Transport layer,
5. Session layer,
6. Presentation layer,
7. Application layer.

Hackers can attack all these layers. We need more information about our systems so that we can better protect them from attacks by cyber attackers and hackers. We need new models of the layers of the cyber world to analyze our environment, considering the potential of cyber attackers and hackers to attack our systems (Fig. 7.18). When we live in virtualized environments, it is not enough to look at just one system without considering its dependence on other systems. We need multilayer analyses method to analyze our systems. Figure 7.18 provides an additional answer and information to this question.

In the future, smart city telecommunications networks will use virtualized, segmented, and sliced network solutions, which must be considered when designing services and defining and analyzing cyber security threats. The EU METIS project sets out the main requirements for future 5G mobile network solutions based on different scenarios, see Fig. 7.19 [12, 30].

We also need to take these new 5G and 6G (7G) mobile network technologies into account when looking at and analyzing dependencies, risks, and cyber threats in different segments of smart city services and infrastructures. Figure 7.20 shows a top-level architectural description of future communication networks and their access interfaces. Figure 7.21, on the other hand, shows the smart city communication network infrastructures for server hotel interfaces, also considered AI/DL/ML issues as well as scenarios and use cases for 5G and 6G implementations.

New mobile network technologies require the expansion and deepening of architectural descriptions due to network virtualization, segmentation, and slicing. New quantum encryption technologies for wireless networks and their implications for user interface solutions must also be considered in order to provide adequate protection against cyber-attacks. Implementing quantum encryption from IoT and sensor devices to data center services is challenging because quantum encryption consumes

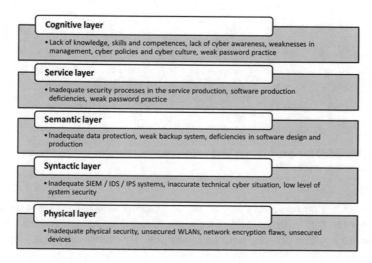

Fig. 7.18 Cyber world layers in hospital perspective [29]

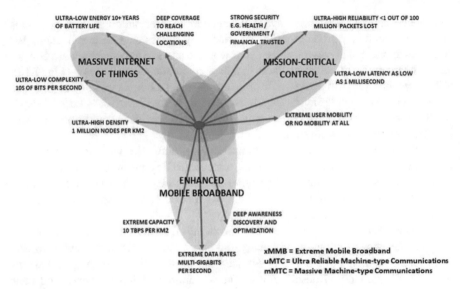

Fig. 7.19 Future scenarios and use cases for mobile networks [17, p. 146]

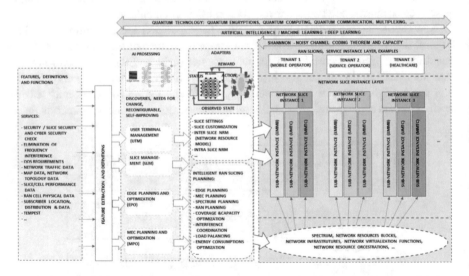

Fig. 7.20 Use of AI/DL/ML in virtualized and sliced smart city communication network infrastructures, access side [31, 32]

a lot of energy and there may not be enough energy inside the sensor devices to encrypt sensor data with this technology. We also often forget Shannon's theorem about the propagation of signals in a noisy channel. This needs to be considered when designing new telecommunication systems with huge data transfer capacity in our

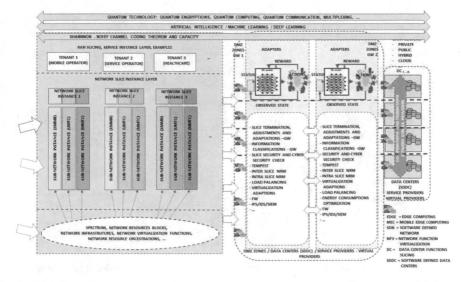

Fig. 7.21 Use of AI/DL/ML in virtualized and sliced smart city communication network infrastructures, data center side [31–33]

smart urban environments. In this way, we can reduce the chances of cyber-attackers attacking our systems.

7.5 Performing and Modelling Threat Analyses

Carrying out a threat analysis of the future society as a whole or of the service sector only as part of a future smart city is a challenging task. This is one reason why the entire environment is divided into sections, service sectors or smaller parts.

The telecommunications arrangements required for services in the service sectors are grouped into different segments. The communications package consists of access, core, drone, satellite, and data center networks. The following threat analysis will be performed for an access network involving many different sensors and IoT devices. It is subject to major changes in the situation shown in Fig. 7.22. First, data from IoT devices and sensors connected to the smartphone is transferred to the smartphone for further transfer. The smartphone is connected to a HUB home system that is connected via an EDGE router to the MEC router or directly to the MEC router. EDGE and MEC routers are virtualized and contain some slices that are used for certain services. In addition, the HUB system may include some firewall functions.

Threat analysis can be done using an attack tree model [34]. In this case, we need descriptions of possible target points in our system to get a sufficiently clear picture of the entire service chain, for which we make probability calculations for attacks. The attack tree model for the access architectures of Fig. 7.22 is shown in Fig. 7.23. The notations in Fig. 7.23 are clarified in Table 7.3.

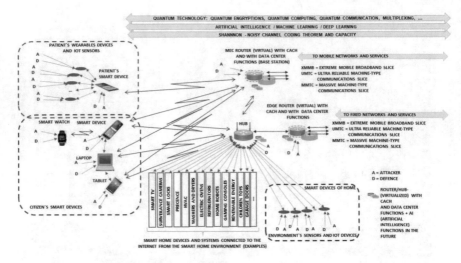

Fig. 7.22 Access network architecture in the home environment (modified from [17, p. 50])

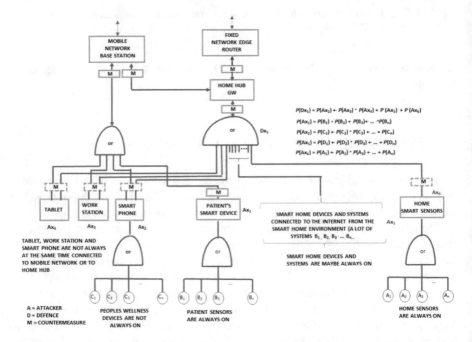

Fig. 7.23 Access network target architecture for calculating probabilities using attack tree model

Table 7.3 Meaning of notations in Fig. 7.23

Action	Examples	Notion
Attack	Sniffing, enumeration, scanning, …,	A
Detection	Port scan, information scan, …,	D
Countermeasure	Analysing of vulnerabilities, safeguards put in place, …,	M

The probabilistic success of attacks ($P(t)$) against a device x in a home can be calculated as follows [34]:

$$PA(t) = pA(1 - pD(t))(1 - pM(t)) \tag{7.1}$$

Probability of successful attacks is calculated as follows:

- $P(Ax1)$ occurs if $B1$ or $B2$ or $B3$,…, Bn occur,
- $P(Ax2)$ occurs if $C1$ or $C2$ or $C3$,…, Cn occur,
- $P(Ax6)$ occurs if $A1$ or $A2$ or $A3$,…, An occur.

A citizen's smart device is connected to different devices on the network, depending on what his/her communication system is, what services he/she uses and provides them. They may also be connected to various network slices. Slices can have different service segments whose services the user is authorized to use:

- Health care slices,
- Mobile broadband slices,
- Internet of Things slices,
- Slices of physical infrastructures,
- …

When we use different services with our smart devices, we also need to check for vulnerabilities and security issues in the applications we use. If there is something wrong with our devices and their protection systems have not been updated, these devices can spread any malware and viruses they may contain widely to smart society's online services, thus contaminating other users' devices and services as well (see Figs. 7.20, 7.21, 7.22 and 7.23). Location and movement information systems inside smart devices need to be checked and protected from attackers, as they can be used against citizens in critical situations.

Bayesian probability calculation can also be used to calculate probabilities of attacks on our systems. By presenting our calculation results in tabular form (Table 7.4), we can see the result more easily. Then we can use enterprise architecture development tools to calculate dependencies, risks, and threats calculate residual risks and threats. From such tables, we can directly see the real situation of our system, its security situation, and what is needed to further protect it [5].

Table 7.4 Threats and risk table, example [1, pp. 523–532]

Ref id	Org	Functions	Category	Threat/Methotology	Threat/Risk	Exixting control	Threat/risk level			Accept/Reduce	Recommended controls	Residual threat/Risk			Check point
	MC						L	C	R			L	C	R	
1-AM		Identify	Access management	Food Printing	– Informa-tion gathering – Target address range – Name space acquisition – Network topology	– Inven-tory of physical assets is made, protec-tive systems like fire-walls, ids, IPS are place, VPN and security are installed	3	3	8	Reduce	Depends on security levels of access networks and its services – Public – Restrected – Confidental – Secret – Top secret … EU directives national recommentations, …	2	2	3	xx

L = Likelihood, C = Consequence, R = Risk

7.6 Conclusions and Future Work

7.6.1 Conclusions

Up-to-date analysis of smart city infrastructures and services with all needs in mind and outlining future prospects was an incredibly challenging task, as smart cities have large and complex environments and structures and should provide services to people, organizations, and governments in real time. A wide range of services are available. That is why I have shared a smart city infrastructure to different segments based on the activity. It was also very difficult to obtain technical specifications for IoT devices, sensors, and actuators. Technical information is needed to check the device's encryption solutions and to decide to which data centers the information coming from the device would be transferred.

In many smart cities, the segment's development work takes place in its own silo and not in cooperation with other segments. It is difficult to obtain sufficient information on, for example, internal communications and information systems in buildings, their development plans, internal networks in buildings, and security and safety mechanisms in buildings. It is therefore challenging to carry out analyses and checks to better consider the new requirements of the security and cyber security directives. All this leads to sub-optimization and not to optimize the development of a smart city as a whole.

Lots of IoT devices, sensors, or actuators are coming into the systems alongside the existing ones and we have using a lot of smart devices. Maybe they are full of vulnerabilities, their security mechanisms are flawed, and the security solutions are not good enough. Hackers and cyber attackers can attack the smart city service system through smart devices, sensors, and IoT devices, even if the attackers are on a different continent. When we use our smart devices for e-health purposes, Security issues are even more critical in the healthcare communications environment.

7.6.2 Future Work

We need a clear picture of the environments of a smart city, because, for example, threats analyses are difficult to carry out and the number of IoT and sensor devices connected to smart city telecommunication networks is growing at a tremendous rate. Technical progress is advancing at an accelerating pace, so the importance of analyzing individual components needs to be considered openly. Albeit dependency and threat assessment analyses could be made comprehensively enough in some areas, it is essential to carry out threat assessments and analysis of future societal services with the support of computer programs and artificial intelligence (AI).

We need to develop and test a high-level architecture description model to obtain more accurate analyses. For example, with Enterprise Architecture (EA), we can describe different segments of a smart city with their services and the factors that affect them [5]. If the vulnerability of OSI layers and the various vulnerabilities associated with the protocols in use are added to this entity, the number of issues to be verified will increase. The inherent defects and vulnerabilities of encryption solutions and cryptographic network solutions also need to be analyzed. The functionality of virtual access network services and the terminals that use them must be investigated in different attack situations. It would be especially important to conduct research on, for example, remote healthcare, rescue, security, and other critical systems. Artificial intelligence (AI) also needs to be investigated and tested for its use to protect various IoT devices and sensors, especially in e-health systems. So that we can better protect these devices against these malicious software and cyber-attacks.

The most critical is a health care system with many vulnerabilities and security challenges. We need to define the architecture and find a safety-compliant solution for healthcare devices and systems so that patients and healthcare professionals can use them safely in the healthcare environment of the smart city of the future.

When IoT devices are placed in the human body and thus the body comes part of the communication network, we must also take care of ethics and moral issues. One research topic is to measure and test various frequency-induced interference in smart city, office, home, and hospital environments, etc., and to check whether interference affects citizens' devices and services used. Radio frequencies also affect the human body in real time, so radiation levels must also be analyzed and efforts made to find the right level so that the patient is not harmed by this radiation.

It is important to study energy efficiency in smart city environments, communication systems, data centers, and smart devices and smart services from communications environments. A research area to consider would be a quantum encryption system that will be used in wireless networks (5G, 6G) and services in the future.

One important research topic should be the use of artificial intelligence (AI) in analysis would speed up work and provide an opportunity to quickly identify and fix vulnerabilities. Thus, security measures could be quickly targeted at the right location to prevent potential penetration into networks and services, minimizing the effects of the attack. Also, one research topic would be the use of artificial intelligence could also be used for real-time tracking and analysis at different gateway (GW) points to analyze network threats in smart city security and management systems.

In addition, we spend millions of euros on the demands of hackers and cyber attackers, so why not develop and make different security zones for hospitals, homes, or other critical environments to protect sensitive systems from radio interference and eavesdropping. If there are no security zones yet, they would be easy to implement. After all, radio frequency interference can prevent the entire operating environment from operating without any physical connection to any part of the network, depending on the location, of course.

References

1. Hummelholm A (2018) Cyber threat analysis in smart city environments. In: ECCWS 2018: proceedings of the 17th European conference on cyber warfare and security. Academic Conferences International, pp 523–532
2. IEC (2014) Orchestrating infrastructure for sustainable smart cities. White paper, international electrotechnical commission (IEC), Geneva
3. SmartCitiesCouncil (2014) Smart cities readiness guide: the planning manual for building tomorrow's cities today. Smart cities council
4. Beecham (2021) M2M sector map. Beecham research limited, http://www.beechamresearch.com/download.aspx?id=18
5. Dragon1-open (2021) Dragon1: enterprise architecture for business & AI strategy operations. Dragon1. https://www.dragon1.com/
6. QFD (2021) What is QFD? QFD institute. http://www.qfdi.org/what_is_qfd/what_is_qfd.htm
7. Limnéll J, Majewski K, Salminen M (2014) Kyberturvallisuus. Docendo, Jyväskylä
8. Hummelholm A (2019a) E-health systems in digital environments. In ECCWS 2019: proceedings of the 18th European conference on cyber warfare and security. Academic Conferences International, pp 641–649
9. CBS (2021) The threats arising from the massive SolarWinds hack. CBS News. https://www.cbsnews.com/news/the-threats-arising-from-the-massive-solarwinds-hack/
10. Gewirtz D (2021) Capitol attack's cybersecurity fallout: Stolen laptops, lost data, and possible espionage. ZDNet. https://www.zdnet.com/article/capitol-attacks-cybersecurity-fallout-stolen-laptops-lost-data-and-possible-espionage/?ftag=TRE49e8aa0&bhid=2883305140448922536407928618284&mid=13233595&cid=2176286718
11. 3GPP (2019) Release 15. 3rd Generation Partnership Project (3GPP). https://www.3gpp.org/release-15
12. 5G-PPP (2016) View on 5G architecture. Version 1.0, 5G PPP Architecture Working Group
13. Hummelholm A (2020) Threat characterization for various layers of infrastructure. In: Cybersecurity and resilience in the Arctic. IOS Press, pp 150–182
14. EC (2012). Directive 2012/27/EU of the European Parliament and of the Council of 25 October 2012 on energy efficiency, amending Directives 2009/125/EC and 2010/30/EU and repealing Directives 2004/8/EC and 2006/32/EC
15. EU (2016a) NIS, directive concerning measures for a high common level of security of network and information systems across the Union. Directive (EU) 2016/1148, European Parliament and Council
16. EU (2016b) GDPR, regulation on the protection of natural persons with regard to the processing of personal data and on the free movement of such data. Regulation (EU) 2016/679, European parliament and council
17. Hummelholm A (2019c) Cyber security and energy efficiency in the infrastructures of smart societies. JYU dissertations, 173. University of Jyväskylä, Jyväskylä
18. EPA (2020) Greenhouse gas equivalencies calculator. U.S. environmental protection agency. https://www.epa.gov/energy/greenhouse-gas-equivalencies-calculator
19. RF-WW (2021) IoT wireless technologies or IoT wireless standards. RF wireless world. https://www.rfwireless-world.com/Terminology/IoT-wireless-technologies.html
20. EU (2017) Regulation on medical devices. European parliament and council of the European union
21. 5G-IA (2015) 5G and e-health. 5G infrastructure association (5G IA). https://5g-ppp.eu/wp-content/uploads/2016/02/5G-PPP-White-Paper-on-eHealth-vertical-Sector.pdf
22. Klosowski T, Murphy D (2020) What is tor and why should i use it? Lifehacker. https://lifehacker.com/what-is-tor-and-should-i-use-it-1527891029
23. Tanase S (2015) Satellite Turla: APT command and control in the sky. Kaspersky. https://securelist.com/satellite-turla-apt-command-and-control-in-the-sky/72081/
24. Cann A, Dimitriou K, Hooley T (2011) Social media: a guide for researchers. Research Information Network, London

25. Mohney G (2017) Hospitals remain key targets as ransomware attacks expected to increase. ABC News. https://abcnews.go.com/Health/hospitals-remain-key-targets-ransomware-attacks-expected-increase/story?id=47416989
26. Palmer D (2018) IoT security warning: Cyber-attacks on medical devices could put patients at risk. ZDNet. https://www.zdnet.com/article/iot-security-warning-cyber-attacks-on-medical-devices-could-put-patients-at-risk/
27. Davis J (2019) Hackers targeting healthcare with financially motivated cyberattacks. Xtelligent healthcare media. https://healthitsecurity.com/news/hackers-targeting-healthcare-with-finan-cially-motivated-cyberattacks
28. Piggin R (2017) Cybersecurity of medical devices: addressing patient safety and the security of patient health information. BSI, London
29. Lehto M (2015) Phenomena in the cyber world. In: Lehto M, Neittaanmäki P (eds) Cyber security: analytics, technology and automation. Springer, Cham, pp 3–29
30. METIS (2020) The METIS 2020 project. https://metis2020.com/
31. Santos JF, Liu W, Jiao X, Neto NV, Pollin S, Marquez-Barja JM, Moerman I, DaSilva LA (2020) Breaking down network slicing: hierarchical orchestration of end-to-end networks. IEEE Commun Mag 58(10):16–22
32. Zhou Y, Tian L, Liu L, Qi Y (2019) Fog computing enabled future mobile communication networks: a convergence of communication and computing. IEEE Commun Mag 57(5):20–27
33. Yao M, Sohul M, Marojevic V, Reed JH (2019) Artificial intelligence defined 5G radio access networks. IEEE Commun Mag 57(3):14–20
34. Wang P, Liu JC (2014) Threat analysis of cyber-attacks with attack tree+. J Inf Hiding Multimedia Signal Process 5(4):778–788
35. Hummelholm A (2019b) Undersea optical cable network and cyber threats. In: ECCWS 2019: proceedings of the 18th European conference on cyber warfare and security. Academic conferences international, pp 650–659

Chapter 8
Cyber Security in Healthcare Systems

Martti Lehto, Pekka Neittaanmäki, Jouni Pöyhönen, and Aarne Hummelholm

Abstract Healthcare is a good example of a structure whose complexity has been increased by the general rapid development of technology and digitalization. Digital development has made it possible to provide services in new ways and on a wider scale, in particular through information networks. Today, the healthcare information environment is a networked entity consisting of various ICT systems, medical devices, and clinical systems with an open system structure. Digitalized functions and services are becoming more common in social welfare and healthcare. Their reliability during incidents and emergencies must be ensured. While the digital world offers good opportunities for improving healthcare systems and enhancing disease analysis, we need to look deeper into this. Devices and information systems may not work well together and there are vulnerabilities in people, processes and technology. Thus, a comprehensive approach to healthcare cybersecurity is needed.

Keywords Healthcare · Cyber security · Hospital · Information systems

8.1 Introduction

Securing the functioning of healthcare information systems is part of ensuring the critical infrastructure and security of supply in society, where preparedness plays a key role. Preparedness means ensuring that all activities and tasks can continue with minimum interruptions and that the required exceptional measures can be performed during disruptions occurring in normal conditions and during emergencies. Vital functions are essential for the functioning of society and they must be maintained in all situations.

The starting point should be that the required client and patient medical information must be available in all situations. Data transmission of diagnostic and other client and patient medical data, digital services and the cyber security of networked social welfare and healthcare equipment must be ensured against cyber-attacks. As

M. Lehto (✉) · P. Neittaanmäki · J. Pöyhönen · A. Hummelholm
Faculty of Information Technology, University of Jyväskylä, Jyväskylä, Finland
e-mail: martti.lehto@jyu.fi

© The Author(s), under exclusive license to Springer Nature Switzerland AG 2022
M. Lehto and P. Neittaanmäki (eds.), *Cyber Security*, Computational Methods
in Applied Sciences 56, https://doi.org/10.1007/978-3-030-91293-2_8

more and more social welfare and health services are provided as home care, the threats to the sector are also spreading outside hospitals and healthcare centers, which must be considered in the contingency and preparedness plans of the healthcare service providers against hybrid operations and different types of cyber threats [1].

In the past two decades, information technology has been widely utilized in healthcare. Electronic Health Records (EHRs), biomedical database, and public health have been enhanced not only on the availability and traceability but also on the liquidity of data. As healthcare-related data is constantly growing, there are challenges for data management, storage, and processing as follows [2]:

1. *Large scale*: As medical informatization improves, particularly with the development of hospital information systems, the volume of medical data has increased. Wearable health devices have also exploded the growth of health care data.
2. *Rapid generation*: Most medical devices, particularly wearable devices, are constantly collecting data. Rapidly generated data needs to be processed promptly to respond immediately to emergencies.
3. *Various structure*: Clinical examination, treatment, monitoring, and other health care devices generate complex and heterogeneous data (e.g., text, image, audio, or video) that is either structured, semi-structured, or unstructured.
4. *Deep value*: The value hidden in an isolated source is limited. However, by combining EHRs and Electronic Medical Records (EMRs), we can maximize the deep value of health care data, such as personal health counseling and public health alerts.

Digital transformation has had a positive impact on healthcare. Telemedicine, artificial intelligence (AI) enabled medical devices, big data analysis, virtual reality patient care, wearable medical devices, and blockchain electronic health records are concrete examples of digital transformation in healthcare. How we interact with healthcare professionals, how data is shared among providers, and how care plans and health outcomes are decided has changed completely [3].

A modern hospital has hundreds—even thousands—of workers using laptops, computers, smartphones, and other smart devices that are vulnerable to security breaches, data thefts and ransomware attacks. Hospitals keep medical records, which are among the most sensitive data about people. Many hospital's electronics help keep patients alive, monitoring vital signs, administering medications, and even breathing and pumping blood for those in the most critical conditions [4, p. 648].

The reason for the interest of criminals is that patient data is well paid for in the black market; typical patient information includes credit card numbers, email addresses, health insurance numbers, employer information, and medical history information. These have value to criminals because they usually last for years. Cybercriminals use information for phishing attacks, fraud, and identity theft. [5, p. 18].

In the health care sector, there are very specific requirements for data processing. The integrity and availability of patient data are extremely important for the safe care of patients. On the other hand, the confidentiality of data must be protected not only to ensure the protection of privacy, but also to prevent the criminal use of personal data. The functionality of the entire hospital environment is critical to patient care. In this case, the digital system and equipment environment of the hospital that connects to the Internet must be considered. An example of the wide scope is the important role of cyber security in building automation in hospital buildings [6, pp. 19–23].

8.2 Hospital as a Cyberspace

Healthcare is a large and diverse sector that provides a huge range of goods and services essential to the health, safety and well-being of the nation. Critical functions in the sector include, but are not limited to:

- Primary health care, specialized hospitals, and ambulatory health care, including doctors, nurses, and occupational health practitioners,
- Health centers and first aid services,
- Health planning organizations, business associates, and health insurance companies,
- Units taking care of the dead,
- Enterprises that manufacture, distribute, and sell medicines, biologicals, and medical devices,
- Biobanks and genomic centers,
- Population-based care and surveillance provided by health agencies at national, regional and local levels.

Digital healthcare tools have the vast potential to improve ability to accurately diagnose and treat disease and to enhance the delivery of healthcare for the individual. Digital tools are giving providers a more holistic view of patient health through access to data and giving patients more control over their health. Digital healthcare offers real opportunities to improve medical outcomes and enhance efficiency.

Different reports show that ransomware, data breaches and other cyber-attacks are on the rise, and healthcare is one of the biggest targets. The healthcare industry increasingly relies on technology that is connected to the internet: from patient records and lab results to radiology equipment and hospital elevators.

8.2.1 Hospital and Cyber World Layers

Martin C. Libicki has created the structure of the cyber world, the idea of which is based on the Open Systems Interconnection Reference Model (OSI). The OSI model groups communication protocols into seven layers. Each layer serves the layer above it and is served by the layer below. The Libicki cyber world model has the following four layers: physical, syntactic, semantic, and pragmatic [7, 8, p. 21; 9, pp. 5–7].

We have modified the model to use five levels:

1. Physical layer,
2. Syntactic layer,
3. Semantic layer,
4. Service layer,
5. Cognitive layer.

The physical layer contains the physical elements of the communication network (fixed and wireless) and medical devices. The syntactic layer consists of various system control and management software and features that facilitate the interaction between devices connected to the network. This layer includes software that provides operating commands for physical devices. The semantic layer contains information and datasets on the user's computer, hospital servers, or cloud service environment, as well as various user-managed functions. The service layer contains all public and commercial services available online. The cognitive layer is the environment of awareness of hospital staff: a world where information is interpreted and contextual understanding of information is created. It can be seen from a larger perspective as the mental layer, including the user's cognitive and emotional awareness.

Hospital cyberspace is more than the internet, including not only hardware, software, data, medical devices, and information systems, but also people and social interaction within these networks and the whole infrastructure. Figure 8.1 illustrates the cyber world layers in hospital perspective.

The International Telecommunication Union describes, that:

Cyber security is meant to describe the collection of tools, policies, guidelines, risk management approaches, actions, trainings, best practices, assurance and technologies that can be used to protect the availability, integrity and confidentiality of assets in the connected infrastructures pertaining to government, private organizations and citizens; these assets

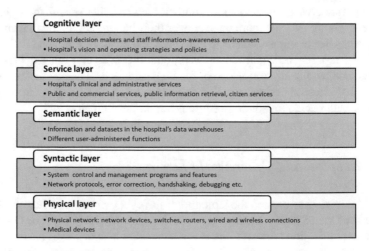

Fig. 8.1 Cyber world layers in hospital perspective

include connected computing devices, personnel, infrastructure, applications, services, telecommunications systems, and data in the cyber-environment. [10]

The ISO/IEC 27032 defines:

cyberspace security as the protection of privacy, integrity, and accessibility of data information in the cyberspace. Therefore, cyberspace is acknowledged as an interaction of persons, software, and worldwide technological services. [11]

8.2.2 Hospital Information Systems

The Hospital Information System (HIS) is an integrated information system designed to comprehensively manage all hospital operations, such as medical, administrative, financial, and legal issues, and the processing of corresponding services. The hospital environment requires the utilization of several different information and automation systems. They are needed in at least four different processes:

1. Administrative information systems,
2. Hospital clinical information systems,
3. Building automation system (Heating, Ventilation, and Air conditioning—HVAC),
4. Physical safety and security information systems.

Figure 8.2 illustrates the generic hospital information system.

A patient record system can be part of a hospital information system. It is a type of clinical information system dedicated to collection, storage, manipulating, and making available of clinical information relevant to patient care. A patient record is a repository of information about an individual patient. This information is generated

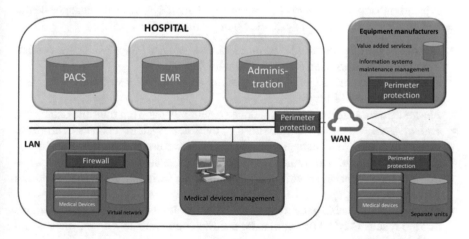

Fig. 8.2 Generic hospital information system [12, p. 21]

by health care professionals as a direct result of interaction with the patient or persons with personal knowledge of the patient, or both.

A health information system refers to a system designed to manage healthcare data. This includes systems that collect, process, store, manage, report, and transmit a patient's electronic medical record (EMR), a hospital's operational management or a system supporting healthcare policy decisions. Health information systems also include those systems that handle data related to the activities of providers and health organizations. EMR/EHR database maintain patient's record by data like contact details, test results, treatment history, and more. It enables the sharing of information to another EMR/EHR system so that different healthcare providers can access the patient's system if compatible data models are used [13, 14].

Hospital information systems must strictly reflect the reality of the organization, its internal formation, care processes, medico-technical means, legal and regulatory environments, and billing procedures. Health information systems consist of six key components [14]:

1. *Resources*: legislative, regulatory and planning frameworks necessary for the functioning of the system, including personnel, financing, logistical support, information and communication technology (ICT),
2. *Indicators*: a complete set of indicators and relevant targets, including inputs, outputs, outcomes, health determinants, and health status indicators,
3. *Data sources*: including both population and institution-based data sources,
4. *Data management*: collection, storage, quality assurance, processing, flow of information, compilation, and analysis,
5. *Information products*: data that has been analyzed and presented as operational information,
6. *Dissemination and use*: the process of making data available to decision-makers and facilitating its use.

A health information system is a complex system of systems that manages health data, including many types of systems. Some of them are [13, 15, 16]:

• *Electronic Health Record (EHR) or Electronic Medical Record (EMR)*: collect, store and share data related to a patient's health;
• *Practice Management System*: manages the daily operations of a practice, such as scheduling and billing;
• *Master Patient Index*: combine separate patient records from multiple databases;
• *Patient Data Repository*: used with a patient data system; enables centralized archiving of electronic patient data as well as active use and storage of data;
• *Pharmacy Management System*: contains all data related to a patient's prescriptions; can be found in several pharmacy settings, including retail, hospital, and long-term care;
• *Patient Portals*: enable patients to access their health data, including medications and lab results (MyData); can also be used to communicate with physicians and track appointments;

- *Clinical Decision Support (CDS)*: analyzes data from clinical and administrative systems; provides an opportunity for physicians to make the best clinical decision;
- *Medical Certificates Sharing System*: transmit electronically the certificates and reports issued by healthcare professionals to those concerned.

Protected Health Information (PHI) is collected or created by a healthcare provider, health plan, employer, healthcare clearing house, or other entity. According to the defining criteria, data contained within a patient's healthcare record is considered to be PHI if there are reasonable grounds to believe that the information can be used to identify an individual. Examples of identifiable data elements include [17, p. 5]:

- Name, address (including postal code), telephone and fax numbers,
- Email addresses,
- Medical insurance or social security/national insurance numbers,
- Identity number,
- Information on designated beneficiaries,
- Any (financial or other) account number, license, vehicle or certificate numbers,
- Medical or otherwise salient device or serial numbers,
- Any associated internet protocol (IP) addresses or URLs/URIs,
- All biometric data (e.g., finger, retina or voice prints and/or DNA),
- Full face photos or images of unique identifiable characteristics,
- X-rays and other diagnostic images.

Hospitals are becoming more reliant on the ability of hospital information systems to assist in the diagnosis, management and education for better and improved services and practices. Separate systems collect research and procedure data during the patient's care chain, which the most important are laboratory systems, through which the necessary tests are ordered, the test results are entered into them and the results are handed over to the requesting unit.

Hospital information systems are among others:

- Radiology Information System (RIS),
- Picture Archiving Communications Systems (PACS),
- Electronic Health Record (EHR),
- Electronic Medical Records (EMR),
- Laboratory Information Systems (LIS),
- Clinical Information System (CIS),
- Pathology information systems,
- Pharmacy information system,
- Intensive care systems,
- Blood bank system,
- Anesthesia information systems,
- Control of surgery operations,
- Remote Patient Monitoring (RPM),
- Imaging systems,
- Maternity ward information systems,

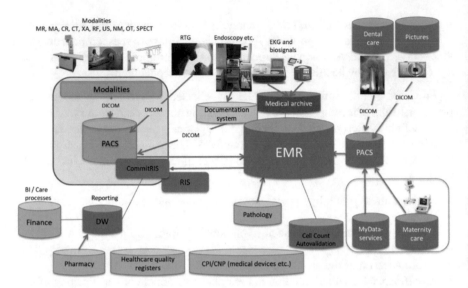

Fig. 8.3 Generic hospital data system; DICOM = Digital imaging and communications in medicine, PACS = Picture archiving and communication system, RIS = Radiology information system), CPI = Clinical physiology and isotope medicine, CNP = Clinical neurophysiology, EMR = Electronic medical records

- Nursing information systems (Nurse call systems),
- Central control systems,
- Financial information system,
- Enterprise Resource Planning (ERP) systems,
- Security systems.

One person generates 1100 terabytes of healthcare data, 6 terabytes of genomics data, and 0.4 terabytes of clinical data in his/her lifetime. The data is fragmented here and there and is not easy to share or analyze. Figure 8.3 illustrates an example of the most significant data sources in the hospital [18].

8.2.3 Medical Devices

Smart devices are connected to fixed or mobile networks and are used to transfer the patient's bio-signal data to hospital systems. In hospital systems, the information is analyzed, and the care staff take the necessary decisions based on analyzed results and gives information on management measures to the patients [4, p. 642].

Medical device is an instrument, apparatus, appliance, implement, machine, contrivance, implant, in vitro reagent, or other similar or related article, including a component part, or accessory intended for use in the diagnosis of disease or other conditions, and/or therapeutic purposes or in the cure, mitigation, treatment, or prevention of disease. These include [19, p. 14, 20, 21]:

- Radiology equipment, radiotherapy, nuclear medicine, operating room or intensive care equipment, surgical robots, electro-medical equipment, infusion pumps, spirometers, medical lasers, endoscope equipment;
- Patient implantable devices (enclosures, pacemakers, insulin pumps, cochlear implants, brain stimulators, cardiac defibrillators, gastric stimulators, etc.) or wearables (external EKG or pressure enclosures, glucose monitors, etc.);
- Devices used

  - for the diagnosis, prevention, monitoring, treatment or alleviation of human diseases,
  - for the diagnosis, monitoring, treatment, alleviation or replacement of human injury or handicap,
  - for investigation, replacement or modification of human anatomy or physiological process,
  - to control human fertilization,
  - in human therapy.

Figure 8.4 shows a generic model of the architecture and key components of medical devices from a cybersecurity perspective.

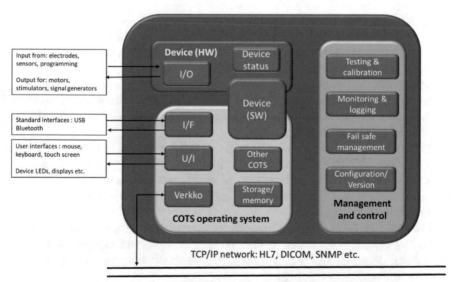

Fig. 8.4 Generic medical device architecture [12, p. 16]

8.3   Cybersecurity Risks Related to Hospital Systems

According to healthcare cybersecurity statistics healthcare would suffer 2–3 times more cyberattacks in 2019 than the average amount for other industries. Ransomware attacks on healthcare organizations were increased to quadruple between 2017 and 2020 and will grow to 5 times by 2021. The HIMSS cybersecurity survey states that nearly 60% of hospital and healthcare IT professionals in the U.S. appraised that the email is the most common point of information compromise. The HIPAA Journal told that claims healthcare email fraud attacks have increased 473% in two years. Over 93% of healthcare organizations have experienced a data breach over the past three years. Gartner predicted that more than 25% of cyberattacks in healthcare delivery organizations will involve the Internet of Things (IoT) which means wirelessly connected and digitally monitored implantable medical devices (IMDs), like cardioverter defibrillators (ICD), pacemakers, deep brain neurostimulators, insulin pumps, ear tubes, and more. Medical devices have an average of 6.2 vulnerabilities each; 60% of medical devices are at end-of-life stage, with no patches or upgrades available [22].

8.3.1   Data Breaches Against Hospital Systems

The report [23] notes that in the year 2020 there was a 25% year-over-year increase in healthcare data breaches. 619 major breaches were reported in 2020, affecting nearly 28.8 million individuals of those 415 were reported as hacking incidents. Some 246 incidents were reported as involving a business associate. After hacking incidents, the next most reported type of breach involved unauthorized access/disclosure. There were about 134 such incidents. Another 28 breaches involved lost or stolen unencrypted computing devices. Table 8.1 illustrates top 10 breaches in 2020 by number of individuals affected in U.S. [23, 24].

2020 was the third worst year in terms of the number of breached healthcare records, with 29,298,012 records reported as having been exposed or impermissibly disclosed in 2020. 266.78 million healthcare records have been breached since October 2009 across 3705 reported data breaches of 500 or more records. Figure 8.5 illustrates that development.

Threats come from a variety of different sources including adversarial, natural (including system complexity, human error, accidents, and equipment failures) and natural disasters. Adversarial groups or individuals have varying capabilities, motives, and resources [25, p. 6].

Threat, vulnerability, and risk form an intertwined entirety in the cyber world. First, there is a valuable physical object, competence or some other immaterial right that needs protection and safeguarding. A threat is a harmful cyber event that may

Table 8.1 Top 10 breaches in 2020 by number of individuals affected in U.S. [23]

Breached entity	Individuals affected	Covered entity type	Type of breach
[a]Trinity health	3.3 Million	Business associate	Hacking/IT incident
MEDNAX services	1.3 Million	Business associate	Hacking/IT incident
[a]Inova health system	1.05 Million	Healthcare provider	Hacking/IT incident
Magellan health inc	1.01 Million	Health plan	Hacking/IT incident
Dental care alliance	1 Million	Business associate	Hacking/IT incident
Luxottica of America	830.000	Business associate	Hacking/IT incident
[a]Northern light health	657.000	Business associate	Hacking/IT incident
Health share of Oregon	654.000	Health plan	Theft
Florida orthopaedic institute	640.000	Healthcare provider	Hacking/IT incident
Elkhart emergency physicians	550.000	Healthcare provider	Improper disposal

[a] Affected by ransomware attack on Blackbaud

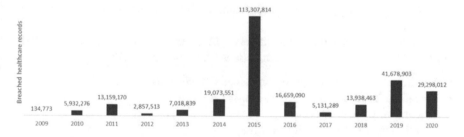

Fig. 8.5 Records exposed in U.S. healthcare data breaches [23]

occur. The numeric value of the threat represents its degree of probability. Vulnerability is the inherent weakness in the system that increases the probability of an occurrence or exacerbates its consequences.

Vulnerabilities can be divided into those that occur in human activities, processes, or technologies. Risk is the value of expected damage. The risk equals the probability times the loss. It can be assessed in terms of its economic consequences or loss of face.

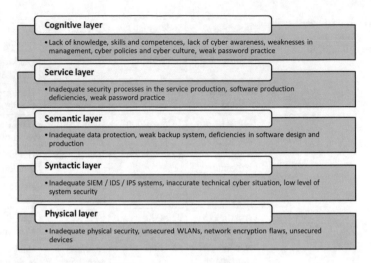

Fig. 8.6 Vulnerabilities in the cyber environment [26, p. 168]

Risk management consists of the following elements: risk assumption, risk allevia-tion, risk avoidance, risk limitation, risk planning, and risk transfer. Countermeasures can be grouped into three categories:

1. regulation,
2. organizational solutions (management, security processes, methods and proce-dures and security culture),
3. security technology solutions.

Figure 8.6 shows typical operational vulnerabilities in a five-layer cyber structure model [7].

In general, the cybersecurity threats and attack vectors in health networks (wireless/fixed) include the following issues [4, p. 645; 25, p. 5]:

*   Monitoring and eavesdropping on patients' vital signs,
*   Threats to information during transmission,
*   Threat routing in networks,
*   Location threats and activity tracking,
*   Distributed Denial of Service (DDoS) threats,
*   Interfering with or inhibiting radio communications from IoT devices and sensors,
*   Exploiting vulnerabilities to access health services,
*   Attack on hospital health information systems,
*   Interfering with wireless communications throughout the hospital and preventing the daily activities of the hospital,
*   Discontinuation of treatment/service (including deaths),
*   Malware and phishing attempts.
*   Misleading staff with spoof emails or fake websites to obtain login credentials or install malware,

- Unintentional or intentional insider threat,
- Loss of patient information, especially Electronic Protected Health Information (ePHI),
- Data breach, data exfiltration and loss of assets,
- Blackmail, extortion, and coercion using exfiltrated sensitive data,
- Intellectual Property (IP) theft.

8.3.2 Cyber Security Risks Related to Medical Devices

Increasing connectivity of medical devices to computer networks and the convergence of technologies has exposed vulnerable devices and software applications to incidents. The medical devices being potentially vulnerable and easy to exploit, so cyber-attack is possible and feasible. The purpose of such attacks can be wide-ranging from the intent to harm a specific patient; or an attack on a specific healthcare provider (e.g., cyber vandalism, crime); or an attack on the larger healthcare system (e.g., cyber terrorism, sabotage), or a military operation to support of a conventional or biological attack. These are serious [12, p. 23].

Cyber security risks are set to increase further with the adoption of the Internet of Things (IoT) by healthcare organizations and consumers. The security risk of medical devices is that they may potentially expose both the data associated with the device and the control of the device itself to an outsider. The convergence of networking, computing technology and software has enabled increasing integration of Hospital Enterprise Systems, Information Technology (IT), Operational Technology (OT) and Clinical Engineering (CE), and suppliers through remote connectivity. This will be revolutionized by cloud-based services and the use of big data analytics. This threat naturally raises the need for consideration between patient safety and cyber security. Therefore, the cyber threat will require increasingly close stakeholder cooperation in the future, especially regarding system/device design and regulation, stakeholder engagement regulators, device manufacturers, healthcare organizations, and IT providers [25, p. 3].

Most medical devices are more vulnerable to cyber-attacks than normal IT endpoints (desktops, laptops, servers), whether it is a specifically targeted attack on the medical device or an unintentional infection of it by common malware. Also, technology convergence has since brought an abundance of commercially off-the-shelf (COTS) technology including common networking infrastructure, operating systems, software, smart mobile devices, computers, and embedded control systems to medical devices. Many medical devices contain configurable embedded computers that might be vulnerable to cybersecurity breaches. Often embedded systems may utilize older, vulnerable operating systems that may be unpatched or even no longer supported.

The common vulnerabilities of the medical devices are among others [25, pp. 13–14]:

- Network-connected/configured medical devices that have been infected or disabled by malware,
- Malware access to the hospital's IT/OT/E system using wireless technology to obtain patient data,
- Uncontrolled distribution of passwords, disabled passwords, hard-coded passwords or default passwords,
- Failure to provide timely security software updates and patches for medical devices and networks,
- Vulnerabilities in older medical devices,
- Poor design practices,
- Incorrectly configured or open ports,
- Lack of encryption and authentication,
- Poor vulnerability management,
- Insecure remote access,
- Poor manufacturing cyber-hygiene.

Several factors make it difficult to protect medical devices, thus causing ongoing insecurity. These factors are the result of technical, administrative and human reasons [21]:

- Providing hackers with vital information about device performance and technical design,
- Vulnerabilities caused by legacy operating systems and software and system incompatibilities,
- Failure to make timely software updates and patches,
- Lack of basic security features in medical devices,
- Favoring online services for connection to existing systems,
- Use of compromised medical devices in attacks on other sections of the healthcare organization's network,
- Lack of awareness of cybersecurity issues and poor security practices,
- Vulnerabilities in the hardware environment caused by balancing cybersecurity, privacy, and effective healthcare processes.

The security risks associated with hospital equipment are also reflected at the system level. From a cybersecurity point of view, the hospital is comprised of a set of critical systems that contain both operational risks and various vulnerabilities, through device vulnerabilities, and are thus exposed to cyber threats.

The predicted rapid growth of connected devices in healthcare applications, the increasing concern over breaches of patient data and more recently the potential risks to patient safety requires security to be a critical feature in medical products and software. An advanced malware-infected device has the potential, in the worst case, to shut down hospital operations, expose sensitive patient information, compromise the operation of other devices, and harm patients [25, pp. 5, 18–19].

8.3.3   Cyber-Attack Vectors Against Hospital

There are a wide range of Internet threats and attacks from virus propagation and worms, such as; distributed denial of service (DDoS) attacks and data theft and manipulation and also hospital systems paralysis. An attack vector is a path or means by which an attacker can gain unauthorized access to a computer, network, or information infrastructure to deliver a payload or malicious outcome. Attack vectors allow attackers to exploit system vulnerabilities, install different types of malware and launch cyber-attacks. Once the attacker has based his/her motivation and objective they choose one or more means to achieve the goal = attack vector [27, 28].

Cyberattacks are committed by a variety of actors. A disgruntled former employee may be aware of vulnerable attack vectors due to their role in the hospital. An individual hacker may be trying to steal personalized information. A hacktivist might initiate a cyber-attack against a hospital to make an ideological statement. Business competitors may try to attack clinical infrastructure to gain a competitive edge. Cyber-criminal groups combine their expertise and resources to penetrate hospital security systems and steal large volumes of data. The intelligence service of a foreign government wants to steal secret information. There are also many known attack vectors that these groups can effectively exploit to gain unauthorized access to the hospital's IT/OT/CE infrastructure.

Inadequately tested hardware upgrades must be considered as one of the most significant risk factors for cyber security in hospital systems and equipment. The threats posed by them can be exploited by both inside and external actors. According to [25, pp. 6–7] threats can be accidental or described because of non-validated changes. Cyberattack vectors among others are:

- *Communication*: network/device communication interference,
- *Database injection*: unauthorized intrusion and data theft,
- *Replay*: data replay to gain access to the system or falsify data,
- *Spoofing or impersonation*: fooling hardware or software in such a way that the communication appears to come from elsewhere,
- *Social engineering*: obtaining information from personnel on the pretext and using it to attack computers, devices or the network,
- *Phishing*: a form of social engineering in which fake emails or websites are used to entice a victim to reveal information,
- *Malicious code*: aimed at gathering information, destroying data, allowing access to the system, falsifying system data or reports, or causing time-consuming irritation to operators and maintenance personnel,
- *Distributed Denial of Service (DDoS)*: reducing the availability of networks and computer resources (e.g., operating systems, hard drives, and applications),
- *Escalation of privileges*: enhancing an attack by gaining access to perform actions not otherwise authorized,
- *Physical destruction*: the destruction or deactivation of physical devices or components through either a direct or indirect network attack (for example, Stuxnet worm).

8.3.3.1 Ransomware

Ransomware is a type of malware that threatens to publish the victim's data or perpetually block access to it unless a ransom is paid. It encrypts the victim's files, making them inaccessible, and demands a ransom payment to decrypt them. In a properly implemented attack, recovering the files without the decryption key is an intractable problem—and difficult to trace digital currencies such as Bitcoin and other cryptocurrencies are used for the ransoms. Ransomware often spreads like other malware as email attachments, requiring the user to open the file. Another infection method is spam, which has links to websites by which malware is downloaded to the computer.

Hospitals have been the target of blackmail malware for many different reasons. One reason is that hospitals often have multiple information systems and also old operating systems in use. It is not possible to frequently update all devices that are in clinical use due to them being constantly needed. Another reason is that the operation of hospitals requires that clinical information systems be available like the patient information system, and without this information, the operation of hospitals will be significantly slowed down, which can cause problems for the patient's health. Special attention was paid in May 2017 to the widespread WannaCry blackmail malware, which spread to 48 National Health Service organizations in the UK.

Example 1. The private mental health services firm Vastaamo has been at the center of a hacking and blackmail scandal after it emerged that highly sensitive information on thousands of patients had been stolen from its database. The total number of compromised data is 40,000.

A blackmailer demanding money from a group of psychotherapy centers has released more highly sensitive personal information about more than 200 of its patients. The perpetrator has threatened to publish more daily unless the Vastaamo psychotherapy center firm pays a ransom of nearly half a million euros. A second batch of about 100 patient reports appeared on the anonymous Tor network, bringing the total to more than 200. They include highly intimate information about Vastaamo customers' personal lives and mental health issues, along with their names, addresses and social security numbers. The unknown extortionist has published messages in English purporting to be in correspondence with representatives of the company. The blackmailer is demanding that Vastaamo pay 40 bitcoins to halt the release of more patient data. The result was the bankruptcy of Vastaamo [29].

Example 2. The WannaCry ransomware (crypto worm) attack was a May 2017 worldwide cyberattack, which targeted computers using the Microsoft Windows operating system by encrypting data and demanding ransom payments in the Bitcoin. It used an EternalBlue exploit of Windows' Server Message Block (SMB) protocol. Microsoft had released patches previously in March to close the exploit, but WannaCry's found and spread in the organizations that had not patched the systems or were using older Windows systems. The WannaCry malware first checks the "kill switch" domain name; if it is not found, then the ransomware encrypts the computer's data, then attempts to exploit the SMB vulnerability to spread out to random computers on the Internet, and "laterally" to computers on the same network. Then the payload displays a message informing the user that files have been encrypted and demands a payment of around US$300 in bitcoin within three days, or US$600 within seven days. Many hospitals world-wide were infected, and some medical devices corrupted specifically hard, like radiology modalities, contract injectors and patient monitoring systems [30, 31].

Looking at the ransomware case that have occurred in hospitals and other health services over the past years, their impact has been significant. Infections have resulted in the inaccessibility of, for example, the patient information system, appointment booking, and X-ray equipment. In some cases, patient data has also been stolen. Only a few hospitals have told the public that they have paid a ransom and many hospitals have had backups that could be used for recovery.

8.3.3.2 Hacking and Data Breach

A data breach is defined as follows:

> A data breach is a security violation in which sensitive, protected or confidential data is copied, transmitted, viewed, stolen or used by an individual unauthorized to do so [32].

Data breaches may involve financial information such as credit card or bank details, personal health information, personally identifiable information (PII), trade secrets of corporations or intellectual property. Most data breaches involve overexposed and vulnerable unstructured data—files, documents, and sensitive information [32].

Patient records contain a large amount of personal information that is of interest to criminals. Information stolen from health care systems are name, address, date of birth, social security number, bank account number, credit card number, medication, treatments/surgeries, insurance information, and much more personal information. The information allows for a wide range of harm to the individual, and criminals can sell the information directly or use the information as part of an attack on an individual.

The US Department of Health and Human Services' (HHS) breach portal containing information about breaches of protected health information. According to the HHS breach portal, data breaches affected 27 million people in 2019 in U.S. Top 10 breaches by number of individuals affected, currently listed on HHS's breach portal, are given in Table 8.2 [32].

8.3.3.3 Distributed Denial of Service Attacks

Distributed denial of service (DDoS) attacks are a popular tactic, technique, and procedure (TTP) used by hacktivists and cybercriminals to overwhelm a network to

Table 8.2 Top 10 breaches by number of individuals affected in U.S. [32]

Name of covered entity	Covered entity type	Individuals affected	Type of breach	Location of breached information	Year
Anthem Inc	Health Plan	78,800,000	Hacking/IT incident	Network server	2015
American medical collection agency	Business associate	26,059,725	Hacking/IT incident	Network server	2019
Optum360, LLC	Business associate	11,500,000	Hacking/IT incident	Network server	2019
Premera blue cross	Health plan	11,000,000	Hacking/IT incident	Network server	2015
Laboratory corporation of America	Health plan	10,251,784	Hacking/IT incident	Network server	2019
Excellus health plan, Inc	Health plan	10,000,000	Hacking/IT incident	Network server	2015
Community health systems professional services corporations	Healthcare provider	6,121,158	Hacking/IT incident	Network server	2014
Science applications international corporation	Business associate	4,900,000	Loss	Other	2011
Community health systems professional services corporation	Business associate	4,500,000	Theft	Network server	2014
University of California, Los Angeles health	Healthcare provider	4,500,000	Hacking/IT Incident	Network server	2015

the point of inoperability. This can pose a serious problem for healthcare providers who need access to the network to provide proper patient care or need access to the Internet to send and receive emails, prescriptions, records, and information. While some DDoS attacks are opportunistic or even accidental, many target victims for a social, political, ideological, or financial cause related to a situation that angers the cyber threat actors [33].

In DDoS attacks, patient data is often not at risk, but patient safety can suffer when they cannot access their data or hospital staff are unable to deliver planned actions because they do not have access to medication data, for example.

Example 3. Anonymous (a well-known hacktivist group) targeted the Boston's Children's Hospital with a DDoS attack after the hospital recommended one of their patients, a 14-year-old girl, that the state takes her custody and withdrawn that from her parents. The doctors believed the child's illness was a psychological disorder and that her parents are using unnecessary treatments for a disorder the child did not have. The custody debate put Boston Children's Hospital in the middle of this controversial case. Some members of Anonymous, viewed this as an infringement on the girl's rights. So, Anonymous launched DDoS attacks against the hospital's network, which resulted in others on that network, including Harvard University and all its hospitals, to lose Internet access as well. In the networks the disorder occurred almost a week, and some medical patients and medical personnel could not use their online accounts to check appointments, test results, and other case information. The hospital must spend more than $300,000 responding to and mitigating the damage from this attack [33].

8.3.3.4   Insider Threats

Organizations are often too preoccupied with defending the integrity of their company and network from external threats to address the very real and dangerous risk that may lie within their own organization—insiders. The insider poses a threat because the legitimate access they have or had to proprietary systems discounts them from facing traditional cybersecurity defenses, such as intrusion detection devices or physical security. They also may have knowledge of the network setup and vulnerabilities, or the ability to obtain that knowledge, better than almost anyone on the outside. While an insider may be simply careless, others cause destruction with malice. The insider threat concept encompasses a variety of employees: from those unknowingly clicking on a malicious link which compromises the network or losing a work device containing sensitive data to those maliciously giving away access codes or purposely selling PHI/PII for profit [34].

Patient data is also at risk when stored, for example, on laptops that are taken outside the hospital. Laptops have been stolen from doctor's offices, cars, and health-care professionals' homes. What makes these situations problematic is that computers often have a password, but the hard drive itself or its data is not encrypted, so a criminal can access all the data if, for example, a password query is bypassed.

Example 4. In Texas hospital an employee built a botnet, using the hospital network, because he wanted to attack rival hacking groups. The individual was eventually caught after he filmed himself staging an infiltration of the hospital network and then posted it on YouTube for public viewing. The video clearly shows when he used a specific key to "infiltrate" the hospital network. His identity as Jesse McGraw, a night security guard of the building could be ascertained. The investigation revealed that McGraw had downloaded malware on dozens of machines, including nursing stations with patient records. He also installed a backdoor in the HVAC unit, that could have caused damage to drugs and medicines and affected hospital patients during the hot Texas summer. McGraw pled guilty to the cyber-attack and is serving a 9-year sentence in addition to paying $31,000 in fines [34].

8.4   Healthcare Cybersecurity

The technical infrastructure underlying the healthcare systems is extremely complex. It must support not only patient records but also a diverse suite of medical devices used in diagnosing, monitoring, and treating patients. Understanding and managing cybersecurity risks for this mission-critical environment is challenging as the healthcare system has a mixture of state-of- the-art applications and devices, as well as older legacy devices that use unsupported operating systems or networking protocols. In addition, it is difficult to make these systems available for updates since they often provide round-the-clock care to patients and cannot be taken out of service. Hospital environments are also challenging because staff and patients Bring their [Your] Own Devices (BYOD) inside the facilities, as well as many different medical devices used in medical research [35, pp. 22–23].

8.4.1   Best Practices

National standardization bodies draw up national standards and participate in the drafting of international standards, taking into account best practice. Organizations use standards and various other guidelines and recommendations on a voluntary basis. The best practices and objectives that emerge from them are usually related to operational development, which are at best, proactive procedures. In the cyber world, they assist their users in improving the reliability, continuity, quality, risk management, and preparedness of an organization's operations. In this case, the functions may be, for example, the management and administration of cyber security or the technical development, maintenance or use of information systems, information networks and ICT services [36].

Many organizations' cyber security activities continue to be characterized by a responsive approach, with hospitals being no exception. A responsive approach means that in the event of a cyber-attack, and the action are quick conclusions and urgent measures. Developing cyber security using best practices creates for the hospital proactive methods instead reactive actions [36].

NIST Framework for Improving Critical Infrastructure Cybersecurity provides a common language for understanding, managing, and expressing cybersecurity risk to internal and external stakeholders. It can be used to help identify and prioritize actions for reducing cybersecurity risk, and it is a tool for aligning policy, business, and technological approaches to managing that risk. It can be used to manage cybersecurity risk across entire organizations, or it can be focused on the delivery of critical services within an organization. The Framework core provides a set of activities to achieve specific cybersecurity outcomes, and references examples of guidance to achieve those outcomes [37, pp. 6–8].

The measures of the NIST framework and their contents are the following [37]:

- *Identify*: Develop an organization's understanding of the management of cyber-security risks in systems, people, assets, data, and capabilities;
- *Protect*: Develop and implement appropriate safeguards to ensure the delivery of critical services;
- *Detect*: Develop and implement appropriate activities to detect a cybersecurity incident;
- *Respond*: Develop and implement appropriate activities in response to a detected cybersecurity incident;
- *Recover*: Develop and implement appropriate activities to maintain resilience plans and restore the performance of services degraded by a cybersecurity incident.

Categories are areas of main activities that can be divided into cybersecurity review groups, such as "access control" or "identification processes." The subcategories are further subdivided into technical sections and/or management activities. Informative references are parts of standards, guidelines and practices [37, pp. 6–8, 13].

Discussion in the U.S. Health Care Industry Cybersecurity (HCIC) Task Force focused on gathering information from external stakeholders and subject matter experts across the healthcare industry and other sectors. The Task Force identified six high-level imperatives that need to be achieved in order to increase security in healthcare industry. The imperatives are [35, pp. 24–44]:

1. Define and streamline leadership, governance, and expectations for healthcare industry cybersecurity,
2. Increase the security and resilience of medical devices and health IT,
3. Develop the healthcare workforce capacity necessary to prioritize and ensure cybersecurity awareness and technical capabilities,
4. Increase healthcare industry readiness through improved cybersecurity awareness and education,
5. Identify mechanisms to protect R&D efforts and intellectual property from attacks or exposure,
6. Improve information sharing of cyber security threats, risks, and mitigations.

Both large and small healthcare organizations struggle with unsupported legacy systems that cannot easily be replaced (hardware, software, and operating systems) and the systems have large numbers of vulnerabilities and few modern countermeasures.

The implementation of key cyber security measures include [38, p. 53]:

- Network segmentation (smart firewalls),
- Network monitoring and intrusion detection,
- Robust encryption,
- Access control,
- Authentication and authorization.

Clinicians in a hospital setting are required to access multiple computers throughout the facility repeatedly (up to 70 times per shift) as they deliver care to patients. To authenticate their identity so that they can perform common tasks (e.g., access a patient's medical record, order diagnostic tests, prescribe medication, etc.), a clinician typically enters his or her username and a unique password. This widely used, single factor approach to accessing information is particularly prone to cyber-attack as such passwords can be weak, stolen, and are vulnerable to external phishing attacks, malware, and social engineering threats. NIST SP 800–6355 adopts alternatives to the use of passwords for user authentication, including items in the user's possession (e.g., a proximity card or token) or biometrics [35, p. 32].

Clinicians also interact with medical devices and the integrity of the devices used in these treatments must be assured from a bioengineering and a cybersecurity perspective. The provider operating the device must be authenticated and authorized to operate it, and the patient needs to be accurately identified as the person authorized to receive the treatment. Moreover, communications between the device and other healthcare technologies should be authenticated (i.e., devices should know what technologies they are communicating with and should only be communicating with technologies with the appropriate credentials) [35, p. 32].

Many hospitals still follow a reactive approach to information security. Measures are frequently taken only after an incident has occurred. In the healthcare context, avoiding incidents is particularly important as trustworthiness is of very high priority. Security incidents may not only threaten personal health information but also patient safety. Hospitals should also be well prepared for the possibility of security incidents by having concrete response and recovery plans in place, like [38, p. 53]:

- Perform a cost benefit analysis for the most important IoT components in the hospital. Smart hospital is expensive to implement, it needs to be adequately protected.
- Create an information security strategy for the smart assets in the hospital. Clear roles and responsibilities as well as regular training and awareness raising activities are key elements of a proactive approach to information security.
- Create a BYOD and mobile device policy for users: as this is a component of a smart hospital ecosystem this needs to become a priority.
- Identify the assets and how these will be interconnected (or connected to the Internet). For some systems, the right move for safety and resilience might be for the manufacturer to refuse built-in network capabilities into the device.
- Define and implement security baselines on all major operating systems.

The healthcare industry must increase outreach for cyber security across all members of the healthcare workforce through ongoing workshops, meetings, conferences, and tabletop exercises. Also, the healthcare industry must develop cyber security programs (including on-line education) to educate decision makers, executives, and board members about the importance of cyber security, as cybersecurity is the responsibility of top management. As part of this holistic cyber security strategy, it is critical that a thorough baseline is established whereby inherent trust can be established between patients and clinicians, technologies, and processes, and ultimately institutions and patients [35, p. 40].

8.4.2 Cybersecurity of Medical Devices

The cybersecurity of medical devices has increasingly become a concern to healthcare providers, device manufacturers, regulators, and patients. Due to their long useful life, unique care-critical use case, and regulatory oversight, these devices tend to have a low security maturity, significant vulnerabilities, and an overall high susceptibility to security threats.

Lifecycle management and procurement of the medical devices need a shared responsibility. As part of their asset and risk assessment processes, actors articulate described cyber security requirements. Figure 8.7 outlines the liability measures between the device manufacturer and the healthcare organization.

Medical device providers should assign priorities and classification to medical devices based on the risk/type of device. The classifications may vary according to the organization's priorities but could follow a model similar to the example in Tables 8.3 and 8.4.

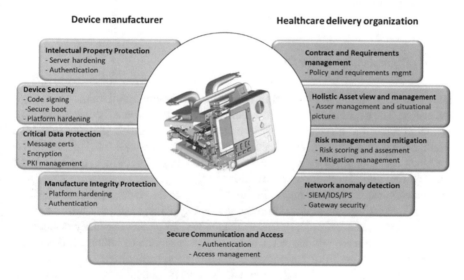

Fig. 8.7 Division of cybersecurity responsibilities for medical devices [39, p. 2]

Table 8.3 Medical device priorities [40, p. 14]

Priority level	Description
1	Lifesaving (defibrillator, pacemaker, ventilator)
2	Curative/Therapeutic (infusion pump, hyperbaric chamber, dialysis)
3	Patient diagnostic (ECG, ultrasound, X-Ray, lab equipment)
4	Analytics (fetal monitors, patient monitors)
5	Miscellaneous (medical cabinet, autoclave, scale)

Table 8.4 Medical device classifications [40, p. 14)

Security classification	Description
A	Over 100,000 records stored, transmitted or processed
B	Between 10,001 and 99,999 records stored, transmitted, or processed
C	Less than 10,000 records stored, transmitted or processed
D	Device does not store, transmit or process PHI

Organizations should not wait for regulators to enforce security standards. Instead, health entities should be active in conversations with regulators about needed data security standards and work collaboratively with the medical device market to ensure appropriate data security measures are factored into product design and implementations. The medical devices should be tiered based on clinical safety considerations, the volume of patient records managed by the device and platform, etc. Medical device security programs must prioritize which devices to secure first and then move on to others over time [41].

8.4.3 New Technology to Help

The technology revolution is creating a unique environment for all industries to grow and evolve. The healthcare and pharmaceutical industries are quickly discovering that the right tech could also be the key to delivering better care to patients. Artificial Intelligence is one of the most prominent examples of Industry 4.0 technology. With machines that can learn from the enormous amounts of data in every healthcare industry, there will be endless opportunities [42].

Cyber-physical systems are ubiquitous and used in many applications, from industrial control systems, novel communication systems, to critical infrastructure. These systems generate, process, and exchange vast amounts of security-critical and privacy-sensitive data, which makes them attractive targets of attacks. Industry 4.0 and its main enabling information and communication technologies are completely changing both services and production worlds. This is especially true for the healthcare domain, where the Internet of Things, Cloud and Fog Computing, and Big Data technologies are revolutionizing eHealth and its whole ecosystem, moving it towards Healthcare 4.0. [43, pp. 1–2; 44]

Industry 4.0 is the combination of the automation process, manufacturing units and smart machines. It consists of digitization, IoT, internal connected network, human resources for supervision, Supervisory Control and Data Acquisition (SCADA), robots for automation of many critical functions, valves, sensors, actuators, PLC system, communication protocols and cyber security. It uses artificial intelligence to help clinical decision-making, share information digitally in hospitals and enables to create smart cyber security to hospital environment [45].

Artificial Intelligence is intelligence exhibited by machines. Any system that perceives its environment and takes actions that maximize its chance of success at some goal may be defined as AI. For example, cognitive computing is a comprehensive set of capabilities based on technologies such as deep learning, machine learning, natural language processing, reasoning and decision technologies, speech and vision technologies, human interface technologies, semantic technology, dialog, and narrative generation, among other technologies. Artificial intelligence and robotics have steadily growing roles in healthcare. Organizations benefit from the ability of cognitive systems to improve their expertise quickly and from sharing it to all those who need it. The knowhow of top experts is quickly made available to all when their subject matter expertise is taught to a cognitive system. Through repeated use, the system will provide increasingly accurate responses, eventually eclipsing the accuracy of human experts. With artificial intelligence, comprehension can be outsourced. As the intelligence of machines improve, they will use deep learning to understand the collective information. With the use of digital sensor data, equipment based on artificial intelligence can used to develop smart advisors, teachers, or assistants. [49, p. 431).

The integrated framework of key security capabilities should form the core of the cyber security solutions. At the core of this structure is security intelligence and analytics. This serves as the key piece, ingesting security data across an IT environment (e.g., logs, flows, incidents, events, packets, and anomalies) as well as information beyond the organization (e.g., blogs, research information and websites) to understand threats and attacks. Security infrastructure uses its own network of integrated security capabilities to intelligently detect the symptoms of a cyber-attack, like a breach on the network, an abnormal login on a high-value server, rogue cloud app usage, and respond appropriately [46].

Figure 8.8 shows an example of IBM's concept of an integrated cybersecurity solution, with analytics capabilities at the heart of the solution. With analytics at the

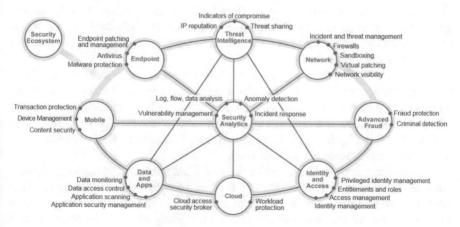

Fig. 8.8 IBM's integrated cybersecurity concept [46]

core, integrated capabilities deliver a level of visibility and defense that no single security solution can provide on its own.

MIT Computer Science and Artificial Intelligence Laboratory (CSAIL) and PatternEx have developed an artificial intelligence AI2 platform to predict cyber-attacks. According to [47], the AI2 platform was able to reach 86% accuracy in detecting cyber-attacks, which is approximately three times better than results of previous studies. Tests were conducted with 3.6 billion data components (log lines), which were generated by millions of users in a three-month research period. To prevent attacks, AI2 identifies suspicious activity by applying clustering algorithms to the input data by utilizing unsupervised machine learning algorithms. Hence, the results will be presented to analysts, who confirm which incidents are real attacks. Analysts also incorporate the outcome into platform models (supervised learning) for the next set of data, which enables further learning. The system is also capable of continuously generating new models within hours, which can significantly improve the speed of its detection ability of cyber-attacks [47, 48, p. 49, 49, p. 435].

Doctors at the University of Tokyo reported that they diagnosed with IBM Watson a 60-year-old woman with rare leukemia that had been identified incorrectly a month earlier. Watson needed only 10 min to compare the patient's genetic changes to the 20 million cancer study publication database. Watson provided an accurate diagnosis, instructions on treatment, and medication to achieve the desired treatment outcomes [50].

8.4.4　Hospital Cybersecurity Architecture

The starting point for a hospital's cyber security architecture can be formed by utilizing the definitions and recommendations of HCIC Task Force to organize policies [35, p. 1]. They relate to organizational leadership and management, hospital resilience, staff competence, and research and information sharing. Measures should also identify cyber security challenges, which consist of the different life-cycle stages of equipment and are reflected at system level in the deployment, management, and maintenance of new devices.

As the pace of development has been very rapid and new technology has been introduced very quickly, the international standard work has not been involved in the development process. We often have manufacturer-specific solutions for IoT devices, different sensors, and data storage systems in some of the service providers' Data Center, shown in Fig. 8.9. This issue in turn leads to a challenge of connecting IoT devices to smart devices [4, pp. 641–642].

The publication of the National Institute of Standards and Technology "Framework for Improving Critical Infrastructure Cybersecurity" can also be applied in a hospital operating environment. The starting point is to identify the processes in the hospital and the devices and systems. This is particularly related to an organization's ability to understand and manage cyber security risks in them. Security measures can then be developed and implemented with appropriate cyber security products and

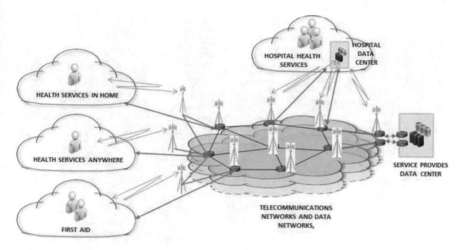

Fig. 8.9 E-Health top level architecture [4, p. 642]

services that specifically address device risks. The situational picture and the situational awareness are the basic resource to create effective cyber security in hospital [51, p. 62; 37]

In cyber-physical systems, networked devices, and their software control physical processes. The operation of the hospital involves a significant number of technical devices and functional entities consisting of cyber-physical systems. The hospital is technically a system of systems and in turn part of a large healthcare complex. Functions are thus networked at many levels of healthcare.

The development of digital methods of treatment with sensors and IoT devices is strongly developed to improve patient care, to look at their condition remotely at home or wherever they are moving in real time. The future healthcare operation environments are presented in Fig. 8.10. So, system-level protection can be developed by applying new technologies to solutions [4, p. 643].

8.5 Conclusion

The most critical components in digital healthcare environments are mainly patient healthcare systems and devices. Their cyber security needs to be analyzed and functionalities monitored as they are critical systems in patient care. Cyber security supports both business and clinical objectives in healthcare and facilitates the delivery of efficient and high-quality patient care. However, this requires a holistic cybersecurity strategy. Organizations that do not adopt a holistic strategy endanger not only their data, organization, and reputation but also the well-being and safety of their patients.

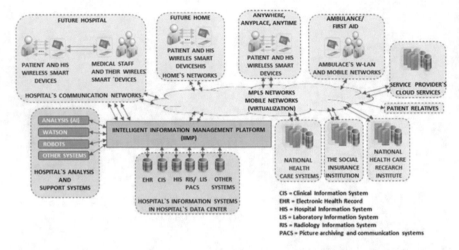

Fig. 8.10 E-Health or m-health operating environment, top level architecture [4, p. 643]

If healthcare operations lack visions, strategies, scenarios, and use cases, we do not have sufficiently precise architectural descriptions of the ICT systems used in healthcare. In this case, it is difficult to perceive the entities and service chains they contain. Risk and threat analyses of the services offered by service providers are difficult and even impossible to perform with sufficient accuracy. It may not be possible to identify and analyse healthcare service chains and to carry out related risk and threat analyses.

The case Vastaamo is a good example of what happens when healthcare services are outsourced (see Example 1). Due to outsourcing, the implementation and auditing of services, the systems and applications required for them are no longer the responsibility of healthcare organizations. In such situations, there may be deficiencies in inspections and audits of critical healthcare systems and applications. If we are unaware of the vulnerabilities in our systems and applications and thus unable to fix them, it will be easy for hackers and cyber attackers to attack our system. In the case of Vastaamo, it is difficult to find the organizations or people who will have to answer in court for these consequences of the information leak. Unfortunately, it is difficult for victims of information leakage to receive adequate and correct compensation, because after a leak, their entire lives may contain long-term suffering and pain.

Often, due to service outsourcing, cloud services may be deployed. Their actual location may not be known, and service chains cannot be defined. In critical services where $1 + 1$ protection and backup are necessary, signal delays are critical, making service chain definitions and their analyses relevant to functionality. In a $1 + 1$ protected operating environment, the exchange of services for the protection system must occur in milliseconds. This means that continuity management must also be considered in the design and development of healthcare systems and services based on different scenarios. Scenario-based design and development also provide better opportunities for analysing cyber security risks and threats to healthcare systems. In addition, there are many integrations between healthcare systems. In the healthcare areas outlined above, we need more research to develop better cybersecurity systems.

There are currently several cybersecurity solutions and tools are available for the needs of healthcare organizations. The challenge is the fragmentation of solutions and tools, as well as the problems of deploying and maintaining new systems, which cause management difficulties and increase complexity throughout the system. The complexity of the systems requires the development of integrated systems that identify both external and internal threats and have comprehensive built-in cybersecurity systems. Integrated and holistic solutions provide the required visibility at all levels of ICT system, which means that the protection and prevention of cyber-attacks can be implemented as a whole rather than as individual procedures.

We recommend that healthcare organizations and hospitals make and develop visions, strategies, and scenarios and collect use cases from their healthcare environment. Based on them, they can develop systems and application architectures. They then have a clear picture of their operating environment, systems, and applications, and can conduct cyber risk and threat analyzes for the systems there.

ANNEX Core Medical Equipment

Core medical equipment is described as follows:

Core medical equipment refers here to technologies that are commonly considered as important or necessary for specific preventive, diagnostic, treatment or rehabilitation procedures carried out in most health care facilities.

Today, there are more than 10,000 types of medical devices available. The selection of appropriate medical equipment always depends on local, regional or national requirements; factors to consider include the type of health facility where the devices are to be used, the health work force available and the burden of disease experienced in the specific catchment area. It is therefore impossible to make a list of core medical equipment which would be exhaustive and/or universally applicable [52].

According to the [52], the core medical equipment are:

- Analyzer, laboratory, hematology, blood grouping
- Anesthesia unit
- Apnea monitors
- Aspirator
- Auditory function screening device, newborn
- Bilirubinometer
- Blood gas/pH/chemistry point of care analyzer
- Blood pressure monitor
- Bronchoscope
- Cataract extraction units
- Clinical chemistry analyzer
- Colonoscope
- Cryosurgical unit
- Cytometer
- Defibrillator, external, automated; semi-automated

- Defibrillator, external, manual
- Densitometer, bone
- Electrocardiograph, ECG
- Electrosurgical unit
- Fetal heart detector, ultrasonic
- Fetal monitor
- Glucose analyzer
- Hematology point of care analyzer
- Hemodialysis unit
- Immunoassay analyzer
- Incubator, infant
- Laser, CO_2
- Laser, ophthalmic
- Mammography unit
- Monitor, bedside, electroencephalography
- Monitor, central station
- Monitoring system, physiologic
- Monitor, telemetric, physiologic
- Peritoneal dialysis unit
- Pulmonary function analyzer
- Radiographic, fluoroscopic system
- Radiotherapy planning system
- Radiotherapy systems
- Remote afterloading brachytherapy system
- Scanning system, CT
- Scanning system, Magnetic Resonance Imaging, full-body
- Scanning system, ultrasonic
- Transcutaneous blood gas monitor
- Ventilator, intensive care
- Ventilator, intensive care, neonatal/pediatric
- Ventilator, portable
- Videoconferencing system, telemedicine
- Warming unit, radiant, infant
- Whole blood coagulation analyser.

References

1. SecurCom (2017) The security strategy for society: government resolution, 2 Nov 2017. The security committee, Helsinki. https://turvallisuuskomitea.fi/wp-content/uploads/2018/04/YTS_2017_english.pdf
2. Zhang Y, Qui M, Tsai CW, Hassan MM, Alamri A (2017) Health-CPS: healthcare cyber-physical system assisted by cloud and big data. IEEE Syst J 11(1):88–95

3. Reddy M (2021) Digital transformation in healthcare in 2021: 7 key trends. Digital authority partners. https://www.digitalauthority.me/resources/state-of-digital-transformation-healthcare/
4. Hummelholm A (2019) E-health systems in digital environments. In: ECCWS 2019: proceedings of the 18th European conference on cyber warfare and security. Academic Conferences International, pp 641–649
5. Lehto M, Limnéll J, Innola E, Pöyhönen J, Rusi T, Salminen M (2017) Finland's cyber security: the present state, vision and the actions needed to achieve the vision. Publications of the government´s analysis, assessment and research activities, 30/2017. (In Finnish)
6. Halonen P (2016) Kyberturvallisuus terveydenhuollossa. Viestintäviraston kyberturvallisuuskeskus. https://docplayer.fi/25743256-Kyberturvallisuus-terveydenhuollossa-perttu-halonen-helsinki.html
7. Lehto M (2015) Phenomena in the cyber world. In: Lehto M, Neittaanmäki P (eds) Cyber security: analytics, technology and automation. Springer, Cham, pp 3–29
8. Libicki MC (2007) Conquest in cyberspace: national security and information warfare. Cambridge University Press, New York
9. Sartonen M, Huhtinen A-M, Lehto M (2016) Rhizomatic target audiences of the cyber domain. J Inf Warfare 15(4):1–13
10. ITU (2018) Guide to developing a national cybersecurity strategy: strategic engagement in cybersecurity. International telecommunication union (ITU), Geneva
11. ISO (2012) ISO/IEC 27032 cyber security trainings. PECB University, Washington, DC. https://pecb.com/en/education-and-certification-for-individuals/iso-iec-27032
12. IHE (2015) Medical equipment management (MEM): medical device cyber security—best practice guide. White paper, integrating the healthcare enterprise (IHE) international. http://www.ihe.net/uploadedFiles/Documents/PCD/IHE_PCD_WP_Cyber-Security_Rev1.1 2015-10-14.pdf
13. Chanchal S (2020) What is hospital information system & our top 15 picks. Software Suggest. https://www.softwaresuggest.com/blog/top-hospital-information-system/#
14. Levin D (2019) What is a health information system? Datica. https://datica-2019.netlify.app/blog/what-is-a-health-information-system/
15. InfoWerks (2020) What is a health information system? InfoWerks. https://infowerks.com/health-information-system/
16. Kanta (2021) What are the Kanta services? Kanta system, https://www.kanta.fi/en/what-are-kanta-services. Retrieved 28 Jan 2021
17. Verizon (2018) Protected health information data breach report. White Paper, Verizon
18. McGovern L, Miller G, Hughes-Cromwick P (2014) The relative contribution of multiple determinants to health outcomes. Health Affairs Health Policy Brief
19. ENISA (2020) Procurement guidelines for cybersecurity in hospitals: good practices for the security of healthcare services. European union agency for network and information security (ENISA). https://www.enisa.europa.eu/publications/good-practices-for-the-security-of-health care-services
20. EU (1993) Council directive 93/42/EEC of 14 June 1993 concerning medical devices. The Council of the European communities
21. Williams P, Woodward A (2015) Cybersecurity vulnerabilities in medical devices: a complex environment and multifaceted problem. Med Devices (Auckl) 8:305–316
22. Herjavec (2019) The 2020 healthcare cybersecurity report: a special report from the editors at cybersecurity ventures. Herjavec group. https://www.herjavecgroup.com/wp-content/uploads/2019/12/Healthcare-Cybersecurity-Report-2020.pdf
23. HIPAA (2021) 2020 healthcare data breach report: 25% increase in breaches in 2020. HIPAA J. https://www.hipaajournal.com/2020-healthcare-data-breach-report-us/
24. McGee MK (2021) Analysis: 2020 health data breach trends—ransomware, phishing incidents, vendor hacks prevail. DataBreachToday. https://www.databreachtoday.com/analysis-2020-health-data-breach-trends-a-15694

25. Piggin R (2017) Cybersecurity of medical devices: addressing patient safety and the security of patient health information. BSI. https://www.bsigroup.com/LocalFiles/EN-AU/ISO%201 3485%20Medical%20Devices/Whitepapers/White_Paper___Cybersecurity_of_medical_dev ices.pdf
26. Lehto M (2014) Kybertaistelu ilmavoimaympäristössä. Teoksessa T. Kuusisto (toim.), Kyber-taistelu 2020, ss. 157–178. Taktiikan laitoksen julkaisusarja 2, No. 1/2014. Maanpuolustusko-rkeakoulu, Helsinki
27. CERT-UK (2015) Annual report 2015/2016. CERT-UK
28. Kovanen T, Nuojua V, Lehto M (2018) Cyber threat landscape in energy sector. In: ICCWS 2018: proceedings of the 13th international conference on cyber warfare and security. Academic conferences international. pp 353–361
29. Yle (2020) Extortionist publishes more sensitive data on psychotherapy centres' patients. Yle. https://yle.fi/uutiset/osasto/news/extortionist_publishes_more_sensitive_data_on_psychother apy_centres_patients/11608960
30. Kusche K (2018) Getting ready for the next international cyber-attack. Presentation at HIMSS 2018 annual conference and exhibition. https://365.himss.org/sites/himss365/files/365/han douts/550237057/handout-CYB2.pdf
31. Symantec (2017) What you need to know about the WannaCry ransomware. Symantec security response. https://symantec-enterprise-blogs.security.com/blogs/threat-intelligence/wannacry-ransomware-attack
32. HHS (2020) Breach portal: notice to the secretary of HHS breach of unsecured protected health information. U.S. Department of health and human services, office for civil rights. https://ocr portal.hhs.gov/ocr/breach/breach_report.jsf
33. CIS (2020a) DDoS attacks: in the healthcare sector. Center for internet security (CIS®). https://www.cisecurity.org/blog/ddos-attacks-in-the-healthcare-sector/
34. CIS (2020b) Insider threats: in the healthcare sector. Center for Internet Security (CIS®). https://www.cisecurity.org/blog/insider-threats-in-the-healthcare-sector/
35. US-GOV (2017) Report on improving cybersecurity in the health care industry. Health care industry cybersecurity task force, U.S. Department of health and human services. https://www.phe.gov/preparedness/planning/cybertf/documents/report2017.pdf
36. Pöyhönen J (2020) Cyber security management and development as part of a critical infrastructure organization: system thinking. Ph.D. thesis, University of Jyväskylä. (In Finnish)
37. NIST (2018) Framework for improving critical infrastructure cybersecurity: version 1.1. National institute of standards and technology (NIST). https://nvlpubs.nist.gov/nistpubs/CSWP/NIST.CSWP.04162018.pdf
38. ENISA (2016) Smart hospitals: security and resilience for smart health service and infrastruc-tures. European union agency for network and information security (ENISA). https://www.enisa.europa.eu/publications/cyber-security-and-resilience-for-smart-hospitals
39. Symantec (2016) Symantec™ industry focus: medical device security. Symantec corpora-tion. https://www.symantec.com/content/dam/symantec/docs/data-sheets/symc-med-device-security-en.pdf
40. Meditology (2017) Hijacking your life support: medical device security. Medi-tology Services. https://www.meditologyservices.com/fullpanel/uploads/files/whitepaper-med ical-device-security-2017.pdf. Retrieved 28 Nov 2018
41. Selfridge B (2018) Healthcare's space junk: medical device & IoT security (part 3 of 5). Meditology services. https://www.meditologyservices.com/healthcares-space-junk-medical-device-iot-security-part-3/
42. Peters J (2020) How is industry 4.0 affecting healthcare. Intetics. https://intetics.com/blog/guest-post-how-is-industry-4-0-affecting-healthcare
43. Sadeghi AR, Wachsmann C, Waidner M (2015) Security and privacy challenges in industrial internet of things. In: DAC'15: proceedings of the 52nd annual design automation conference, article 54. ACM, New York, pp 1–6
44. Aceto G, Persico V, Pescapé A (2020) Industry 4.0 and health: internet of things, big data, and cloud computing for healthcare 4.0. J Ind Inf Integr 18:100129

45. Javaid M, Haleem A (2019) Industry 4.0 applications in medical field: a brief review. Curr Med Res Pract 9(3):102–109
46. Falco C (2016) Unleashing the immune system: How to boost your security hygiene. Security Intelligence. https://securityintelligence.com/news/unleashing-the-immune-system-how-to-boost-your-security-hygiene/
47. Conner-Simons A (2016) System predicts 85 percent of cyber-attacks using input from human experts. MIT News. http://news.mit.edu/2016/ai-system-predicts-85-percent-cyber-att acks-using-input-human-experts-0418
48. Veeramachaneni K, Arnaldo I, Korrapati V, Bassias C, Li K (2016) AI^2: training a big data machine to defend. In: 2016 IEEE 2nd international conference on big data security on cloud (BigDataSecurity), IEEE international conference on high performance and smart computing (HPSC), and IEEE international conference on intelligent data and security (IDS). IEEE, pp 49–54
49. Vähäkainu P, Lehto M (2019) Artificial intelligence in the cyber security environment. In: ICCWS 2019: proceedings of the 14th international conference on cyber warfare and security. Academic conferences international, pp 431–440
50. Fingas J (2016) IBM's Watson AI saved a woman from leukemia. Engadget https://www.eng adget.com/2016-08-07-ibms-watson-ai-saved-a-woman-from-leukemia.html
51. Lehto M, Limnéll J, Kokkomäki T, Pöyhönen J, Salminen M (2018) Strategic leadership of cyber security in Finland. Publications of the government´s analysis, assessment and research activities, 28/2018
52. WHO (2011) Core medical equipment. World health organization (WHO). https://apps.who. int/iris/bitstream/handle/10665/95788/WHO_HSS_EHT_DIM_11.03_eng.pdf?sequence=1

Chapter 9
Cyber Security of an Electric Power System in Critical Infrastructure

Jouni Pöyhönen

Abstract The functioning of a modern society is based on the cooperation of several critical infrastructures, whose joint efficiency depends increasingly on a reliable national electric power system. Reliability is based on functional data transmission networks in the organizations that belong to the power system. This chapter focuses on the procedures applied to cyber security management in the processes of an electricity organization, whereby different standards will also be utilized. The major contributions of the chapter are that it integrates cyber security management into the process structures of individual electricity production organization and that it utilizes the PDCA (Plan, Do, Check, Act) method in developing an organization's cyber security management practices. In order to put the measures into practice, the leadership of an electricity organization must regard trust-enhancing measures related to cyber security as a strategic goal, maintain efficient processes and communicate their implementation with a policy that supports the strategy.

Keywords Critical infrastructure · Power system · Electricity organization · Cyber security management · Trust

9.1 Introduction

The critical infrastructure consists several systems and services that are essential for national society. The functioning of a modern society is based on the cooperation of several operators, whose joint efficiency depends increasingly on a reliable national electric power system. Crucial in the cyber environment are also functional data transmission networks and the usability, reliability and integrity of systems and services in the operating environment, whose cyber security risks are continuously augmented by threatening scenarios of the digital world.

J. Pöyhönen (✉)
Faculty of Information Technology, University of Jyväskylä, Jyväskylä, Finland
e-mail: jouni.a.poyhonen@jyu.fi

Fig. 9.1 Simplified
composition of critical
infrastructure

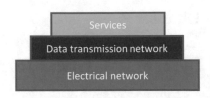

Security of supply means the ability to uphold society's vital functions in state of emergency and, in that sense, it is relevant to emphasize the importance of safeguarding the basic national structures and services. It is essential for the vital functions of society, including both physical facilities and structures as well as electronic functions and services. Collectively, they are called the critical infrastructure. Energy supplies, digital services, and logistics and transport must be safeguarded in the event of serious incidents and emergencies. In that sense it is also important to improve cyber security preparedness [9].

Based on the previous research results, the concept of national critical infrastructure can be simplified in accordance with Fig. 9.1. An energy supplies operator of electric power system can position its own strategic role and identify its operation as part of an entity whose other parts depend on a reliably functioning electrical network. This also facilitates the identification of cyber dependencies within the services of the service layer so that they can be secured with the most efficient and practical measures [24].

Finland's electric power system—comprising power plants, a nationwide transmission grid, regional networks, distribution networks and electricity consumers—is part of an inter-Nordic power system together with the systems of Sweden, Norway, and Eastern Denmark. In addition, there are direct current transmission links to Finland from Russia and Estonia in order to connect the Nordic system to the power systems of Russia and the Baltic countries. The inter-Nordic system is furthermore connected to the system in continental Europe via direct current transmission links [8].

Electricity is produced at Finnish power plants in various ways, using several energy sources and production methods. The major sources of energy include nuclear power, waterpower, coal, natural gas, wood fuels and peat. In addition to the sources of energy, production can be classified according to the production method. In Finland there are about 120 enterprises that produce electricity as well as around 400 power plants, over half of them hydroelectric power plants. Nearly a third of electricity is produced in connection with heat production. Compared with many other European countries, Finland's electricity production is decentralized. A diverse and decentralized electricity production structure increases the security of the national energy supply [6].

The national significance of an electric power system is very similar irrespective of the country. For example, in the USA the power system is considered to be a critical infrastructure and a key resource for the functioning of the entire society. In the USA it can be seen that the grid represents a technologically highly advanced system entity and that its solutions call for the use of the most demanding technologies. Grid technology and its control procedures constitute the principal areas in examining cyber security [17].

The electric power system with all its components belongs to critical national infrastructure: it is vital for the operations of the country and its outage or destruction would weaken national security, the economy, public health and safety as well as make the operations of state administration less effective. Even one second power failure can cause harm for sensitive industrial processes may stop. Data in information systems may be lost, fifteen minutes failure may harm people's daily activities and cause traffic delays, after few hours industrial processes may undergo significant damage, mobile phone networks will face problems, domestic animal production will be disturbed and finally after days the operations of society will be seriously harmed [12].

The global threats within the cyber environment have remained at a high level over the past few years, as stated in the annual international business world surveys by the World Economic Forum. They are seen to be among the major global risks based on the probability and impact of their realization [29].

This chapter focuses on factors related to cyber security management in an electricity production organization of energy supplies that is typical part of power system. It will also examine the answer to the question: How the cyber security factors should be considered in the organization process structures while creating continuity in its operation within a dynamic cyber environment?

9.2 Organization's Cyber Structure

9.2.1 Structure of an Organization's Cyber World

According to EU commission "Information and Communication Technologies (ICTs) are increasingly intertwined in our daily activities. Some of these ICT systems, services, networks and infrastructures (ICT infrastructures) form a vital part of European economy and society, either providing essential goods and services or constituting the underpinning platform of other critical infrastructures" [7]. Information and Communication Technology (ICT) systems are part of the organization's critical infrastructure and thus constitute a significant part of the operations that support an organization's core processes. Organization-level ICT systems are related to administration and to the management of information and material flows in the network. The production level includes industrial automation called Industrial Control Systems (ICS).

Martin C. Libicki has created a structure for the cyber world, his idea is based on the Open Systems Interconnection Reference Model (OSI). The OSI model groups communication protocols a broken down into seven layers. Each layer serves the layer above it and is served by the layer below it. The Libicki cyber world model has the following four layers: physical, syntactic, semantic and pragmatic [18].

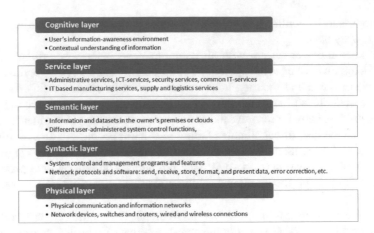

Fig. 9.2 Five-layer structure for the cyber world (modified from [16])

Martti Lehto, cyber security professor at the University of Jyväskylä, has updated the Libicki's four layers cyber world model by adding the fifth layer in order to consider an organization's networking needs. The structure is described in Fig. 9.2.

In the case of the five-layer model structure, the physical layer contains the physical elements of the ICT and ICS devices, such as computers, control devices, communications network devices (switches and routers) as well as wired and wireless connections. The syntactic layer is formed of various system control and management programs and features, which facilitate interaction between the devices connected to the different kind of networks, such as networks protocols, error correction, hand-shaking, etc. The semantic layer contains the information and datasets in the user's computer terminals as well as different user-administered and controlled functions. The service layer is the heart of the entire network. It contains administrative services, ICT-services, security services, IT based manufacturing services, supply and logis-tics services. The cognitive layer portrays the user's information-awareness environ-ment: a world in which information is being interpreted and where one's contextual understanding of information is created.

9.2.2   Structure of an Electricity Organization's Cyber Environment

Organization's operational systems and supply chains are complex systems of systems characterized by a conglomeration of interconnected networks and depen-dencies. The general networks and working processes involved in the operation of an organization can be illustrated with a logistics framework that comprises a supplier network, a production process, a client network, and information and material flows that connect them [24].

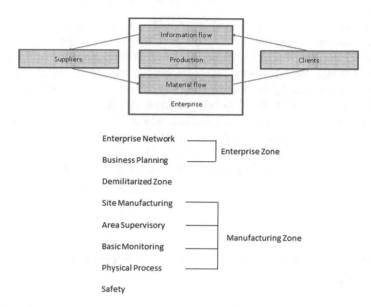

Fig. 9.3 Logistics framework of an electricity company (adapted) and common IT and industrial automation systems [3] (adapted from [14])

Information Technology (IT) systems are part of a company's infrastructure and thus constitute a significant part of the operations that support a company's core processes. Organization-level IT systems are related to administration and to the management of information and material flows in the network. The production level includes industrial automation and control systems (ICS). Figure 9.3 presents the structure of an organization's logistics framework and common ICT/IT and ICS systems.

The highest levels of IT system hierarchy include the general information systems of administration and the enterprise resource planning (ERP) system (Enterprise Zone). The top level of a typical ERP system includes overall process management by, for example, guiding the production volume. It also covers the restocking of raw materials, storing, distribution, payment traffic and human resources. If needed, between ERP software and control rooms there may be a manufacturing execution system (MES), which makes it possible to transfer the information obtained from the control room to the ERP system.

The ICS systems of production within an electricity organization comprise their own hierarchy levels (Manufacturing Zone). Based on their control systems and network structure, the Finnish Automation Society roughly classifies ICS systems into the following groups [2]:

1. Supervisory Control and Data Acquisition Systems (SCADA),
2. Programmable Logic Control (PLC),
3. Distributed Control Systems (DCS).

The transmission network of the Finnish national grid is owned by Fingrid Oyj. The distribution network consists of dozens of enterprises, and electricity is produced by about 120 enterprises and 400 power plants in different parts of the country. The power system structure is thus highly decentralized. Every organization of it is responsible for managing its own working processes. From the perspective of the entire power system of the critical infrastructure, the major threats to cyber security concern the transmission and distribution networks, switching and transforming substations, and power plants. A decentralized structure limits the potential consequences of these threats in the power system. On the other hand, decentralized power system requires good overall cyber security management for all of the electricity organizations as well as the capability to manage and control continuity of business processes in the cyber environment.

9.3 Main Cyber Security Threats in an Electricity Organization

When evaluating the role of electricity production systems in the cyber world as well as the factors that affect their cyber security, it is of primary importance to be aware of the most central features of the systems. For instance, the distributed industrial automation systems (DCS) used in controlling production processes can be characterized by saying that their operation is highly established and that their life cycles are long compared with other IT systems in a company. The life cycles of ICS systems can even be several decades, as far as the basic systems are concerned. Moreover, the structure of the basic systems is changed infrequently. The changes are mainly carried out as system life-cycle updates in connection with larger maintenance or alteration works.

The resources of ICS systems are also restricted, which is why it has not been possible to use typical technological information security solutions or cryptographies in them. Their user organizations are properly trained for their tasks and thus familiar with the devices as well as with the operating principles and operating environments of these devices. The data warehouses of ICS systems chiefly include process data, whereas administrative IT systems commonly include confidential business information. Unlike in administrative IT systems, no direct connection to the internet is usually needed in ICS systems. In the latter systems, IT devices are not used for purposes other than their decentralized tasks within the production process, its measurement and control tasks, and security functions. The monitoring of operations and staff in ICS systems is strictly controlled because of, for example, the availability and safety requirements of process operation [2].

Regarding the threats of information in ICT/IT and ICS environment terms such as, denial, loss, and manipulation are descriptive. Denial of useful information when it is needed is a condition which occurs only while the attack is active. Loss of information refers to sustained loss of an asset that continues after the active malicious interaction has ceased. Manipulation alters the information asset and can be either loud and easy to detect or subtle and longer sustained.

The aforementioned ICT/IT and ICS systems are part of the common cyber world, in which the primary risks are related to the loss of money, sensitive information and reputation as well as to business hindrance. Security solutions are hereby the key elements in risk management. The vulnerabilities behind the risks can be analyzed as insufficient technology in relation to attack technology, insufficient staff competence or inappropriate working methods, deficiencies in the management of organizations, and lacks in operating processes or their technologies. The most common motives of attackers are related to the aim of causing destructive effects on processes, making inquiries about process vulnerabilities, and anarchism or egoism. These attacks can even be carried out by state-level actors, but perhaps most commonly by organized activists, hackers or individuals acting independently [15].

Harmful measures to the systems of an electricity organization can be implemented by foisting mal- and spyware into the systems utilizing the staff; or they can include intruding or network attacks via wireless connections or the internet. The intruders' goals may be related to the prevention of network services, the complete paralyzation of operations, data theft or distortion, and the use of spyware. Components pre-infected with so-called backdoors or the programming of components intentionally for the purposes of attackers is also increasingly common in today's cyber world [15].

In the USA the security threats to the electric power system concern power plant logistics. They involve interfering and harming raw material supply routes, doing physical damage to transmission and distribution networks as well as to the transformer and switching substations between them, or performing cyberattacks to the control and regulation systems of the power grid [17].

Protecting the power system against threats implies measures taken based on risk assessment, and they ensure the availability of primarily digital information in the operating processes being examined. The measures are highly significant for the overall availability of the systems that support the processes. Availability plays a key role in achieving business results and promoting the reliability of activities. Further central goals include the reliability and content integrity of information within the processes and used by the processes. Overall trust should be built from these starting points, based on the target organization's realistic idea of its own capabilities to reliably manage the challenges involved in operations within the cyber world.

An organization can use their own capabilities to develop security in their cyber domain. By enhancing the capabilities of people, processes and technology, outcomes or effects applicable to the operational domain can be achieved [11].

9.4   Organization's Decision-Making Levels and System View

9.4.1   System-Level View of Organization's Cyber Security

In practice, all organizations of the power system operate in very complex, inter-related cyber environments, in which the new and long used information technical system entities (e.g., system of systems) are utilized. Organizations are dependent on these systems and their apparatus in order to accomplish their missions. The management must recognize that clear, rational and risk-based decisions are necessary from the point of view of business continuity. The risk management at best combines the top collective risk assessments of the organization's individuals and different groups related to the strategic planning, and also to the operative and daily business management. The understanding and dealing of risks are an organization's strategic capabilities and key tasks when organizing the operations. This requires, for example, the continuous recognition and understanding of the security risks at the different levels of the management. The security risks may be targeted not only at the organization's own operation but also at individuals, other organizations, and the whole society [23].

The Joint Task Force Transformation Initiative [23] recommends implementing an organization's cyber risk management as a comprehensive operation, in which the risks are dealt with from the strategic to the tactical level. That way, risk-based decision-making is integrated into all parts of an organization. In Joint Task Force Transformation Initiative's research, the follow-up operations of the risks are emphasized in every decision-making level. For example, in the tactical level, the follow-up operations may include constant threat evaluations about how the changes in an area can affect the strategic and operational levels. The operational level's follow-up operations, in turn, may contain, for example, the analysis of new or current technologies in order to recognize the risks to business continuity. The follow-up operations on the strategic level can often concentrate on an organization's information system entities, the standardization of the operation and, for example, on the continuous monitoring of the security operation [23].

In order to comprehensively build organization cyber security, organization leadership must define and guide actions at the strategic, operational and technical-tactical levels. The strategic level provides answers to 'why' and 'what' questions. The operational and tactical levels answer the 'how' question. The approach guided by questions ensures that the right things are done and that they are done in line with the set goal. The technical-tactical level must implement the goal-oriented activities defined at the strategic level, not create it. The organization's organizational capability in implementing the cyber security measures required by the technical-tactical level ultimately determines how the organization manages potential disturbance situations [20].

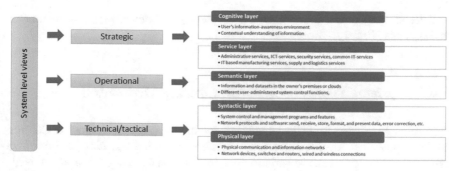

Fig. 9.4 System-level view of organization's cyber security

All three organization´s main decision-making levels can be integrated into the five-layer cyber structure in order to have a comprehensive system view of the organizations cyber security environment. It is a system thinking approach to the subsect of organization's cyber security. The principle is described in Fig. 9.4 [25].

The three decision-making levels are added to the five-layer cyber structure in order to have a comprehensive system view of the organizations cyber security environment. The following standards were utilized to support the idea to create a trust based cyber security architecture framework that is based on a comprehensive system view of the organizations cyber world.

NIST 800-39 publication places information security into the broader organizational context of achieving mission/business success. The objectives are as follows [23]:

1. Ensure senior leaders/executives recognize cyber security risks and manage such risks.
2. Ensure the risk management process is being conducted across the three tiers of organization, mission/business processes, and information systems.
3. Foster the awareness of cyber security risks so that mission/business processes are designed within designed within comprehensive enterprise architecture, and system development life cycle processes.
4. Help people in system implementation or operation understand how cyber security risks in systems affect the mission/business success (organization-wide risk).

The ISO/IEC 9000 standard family of quality management systems helps organizations ensure that they meet the stakeholders needs related to products or services. The main goal is the customers satisfaction. The fundamentals of quality management systems, including the seven quality management principles (customer focus, leadership, the engagement of staff, a process approach, continuous improvement, evidence-based decision-making, and relationship management) are the basic principles of the standard family [28].

The ISO/IEC 27,000 standard family provides recommendations for information security management system (integrated elements of an organization to establish policies and objectives and processes to achieve those objectives), risk treatments and controls [10].

9.4.2   Systems Views and Trust-Enhancing Measures

According to process statistical management theory (Statistical Process Control, SPC) all processes involve operational variation, variation was classified into two types according to its causes: variation due to common causes (or the system itself) and variation due to special causes (i.e. named and assignable causes). Special causes come from outside of the process and usually generate more variation in the process than the common causes. In uncontrolled processes, deviation as a result of both types occurs simultaneously [19].

In principle, Lillrank's theory on the causes of process variation can also be generalized to the processes of an electricity organization. The measures taken by organization leadership can be targeted at reducing variations resulting from both aforementioned types of causes. Proper planning and control of process performance reduce variation generated by random causes. At a general level, it is always recommended to aim at reducing this variation. If corporate leadership, in particular, concentrates too much on process changes resulting from random causes, it can lead to overreactions in process control due to the measures chosen. At its worst, this can lead to loss of control in managing the overall process. The actions of corporate leadership should indeed be targeted primarily at proactively preventing variation generated by special causes. Almost without exception, serious cyber security disturbances occurring in the operating process cause blackouts. They are not in the normal range of variation. Taking these special causes into account in planning and proactively implementing security activities reduces related risks and improves the overall reliability of the organization's operations.

Excellence of an organization can be measured in its ability to recognize both causes of variation in operational processes at the right time and with the right attitude. In cyber security operations, this is mandatory in order to handle complicated technologies and complex system environment. To promote the most significant cyber security culminates in systems thinking and the point of views of strategy, operational and technical/tactical measures need to have comprehensive cyber trust.

One of the most fundamental cybersecurity tasks of the organization's uppermost management is the continuous development and maintaining of the trust in operation as part of the national critical infrastructure. The strategic choices relate to the reputation of an organization. The management is required to make concrete strategic choices and to support and guide the performance of the chosen operations through the whole organization. An important task of the management is to take care of the adequate resourcing of operations. The chosen operations must be communicated extensively with the organization's personnel and other interest groups. It is important to create a cybersecurity assessment model for the needs of the uppermost management. With the help of that model, for example, other organizations can evaluate their cybersecurity level, become aware of their weaknesses and insufficient contingency planning, and take care of, at a minimum, the basics. The operations require strategic level decisions from the organization's uppermost management.

The operational level operations are used to advance strategic goals. Comprehensive security- and trust-adding operations require comprehensive cybersecurity management and maintaining the situation awareness of the cyber operational environment. Its starting point has to be the target's risk assessment, and the operation analyses carried out based on it. The operational level's concrete hands-on operations must be targeted at the confirmation of information security solutions and the composition of the organization's continuity and disaster recovery plans. The goal must be the continuous monitoring of the operational processes' usability, and the decision-making support in case of incidents that require analyzing and decisions.

At the technical/tactical level, the organization runs all operational processes and use such protection techniques in their ICT/IT and ICS systems that extend from the interface of the Internet and the organization's internal network right up to the protection of a single workstation or apparatus. These technical solutions make it possible to verify different harmful or anomalous observations. The typical technologies are related to security products such as network traffic analysis and log management (Security Information and Event Management, SIEM), firewall protection, intrusion prevention and detection systems (IPS and IDS) and antivirus. The situation awareness builds up to centralized monitoring rooms (Security Operations Centre, SOC). These technical solutions can be under the organization's control, or the service can be outsourced to the information security operator. A crucial goal is the situational awareness and protection of the business processes.

The key words of an organization's excellence are leadership, management, capabilities, and the measures to have continuous improvement actions. The cyber trust is related to the reliability of an organization. The summarized measures should be taken in all decision levels to increase an organizations cyber trust as illustrated in Fig. 9.5 [25]. The content of Fig. 9.5 is derived from the organization-wide risk management standard, NIST 800-39 [23] and perspectives from ISO/IEC 9000 standard family (seven quality management principles) and ISO/IEC 27000 standard family (information security management).

Building organization cyber security measures begins from the level of vision and strategy work. The visions created by leadership to enhance cyber trust are translated into strategic goals, operational-level actions, guidelines, and policy. The practical measures derived from the strategy are realized at the technological-tactical level. Organizational capability factors enable the success of the measures. Establishing measures that increase cyber world security and trust is primarily the responsibility of corporate leadership. Integrating the necessary measures with the idea of ensured business activities increases their significance and benefits through better processes for the entire organization, interest groups and society.

The continuous improvement of activities related to cyber security enhance the organization's capability to proactively prevent disturbances and tolerate potential changes to the operational processes. The competence and the possibilities open to fully influence the organization will help develop the overall operations of the

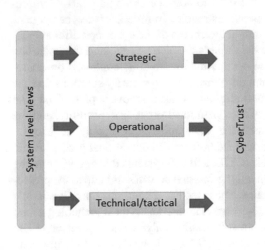

Leadership, management

- Visio, strategy, culture, values
- Reputation, responsibilities
- Risk level approval
- Relationship
- Resourcing, commitment

Cyber security management

- Risk management
- Policy
- Situational awareness
- Continuity/recovery plans

Cyber secure processes

- Protected processes
- Control mechanisms
- CS-services
- CS-products
- Availability, reliability, integrity

Continuous improvement

- Performance measurement
- Feedback
- QM tools
- Benchmarking

Capabilities

- People
- Processes
- Technology

Fig. 9.5 Measures increasing an organization cyber trust

organization. The continuous development of activities and staff competence support the measures taken at the strategic, operational and technological-tactical levels.

The major cyber security excellence of the organization within the system thinking principles of leadership, management, process, and measures support that trust is enhanced and maintained at all levels of business activity. Comprehensive attitude to increase cyber trust, together with the development of capabilities related to cyber activity, also improve a company's competitive edge.

9.5   Implementing Measures to Enhance Cyber Trust

In this section, enhancing cyber trust is underlined. It is needed to understand risks, risk categorization, and system thinking approach in an organization. System thinking enables a trust-based cybersecurity architecture framework for an organization. It is then possible to make risk assessment for the whole organization.

There are still unidentified risks and therefore it is necessary to implement opera-tions that add the resilience of the organization. They can be developed by utilizing preparedness process. At the end of the section, the PDCA (Plan, Do, Check, Act) method is recommended for developing cyber security activities and capabilities in an organization.

9.5.1 Trust-Enhancing Measures

In general risk assessment or classifications, it is assumed that every event is already identified, but it does not help to characterize unidentified risks. In the cyber world it is obvious that ICT/IT and ICS systems vulnerabilities can cause various risks. It is supposed that there is a very rare but well-known event. People know its identity but do not know if it will really happen. This event should be classified as unknown, because the occurrence and also the impact are uncertain.

In order to distinguish identified risks from unidentified risks, the level of knowl-edge about the risk occurrence should be about being able to identify the risk in advance or not. The level of knowledge about the impact of risk should include occurrence as well as impact since either occurrence or impact of a risk can be uncertain. A schematic structure of the risk categorization is shown in Table 9.1. In this model, events are categorized by "identification" and "certainty" [13].

As a solution to response to all listed challengers, a model of comprehensive system-level view in cybersecurity management is needed. It would be consisting of an organization's five-layer cyber structure and the strategic, operative and tech-nical/tactical level approaches. These approaches include a survey of measures adding trust in the organization. An organization's architectural cybersecurity frame-work is constructed of these components and can be put to use in developing further steps in cybersecurity management on all levels of decision-making (strategic, operative and technical/tactical). Three practical measures for development are:

1. First, embedding high level and new technological solutions into the organiza-tion's piers cyber security structure (known known, unknown known),
2. Second, drafting comprehensive cyber security risk assessments (known unknown) and

Table 9.1 Schematic structure of the risk categorization [13]

	Certain (known)	Uncertain (unknown)
Identified (known)	Known-known (identified knowledge)	Known-unknown (identified risk)
Unidentified (unknown)	Unknown-known (untapped knowledge)	Unknown-unknown (unidentified risk)

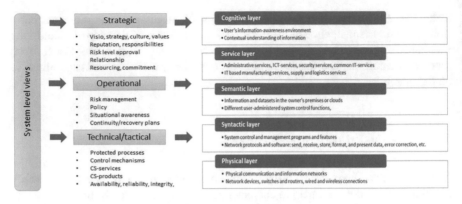

Fig. 9.6 Trust based cyber security architecture framework

3.   Third, preparing contingency plans in order to improve an organization's resilience (unknown unknown).

Integration of the measures increasing an organization cyber trust (Fig. 9.5) and the system thinking approach to organization five-layer cyber structure (Fig. 9.2) makes it possible to have a trust based cyber security architecture framework (Fig. 9.6).

BusinessDictionary.com defines a capability in general as the "measure of the ability of an entity (department, organization, person, system) to achieve its objectives, especially in relation to its overall mission", and in quality perspective as the "total range of inherent variations in a stable process" [4]. Dickenson and Mavris [5] define a capability as "the ability to achieve a desired effect under specified standards and conditions through combinations of ways and means to perform a set of tasks" and also "the ability to execute a specified course of action". Thus, the capability of an organization can be seen also as an ability to learn from its experiences and use relevant information to improve the cyber security processes. Organization capabilities support actions of the architecture framework.

The process approach promoted by ISO/IEC 9001 systematically identifies processes that are part of organization quality system. Related to the quality management system, the PDCA cycle is a dynamic cycle that could be implemented in each process throughout the organization. It combines planning, implementing, controlling and continual improvement. That way an organization would achieve continual improvement once it implements the PDCA cycle [1].

9.5.2   Risk Analysis

The vision for achieving a company's goals is the point of departure for trust-enhancing measures. The definition of strategy derived from the vision guides the actions taken in order to achieve the goals. At the first stage, it is most practical to facilitate the definition of strategy by performing risk analysis on cyber threats.

When examining an electricity company, the targets of risk analysis are determined by the company's logistics framework and its IT processes. An electricity company's systems include a fuel logistic and feed system, a production system and its support processes, and the electricity distribution system. Due to all the aforementioned components being needed in the operation of an electricity company, their mutual dependence as well as operations management and monitoring are crucial for the success of overall production. In managing cyber security, the different functions of the logistics framework must be treated as subjects of equal value.

If an organization is familiar with the factors affecting the operation of processes, their most vulnerable points in the cyber world and the cyberattack methods most probably threatening the processes, it possesses the most relevant information for creating protective plans for potential treats. Vulnerability analysis against attack methods is a systematic tool for identifying and assessing risks related to process operations as well as for choosing the most suitable measures to enhance cyber security trust. The analysis provides a comprehensive overall picture of the needs to develop the processes.

The risk management standard ISO/IEC 27,005 of the ISO/IEC 27,000 standard family includes the risk management process (presented in Fig. 9.7), which can be utilized in analyzing the risks involved in the electricity production process. Risks can be classified in a treatment process according to Fig. 9.7. The aim should be to reduce or completely eliminate the highest risks using different measures. Corporate leadership prioritizes the highest risks to the processes based on risk identification and chooses the measures that best suit risk management and development of proactive measures in the cyber environment. Less significant risks can be retained, aiming to

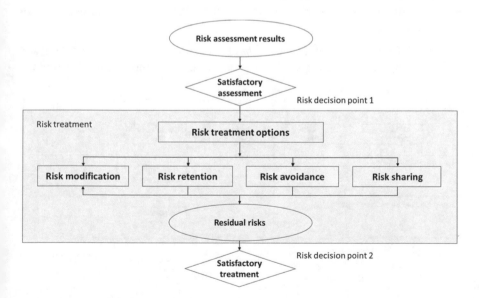

Fig. 9.7   ISO/IEC 27,005: risk management [27]

manage them. Risk transfer in the cyber environment of an electricity company can be possible through its logistics network. This means that responsibility questions must be resolved using a clear internal operations model within the network.

9.5.3 Resilience Adding Operations

Resilience adding operations can be developed by utilizing preparedness planning. Linkov et al. [21] introduce a resilience matrix framework (later the "Linkov model") that can be used for this planning. It combines the four stages of a system (1) plan/prepare, (2) absorb, (3) recover and (4) adapt with the four domains of a system (1) physical, (2) information, (3) cognitive and (4) social. Later on, Linkov et al. [22] apply their model further to cyber systems. Their purpose is to develop efficient metrics to measure the resilience of cyber systems [21, 22].

The resilience management process (Fig. 9.8) was developed for the electricity company in the paper [26]. It can be linked to the management system of an organization. When creating the resilience management process, the following procedures would be utilized:

- the definition of the target organization's cyber-physical systems (ICT/IT and ICS systems),
- SWOT analysis (Strengths, Weaknesses, Opportunities and Threats),
- the Linkov model,
- Open Source Intelligence (OSINT), and
- the electricity organization's strength in utilizing its own operating networks for data collection.

After defining the target organization, the cyber-physical systems (ICT/IT and ICS systems) related to its operational processes are recognized and placed in the

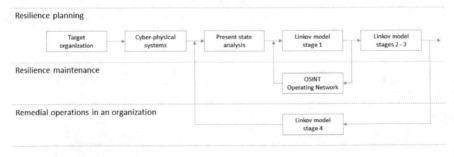

Fig. 9.8 Implementation process of the resilience operations

systems' cyber structure described in Sect. 2. After that, SWOT analysis can be applied to the organization as a theme interview by taking into consideration the cyber structure in order to describe the present state of the organization's cyber security situation. This enables the drafting of the resilience basic plan during the normal conditions (stages 1–3 of the Linkov model) for all the domains (physical, information, cognitive and social). Stage 4 of the Linkov model includes all the aforementioned domains also, but their final content must be defined based on the aftermath of the possible disturbance situation. The purpose is that the organization learns from the disturbance situations as efficiently as possible. The operations of the organization are developed by repeating SWOT analysis. As a result, the plans are updated for each stage of the Linkov model as part of a repairing operation. The preparedness planning of the normal conditions (stage 1) should continuously be maintained with the help of the OSINT model and by utilizing the company's own data collection channels, such as operating networks, in the update process.

The implementation process of the resilience operations serves all the decision-making levels of an organization. In SWOT analysis, the analysis of the organization's performance and its operational environment on the whole, supports strategic planning. It also produces information to other decision-making levels in learning and problem recognition, evaluation, and development of operational processes. The Linkov model serves the planning and maintenance of the organization's operational continuity management, which supports the operation of the operational and technological-tactical levels.

The basis for trust adding operations is the envisioning of the company's operations in order to achieve its goals. It is made possible with the strategy definition derived from the vision. The electricity company's operational business processes include systems such as fuel logistics and input system, production system and its support processes, and electricity distribution operation. Due to all the aforementioned components being needed in the operation of an electricity company, their mutual dependence, and the control and supervision of functionality solve the succeeding of the whole operation. In order to achieve a successful cyber security management, the different operations should be considered as equal.

The Linkov model and its different stages especially suit the operational and technical-tactical level preparedness planning, and ensures the continuity of operation. Considering the structure of the previously described cyber-physical systems, it is possible to find those targets from the operation of an electricity company that are in a central position in preparedness planning. The company-specific content of the operations has to be based on the present state analysis carried out before using the Linkov model, and on the situational awareness got from that in the form of target organization's strengths, weaknesses, possibilities, threats and their mutual relations. SWOT analysis from organizations cyber security capabilities gives a good overview of targets to be achieved in operations continuity enchantment and management against threats. Based on the analysis, the related needs of each power systems (electricity production) organization can be planted on the planning stages of the Linkov model (see Table 9.2).

Table 9.2 Research results planted on the Linkov model

	Plan/Prepare	Absorb	Recover	Adapt
Physical	Technical situational awareness Segmentation Alternative resources	Recognition of disturbances, their scope and impacts Protection of sensitive information Deployment of alternative resources Isolation of disturbance	Maintenance of situational awareness Ramp-up Testing	Updates
Information	Classification and prioritization of critical systems Business impacts Preparation of sensitive information protection Communication plans	Documentation Informing of authorities and stakeholders	Documentation Informing of the press	Aggregation of documents
Cognitive	Perception of situational awareness Scenarios and models Situational management Resourcing Training and benchmarking Feedback system	Analysis of situational awareness Additional resources Prioritization Censor information	Allocation of expertise Collection of data and log information	Log analysis Impact analysis Situation analysis Feedback analysis System updates Continuous improvement
Social	Naming of stakeholders' contact persons Training for exceptional situations	Informing about operations	Informing about operations	Staff training Informing about development operations Update of stakeholder information

The following operations of the plan/prepare and absorb stages within the physical domain of the Linkov model were recognized:

- taking care of the functionality, supervision, and control of the technology,
- planning of the system isolation and needed operational segments, and
- planning of the alternative networks and routes.

In case of a disturbance situation, firstly, the situational awareness of the incidence, its nature, distribution, and scope are clarified, as well as its impact. After that, the plans are used for their needed parts. In the recovery stage, the cleanliness and functionality of the systems is ensured for all their parts. Then, the comprehensive ramp-up of the machines is guided through. The adaptation stage is determined by the experiences received from the incident, but at least the technical protection operations must be considered carefully.

The documentation planning is emphasized in the operations of information domain, by paying attention to the situation-specific documentation itself, and the critical operations and related requirements has to be documented already in the planning stage. The documentation both serves the operation in a disturbance situation and enables the information documentation during the disturbance situation and in a recovery stage, so that the utilization of situation-specific experiences and learning in the adaptation stage is made possible. The informing of essential stakeholders and different authorities must also be included in each stage.

In our case study, the plan of cognitive domain grew the most of all domains. Thus, it can be seen as significant for management, in building the situational awareness, in continuity management, in prioritizing the operations, and in managing and controlling different resources, including services. All these operations play a decisive role in a disturbance situation, in the recovery stage and in the adaptation stage when utilizing the knowledge gained from the previous stages.

The planning stage of the social domain consists of more specific communication plans than in the information domain, including the named contact persons, and both internal and external interest groups. The wide-scale situation-specific informing in the different stages results from the planning of the social domain. In addition, the planning of the social domain includes the whole staff training in managing all the different stages.

9.5.4 PDCA Method as a Tool for Developing Activities

An organization's policy demonstrates that its leadership is committed to implementing strategic measures. In the business world, general strategic measures are mainly targeted at promoting the business activities, which means that taking cyber security into account as part of the overall strategy supports the business development targets. Cyber security as part of company policy is a way of communicating to staff and interest groups on the necessity and significance of development projects. Operational goals are formed as processes derived from the policy, whereby risk analysis has been considered. In order to create the measures, the organization must have a systematic approach to developing its operations.

The ISO/IEC 9000 Standard recommends the PDCA (Plan, Do, Check, Act) method for a systematic development of an organization's activities. The method is based on a cycle of four development phases. The first phase (Plan) comprises planning, during which the subject is analyzed, and alternative measures are created based on the analysis. In the realization (Do) phase, the chosen measures are put into practice. Thereafter the functionality, efficiency and appropriateness of the chosen measures are checked in practice (Check). At the last (Act) phase of the cycle, the chosen measures are improved, if necessary, and established as standard practice. After the cycle has been implemented once, one will return to the first phase and start a new cycle with improvement actions based on a new situation analysis. Development can thus proceed as an endless process, in which a new level of activities is achieved after each cycle. The method is based on the idea of continuous learning and continuous improvement of activities.

The measures during one round of the cycle usually require a lot of planning, so sufficient time should be reserved for them. It is important to select the measures in relation to the resources needed for their implementation. The maturity level of the organization's development activities affects the evaluation of the implemented measures. When developing cyber security, at an initial stage the aim can be to recognize the need for cyber security management and to define cyber security risks for business. Hereby, the PDCA cycle may comprise the administrative actions most necessary according to risk assessment, such as a coherent information security policy in production, practical guidelines for maintaining information security in production, and potential preliminary system-specific cyber security checks. The targets for development must later be chosen according to risk prioritization.

The following is one possible process model for developing cyber security management with the PDCA method:

PLAN, planning phase

1. Choose the target for development based on risk assessment

   - present state
   - schedule and goal

2. Create a picture of the current cyber security situation

   - earlier measures, knowledge from partners
   - disturbances in the branch resulting from special causes

3. Analyze the problems and define corrective actions

   - identify relevant potential harms caused by the disturbances
   - choose the measures available to anticipate and manage the situation.

   DO, implementation phase

4. Implement the chosen measures

   - choose the actors responsible for implementation
   - organize information and training for staff

   CHECK, checking phase.

5. Check the impact of the measures

   - compare the results with the goals

- return to phase 3 if the goals have not been achieved

ACT; regularize the measures.

6. Regularize the chosen development measures

    - update necessary guidelines, technological solutions and services
    - continue staff training

7. Draw conclusions and make plans for the future

    - continue development according to new goals
    - update threat and risk analyses.

In this section, we have described one way of launching primary basic solutions related to cyber security management in an electricity production organization. These first steps provide a basis for later development activities and continuous improvement in a dynamic cyber environment.

9.6   Conclusion

The national power grid and its electricity production are part of a country's critical infrastructure—the operation of a modern society is based on a reliable electric power system. Ensuring the availability and reliability of processes in electricity organizations in all environments is vital for the efficient functioning of critical infrastructure. Therefore, the measures taken in electricity organizations in order to manage and control the cyber security of processes are an essential component of the reliability of production.

The major cyber environment risks within the processes of an electricity organization require that trust be enhanced and maintained at all levels of business activity. Comprehensive measures to increase cyber trust, together with the development of cyber security excellence of organization related to cyber activity, also improve a company case its competitive edge.

The initial measures taken to develop cyber security management and trust can be summarized and prioritized as follows:

1. It is ensured that the organization sees cyber security measures as strategic goals and that sufficient resources are allocated to the chosen measures.
2. Risk assessment is performed, the organization's policy is updated to meet the requirements of cyber security and the resilience management process is implemented.
3. The primary trust-enhancing development measures needed based on risk assessment are taken at the first development phase, using the PDCA method.
4. A continuous process is formed of the development actions by choosing the subjects of the next cycle, and the PDCA development cycle is repeated. This procedure will provide the organization with a culture of continuous learning

and improvement. The organization's capabilities and competitive advantage are enhanced.

5. The impact of the measures is monitored as part of the organization's audit and management procedures (e.g., as a part of the ISO/IEC 9001 Standard procedures).

References

1. 9001Quality (2020) The Plan Do Check Act (PDCA) cycle. 9001Quality http://9001quality.com/plan-do-check-act-pcda-iso-9001/. Accessed on 27 Jan 2020
2. Automaatioseura (2010) Teollisuusautomaation tietoturva: Verkottumisen riskit ja niiden hallinta. Suomen Automaatioseura ry, https://www.automaatioseura.fi/site/assets/files/2157/sas29_teollisuusautomaation_tietoturva.pdf
3. Bowersox D, Closs D, Helferich O (1986) Logistical management: a systems integration of physical distribution, manufacturing support, and materials procurement, 3rd edn. Macmillan, New York
4. BusinessDictionary (2020) BusinessDictionary.com. Online business dictionary. Available at http://businessdictionary.com/
5. Dickerson CE, Mavris DN (2010) Architecture and principles of systems engineering. CRC Press
6. ET (2020). Sähköntuotanto. Energiateollisuus ry (ET), https://energia.fi/energiasta/energiantuotanto/sahkontuotanto
7. EU (2009) Protecting Europe from large scale cyber-attacks and disruptions: Enhancing preparedness, security and resilience. 52009DC0149, Communication from the Commission to the European Parliament, the Council, the European Economic and Social Committee and the Committee of the Regions, EU Commission, https://eur-lex.europa.eu/LexUriServ/LexUriServ.do?uri=COM:2009:0149:FIN:EN:HTML
8. Fingrid (2020) Electricity system of Finland. Fingrid Oyj, https://www.fingrid.fi/en/grid/power-transmission/electricity-system-of-finland/
9. Finlex (2018) Valtioneuvoston päätös huoltovarmuuden tavoitteista. Finnish Government, https://www.finlex.fi/fi/laki/alkup/2018/20181048
10. ISO/IEC (2018) ISO/IEC 27000:2018: Information technology, security techniques, information security management systems, overview and vocabulary. International Organization for Standardization (ISO), https://www.iso.org/standard/73906.html. Accessed 27 Jan 2020
11. Jacobs PC, von Solms SH, Grobler MM (2016) Towards a framework for the development of business cybersecurity capabilities. Business Manage Rev 7(4):51–61
12. Kananen I (2013) Sähköjärjestelmä yhteiskunnan toimivuuden perustana. Seminar presentation, Fingridin käyttövarmuuspäivän seminaari 2.12.2013, Fingrid, http://wms.magneetto.com/webcasts/hd1/fingrid/2013_1202_kayttovarmuuspaiva_02_Kananen/Attachment/02_Kayttovarmuuspaiva_021213_Kananen.pdf
13. Kim SD (2012) Characterizing unknown unknowns. In: Paper presented at PMI® Global Congress 2012—North America (Vancouver, 2012). Project Management Institute, Newtown Square, PA
14. Knowles W, Prince D, Hutchison D, Disso JFP, Jones K (2015) A survey of cyber security management in industrial control systems. Int J Crit Infrastruct Prot 9:52–80
15. Lehto M (2015) (2015). Phenomena in the cyber world. In: Lehto M, Neittaanmäki P (eds) Cyber security: analytics. Technology and Automation. Springer, Cham, pp 3–29
16. Lehto M (2018) The modern strategies in the cyber warfare. In: Lehto M, Neittaanmäki P (eds) Cyber security: power and technology. Springer, Berlin, pp 3–20

17. Lewis TG (2015) Critical Infrastructure protection in homeland security: defending a networked nation. eBook, 2nd ed, Wiley
18. Libicki MC (2007) Conquest in cyberspace: national security and information warfare. Cambridge University Press, New York
19. Lillrank PM (1998) Laatuajattelu: Laadun filosofia, tekniikka ja johtaminen tietoyhteiskunnassa. Otava
20. Limnell J, Majewski K, Salminen M (2014) Kyberturvallisuus. Docendo, Jyväskylä
21. Linkov I, Eisenberg D, Bates M, Chang D, Convertino M, Allen J, Flynn S, Seager T (2013) Measurable resilience for actionable policy. Environ Sci Technol 47:10108–10110
22. Linkov I, Eisenberg D, Plourde K, Seager T, Allen J, Kott A (2013) Resilience metrics for cyber systems. Environ Syst Decis 33(4):471–476
23. NIST (2011) Managing information security risk: organization, mission, and information system view. Joint Task Force Transformation Initiative, NIST Special Publication 800-39, National Institute of Standards and Technology (NIST), Gaithersburg, MD
24. Pöyhönen J, Lehto M (2017) Cyber security creation as part of the management of an energy company. In: ECCWS 2017: proceedings of the 16th European conference on cyber warfare and security, pp 332–340. Academic Conferences and Publishing International, Reading
25. Pöyhönen J, Lehto M (2020) Cyber security: trust based architecture in the management of an organization security. In: ECCWS 2019: proceedings of the 18th European conference on cyber warfare and security, pp 304–313. Academic Conferences and Publishing International, Reading
26. Pöyhönen J, Nuojua V, Lehto M, Rajamäki J (2018) Application of cyber resilience review to an electricity company. In: ECCWS 2018: proceedings of the 17th European conference on cyber warfare and security, pp 380–389. Academic Conferences and Publishing International, Reading
27. SFS (2012) SFS-käsikirja 327: Informaatioteknologia, turvallisuus, tietoturvallisuuden hallintajärjestelmät. Suomen standardoimisliitto SFS ry, Helsinki
28. SFS (2016) Johdanto laadunhallinnan ISO 9000 —standardeihin: Kalvosarja oppilaitoksille. Suomen standardoimisliitto SFS ry, Helsinki, http://slideplayer.fi/slide/11133323/
29. WEForum (2020) The global risks report 2020. Insight report, 15th ed, World Economic Forum, http://www3.weforum.org/docs/WEF_Global_Risk_Report_2020.pdf

Chapter 10
Maritime Cybersecurity: Meeting Threats to Globalization's Great Conveyor

Chris Bronk and Paula deWitte

Abstract This chapter addresses the issue of cybersecurity in the global maritime system. The maritime system is a set of interconnected infrastructures that facilitates trade across major bodies of water. Covered here are the problem of protecting maritime traffic from attack as well as how cyberattacks change the equation for securing commercial shipping from attack on the high seas. The authors ask what cyberattack aimed at maritime targets—ships, ports, and other elements—looks like and what protections have been emplaced to counter the threat of cyberattack upon the maritime system.

Keywords Cybersecurity · Maritime commerce · International security

10.1 Introduction

International maritime operations remain a primary vehicle of globalization. More than 80 percent of the world's cargo is carried by ship. While mobile phones and other small, lightweight, highly-valuable items may go by air, almost everything else traveling from continent-to-continent is transported by maritime vessels. Shipping remains a fundamental component to global trade, wherein ports large and small serve as the departure and arrival point for containerized, bulk, and liquid cargo.

Transport by ship has become a deeply automated process in which computers are employed in everything from navigation and propulsion to cargo handling and customs. Increasingly, the computers involved in maritime cargo operations are also networked, largely employing the same protocols as other Internet-based forms of communication [1]. This rise in networked computerization in ships and in systems that support shipping from onshore present new opportunities for malicious parties to disrupt maritime commerce in ways that piracy and open naval hostilities cannot.

C. Bronk (✉)
University of Houston, Houston, TX, USA
e-mail: rcbronk@uh.edu

P. deWitte
Texas A&M University, College Station, TX, USA

© The Author(s), under exclusive license to Springer Nature Switzerland AG 2022
M. Lehto and P. Neittaanmäki (eds.), *Cyber Security*, Computational Methods in Applied Sciences 56, https://doi.org/10.1007/978-3-030-91293-2_10

Cyberattacks may be launched across global distances and can have potentially devastating impact. They can't necessarily be steered around as with threats like regional conflict or piracy. Nonetheless, we argue the threat of cyberattack is real and prompts us to answer several questions. First, we ask how does cyberattack threaten the global system of maritime enabled commerce? Second, we investigate the cyber threat to maritime system. In our third and last thrust of inquiry, we attempt to identify what norms, standards, practices, and law may be needed to protect the system of global maritime commerce from cyberattack as well as practical prescriptions for US public policy as well as international policy.

Before moving on to discussion of international security antecedents to cyber issues found in this area, there is a matter of definitional housekeeping. The authors prefer to use the term *maritime system* to define the operational space in which shipborne and port activities take place, principally for commercial purposes. US Coast Guard (USCG) and Department of Homeland Security (DHS) documents describe a Maritime Transport System (MTS) that encompasses much area where cybersecurity issues are to be found in the maritime system, but not necessarily all of it. DHS's definition extends to ports and coastal authorities but not necessarily ships plying the seas far from US territory.

For centuries, states have pursued control of the seas, often in competition or conflict with one another. While two great powers, the United Kingdom and the United States, have exerted much effort to control the seas and allow for the free flow of trade on the world's oceans, other powers have contested their (mostly) benevolent hegemony for the seas [2]. Ahistorical perspectives on maritime security are likely foolish while thinking about cybersecurity issues as cyberattacks may well achieve results previously ascribed to warships, privateers, or pirates on the high seas.

10.2 Seaborne Commerce and Sea Control: Lessons from the Last Century

Disruption of shipping activity is often fundamental component of naval conflict [3]. In both world wars, submarine campaigns represented a mortal threat to multiple powers, but not least the island nations of the Axis and Allied coalitions. In the Second World War, it could be argued that one submarine campaign, Germany's in the Atlantic, ultimately failed, though at great cost to the Allies, while another, the United States' campaign against Japanese merchant shipping, was a success. For decades after the war, the United States and its NATO allies prepared for a clash of naval forces and doctrine in the North Atlantic.

There the issue was to what degree NATO's naval forces—surface ships, submarines, and aircraft—could protect a massive reinforcement from North America to Europe. It was assumed that the Soviet Union would sortie hundreds of submarines and surface warships to disrupt the Alliance's maritime link. How well the respective strategies of NATO and the Soviet Union would have fared remains a well-educated guess, but few estimates were particularly rosy with regard to the fortunes of merchantmen on the North Atlantic in a potential war with the Soviet Union [4]. Nonetheless, in 1986 Mearsheimer argued, "the Navy's main value for deterrence lies in the realm of sea control, where protection of NATO's sea lines of communication (SLOCs) might matter to Soviet decision-makers contemplating war in Europe" [5].

While no major war between East and West came to pass between 1945 and 1989, regional conflicts did have an impact on international maritime commerce. Perhaps most important of them was the closure of the Suez Canal from 1967 to 1975. Shut at the onset of the June 1967 Six Day War, Israeli and Egyptian troops faced off across the 120 mile-long waterway between the Mediterranean and Red Seas until 1973s Three Day War. The canal was ultimately reopened as relations improved between Cairo and Tel Aviv in 1975. The canal's closure increased the distance of a sea journey from Mumbai to London from 6200 nautical miles to more than 10,800 nm. Feyrer argues persuasively how closure of Suez led to significant reduction in trade between nations on either side of it [6, 7].

Despite being the last naval conflict of its kind, the 1982 War over the Falkland Islands had minimal impact on international seaborne commerce, during the war between Iran and Iraq from 1980 to 1988 merchant ships involved in the export of oil from both belligerents were attacked more 450 times [8]. Both sides sought to interdict their opponent's capacity to sell oil internationally thereby acquiring funds to continue the war effort. US intervention in the Persian Gulf during the conflict ultimately led to the crippling of two warships, the *Stark* (hit by Iraqi missiles) and *Samuel B. Roberts* (which struck an Iranian mine). US protection of commercial shipping illustrated that such duty remained dangerous and unpredictable, however punitive attacks on Iranian forces after the damaging of the *Roberts* largely curtailed Iran's capacity to harm US or allied commercial vessels.

Absent major international conflict, disruption to maritime commerce has arisen in new forms. Somalia's incapacity to exert control over her littorals during the country's slide to largely ungoverned status in the 1990s led to a resurgence in maritime piracy in the 2000s. Regional warlords and bandits engaged in a significant piracy campaign, involving the hijacking of dozens of vessels, some held for periods of years for ransoms in excess of $1 million. However, coordinated international response as well as military operations onshore have had a desired result of reducing the Somali pirate problem to a negligible one [9].

10.3   Cybersecurity and the Maritime System

When we think of piracy on the high seas, it is a mostly unsophisticated endeavor. A few men, armed with rocket propelled grenades and Kalashnikovs, possessing boarding gear and a fast boat are usually all that is needed to highjack a vessel displacing 50,000 tons or more (naval vessels excepted). Ransoms for these hijacked ships has reached well into the millions of dollars.

How cyberattack may disrupt shipping is different. To get our arms around cyber threats, we need to begin using some imagination as to what is requisite for a pulling off a cyberattack that either steals something of value or does damage to a maritime vessel or other piece of infrastructure. The authors like to consider the beginning point of thinking about such attacks as the *bad guy-ology* of the attacker.

What does this mean? When we speak of bad guys in cyberspace, we are talking about people who can act alone, in small groups or large ones, supported or deployed by nation states or not. They craft source code for sophisticated tooling, penetrate computer networks, and do a lot of the same data management work as most Internet enterprises (servers, databases, means of communication, etc.) also toil in [10].

We have witnessed reports of computer security breakdown in the face of increasingly sophisticated attack for more than 20 years now. This has been going on for a long time. Hackers and, equally importantly, hacker groups have been around for a while and they have evolved within both domestic and international political spheres. They have power. A former member of the Cult of the Dead Cow (cDc) hacker organization ran a Democratic campaign for one Texas's US Senate seats in 2018.

Concurrently, there has been a convergence of politics and cyberattack that extends from "kinetic" hacks like the *Stuxnet* campaign launched against Iran's nuclear enrichment program and the information warfare operation exemplified by the email breach at the US Democratic National Committee by foreign, state-supported hackers. Those individuals, in the employ of Russia leaked stolen data to the Wikileaks organization during the 2016 US Presidential election. Both these episodes illustrate how important or impactful cyberattacks have been and what breaks when they occur.

Thus when we begin thinking about cyber vulnerabilities in the maritime sector, we need to focus firstly on what happens when things break [11]. There is an exercise afoot in which mapping vulnerabilities to components are linked to pieces of information and computing infrastructure. We may not need to worry about a pump that can only be turned on by a human being, but one operated by computer and interconnected by network, we do worry about.

Where cybersecurity concerns come into play is after identifying things that could go wrong, i.e. that also are very detrimental to safety or continuation of operation. There need to be many people thinking about what can go wrong in shipping as with any piece of critical infrastructure. It may seem simple, but the computerization of it is not.

Furthermore, it must be reminded just how important maritime trade is to the global economy and what disruptions to it may produce in global manufacturing or energy supply chains. Hopefully this answers the question of why cybersecurity in the maritime system is important. It's important because of how closely seaborne trade tracks with world GDP and other economic indicators. Trade on the oceans exceeds 10 billion tons per year [12]. With many nations highly dependent on forms of import or export, disruption of those flows could be potentially useful to adversaries or enemies. In a time of increased economic conflict, could the cyber weapon not be employed against the maritime system? Of course, and it already has.

The *Stuxnet* or *Shamoon* of the maritime system, thus far is the cyberattack against Denmark's Møller-Maersk, the world's largest container ship operator. But Maersk is not just the biggest in container shipping, it also operates the ports themselves, including the Port of Los Angeles, the busiest port by container volume in the US. Maersk was also the victim of the most expensive and destructive cyberattack against any form of logistics company in June 2017.

The company's IT infrastructure was walloped by the propagation of the *NotPetya* malware across its computer networks. It was crippled by the attack, which shut down port operations—cranes, gates, freight forwarding instructions, and many, many other processes, at 17 of the company's 76 ports. After the attack, "For days to come, one of the world's most complex and interconnected distributed machines, underpinning the circulatory system of the global economy itself, would remain broken" [13].

With Maersk's woes as a backdrop, thinking about the bad guy-ology of cyber-attack in the maritime system is shaped by two avenues for action. First is beginning with a desired impact of an attack, perhaps misidentifying cargo containers to facilitate smuggling. The second relates to systems' exposure to attack and how vulnerabilities may be exploited to produce a desired effect. So, we can start with two general types of questions. One is, "If I want to disrupt x with some form of cyberattack, how do I do it?" But also important is, "If I can see a vulnerability on resource y, what can I do with it?".

Returning to the Maersk case, it has been largely judged to be a victim of a cyber-attack spilling over from the years' long conflict between Russia and her former sister republic, Ukraine. So, the enormously costly attack on Maersk was the collateral damage of a Russian-sponsored attack on a country more than 1500 km from Maersk's headquarters in Copenhagen. So, for as much damage and distress as *NotPetya* visited upon Maersk, it wasn't the intended target. We are left to wonder what damage an attack with some intent and planning might do to another major shipper and operator of ports.

Moving forward, we need to chronicle the places in which bad things can happen by cyber means and categorize them to some degree. The apparent dichotomy for maritime cybersecurity is a divide between operations at sea and those undertaken while in port. This is a useful distinction as the level of data connectivity for ships at sea is far more constrained than for other pieces of the maritime system functioning at pier-side and further inland. While ports and their IT infrastructure largely benefit from connectivity to high-speed, backbone Internet networks, ships at sea do not. They rely almost exclusively on satellite connectivity to transmit and receive data, and that connectivity is vastly expensive. But let us begin with the cyber issues faced by ships at sea.

10.3.1 Cyber Issues for Maritime Vessels

Navigation by stars and sextant has been largely abandoned by the world's mariners. Most ships ply the world's sea lanes with the aid of three computer-driven systems: the automatic identification system (AIS); the global positioning system (GPS); and the Electronic Chart Display Information System (ECDIS). These three systems are the pillars of computerized navigation for merchant shipping today.

"AIS is a non-encrypted transponder responsible for transmitting course, speed, type of vessel, type of cargo, at-anchor or underway status; and other information for safety at sea" [14]. AIS transponders have been required of ocean-going vessels since 2002, however the functionality of AIS has been subverted for a variety of purposes. Substantial evidence exists that Iran switches off AIS transponders to facilitate sanctions evading behavior in its export of crude oil. North Korea also allegedly disables AIS ostensibly to allow its merchant vessels a greater degree of latitude in avoiding sanctions.

Also important to maritime navigation is GPS. Its use makes navigation on the high seas far more accurate and simpler than ever before. As long as a merchant vessel can communicate with satellites of GPS system, its location can usually be pinpointed within a few meters. GPS is also employed in military targeting, and as a result, measures able to confuse, block, or spoof GPS signals have appeared. The US Coast Guard issued an alert regarding a 2015 incident in which a loss of GPS connectivity to multiple ships departing a non-US port occurred. In 2017, multiple vessels observed degradation and loss of GPS connectivity while sailing in the Black Sea.

Of all the systems of concern with regard to cyberattack, perhaps none is more worrisome than ECDIS. As it is a system that interfaces with navigational gear, sensors, and control systems for driving the ship, ECDIS represents a highly-dangerous target to cyberattack. Even bad ECDIS data is a significant issue. The US Navy minesweeper *Guardian* was severely grounded off the Philippines in 2013 largely due because, "leadership and watch teams relied primarily on an inaccurate Digital Nautical Chart (DNC) coastal chart during planning and execution of the navigation plan" [15]. In addition, multiple cybersecurity and maritime publications have reported on ECDIS's susceptibility to manipulation by unauthorized parties, possibly leading to grounding or collision.

In addition to the major navigational systems present aboard contemporary merchant vessels, there is an enormous amount of automation in shipboard operations. Contemporary cargo vessels, including the largest ones, have automated away large numbers of crew. Large merchant vessels displacing upwards of 100,000 tons are now operated by crews as small as 10 persons or less. The computer systems that replace crew members are process control systems, often provided by automation firms servicing multiple sectors.

One of them is Schneider Electric, a French firm that offers products in no less than 11 merchant shipping applications. Schneider's products are germane to this paper as its Triconex® brand of process control software is widely-utilized in industrial applications in a variety of sectors. Unfortunately, it was also allegedly compromised by a cyberattack in a petrochemical facility in Saudi Arabia. Shipboard systems likely contain a significant number of vulnerabilities, and while they can't be attacked in the way cable- and fiber-based networks are, there are plenty of other avenues for attack, including by insiders in a constant churn of crew turnover.

10.3.2   Cyber Issues in Port Operations

While ships at sea present a peculiar case in what may be considered operational technology (OT) cybersecurity, operations on land are quite different. While shipboard systems may largely be disconnected while at sea, port systems are largely interconnected and often widely exposed to the Internet. And what complicates their cybersecurity even more is that ports are incredibly heterogenous in ownership, operational, and technological composition. Coast Guard port inspectors reputedly quip, "If you've seen one port, you've seen a port."

Ports are often owned by local or regional governments, operated by a commercial operator, and served by myriad firms and offices who make the port work. Consider the Port of Houston, one of the nation's largest, and the most energy-related port in the United States (more on that later). Along the 52-mile Houston Ship Channel is the Port of Houston and its Port Authority (PHA), a mix of publicly- and privately-operated shipping terminals, and other port facilities, 150 different ones in total. It is home to the second and third largest oil refineries in the US and considered the primary energy port in the country. Some 260 million short tons of cargo and more than two million twenty-foot equivalent cargo containers passed through it in 2018.

It is also a very highly automated and networked port. And at the core of the digital operations is something called Navis. Navis is an interconnected suite of software:

> Designed to manage all facets of terminal and cargo operations; it employs, among other things, optical character recognition to scan cargo and manage its movement. When cargo exits the port by truck or rail, not only does NAVIS [sic] electronically log the cargo out and thus simultaneously functioning as part of PHA's security access control system, it also generates billing invoices for PHA. PHA's gantry cranes, fuel farms, and even its HVAC systems are networked [16].

Thinking like a good bad guy, if so much of the Port of Houston's daily operations are largely dependent on the Navis software, then that is probably also an excellent target if the aim is to steal from or disrupt the port. Has Navis been compromised or been found vulnerable? Yes, in 2016, a SQL-injection flaw (a vulnerability found in a database service) was found in Navis software. The US Department of Homeland Security's now defunct Industrial Control Systems–Computer Emergency Response Team (ICS-CERT) reported a previously unknown vulnerability and Navis released a patch for it. The vulnerability could have been exploited by a novice attacker [Q].

Navis has published a library of white papers on enhancing port efficiency. They have titles like *A New Frontier: Business Intelligence, Big Data and the Impact on the Global Supply Chain* and *Port of the Future: A Sense of Wonder*. None of its white papers cover the topic of cybersecurity.

Although Navis and other port system software may have a central role in operations, the systems of many companies and government agencies also interconnect at major ports like Houston. These organizations run email systems, web servers, databases, and all manner of OT systems having to do with port operations. Some of the firms participating in port operations are among the largest corporations or conglomerates in the world, but others are far smaller.

What this means is that getting all the actors involved in the operation of a large US cargo port to adopt a framework or set of practices regarding cybersecurity is difficult. As the Maersk cyberattack illustrated, the loss of even one major firm's system at a large port may bring operations to a screeching halt. Of course, there are many things that may occur to disrupt port operations.

Again, port cybersecurity is different from ship cybersecurity. The targets aboard ships that bad guys care most about are likely those related to navigation and propulsion, both highly automated in contemporary merchant vessels. But in ports, there are many more points of entry to interconnected port systems. Modern port systems talk to railroad systems, and Navis has software, "to automatically route railcars to hub assignments and plan train load sequences" [17].

What this amount to is a scenario in which the purveyors of port operations computer software and automation drive to enhance interoperability and operational efficiency as their primary activity. This drive for efficiency is acceptable, however, automation rife with cyber vulnerabilities may be exploited by malicious actors. Such exploits must be countered by law, policy, and technology. How government and the private sector cooperate on preventing cyberattack is critical to the ongoing function of the global maritime system.

10.4 Law, the Sea, and Cyberspace

A fundamental issue pertaining to the law in sea is the concept of jurisdiction or the *power of a court or locale to regulate persons, objects, or conduct under their law.* Because the world's oceans are international, there is an issue of who has jurisdiction in matters occurring on the oceans. The United Nations Law of the Sea Convention (UNCLOS) attempts to establish a legal framework for the peaceful, cooperative use of the seas. UNCLOS replaced other UN initiatives with this framework. UNCLOS binds only those member countries of the UN and establishes jurisdiction for each country as 12 nautical miles (13.8 miles) from the coastline with a 200-mile exclusive economic zones.

However, multiple countries claim jurisdiction based on their own laws. United States Law, for example, claims that the:

> Special territorial and maritime jurisdiction of the United States includes: (1) The high seas, any other waters within the admiralty and maritime jurisdiction of the United States and out of the jurisdiction of any particular State, and any vessel belonging in whole or in part to

the United States or any citizen thereof, or to any corporation created by or under the laws of the United States or of any State, Territory, District, or possession thereof, when such vessel is within the admiralty and maritime jurisdiction of the United States and out of the jurisdiction of any particular state. [18]

The issue of jurisdiction is especially problematic when it comes to cyberattacks. Does jurisdiction refer to the originating nation of the attacker? The nation of the target? What is a nation is used as an intermediary in the attack? Can multiple nations claim jurisdictions? Unfortunately, the current status of the law remains fragmented with attempts to re-use existing laws and regulations into cyber attack scenarios the challenges to our current civil law framework and in more particular our maritime law legal framework center upon the application of existing legal concepts. This general lack of jurisdiction over hackers presents another issue. What if the damage from the cyberattack is not physical and the lack of physical damage arising from a successful Information Technology (IT) environment cyberattack are legal issues difficult to place within our current civil law framework? In short, the lack of physicality in an IT environment cyberattack presents challenges to our existing civil law framework.

Another attempt to regulate internationally is with the Tallinn 2.0 Manual for International Law Regarding Cyber Operations [19]. The title of this document is problematic. First, it is not international law but rather an attempt by NATO to define rules regarding cyber operations binding among NATO countries. Secondly, the term "cyber operations" is misleading as, on its face, it seems to mean transactions related to cyberspace, but in reality, is synonymous with cyberwar.

The Tallinn Manual establishes a basis for sovereignty, due diligence, jurisdiction, and international responsibility and these uses this basis to prescribe laws for air, sea, and space. Its chapter on the Law of the Sea promulgating ten rules based on the recognized 200-mile economic zone. Both the Tallinn Manual and UNCLOS are limited based on their ability to control the members of their respective groups. As cyberattacks become more common against maritime assets, it will be up to the international courts to determine the effect of regulations and laws, and if these courts actually have the power to regulate.

10.5   Relevant Public Policy

As mentioned above, protection of the maritime system in the wake of the September 11 attacks on the United States and elsewhere has largely been aimed at protecting the physical security and integrity of cargo operations. Planning in port and shipboard security has largely been aimed at thwarting terror threats (smuggling of nuclear weapon or radiological components, other weapons, piracy, etc.) not cyber ones. That said, cybersecurity, or at least cybersecurity risk management has received attention from US national policymaking bodies as well as international organizations and associations.

10.5.1 US Cyber Security Policy Guidance

In the United States, there are sixteen critical infrastructure sectors. These sectors cover cyber as well as physical security. The cybersecurity of ships and ports falls under the DHS's Transportation Systems Sector (TSS). That sector covers not only maritime issues, but also highways, rail, aviation, pipelines, and postal operations. The TSS plan was released by DHS in 2015. It covers a great number of industries and identifies the Coast Guard as the lead agency for maritime safety and security, including cybersecurity. This status is the point of origin cybersecurity strategy produced by the USCG. In addition, the US Maritime Administration (MARAD) maintains an Office of Maritime Security which has added cybersecurity to its portfolio.

Establishing the path for securing systems relevant to maritime operations from cyberattack has become a priority in the US. US policy on cybersecurity for the MTS is still developing but was outlined in the *US Coast Guard Cyber Strategy*. The strategy rests on three pillars: defending cyberspace; enabling operations; and protecting infrastructure. That final piece is where the Coast Guard places the MTS mission, stating:

> Maritime critical infrastructure and the MTS are vital to our economy, national security, and national defense. The MTS includes ocean carriers, coastwise shipping along our shores, the Western rivers and Great Lakes, and the nation's ports and terminals. Cyber systems enable the MTS to operate with unprecedented speed and efficiency. Those same cyber systems also create potential vulnerabilities. As the maritime transportation Sector Specific agency (as defined by the national infrastructure protection plan), the Coast Guard must lead the unity of effort required to protect maritime critical infrastructure from attacks, accidents, and disasters. [20]

The US Coast Guard's strategy heavily emphasizes risk management. This makes a great deal of sense, as shippers and other operators in the maritime system have a long history of managing risk and employing insurances to mitigate risk of loss (UK insurer Lloyd's has been in operation since 1686).

The Coast Guard's strategy rests on two legs: (1) assessment of risk through promotion of cyber risk awareness and management; and (2) prevention via the reduction of vulnerabilities in the MTS. This strategy is likely in need of revision, it was released in 2015, and it's concrete objectives—risk assessment tools and methodologies; cybersecurity information sharing; cyber vulnerability reduction; and cybersecurity education and training—align with the early stage of cybersecurity development found in the maritime system.

10.5.2 International Cybersecurity Guidance

Beyond US policy, the International Maritime Organization (IMO) also has begun to stir in approaching the issue of how cybersecurity impacts its role as the UN specialized agency concerned with "the global standard-setting authority for the safety, security and environmental performance of international shipping." The IMO issued guidance on maritime cyber risk management in 2017 [21]. It detailed eight areas where vulnerable systems can be found:

1. Bridge systems;
2. Cargo handling and management systems;
3. Propulsion and machinery management and power control systems;
4. Access control systems;
5. Passenger servicing and management systems;
6. Passenger facing public networks;
7. Administrative and crew welfare systems; and
8. Communication systems.

The IMO's primary tools for guidance emanate from other bodies including: The *Guidelines on Cyber Security Onboard Ships*; the International Organization for Standardization and International Electrotechnical Commission ISO/IEC 27,001 standard on security techniques; and the US National Institute for Standards and Technology's *Framework for Improving Critical Infrastructure Cybersecurity*. While the latter two documents are applied broadly to many areas of commercial activity, the Guidelines on Cyber Security Onboard Ships (GCSOS) is a much more specific one and deserves greater attention.

Where the USCG has hung its hat on a strategy for cybersecurity in the MTS, GCSOS is an attempt to move toward an industry guidebook for securing ship-board systems. Therefore, it draws significant attention on a set of initiatives that can protect maritime activity. It represents the combined work of nine major associations involved in maritime shipping and transport. Furthermore, it is focused on the cybersecurity of ships, *not ports*.

The GCSOS is a seven-part document that may be best described as a handbook on cybersecurity related to ships engaged in commercial activity. It identifies the primary concern regarding cybersecurity to be found in this area:

As technology continues to develop, information technology (IT) and operational technology (OT) onboard ships are being networked together – and more frequently connected to the internet. [22]

The document also identifies the two major areas of concern regarding a cyberattack upon a ship: its navigation and propulsion systems. Without those functioning properly, safe shipboard operations can't be guaranteed.

Because the GCSOS is essentially a handbook or perhaps even a primer, it covers the full gamut of cybersecurity issues from threats to response and recovery in a relatively brief document. Nonetheless, it stands as significant contribution to cybersecurity in the maritime system. Moving beyond the primer phase of cybersecurity in the maritime system will necessitate new approaches and investments, detailed in the final section of this paper.

10.6   Conclusion and Prescriptions

Maritime cybersecurity has been identified as an issue of some importance in the global cybersecurity agenda. It does not rank as high as energy or power issues, nor have the maturity of corporate and government response found in the financial sector, but it is on the agenda.

We see the state of maritime cybersecurity as this. There is some emphasis on ships, but less on ports, and less still on things connected to ports. All matter and with many, many points of connection to port systems, establishing international, industry-wide standards will likely require extensive coordination and expenditure of intellect. Nonetheless, activity can be undertaken to secure the maritime system by policy and through educational endeavor.

10.6.1   Directions for Public Policy

Obviously maritime cybersecurity issues are inherently international or global in nature. Their remedy will require an investment by stakeholders in both government and the maritime industry with significant input from players in shipbuilding, maritime operations, port activities, and other functions that may be found in the maritime system.

If mere regulation was the answer to cybersecurity issues in this area of endeavor or any other, the job would be one from policymakers alone. Regulation will be only a part of the process of increasing cybersecurity capacity. Nonetheless, when useful frameworks, guidance, rules, and international law may be promulgated, they should be. We just need to be cognizant of the rapid change that may occur as a result of technological innovation. It may be difficult to forecast the future vulnerabilities produced, but certainly this does not constitute a pass for policy action.

Policymakers concerned with addressing the cybersecurity issues to be found in the maritime system must recognize that a workforce of experts in cybersecurity able to address the issues faced by shipping lines, naval architects, automation software developers, or port operators will need to be created and grown. Its beginnings will stem from the tiniest of cadres now extant.

The maritime cybersecurity workforce will be composed of professionals who understand the programming and operation of computer systems as well as having an understanding of the multiple areas of expertise found across the maritime system. For instance, addressing issues in ship propulsion systems requires skills in both the operations of those systems as well as the cybersecurity problems that arise in their development and operation. The same would be true of systems for tracking cargo or navigation.

10.6.2   Research and Education

The workforce issue will necessitate training and education of varying depths. Some professionals will no doubt receive cybersecurity education and training at mid-career while others, if demand is sufficient, will enter the workforce with specialist degrees combining maritime and cybersecurity curriculum. At a deeper level, experts from industry, government, and academia may well need to collaborate around centers for interchange of expertise and research activity. This is already present in cyber activities for everything from the power grid to the banking system.

In the United States, a maritime cybersecurity research and development capability should be established along the lines of Department of Energy (DOE) cybersecurity organizations across its infrastructure of national labs. Considerable investment has been undertaken by the DOE in cybersecurity for the electricity power grid as well as other process control systems. DOE has made considerable investment at its Idaho National Lab (INL) in cybersecurity for Supervisory Control and Data Acquisition (SCADA) systems, found in all manner of industrial applications.

Both DHS and MARAD have grants programs in place for enhancing security of the MTS and ports. One official with whom we discussed this paper described one of the DHS program's outcomes being multiple sales of updated fireboats to major ports. This was verified in our research of DHS granting activity. How government funding can be coupled with industry initiatives should be another area for activity in the cybersecurity of the maritime system.

Few areas of critical infrastructure are riper for strategy and investment related to cybersecurity protection than the maritime system. In addition, research should be undertaken on the protection of computer systems in both shipboard and port operations so that cyberattacks will be less damaging or debilitating to maritime trade.

References

1. Øvergård KI, Sorensen LJ, Nazir S, Martinsen TJ (2015) Critical incidents during dynamic positioning: operators' situation awareness and decision-making in maritime operations. Theor Issues Ergon Sci 16(4):366–387
2. Bueger C (2015) What is maritime security? Mar Policy 53:159–164
3. Mahan AT (2013) The influence of sea power upon history 1660–1783. Read Books Ltd
4. Wood RS, Hanley JT Jr (1985) The maritime role in the North Atlantic. Naval War College Rev 38(6):5–18
5. Mearsheimer JJ (1986) A strategic misstep: The maritime strategy and deterrence in Europe. Int Secur 11(2):3–57
6. Feyrer J (2009) Distance, trade, and income: the 1967 to 1975 closing of the Suez Canal as a natural experiment. In: Working paper 15557, National Bureau of Economic Research
7. Parinduri R (2012) Growth volatility and trade: evidence from the 1967–1975 closure of the Suez Canal. MPRA Paper No. 39040, University of Munich
8. O'Rourke R (1988) The tanker war. Proceedings 114(5)

 9. Murphy MN (2008) Small boats, weak states, dirty money: the challenge of piracy. Columbia University Press, New York
10. Coleman G (2014) Hacker, hoaxer, whistleblower, spy: the many faces of anonymous. Verso books
11. Nicholson A, Webber S, Dyer S, Patel T, Janicke H (2012) SCADA security in the light of cyber-warfare. Comput Secur 31(4):418–436
12. UNCTAD (2017) Review of maritime transport. In: Geneva: United Nations Conference on Trade and Development (UNCTAD)
13. Greenberg A (2018) The untold story of NotPetya, the most devastating cyberattack in history. Wired, September
14. Hayes CR (2016) Maritime cybersecurity: the future of national security. Master's thesis, Naval Postgraduate School, Monterey, CA
15. US. Command investigation into the grounding of USS guardian (MCM 5) on Tubbataha Reef, Republic of the Philippines that occurred on 17 January 2013. United States Pacific Fleet (2013)
16. Kramek J (2013) The critical infrastructure gap: U.S. port facilities and cyber vulnerabilities. Brookings
17. NIST (2016) CVE-2016-5817 detail. National Institute of Standards and Technology (NIST)
18. US. Maritime jurisdiction. U.S. Department of Justice, archived, https://www.justice.gov/arc hives/jm/criminal-resource-manual-670-maritime-jurisdiction
19. Schmitt MN (ed) (2017) Tallinn manual 2.0 on the international law applicable to cyber operations, 2nd ed, Cambridge University Press
20. US. United States coast guard cyber strategy. US Coast Guard (2015)
21. IMO (2017) Guidelines on maritime cyber risk management. International Maritime Organization (IMO)
22. ICS (2018) The guidelines on cybersecurity onboard ships, version 3. International Chamber of Shipping (ICS) and other organizations

Chapter 11
Cyberattacks Against Critical Infrastructure Facilities and Corresponding Countermeasures

Petri Vähäkainu, Martti Lehto, and Antti Kariluoto

Abstract Critical infrastructure (CI) is a vital asset for the economy and society's functioning, covering sectors such as energy, finance, healthcare, transport, and water supply. Governments around the world invest a lot of effort in continuous operation, maintenance, performance, protection, reliability, and safety of CI. However, the vulnerability of CI to cyberattacks and technical failures has become a major concern nowadays. Sophisticated and novel cyberattacks, such as adversarial attacks, may deceive physical security controls providing a perpetrator an illicit entry to the smart critical facility. Adversarial attacks can be used to deceive a classifier based on predictive machine learning (ML) that automatically adjusts the heating, ventilation, and air conditioning (HVAC) of a smart building. False data injection attacks have also been used against smart grids. Traditional and widespread cyberattacks using malicious code can cause remarkable physical damage, such as blackouts and disruptions in power production, as attack vectors to manipulate critical infrastructure. To detect incoming attacks and mitigate the performance of those attacks, we introduce defensive mechanisms to provide additional detection and defense capabilities to enhance the inadequate protection of a smart critical facility from external.

Keywords Adversarial attacks · Critical infrastructure · Cyberattacks · Cyber-physical system · Defensive mechanisms

11.1 Introduction

Cyber-physical systems (CPS) are sociotechnical systems seamlessly integrating analog, digital, physical, and human components engineered for function through integrated physics and logic [49]. Cyber-physical systems can be considered as integrations of computation, networking, and physical processes. CPSs can be implemented as feedback systems that are adaptive and predictive, intelligent, real-time, networked, or distributed, possibly with wireless sensing and actuation. In

P. Vähäkainu (✉) · M. Lehto · A. Kariluoto
Faculty of Information Technology, University of Jyväskylä, Jyväskylä, Finland
e-mail: petri.vahakainu@jyu.fi

© The Author(s), under exclusive license to Springer Nature Switzerland AG 2022
M. Lehto and P. Neittaanmäki (eds.), *Cyber Security*, Computational Methods
in Applied Sciences 56, https://doi.org/10.1007/978-3-030-91293-2_11

CPSs, physical processes are controlled and monitored by embedded computers and networks with feedback loops where physical processes influence computations and contrarily. These kinds of systems provide the foundation of critical infrastructures (CI), providing means to develop and implement smart services of the future, and improving quality of life in various areas. Cyber-physical systems interact directly with the physical world, thus, they are able to provide advantages to our daily lives in the form of automatic warehouses, emergency response, energy networks, factories, personalized health care, planes, smart buildings, traffic flow management, etc.

Critical infrastructure refers to infrastructure that is vital in providing community and individual functions. It can include buildings, e.g., airports, hospitals, power plants, schools, town halls, and physical facilities as roads, storm drains, portable water pipes, or sewer systems [34]. CI can be considered a subset of the cyber-physical system, which includes smart buildings [72]. Smart buildings utilize technology aiming to create a safe and healthy environment for its occupants. Smart building technology, which is still in the early stages of growth and adoption, increases moderately and is becoming a significant business around the world.

Cyber threats against critical infrastructures raise concerns these days, and cyber-physical systems must operate under the same assumption that they might become a target. For example, in the case of an adversarial attack, a perpetrator could fool the Machine Learning (ML) model and gain entry to a building causing significant security threats. The perpetrator may also use the predictive deep learning neural network (DNN) used to adjust the HVAC system by conducting adversarial attacks to cause challenging situations in the form of energy consumption spikes causing high costs. The impact is not negligible as the cost of power spikes has a long payback time; in some cases, several years. Defensive countermeasures against these kinds of attacks are not always straightforward, but adversarial training, defensive distillation, or defense-GAN methods can be utilized in certain cases.

DoS/DDoS, Malware, and Phishing, are traditional attacks capable of causing a considerable threat to critical infrastructure sectors, such as energy and transportation. Perpetrators have utilized DDoS attacks in disrupting the heating distribution system by incapacitating the controlling computers used to heat buildings. This type of attack has also been used when attacking transportation services to cause delays and disruptions over travel services, such as communications, internet services, ticket sales, etc. [54]. Perpetrators may also conduct False Data Injection Attacks (FDIA) to cause a significant threat to, for example, smart grids (SG). They may disrupt energy and supply figures to cause false energy distribution resulting in additional costs [29] often with destructive consequences, or they may conduct the attack towards the smart meter of the power grid to lower one's own electricity bills [76]. If the perpetrator initiates an attack against the power-line connections of the power grid, he or she may be able to separate nodes from the power grid to fool the energy distribution system, which may result in power defects or increased energy transmission costs. In order to provide efficient countermeasures against FDIA attacks, detection methods, such as blockchain, cryptography, and learning-based methods, can be considered.

In the past years, the utilization of malicious software (malware) when conducting attacks towards critical infrastructures have increased. In 2012, Shamoon malware was used to attack the Saudi Arabian national petroleum company, Aramco, by wiping hard disk drives [7]. In 2016, BlackEnergy malware was used to cause disruptions to the Ukrainian electrical grid [81]. Petya malware infected websites of Ukrainian organizations, banks, ministries, newspapers, and electrical utilities [83]. Phishing attacks bring in a human component in which a perpetrator exploits human error and manipulates user behavior, for example, to obtain access to a target system. These kinds of attacks could be detected with deep learning (DL) methods.

In this chapter, the authors briefly introduced the concepts of critical infrastructure, cyber-physical systems, and topical attack vectors against critical infrastructure and countermeasures, respectively. In Sect. 11.2, the authors explain critical infrastructure and resilience concepts in more detail. Section 11.3 addresses the cyberphysical system and presents some relevant CPS sectors these days. Section 11.4 defines cybersecurity and explains the intertwined concepts of cybersecurity, threat, vulnerability, and risks in more detail. Section 11.5 describes artificial intelligence and machine learning and discusses the most common and sophisticated deep learning methods. In Sect. 11.6, we showcase well-known cyberattacks utilized against critical infrastructure facilities, such as smart buildings. Section 11.7 focuses on reviewing the defense mechanisms utilized in combating cyberattacks towards critical infrastructure facilities. Lastly, Sect. 11.8 concludes the study.

11.2 Critical Infrastructure and Resilience

Critical infrastructure (CI) is the body of systems, networks, and assets that are so essential that their continued operation is required to ensure the security of the state, nation, its economy, and the public's health and safety [35]. Critical infrastructure provides services crucial for everyday life, e.g., banking, communication, energy, food, finance, health, transport, and water (Table 11.1). Infrastructure, which is resilient and secure, is a backbone in supporting productivity and economic growth. Disturbances in critical infrastructure can cause harmful consequences for businesses, communities, and governments affecting service continuity and supply security. GOV-AU [46]. Disruptions to critical infrastructure can be caused by, for example, real-world cyber-attacks, which may include environmental damage, financial loss, and even substantial personal injury.

In Finland, critical infrastructure has not been defined in legislation, but the Finnish Government discussed Finnish supply security objectives in 2013. The Finnish Government's decision on supply security objectives contains information about integral threats against the performance of society's vital functions. The decision divides critical infrastructure protection as follows [115]:

1. Energy production, transmission and distribution systems,
2. Information and communication systems, networks and services,
3. Financial services,
4. Transport and logistics,

Table 11.1 Critical infrastructure sectors in Finland [115], EU [40] and United States CISA [23]

Finland	EU	United States
Energy production	Energy	Chemicals
Transmission and distribution systems	Transport	Business
Information and communication systems, networks and services	Banking	Communications
	Financial market infrastructures	Critical infrastructure manufacturing
	Health sector	Damns
Financial services	Drinking water supply and distribution	Defense industry
Transport and logistics	Digital infrastructure	Emergency services
Water supply		Energy
Infrastructure, construction and maintenance		Financial services
Waste management in special situations		Food and agriculture
		Government facilities
		Healthcare and public health
		Information technology
		Nuclear reactors, materials, and waste
		Transportation systems
		Water and wastewater systems

5. Water supply,
6. Infrastructure construction and maintenance,
7. Waste management in special situations.

European Parliament adopted the directive on security of network and information systems NIST [78] on 6 July 2016, aiming to bring cybersecurity capabilities at the same level of development in all the EU Member States and ensure that exchanges of information and cooperation are efficient, including at the cross-border level. The directive increases and facilitates strategic cooperation and the exchange of information among the EU Member States. EC [38].

The core idea of the NIS directive is that relevant service operators and digital service providers must ensure the security of their information infrastructure is secure, ensure business continuity in case of adverse information security disruptions, and report any substantial information security breaches to authorities [104]. According to ([40], ANNEX-II), NIS sectors are the following:

1. Energy,
2. Transport,
3. Banking,
4. Financial market infrastructures,
5. Health sector,
6. Drinking water supply and distribution,
7. Digital infrastructure.

In the United States, there are 16 critical infrastructure sectors whose assets, systems, and networks are so vital to the country that operational incapability or destruction would have a harmful impact on security, economic security, public health, or safety. These 16 sectors are the following CISA [23].

1. Chemicals,
2. Business,
3. Communications,
4. Critical manufacturing,
5. Damns,
6. Defense industry,
7. Emergency services,
8. Energy,
9. Financial services,
10. Food and agriculture,
11. Government facilities,
12. Healthcare and public health,
13. Information technology,
14. Nuclear reactors, materials, and waste,
15. Transportation systems,
16. Water and wastewater systems.

Critical infrastructure is facing various threats that may lead to the appearance of disruptive events causing disruption or failure of the services provided. Minimizing the impact of disruptions and ensuring continuity of services is often cost-effective and the most resilient way, which can be approached with strengthening the resilience. Resilience in the CI system can be seen as a quality that mitigates vulnerability, minimizes the effects of threats, accelerates response and recovery, and facilitates adaptation to a disruptive event [100]. According to [18], resilience is a fundamental strategy that makes the business stronger, communities better prepared, and nations more secure. Hence, resilience is an ability to absorb, adapt to, and quickly recover from a disruptive event [100].

In cybersecurity, (cyber) resilience denotes the ability to plan, respond, and recover from cyber-attacks and possible data breaches and continue to operate efficiently. An organization can be cyber resilient if it can safeguard itself against cyberattacks, provide expedient risk control for information protection, and assure continuity of operation within and after a cyber incident. For an organization, cyber resilience aims to preserve the ability to deliver goods and services concerned, such as the ability to restore common mechanisms, change or modify mechanisms according to the need during a crisis or after a security breach [79]. These kinds of attacks, such as cybersecurity breaches or cyberattacks, are able to cause companies significant damage attempting to destroy, expose, or obtain unauthorized access to computer networks, personal computer devices, or computer information systems [97].

Cyber resilience consists of four elements [79], which are the following:

1. Manage and protect,

2.   Identify and detect,
3.   Respond and recover,
4.   Govern and assure.

Manage and protect consists of the capability to identify, analyze, and handle security threats associated with networks and information systems; third and fourth-party vendors included. Identify and detect consists of continuous security monitoring and surface management of threats to detect anomalies and data breaches in addition to leaks before they cause significant problems. Respond and recover concerns incident response planning in order to assure continuity of functions (e.g., business) even in case of a cyberattack. Govern and assure confirms that the cyber resilience scheme is supervised as usual through the whole organization.

11.3   Cyber-Physical Systems

NIST [78] described cyber-physical systems (CPS) as "smart systems that encompass computational (i.e., hardware and software) and physical components, seamlessly integrated and closely interacting to sense the changing state of the real world." Rajkumar et al. [99] instead characterized cyber-physical systems as "physical and engineered systems whose operations are monitored, controlled, coordinated, and integrated by a computing and communications core." While according to [49], cyber-physical systems are sociotechnical systems seamlessly integrating analog, digital, physical, and human components engineered for function through integrated physics and logic.

These definitions have many similarities, especially; they agree on CPS systems having a physical part, seamless integration of the devices, and controlling software. Compared to the NIST definition, on the one hand, the definition by Rajkumar et al. [99] impress the need for monitoring, controlling, and coordinating the functioning of the engineered system. On the other hand, the definition by [49] includes the human aspect and the need for the system to have a reason to exist in the first place. However, the most general definition the authors have come across is the one by [63], all CPSs include both computational (cyber) part, which controls the system, and a physical part, which includes sensors, actuators, and the frame.

There are various definitions of cyber-physical systems as introduced above. Therefore, the authors settle for defining a cyber-physical system as a cohesive group of computational devices capable of communication; and controlling, coordinating, and monitoring software, engineered and closely integrated aiming to solve the common problem the physical frame or the users of the physical frame might come across during operation of the entire system under uncertainties related to the physical frame and agents. The agents refer to hardware (e.g., sensors, actuators, or other devices) and software (e.g., ML-based access control, energy consumption control programs, etc.) that generate or process the data in any way, including humans. One should understand that different definitions of CPS serve a specific need, and every cyber-physical system might not fit the said definition even though it might be a cyber-physical system.

CPSs can be implemented as feedback systems that are adaptive and predictive, intelligent, real-time, networked, or distributed, possibly with wireless sensing and actuation. In CPSs, physical processes are controlled and monitored by embedded computers and networks with feedback loops where physical processes influence computations and contrarily. CPSs are data-intensive, generating a lot of data during their use. For example, sensors may be able to collect air pressure, CO_2, humidity, motion detection, temperature, etc. These kinds of systems provide the foundation of critical infrastructures (CI), providing means to develop and implement smart services of the future, and improving quality of life in various areas. Cyber-physical systems interact directly with the physical world; thus, they are able to provide advantages to our daily lives in the form of automatic warehouses, emergency response, energy networks, factories, personalized health care, planes, smart buildings, traffic flow management, etc.

Feedback system refers to programs having the capacities to accept and use data both from previous time steps and current time step in the calculation of how the program should change the state of its comprising components or, in other words, how the actuators should be adjusted to implement changes to the system's flow. For example, the program might try to decide how the valve of the HVAC cooling device should be adjusted to save the maximum amount of energy with the least number of changes made to the device's state. Without this knowledge of previous events or data by the system, it can be difficult to make intelligent choices that affect the future state of the network.

CPS can utilize, for example, the interconnected network of various embedded Internet-of-Things (IoT) sensors, devices, and actuators, which observe a small portion of the physical world, and based on the decisions made by the guiding program, change the actuators behavior and thus, cause change to the behavior of the surroundings. The change in physical surroundings might have large scale effects for the whole system's operation, such as advancements of indications to impending and unavoidable service breaks. Therefore, the software program attempts to harmonize the totality of the ensemble of sensors and actuators under the challenges brought upon by the system and the real-world. One of these challenges can be, for example, the replacement of an old actuator with a new one. If the new actuator has capacities beyond the old device, recognizes a different protocol, or stores data in some other format than the old one, then the program might not be able to communicate with the device, and it may cause an error to the system holistically, and thus, the CPS may need calibration or human intervention to correct.

Cyber-physical systems are becoming more and more widespread in the future. For example, even though smart building technology is still in the early stages of growth, its adoption throughout the world is increasing, and it is becoming a remarkable business. For example, the value of smart cities (another embodiment of CPS) is expected to reach over USD 820 billion in the year 2025 Markets [68]. The same could be said about smart grid technology used to manage energy consumption in energy networks. According to a whitepaper by Business Finland [22], the energy clusters' yearly turnover just in Finland has reached EUR 4.4 billion.

A smart building concept can be defined as a set of communication technologies enabling different objects, sensors, and functions within a building to communicate and interact with each other and be managed, controlled, and automated in a remote way EC [39]. It can measure information, such as the temperature of a room or state of windows (open or closed), by utilizing sensors located in the building. The building can become smart if it can obtain such information. An actuator can be used to open a door or to increase the heating temperature of buildings. Intelligent sensors provide significant amounts of information, which must be gathered, processed, and utilized to enable smart functionalities. CPS provides means to utilize sensors to collect data from smart buildings to adjust and control automatically, for example, heating, ventilation, and air conditioning (HVAC) systems. Relevant variables, such as energy, electricity, water consumption, inside and outside temperature, humidity, carbon dioxide, and motion detection, can be utilized in controlling the functions of smart buildings.

Automation and digitalization have become important topics in the energy sector these days, as modern energy systems (e.g., smart grids) increasingly rely on communication and information technology to combine smart controls with hardware infrastructure. The smart grid is another complex example of a cyber-physical system, which continuously evolves and expands. These technologies leveraged the intelligence level of the SG by enabling the adoption of a wide variety of simultaneous operation and control methods into it, such as decentralized and distributed control, multi-agent systems, sensor networks, renewable energy resources, electric vehicle penetration, etc. [75]. In brief, smart grids are electric networks that employ advanced monitoring, control, and communication technologies to deliver reliable and secure energy supply, enhance operational efficiency for generators and distributors, and provide flexible choices for prosumers by integrating the physical systems (power network infrastructure) and cyber systems (sensors, ICT, and advanced technologies) [121].

11.4   Cybersecurity

The history of cybersecurity dates back to the 1970s when ARPANET (The Advanced Research Projects Agency Network) was developed during a research project. At this time, concepts of ransomware, spyware, viruses, or worms did not yet exist. These days due to active cybercrime, these concepts are frequently mentioned in the headlines of newspapers. Cybersecurity has become a preference for organizations worldwide, especially concerning critical infrastructure. The question is not if the system will be under attack, but the question is when it will happen. Hence, proper measures to detect and prevent malicious cyberattacks are required in order to secure essential assets for the functioning of a society or economy.

The concept of cybersecurity can be defined in various ways. Cambridge dictionary defines cybersecurity as follows: "things that are done to protect a person, organization, or country and their computer information against crime or attacks carried out using the internet" [25]. Gartner Glossary defined cybersecurity as the combination of people, policies, processes, and technologies employed by an organization to protect its cyber assets [47]. Cybersecurity can also be thought of as a practice of protecting systems, networks, and programs from digital attacks [32]. Furthermore, cybersecurity can be defined subsequently: "cybersecurity refers to the preventative techniques used to protect the integrity of networks, programs, and data from attack, damage, or unauthorized access" [88].

The main purpose of cybersecurity is to ensure information confidentiality, integrity, and availability, which form the well-known CIA triangle. Confidentiality means that data should not be exposed to unauthorized individuals, entities, and processes or to be read without proper authorization. Integrity means that the data concerned is not to be modified or compromised in any way; therefore, maintaining the accuracy and completeness of the data is crucial. The data is assumed to be accessed and modified by authorized individuals, and it is anticipated to remain in its intended state. Availability means that information must be available upon legitimate request, and authorized individuals have unobstructed access to the data when required [82].

In the field of cybersecurity, threat, vulnerability, and risk are intertwined concepts. The risk is located in the intersection of an asset, threat, and vulnerability, being a function of threats exploiting vulnerabilities to obtain, damage, or destroy assets. Threats may exist, but if there are no vulnerabilities, there is no risk, or the risk is relatively small. The formula to determine risk is the following: risk = asset + threat + vulnerability [45]. The generic definition of risk is the following: "risk is a description of an uncertain alpha-numeric expression (objective or subjective), which describes an outcome of an unfavorable uncertain event, which might degrade the performance of a single (or community of) civil infrastructure asset (or assets)" [42]. Assets denotes what to be protected, a threat is a target to be protected against, and vulnerability can be experienced as a gap or weakness in protection efforts. Threats (attack vectors), especially in cybersecurity alludes to cybersecurity circumstances or events with prospective means to induce harm by way of their outcome. Attack surface sums up all attack vectors (penetration points), where a perpetrator can attempt to gain entry into the target system. Common types of intentional threats are, for example, DoS/DDoS attacks, malware, phishing attacks, social engineering, and ransomware. General vulnerabilities are, e.g., SQL injections, cross-site scripting, server misconfigurations, sensitive data transmitted in plain text, respectively.

Measures in the field of cybersecurity are associated with risk management, vulnerability patching, and system resiliency improvements [64]. Cybersecurity risk management uses the concept of real-world risk management and applies it to the cyber world by identifying risks and vulnerabilities and applying administrative means and solutions to sufficiently protect the organization. Reducing one or more of the following components [103] is an integral part of the risk management process: threat, vulnerability, and consequence. In order to improve system resiliency, improving one or more of the following components is required to be improved: robustness, resourcefulness, recovery, and redundancy. Robustness includes the concept of reliability and alludes to the capability to adopt and endure disturbances and crises. Redundancy involves having excess capacity and back-up systems, enabling the maintenance of core functionality in case of disturbances. Resourcefulness denotes the capability to adjust to crises, respond resiliently, and, when possible, to change a negative impact into a positive one. Response means the capability to mobilize quickly prior to crises, and recovery denotes the capability to regain a degree of normality after a crisis or event.

The important question is to detect the challenges of cybersecurity and to counter them expediently. Cyberattacks cannot be prevented entirely. Hence, an integral part of cybersecurity is to preserve the capability to function under a cyberattack, stop the attack and restore the organization's functions to the previous regular state before the incident took place [65]. In order to counter cyber threats, appropriate measures are important to be taken care of in addition to building adequate protection against the harmful impact of the threats. For example, organizations may utilize an incident response plan (IRP) to detect and react to computer security incidents, determine their scope and risk, respond appropriately to the incident, communicate the results and risks, and reduce the likelihood of the incident from reoccurring [27].

11.5   Artificial Intelligence and Machine Learning

Artificial intelligence is a mathematical approach to estimate a function, and it can be expressed with mathematical terms as $f(x) : R^n \rightarrow R^m$, where $f(x)$ is the function to model, R^n represents the real multidimensional input values, and R^m represents the possible real multidimensional output values. The machine learning research field is needed to make AI models and systems more capable of handling new situations [55] because resources might have been limited during initial training, and the occurring circumstance might be from outside the original input or output domain that was used for training of the model. Deep Learning (DL) is a subfield of ML, where the learning is done with models that have multiple layers within their structure. The additional depth can help the models to learn more complex associations within the given data than regular AI models [62], hence DL models are called deep.

Artificial intelligence is a very enticing choice for many different use cases, where the function to be estimated either unknown or difficult to implement in practice, such as machine translations. In practice, the quality and quantity of data, the structure of the model, and training time, as well as the training method, affect how any AI learns to make its choices. Especially, the data quality is an important aspect of the training of an AI. In a case where there is no connection between given inputs and expected outputs, the outcome of the trained model will not reflect reality. In other cases, the poor quality of data may cause the model to gain no insights into the intended use. In a worse case, the model passes the production inspections and winds up in a live situation where it just does not function properly. The malfunction is even worse if it hides itself to take place only under certain specific situations or if the model's use case is of high importance. Therefore, the implementation of artificial intelligence requires, if not expert knowledge of the field where it is intended to be applied to, but rather clear, innate relation between the inputs and the outputs, and rigorous documenting, testing, and follow up after the implementation.

Ensemble methods refer to grouping different ML models together to process inputs, or according to [114], to the manner, the data is to be used in the training phase of these models. Either the definition, both typically consider the ensemble as some version of two different structures, which either process the inputs in sequence or in parallel (that in the case of model training are both resource inefficient and inaccurate, respectively [114]. With the utilization of ensembles, it is possible to improve ML models' performance. Imagine that you have similar ML models, which have been trained for the same problem domain, but the data they have been trained with were from different patches or data sources. Hence, it is not probable that these models have had the same learning experience and that they would calculate exactly the same predictions with the same prediction confidences based on the same inputs. In an ensemble, the performance scores may rise as the result of the ensembled models' outputs, and confidence scores are compared against each other. The errors stemming from individual models' states get mitigated, thus lessening the effect of any bias within the models. The process can be thought of like voting, where the most endorsed output becomes the actual final output, or more commonly, the final output is some weighted combination of the predicted outputs.

Decision trees (DTs) represent the more traditional algorithms used in artificial intelligence development, and their popularity is mostly related to the ease of interpretation of the results. The interpretation is simpler because these models' behavior is well defined, forming decision rules or paths from the data systematically. A decision tree is a flowchart-like tree structure where an internal node represents a feature or attribute, the branch represents a decision rule, each leaf node represents the outcome, and the first node in a DT is known as the root node. It learns to partition based on the attribute value partitioning the tree recursively and providing the tree classifier a higher resolution to process different kinds of numerical or categorical datasets [108]. Depending on the decision criteria, the algorithm chooses which part of the input data is most significant at each iteration until the conclusion criteria have been filled. It can model nonlinear or unconventional relationships. In other words, DTs can be used to explain the data and their behavior. In addition, many coding libraries have visualization capacities of these paths. However, the decision tree's performance suffers from unbalanced data, overgrowing decision paths, which may also hinder the model's interpretation, and updating a DT by new samples is challenging [108].

Random Forest (RF) includes a significant number of decision trees forming a group to decide the output. Each tree specifies the class prediction resulting in the most predicted class in DTs. RF trees protect each other from distinct errors, and if a single tree predicts incorrectly, other trees will correct the final prediction. RFs can reduce overfitting, deal with a huge number of variables in a dataset, estimate the lost data, or estimate the generalization error. RFs experience challenges in repro-ducibility and interpreting the final model and results. RFs are swift, straightforward to implement, extremely accurate, and relatively robust in dealing with noise and outliers. RFs are not fit for all the datasets as they tend to induce randomness into the training and testing data [108].

Neural network (NN) is a popular base model used in the development of AI solutions. The model has three layers: an input layer, a hidden layer, and an output layer, where data flows from the input layer through the hidden layer consisting of multiple layers, and the result is produced to the output layer. NNs are a collection of structured, interjoined nodes whose values are comprised of all the weights of the connections coming to each node. Every value of a node is inputted to an activation function, such as a rectified linear unit (ReLu). The activation function is typically the same for all the nodes in the same layer.

NN may require a lot of quality data. The need is formed based on the difficulty of the problem, suitability of the data, and the chosen structure and size of the model. In case there are a limited amount of quality data available, it can be beneficial to attempt using two competing neural networks to generate the missing training data. According to [95], the general way is to have the first model to generate new values based on the original data, and the second model tries to classify the original and generated inputs (the outputs of the first model) from each other. The results of the classifier are then used as feedback for improving the generator and the classi-fier. Eventually, the generated outputs' distributions move closer and closer to the real inputs. This machine learning method is called Generative Adversarial Neural networks (GAN) [95].

Long-Short Term Memory neural network (LSTM) is a special case of Recur-ring Neural Network (RNN) [66], which retains output information from previous timesteps as part of the input information. The extra information can be helpful, i.e., when forecasting with sequential data. Because NNs can suffer from the problems of vanishing and exploding gradients, which likely will increase with the growth of sequence size, LSTMs have three gates within each node that are used to control the information going through them [66]. These logical gates use sinh and tanh activa-tion functions to control the flow and size of internal representations of the inputs and outputs. RNN, LSTM, and their various variants have been used, for example, in machine translation tasks [123], predicting the smart grid stability [5], and classifying malware [11].

Even though NN models suffer from data issues and it can be more difficult to interpret how models have reached their conclusions, they are perceived to attain more accurate results than some of the traditional algorithms, such as decision trees. In addition, [122] used DTs to interpret the predictions of a Convolutional Neural Network (CNN) model, thus explaining the model's behavior. A convolutional neural network is a neural network that has special layers within its hidden layers. These layers group the inputs systematically from the previous layer and calculate a value for each of these groups, which they then output for the next layers as inputs [6], consequently, reducing the layer's dimensions. The field of research focused on explaining and interpreting these malleable algorithms for human experts in an easily understandable form is called explainable artificial intelligence [14].

11.6 Cyberattacks Against Critical Infrastructure Facilities

This section introduces and discusses well-known cyberattacks, such as adversarial, DoS and DDoS, False data injection (FDI), malware and phishing attacks from a critical infrastructure perspective and illustrates utilization of attacks mentioned with real world case-examples.

11.6.1 Adversarial Attacks

An adversarial attack is an attack vector created using artificial intelligence. These attacks are adversarial disruptions constructed purposely by the attacker. The disruptions are imperceptible in the human eyes but generally adversely impact neural network models. These days, adversarial attacks towards machine learning models are becoming more and more common, bringing out noticeable security concerns. For example, in the context of smart building (CPS), an attacker may have a chance to deceive the ML model into causing harm, such as to create conditions for consumption spikes, when attacking the heating system guided by predictive machine learning-based feedback system.

An adversarial attack happens when an adversarial example is sent as an input to a machine-learning model. An adversarial example can be seen as an instance to the input with features that deliberately cause a disturbance in an ML-model to deceive the ML-model into acting incorrectly and into making false predictions [52]. Deep learning applications are becoming more critical each day, but they are vulnerable to adversarial attacks [113] argue that making tiny changes in an image can allow someone to cheat a deep-learning model to classify the image incorrectly. The changes can be minimal and invisible to the human eye and can eventually lead to considerable differences in results between humans and trained ML-models.

The effectiveness of these attacks is determined based on the amount of information the perpetrator has concerning the model. In a white-box attack, a perpetrator has total knowledge about the model (f) used in classification, and she knows the classifier algorithm or training data. She is also aware of the parameters (θ) of the fully trained model architecture. The perpetrator then has a possibility to identify the feature space where the model may be vulnerable (e.g., where the model has a high error rate). The model can then be exploited by modifying an input using an adversarial example crafting method [28].

There may be indirect ways to obtain an adequate amount of knowledge about a learned model to apply a successful attack scenario. For example, in case of a malware evasion attack, a set of features may be public through published work.

Datasets used to train the detector might be public, or there might be similar ones publicly available. The learner might use a standard learning algorithm to learn the model, such as deep neural networks, random forest, or Support Vector Machine (SVM), by using standard techniques to adjust hyperparameters. This may lead to the situation that the perpetrator can get a similar working detector as the actual one [116].

In the case of Black-box attacks, the perpetrator does not know the type of the classifier, detector's model parameters, classifier algorithm, or have any knowledge about the training data in order to analyze the vulnerability of the model [20]. For example, in an oracle attack, the perpetrator exploits a model by providing a series of carefully crafted inputs and observing outputs. In model inversion type of an attack, the perpetrator cannot directly access the target model, but she can indirectly learn information, such as model structure and parameters, about the model by querying the interface system and gather the responses [28, 91]. presented a strategy (Papernot-attack) to produce synthetic inputs by using some collected real inputs. Many studies are focusing on research utilizing images as datasets (MNIST or CIFAR). In such a case, the perpetrator can, for example, fetch several pictures of the target dataset and use the augmentation technique for each of the pictures to find new inputs that should be labeled with the API. The next step is to train a substitute by sequentially labeling and augmenting a set of training inputs. After the substitute is accurate enough, the perpetrator can launch white-box adversarial attacks, such as FGSM (Fast Gradient Sign Method) or JSMA (Jacobian Saliency Map Approach), to produce adversarial examples to be transferred to the targeted model [48].

Jacobian-based saliency map algorithm (JSMA) was presented by Papernot et al. [89] to optimize L0 distance. JSMA attack can be used for fooling classification models, for example, neural network classifiers, such as DNNs in image classification tasks. The algorithm can induce the model to misclassify the adversarial image concerned as a determined erroneous target class [119]. JSMA is an iterative process, and in each iteration, it saturates as few pixels as possible by picking the most important pixel on the saliency map in a given image to their maximum or minimum values to deceive the classifier [92]. Even though the attack alters a small number of pixels, the perturbation is more significant than L∞ attacks, such as FGSM [67]. The method is reiterated until the network is cheated or the maximal number of altered pixels is achieved. JSMA can be considered as a greedy attack algorithm for crafting adversarial examples, and it may not be useful with high dimension input images, such as images from the ImageNet dataset [67].

The JSMA attack can cause the predictive model to output more erroneous predictions, which can, eventually, make the controlling model either complacent or too reactive. Both choices could be monetarily crippling. For example, [89] were able to perturb both categorical and sequential RNNs with JSMA adversarial attack. Therefore, the chance exists that the perpetrator could, if given enough time and resources, afflict damage to both AI models, namely the cybersecurity AI model and the controlling AI model.

A white-box attack uses the target model's gradients in producing adversarial perturbations. FGSM was introduced by [48] to generate adversarial examples against

NN. FGSM can be used against any ML-algorithms using gradients and weights, thus providing low computational cost. The gradient needed can be calculated by using backpropagation. If internal weights and learning algorithm architecture is known, with backpropagation FGSM is efficient to execute [33]. FGSM fits well for crafting many adversarial examples with major perturbations, but it is also easier to detect than JSMA, therefore, JSMA is a stealthier perturbation, but the drawback is higher computational cost than FGSM. Defense mechanisms can prevent a relatively considerable number of FGSM and JSMA attacks [48].

Carlini and Wagner [26] has been presenting C&W attack, one of the most powerful iterative gradient-based attacks towards Deep Neural Networks (DNNs) image classifiers due to its ability to break undefended and defensively distilled DNNs on which, for example, the Limited-Memory-Broyden-Fletcher-Goldfarb-Shanno (L-BFGS) and DeepFool attacks fail to find the adversarial samples. In addition, it can reach significant attack transferability. C&W attacks are optimization-based adversarial attacks, which can generate L0, L2, and L∞ norm measured adversarial samples, also known by CW0, CW2, and CW∞, respectively. The attack attempts to minimize the distance between a valid and perturbed image while still causing the perturbed image to be misclassified by the model [109]. In many cases, it can decrease classifier accuracy near to 0%. According to [101], C&W attacks reach a 100% success rate on naturally trained DNNs for image datasets, such as MNIST, CIFAR-10, and ImageNet. C&W algorithm is able to generate powerful adversarial examples, but computational cost is high due to the formulation of the optimization problem.

Gradient-based and gradient-free adversarial attacks mentioned in this chapter, such as C&W, FGSM, and JSMA, can perturb the input data in such a way that the inputs seem valid for a human but mess maliciously with, e.g., a machine-learning model that can automatically adjust HVAC and other heating devices of smart buildings. This kind of model may gather data from local measurement units (IoT sensors) and external data from the weather database, including data from social media accounts. Data can then be properly merged and cleaned to be utilized in training the predictive model. The predictive model may use, e.g., LSTM neural networks to perform energy load forecasts and calculate the need for new commands to be sent to the actuators.

This kind of a classification-oriented LSTM neural network can be attacked, for example, by using the mentioned JSMA attack method. It then perturbs the input in the desired direction to selectively make the model misclassify to an appropriate output class [9]. Deep neural networks can be deceived by adding even minor perturbations, such as flawed pixels, to form an image classification problem and to be used to deceive sophisticated DNNs in the testing or deploying stage. The vulnerability of adversarial examples is an ample and ever-growing risk, especially when the field of critical infrastructure is concerned. Fooling the predictive deep neural network used to adjust the HVAC system of a cyber-physical system can cause challenging situations in the form of energy consumption spikes causing increasing operational costs.

11.6.2 DoS and DDoS Attacks

Denial of service (DoS), and its variant (DDoS), is one of the major threats, and it can cause disastrous consequences because of its distributive nature. These attacks conducted by a perpetrator may use single or even multiple computers known as zombies in order to consume the victim's resources so that the server cannot provide a requested service to a legal or legitimate user. The perpetrator utilizes the advantage of the internet, network bandwidth, and connectivity to target the open points and initialize floods of thousands or even millions of packets to knock off the victim's server. The server either crashes or becomes incapable of serving all of the incoming requests, and it cannot serve the legitimate clients who are trying to use the service provided by the server concerned. These attacks' main targets can be, for example, default gateways, personal computers, web servers, etc.

Perpetrators aim to look for the path they can use to gather the secret information they are after. This denotes compromising confidentiality. The second phase, which compromises the integrity, is to gain access to the confidential information to alter it. The third phase is to compromise the availability, which is the main target of perpetrators as compromising confidentiality and integrity are more challenging, requiring more advanced technical skills in order to succeed. Administrative privileges on the target system are not needed when availability is compromised. Perpetrators can compromise the service's availability by exhausting the resources to make the service unavailable for legitimate users, as mentioned earlier.

DoS/DDoS attacks can be conducted in many ways using different kinds of program codes and tools, and they can be initiated from different OSI model layers. OSI has seven layers, which are physical (layer 1) covering transmission and reception of the unstructured raw bit stream over a physical medium, data-link (layer 2) responsible for conducting an error-free transfer, network (layer 3) handles routing of the data, transport (layer 4) responsible for the packetization and delivery of data, session (layer 5) taking care of establishment, coordination, and termination of sessions, presentation (layer 6) handles data translation and sending it to the receiver, and application (layer 7) where communication partners are identified. All the messages and creating packets initiate at this level [84].

DDoS attacks may cause physical destruction, obstruction, manipulation, or malfunction of physical assets on the physical layer. MAC flooding attack floods the network switch with data packets, which usually happens on the Data-link layer. Internet Control Message Protocol (ICMP) flooding utilizes ICMP messages to overload the targeted network's bandwidth, a network layer 3 infrastructure attack method. SYN flood and Smurf attacks are transport layer 4 attack methods. In an SYN attack, series of "SYN" (synchronize) messages are sent to a computer, such as a web server, after communication between two systems over TCP/IP has been established [31]. A smurf attack is an old DoS attack, which uses a great number of ICMP packets to flood a targeted server. SYN attack utilizes TCP/IP communication protocol to bombard a target system with SYN requests to overwhelm connection queues and force a system to become unresponsive to legitimate requests. On the session layer, a perpetrator can use DDoS to exploit a vulnerability in a Telnet server running on the switch, forcing Telnet services to become unavailable. On the presentation layer, the perpetrator can also use malformed SSL requests as inspecting SSL encryption packets is resource intensive. Vulnerabilities to DDoS attacks on the application layer are, e.g., use of PDF GET requests, HTTP GET, HTTP POST methods on website forms, when logging in, uploading photo/video or submitting feedback, etc. [96].

Perpetrators may utilize botnets, which can be described as a network of several or a large number of computers or internet-enabled devices that have been taken over remotely, to launch numerous types of attacks, such as DDoS, spamming, sniffing and keylogging, identity theft, ransom and extraction attacks, etc. Botnet (zombies) target vulnerabilities in different layers of the open systems interconnection. These attacks can be divided, e.g., in the following way:

1.   Application layer attacks,
2.   Protocol attacks,
3.   Volumetric attacks.

Application layer attacks are the most primitive form of DDoS mimicking normal server requests. This type of attack was explained in detail at the beginning of this chapter. Protocol attacks exploit the way servers process the data to overload and overwhelm the intended target. One way to conduct this type of attack is to send data packets, which cannot be reassembled, resulting in overwhelming the server's resources. Volumetric attacks are similar to application attacks, but in this type of DDoS attack, the whole server's available bandwidth is used by botnet requests. A high amount of traffic or request packets to a targeted network will be sent in order to slow down or stop the target services [94].

DDoS attacks are able to cause a significant threat to critical infrastructure sectors, such as energy and transportation. The DDoS attack disrupted the heating distribution system, at least in two properties in the city of Lappeenranta, eastern Finland, in 2016. In the incident, attacks incapacitated the controlling computers of heating in the buildings concerned. The attack lasted from late October to November 3, causing inconvenience and potentially hazardous situations as the outside temperature was below freezing. During the attack, the system tried to respond by rebooting the main control circuit, which was then continuously repeated, making heating incapable of working. Unfortunately, building automation security is often neglected, and housing companies are often reluctant to invest in firewalls and other security measures in order to improve the general security situation [54].

DDoS attacks have been conducted against transportation services, causing train delays and disruption over travel service. Swedish transportation system experienced such an attack on October 11 in 2017, via two internet service providers, TDC and DGC. The DDoS attack crashed the train location monitoring IT system, guiding operators to go and stop the train. The attack also knocked out the federal agency's email system, road traffic maps, and website services. As a result of the attack, train traffic and other services had to be operated manually by utilizing back-up processes [15]. In 2018 Danish rail travelers experienced trouble while buying tickets due to a paralyzing DDoS attack on Denmark's largest DSB railway company's ticket system.

The attack made it impossible to buy a ticket via the DSB app, on the website, ticket machines, and kiosk stations. Additionally, the attack also restricted communications, telephone systems, and internal mail was also affected [87]. In order to communicate delays to customers, the company had to utilize social media and ground staff [70]. The Freedom of Information Data states that up to 51% of critical infrastructure organizations in the UK are potentially vulnerable to these attacks due to incapability of detecting and mitigating short-duration DDoS attacks on their networks, and as a result, 5% of these operators experienced DDoS attacks in 2017 [102]. CI operators, such as transport agencies, cannot leave DDoS attack protection at the chance, they are required to build and improve resilience in combatting these attacks.

11.6.3 False Data Injection (FDI) Attacks

False data injection (FDI) attack poses a significant threat towards the traditional power grid (PG), and in these days, smart grid, technologies that provide power to be used, for example, in cyber-physical systems, such as smart buildings. Smart grids are electrical grids, which utilize information and communication (ICT) technology in providing reliable, efficient, and robust electricity transmission and distribution. Hence, smart grids are not solely well-known power lines in traditional "dumb" energy infrastructures, but they represent a relatively new type of energy distribution system standing among the key relevant concepts in supporting sustainable energy city. SGs are connected to smart meters, which can be installed in entities, such as smart factories, hospitals, schools, etc. include components, which enable predictive analytic services in order to balance the production and consumption in the grid system. Advanced services, such as real-time pricing, provide consumers and suppliers relevant information to manage their energy demands and supplies. The service allows energy distribution to be performed in a dynamic and effective manner. [29] In addition, SGs merge the non-renewable and renewable energy resources into each other, reducing environmental problems [44].

FDI attacks are typically utilized when conducting attacks against the functionality of smart grids in order to disrupt, for example, real energy and supply figures causing erroneous energy distributions resulting in additional costs or destructive consequences [29]. According to [76], the perpetrator can, for instance, use these attacks to modify the smart meter data to lower her electricity bill or target remote terminal unit (RTU) to inject false data to the control center resulting in an increased outage time. FDI attacks can be considered as a type of integrity violation aiming to pose arbitrary errors and distortion to the device's measurements, influencing the state estimate (SE) precision. SE is a vital service for system monitoring in ensuring reliable operation in the power system and in addition to the energy management system (EMS), which processes real-time data gathered by the SCADA system. Smart meters are able to further infer state estimations (e.g., energy demands and supplies) and to make initial decisions, for example, concerning data fusion before the estimations reach control centers. The information provided can be utilized to optimize energy distribution with regard to power grid performance metrics in order to maximize the network utility and energy efficiency while minimizing energy transmission costs. Hence, FDI attacks violate SE's integrity making the smart grid system unstable in the worst-case scenario.

The perpetrator may inject the false monitoring data into the smart grid by using, e.g., the following ways:

1. Compromising the smart meters, sensors or RTUs,
2. Capturing the communication between sensor networks and SCADA system,
3. Penetrating the SCADA system resulting in an incorrect estimate of the smart grid state, which may eventually lead even to large-area power failure accidents.

According to Sargolzaei et al. [106], the perpetrator's aim is not solely to inject false information to distract the solid operation of the target system but also to inject incorrect data, which keeps the system's controller and detection mechanism in the shadows concerning the incident. The perpetrator may also utilize means to gather side-information, such as to perform particular analyses and techniques to collect knowledge about the nominal state values of the agents, concerning the structure of the target system to conduct FDI attacks to increase the destructive power of the attack. In order to conduct the malicious attack, the perpetrator may need to inject "realistic" false data, which is close enough to the nominal states and parameters of the system to various sensors at the same time. This procedure makes FDI difficult to detect, especially if system architecture is known.

The perpetrator can conduct attacks against one or multiple of the following FDI attack surfaces: energy demand, energy supply, grid-network states, and electricity pricing. Attacking against energy demand can cause fraudulent values of the state estimation raising financial costs to both the energy users and providers due to extra cost of power transmission or waste energy. It may also lead to power outage situations, in which energy requests to the smart grid is less than the energy demand that nodes (representing the average energy demand/supply, e.g., a town) of the grid require. Energy-supply nodes provide the value of SE, and an FDI attack can secretly mitigate the amount of energy supplied, leading to an energy shortage situation of energy-demand nodes as the nodes cannot receive the required energy. In the opposite situation, an increase in wasted energy can occur [29].

Grid-network states represent the configurations and conditions of power grids, for example, grid topologies and power-lines capacities. The perpetrator can use FDI to attack power-line connections in order to isolate nodes from the power grid deceiving the energy distribution system and leading to power shortages or energy transmission costs. Dynamic electricity pricing helps in balancing the power loads between peak and off-peak periods and reduce consumer electricity bills. The perpetrator can lower her electricity price causing loss of company revenue or lower prices during peak hours, leading to the grid system eventually overloading. Hence, fake pricing causes remarkable damage to the financial and physical subsystem, obliterating the advantages of optimum supply efficiencies [29].

11.6.4 Malware Attacks

Malware and software-enabled crime is not a new concept but dates back to the year 1986, when the first malware, Brain. A., appeared for a PC computer. The appearance of malware proved that PC is not a secure platform, and safety measures should be considered. Malware or malicious software is software created and possibly used by perpetrators to disrupt computer functions, collect sensitive information, damage the target device, or obtain access to a private computer system. The form of malware can be, for example, active content, code, scripts, or another kind of software. Malware incorporates adware, computer viruses, dialers, keyloggers, ransomware, rootkits, spyware, trojan horses, worms, and other types of malicious computer programs. In general, most of the common malware threats are worms or trojans instead of regular and ordinary computer viruses [73]. Since 2018 Ransomware attacks have been showing signs of growth. Malware attacks can occur on all kinds of devices and operating systems, such as Android, iOS, macOS, Microsoft Windows, etc.

Malware attacks against critical infrastructure have been increased during the past several years. In 2012 Iran conducted a destructive retaliation wiper Shamoon malware attack towards Saudi Arabia's national oil conglomerate, Saudi Aramco. The functionality of Shamoon is to wipe out all data from hard disks, and it was used to overwrite hard drives of 30,000 computers in the Aramco-case [7]. In 2016, a trojan type of malware called BlackEnergy was used to cause disruptions to the Ukrainian electrical grid. BlackEnergy is a modular backdoor that can be utilized to conduct DDoS, cyber espionage, and information destruction attacks towards ICS/Scada, government, and energy sectors worldwide. BlackEnergy malware family has been present since 2007, and initially, it started as an HTTP-based botnet for DDoS attacks. Later on, the second version, BlackEnergy2, was developed, which was a driver component-based rootkit installed as a backdoor. The above mentioned version of the backdoor predominantly spread via targeted phishing attacks by email, including the malware installer. The later version is BlackEnergy3, which was used to attack against Ukrainian electrical power industry. This version can be used when conducting phishing attacks containing Microsoft Office Files packed with malicious obfuscated VBA macros to infect target systems [81].

Another type of malware that appeared in 2015 and which have been used in attacking healthcare sector critical infrastructure facilities is known as DragonFly. The malware specifically targets industrial control system (ICS) field devices in the energy sector in Europe and in the US. Utilization of the DragonFly remarkably grew during the year 2017. Perpetrators have been interested in learning how energy facilities operate and also how to gain access to operational systems themselves. The malware uses different sorts of infection vectors to obtain access to a victim's network. These vectors include malicious emails, trojan software, and watering hole attacks to leak the victim's network credentials and exfiltrate them to an external server. Hijacked device contacts a command-and-control server, which is controlled by perpetrators providing a back door to the infected device [19].

Stuxnet malware (worm) increased awareness of cybersecurity and related issues in the world after it was detected in 2010. The worm was targeting centrifuges used in the uranium enrichment process in a nuclear plant in Natanz in Iran. Governments around the world had to face the fact the critical infrastructures were vulnerable to cyberattacks with a possibility to cause catastrophic effects. The aim of this malware was to sabotage centrifuges in the power facilities in order to stop or delay the Iranian nuclear program. It is believed that the malware was uploaded to the power plant's network by using an infected USB drive [13].

Stuxnet is larger than other comparable worms, and it is implemented by using various programming languages with encrypted components. It used four zero-day exploits when infecting computers, which are a connection with shared printers, and vulnerabilities concerning privilege escalation, allowing the worm to run the software in computers during lock-down. The worm caused damage to the centrifuges by making them alternate between high and low speeds and by masking the change of speed to look normal. Due to the procedure, Iran had to replace 10% of its centrifuges yearly. The incident showed critical infrastructure could be targeted by cyber threats, and even networks separated from each other did not protect against the malware. It is integral to increase protection against this kind of malware and, in addition, to improve resilience during cyberattacks [13].

Duqu followed the well-known Stuxnet malware worm and was detected by the Laboratory of Cryptographic and System Security at the Budapest University in Hungary in 2011. The similarity of the malware structure to Stuxnet is so, which indicates that it was developed and implemented by Stuxnet authors or developers who have had access to the source code. Unlike Stuxnet, Duqu was mainly implemented for cyber espionage purposes to obtain a deeper understanding of network structures in order to detect vulnerabilities to exploit and develop better attack methods to penetrate the defenses [17]. Duqu is an information stealer rootkit targeting MS Windows-based computers collecting keystrokes and other relevant information, which could be used when conducting attacks against critical infrastructures, such as power plants or water supply around the world. After penetrating the defenses, Duqu injects itself into one of four general Windows processes: Explorer.exe, IEExplore.exe, Firefox.exe, or Pccntmon.exe, downloads and installs an information-stealing component to gather information from the infected target system, encrypts the data, and uploads it to the perpetrator's system. Smart grid with smart meters, substations, intelligent monitors, and sensors provide an attractive attack surface to perpetrators' exploitation of critical infrastructure systems in their minds [118].

Triton is among the most hazardous malware spreading over the networks worldwide, targeting critical infrastructure facilities utilizing automated processes. The malware was first detected in 2017 during the malicious attack towards Tasnee-owned petrochemical plant facility using Schneider Electric's Triconex Safety Instrumented System (SIS), which then experienced a sudden shutdown. The malware was deployed in emergency safety devices, which are required to be started in case of plant toxic gas leaks and during emergency situations. Triton, among other dangerous malicious attacks, can cause safety mechanisms to experience physical damage due to the incapability of operating during emergency situations. It can be used to target industrial control systems (ICS) and to use a secure shell (SSH) based tunnel to deliver attack tools to the victim system and running remote commands of the malware program. A perpetrator accesses information technology (IT)- and operational technology (OT)-networks, installs back doors in the computer network, and accessing the safety instrumentation system (SIS) controller in the OT network in order to secure and maintain the target's networks using attack tools [77].

11.6.5 Phishing Attacks

Phishing is a social engineering technique that can be utilized to override technical controls designed and implemented to mitigate security risks in information systems. Social engineering is a manipulation technique exploiting human error to obtain sensitive private information, access, or valuables. The weakest link in the security program is us, the humans. In cybercrime, perpetrators exploit the human component to deceive end-users of the system by manipulating user behavior to expose data, spread malware infections, or provide entry to the restricted system. Attacks can be conducted online, in-person, or via other means. In addition to manipulation of user behavior, perpetrators can exploit a user's lack of knowledge, e.g., "drive-by-download," which infers to installing malicious programs to devices without the user's approval [58].

Phishing takes advantage of this weakness and exploits the vulnerability of human nature to obtain access to a target system [98]. Even though organizations have been long increasing employee awareness of cybersecurity threats, phishing is still among the starting points for various cyberattacks. According to surveys, up to 46% of successful cyber attacks started with a phishing email sent to an employee [36]. According to [2], the attack can be used to steal user's confidential information, such as passwords, social security numbers, and banking information, and takes place when cybercriminals disguise as a trusted entity and fool users to click on fake links included in the email received. In addition, cybercriminals also target organizations belonging to the target country's critical infrastructure sector (e.g., telecommunications or defense subsector) by utilizing the special form of phishing, a spear-phishing.

Spear phishing is a certain type of phishing, in which the context and victim are examined, and which utilize custom-made email message that can be sent to the victim. As mentioned before, received email messages can include a malicious link or email attachment to deliver malware payload to direct a benevolent individual to counterfeit websites. These websites can then be used to inquire, e.g., login credentials or ask to download malicious (malware) software to the victim's device. The perpetrator is then able to utilize the credentials or infected devices in order to obtain entry to the network, steal information, and in many cases, stay inconspicuous for a prolonged amount of time [21].

Spear phishing attacks used to conduct attacks towards critical infrastructure occurred in 2014 when a perpetrator initiated a spear-phishing attack against Korea Hydro and Nuclear Power (KHNP). The attack resulted in the leak of personal details of 10,000 KHNP workers, designs and manuals, nuclear reactors, estimates of radiation exposures among residents, etc. During only a few days, the perpetrator managed to send almost 6000 phishing email messages, which included malicious codes to more than 3000 employees. The catch was to demand money for not leaking sensitive classified information to other countries or not to be published in social media on the internet. Luckily, the server containing the information was isolated from the intranet; therefore, the perpetrator managed to cause only confusion in Korean Society. However, cyberattacks towards nuclear power plants may pose a significant risk and damage to all living organisms and the environment over a wide area. Hence, extensive security countermeasures should be developed to mitigate these risks [85]. Additionally, it is suspected that the Ukrainian power grid was initially attacked with a phishing attack followed by BlackEnergy malware, leaving hundreds of thousands of homes without electricity for six hours [8].

11.7 Defensive Mechanisms Against Cyberattacks

This section focuses on reviewing possible detection and prevention mechanisms that could be utilized in combating previously mentioned cyberattacks threatening critical infrastructure facilities.

11.7.1 Defending Against Adversarial Attacks

Adversarial examples are maliciously perturbed inputs designed to deceive a machine learning model at test time, posing a significant risk to the ML models. These inputs can transfer across models meaning that the same adversarial example is generally misclassified by various models. Adversarial examples can be countered with adversarial training of ML model classifier, which is one of the earliest and well-known defense methods in combatting adversarial example crafting (e.g., FGSM). The adversarial training method has reached the de-facto standard status in providing robust models [112]. Robustness can be improved by augmenting the ML model training dataset with perturbed inputs in case of the training set is the same as the perpetrator uses [105]. Robustness can be reached by adversarial training based on the strength of the adversarial examples utilized. Hence, training a model by using fast non-iterative FGSM produces robust protection towards non-iterative attacks, such as JSMA. Defending against iterative adversarial examples also requires training to be done with iterative adversarial examples [107]. If a perpetrator uses a different kind of attack strategy, the efficiency of the adversarial training will decrease [105].

This method can be applied to large datasets when perturbations are crafted using fast single-step methods. Adversarial training generally attains adversarial examples by utilizing an attack, such as FGSM, and tries to build adequate defense targeting such an attack. The trained model can indicate poor generalization capability on adversarial examples originated from other adversaries. When combining adversarial training on FGSM with unsupervised or supervised domain adaptation, the robustness of the defense could be improved. Unfortunately, the robustness of adversarial training is possible to evade by applying a joint attack with indiscriminate perturbation from other models [111]. In addition, utilization of adversarial training as a robust defense method is limited in real-life situations due to extensive computational complexity and cost [107].

Defensive distillation can be considered as an adversarial defense method to counter adversarial attacks, such as FGSM or JSMA. The method is one of the adversarial training techniques, which provides flexibility to an algorithm's process, making it less susceptible to exploitation. According to [122], the idea behind defensive distillation is to generate smooth classifiers that are more resilient to adversarial examples by mitigating the sensitivity of the DNN to the input perturbation. The technique also improves the generalization ability as it does not alter the neural network architecture, and in addition, it has low training overhead and no testing overhead.

Papernot et al. [90] investigated the defensive distillation and introduced a method that can reduce the input variations making the adversarial crafting process more challenging, providing means to DNN to generalize the samples outside the training set and mitigating the effectiveness of adversarial samples on DNN. The defensive distillation reflects a strategy to pass the information from one architecture to another by reducing the size of DNN. The distillation method provides a dynamic method demanding less human intervention and the advantage of being adaptable with yet not known threats. In general, effective adversarial defense training requires a long list of known vulnerabilities of the system and possible attack vectors. Utilization of defensive distillation decreases the success rate of the adversarial crafting process and is also effective against adversarial attacks, such as JSMA.

As a disadvantage, if a perpetrator has a lot of computing power available and the proper fine-tuning, she can utilize reverse engineering to find fundamental exploits. Defense distillation models are also vulnerable to poisoning attacks in which a malicious actor corrupts a preliminary training database [37]. Defensive distillation can be evaded by the black-box approach [89] and also with optimization attacks [113]. Carlini and Wagner [26] proved that defensive distillation failed against their L0, L2, and L∞ attacks. These new attacks succeed in finding adversarial examples for 100% of images on defensively distilled networks. Previously known weaker attacks can be stopped by defensive distillation, but it cannot resist more powerful attack techniques.

Defense-GAN (Generative Adversarial Networks) is a feasible defense strategy providing advanced defense mechanisms against white-box and black-box adversarial attacks posing a threat towards machine learning classifiers. Defense-GAN is trained to model the distribution of unperturbed images, and before sending the given image to the classifier, the image is projected onto the generator by minimizing the reconstruction error and passing the resulting construction to the classifier. Training the generator to model the unperturbed training data distribution reduces potential adversarial noise. Defense-GAN can be used in conjunction with any ML classifier without a need to alter the classifier structure or re-train it, and utilization of the Defense-GAN mechanism should not significantly decrease the performance of the classifier. The mechanism can be used to combat any attack as it does not presume an attack model, but it can utilize the generative efficiency of GANs to reconstruct adversarial examples [105].

Defense-GAN overcomes adversarial training as a defense method, and when conducting adversarial training using FGSM in generating adversarial examples against, for example, the C&W attack, adversarial training efficiency is not sufficient. In addition, adversarial training does not generalize well against different attack methods. Increased robustness gained by using adversarial training is reached when the attack model used to generate the augmented training set is the same as that used by the perpetrator. Hence, as mentioned, adversarial training endures inefficiently against the C&W attack; therefore, a more powerful defense mechanism should be utilized. Training GANs is a remarkably challenging task, and if GANs are not trained correctly and hyperparameters are chosen incorrectly, the performance of the defensive mechanism may significantly mitigate [105].

11.7.2 Defending Against DoS and DDoS Attacks

Distributed Denial of Services (DDoS) attacks have been increasing, contributing to the majority of overall network attacks. Detecting and preventing DDoS attacks is a challenging task, and practically designing and implementing a DDoS defense is incredibly difficult. DDoS attack and defense issues have been under intensive research, and various research has been conducted in the field of the subject concerned. The purpose of a traditional DDoS detection system is to separate malicious packet traffic from abnormal traffic [74]. Under the traditional network environment, methods for defense against DDoS attacks mainly consist of attack detection and attack response. Attack detection bases on attack signatures, congestion patterns, protocols, and source addresses, forming an efficient DDoS detection mechanism [30].

The detection model has two categories: misuse-based detection and anomaly-based detection. Misuse-based detection utilizes feature-matching algorithms and matches the gathered and extracted user behavior features with the known feature database of DDoS attacks to detect if an attack has been conducted earlier. An attack in a system is detected wherever the sequence of activities in the network matches with a known attack signature. Anomaly-based detection has been used with monitoring systems in order to determine if the states of the target systems and user's activities differ from the normal profile, and it can then deduct if an attack is taking place. The following step is for an attack response to appropriately filter or limit the network traffic as much as possible after the DDoS attack has been commenced [30].

Artificial intelligence and its subfield of machine learning have been applied to cybersecurity in recent years, and it has affected the development of an ML-based attack detection model. Machine learning is able to gather relevant information from the data and integrate previously collected knowledge to discriminate and predict new data. Hence, ML-based methods can provide better detection accuracy in comparison to traditional detection methods. As a drawback, data generated by the DDoS attacks are usually burst and diverse. In addition, background traffic size may also have an impact on the detection model, mitigating the model's detection accuracy [30].

Various studies have been conducted to address the prevention and detection of cyberattacks, such as DDoS attacks, and numerous of them are utilizing ML-based methods, such as support vector machine (SVM), Random Forest, and Naïve Bayes. As an example, [93] conducted research in order to detect DDoS attacks by using Random Forest and SVM ML-methods. Authors of the research trained random forest model with the training data set and mixed the remaining set of attack data packets with the normal traffic as the test set of the model, cross-sampled normal traffic and attack traffic, calculated behavior of each sample, and controlled the sampling flow period to control the ration of normal traffic to attack traffic. LIBSVM library was then utilized to detect the data of the SVM algorithm and compared it with the random forest model detection results. The research results showed that both Random Forest and SVM methods provided significant (93–99%, depending on the sampling period) DDoS attack detection accuracy against TCP, UDP, and ICMP flood attacks.

He et al. [51] proposed a prototype DOS attack detection system on the source side in the cloud, based on machine learning techniques. The prototype was implemented under a real cloud setting, and it included six servers (S0...S5), each server running multiple virtual machines. The authors launched four different kinds of DDoS attacks (SSH brute-force, DNS reflection, ICMP flooding, and TCP SYN attacks) on virtual machines from the S0 server. The victim was a virtual machine on another server S1 running web service. Authors deployed their defense system on the server launching virtual machines running the attacks. Other virtual machines on servers (except S0 and S1) request web service, simulating the legitimate users. The data utilized in the experiment was gathered of network packages coming in and going out of the attacker virtual machines for nine hours. Supervised learning algorithms, such as Linear Regression (LR), SVM (linear, RBF, or polynomial kernels), Decision Tree, Naïve Bayes, and Random forest, were evaluated. For unsupervised algorithms, such as k-means, Gaussian Mixture Model for Expectation–Maximization (GMM-EM), were evaluated, respectively. Supervised algorithms all achieved over 93% accuracy (Random Forest had the best accuracy with 94.96%), but unsupervised ones reached only 63–64% accuracy.

Haider et al. [50] presented a novel deep learning framework for the detection of DDoS attacks in Software Defined Networks (SDNs), which is a prevalent networking paradigm decoupling the control logic from the forwarding logic. SDNs consist of applications (applications running on physical or virtual hosts), control (operating system), and forward planes (network constructed through programmable switches). The framework utilizes ensemble CNN models for improved detection of Flow-based data being critical attributes to SDNs. The authors evaluated the proposed framework with the Flow-based dataset CICIDS2017, which is a public, fully labeled dataset comprised of at least 80 features of network traffic, including both benign and multiple types of attack traffic. The proposed approach provided 99.45% detection accuracy and minimal computational complexity in detecting DDoS attacks with reasonable testing and training time.

11.7.3 Defending Against False Data Injection (FDI) Attacks

FDI attack was introduced in the smart grid domain causing remarkable security challenges to the operation of power systems and can be utilized to circumvent conventional state estimation bad data detection security measures implemented in the power system control room [12]. FDIA detection problem has been attempted to solve by using various kinds of optimization methods, such as sparse matrix optimization problem, which can be solved by using the combination of a nuclear norm minimization and low-rank matrix factorization methods. In order to mitigate the resources required in the FDIA detection process, threshold-based comparisons have been commonly utilized. An experimental study shows that the usage of the Euclidean distance metric with a Kalman filter with the selected threshold helps to identify FDIA better than many other metrics. In addition, comparing residual signals with a predefined threshold can be used to detect the FDIA in a networked cyber-physical system. Nonetheless, a progressive number of FDIA attacks have been able to override threshold-based detection methods [117]. In order to efficiently combat FDIA attacks, more advanced detection methods, such as blockchain, cryptography, and learning-based methods, can be utilized.

Addallah and Shen [1] presented a prevention technique for FDI attack, which guarantees the integrity and availability of the measurement units (measuring the smart power grid's status) and during their transmission to the control center even with the existence of compromised units. McEliece public-key cryptography system is able to guard the integrity of the smart power grid data measurements and prevent the impact of FDIA. As a drawback, cryptographic algorithms require a substantial amount of computing resources due to computational complexity. One of the common buzzwords these days, a blockchain, has been examined by Ahmed et Pathan [4] to generate a shield and protect the data authenticity. The authors empirically demonstrated that the blockchain-based security framework is capable of securing healthcare images from false image injection attacks. The blockchain-based security framework introduced by the authors is decentralized as in nature, provided cryptographic authentication and consensus mechanism in order to counter FDIA attacks more efficiently than other previous methods.

Learning-based methods provide a novel and more sophisticated way of countering FDIA attacks. Esmalifalak et al. [41] proposed an FDIA detector mechanism by utilizing the principle component analysis (PCA) and supervised learning -based support vector machine (SVM) model to statistically separate normal operations of power networks from the case under stealthy attacks. Methods mentioned were utilized to combat a new type of FDIA attacks, such as stealth attacks, which cannot be detected by conventional bad data detection using state estimation. The detection performance of the SVM-based method was relatively high, with 90.06% accuracy in comparison to Euclidean detector's 72.68% and Sparse Optimization 86.79% [117]. Wang et al. [117] utilized wide and recurrent neural networks (RNN) model to learn the state variable measurement data and identify the FDIA. The wide component consists of a fully connected layer of neural networks, and the RNN component includes two LSTM layers. The wide component is able to learn the global knowledge and the RNN component has a capability to catch the sequential correlations from state variable measurement data. Wide component accuracy reached 75.13% and RNN model 92.58%, respectively. The proposed combination of Wide and RNN models detection performance reached up to 95.23% accuracy, which outperforms the previously mentioned learning-based detection methods.

He et al. [51] presented Conditional Deep Belief Network (CDBN) in order to analyze the temporal attack patterns that are presented by the real-time measurement data from the distributed sensors/meters. The aim is to efficiently reveal the high-dimensional temporal behavior features of the unobservable FDI attacks, which are able to bypass the State Vector Estimator (SVE) mechanism. According to Niu et al. [80], no prior studies have been conducted on the dynamic behavior of FDI attacks. Detecting FDI attacks is considered a supervised binary classification problem, which is not able to detect dynamically evolving cyber threats and changing the system configuration. The authors developed an anomaly detection framework based on a neural network in order, to begin with, the construction of a smart grid specific intrusion detection system (IDS). The framework utilizes a recurrent neural network with LSTM cell to capture the dynamic behavior of the power system and a convolutional neural network (CNN) to balance between two input sources. In case a residual between the observed and the estimated measures is greater than a given threshold, an attack is launched.

11.7.4 Defending Against Malware Attacks

Malware infections have been significantly increasing in the past years, and large quantities of malware are automatically created each day. According to [10], almost 10 million malware infection cases have occurred per day during the first quartal in 2020, and 64% of the malicious attacks were targeting educational institutions. These days, 17 million malware programs are registered montly, and up to 560,000 new pieces of malware are detected each day [56]. The number of cybercriminals conducting vicious acts such as malicious attacks has been increasing quickly. The exponential growth of malware has been causing a remarkable threat in our daily life, sneaking in stealth to the computer system without revealing an adverse intent to disrupt the computer operations. Due to the enormous number of malwares, it is impossible to deal with the malware solely by human engineers and security experts, but advances and sophisticated detection methods are required.

Malware detection methods can be categorized in various ways depending on the point of view. One possible way is to divide malware detection methods into signature-based and behavioral (heuristic) -based methods. Signature-based detection has been the most widely utilized way method in antivirus programming. This method extracts a unique signature from a malware file and utilizes it in order to detect similar malware [120]. Signature-based detection can be efficiently used to detect the already known type of malware, but it has challenges in detecting zero-day malware and can also be easily defeated by malware that uses obfuscation techniques. Obfuscation techniques include, for example, dead code insertion, register reassignment, instruction substitution, and code manipulation [110]. Additionally, signature-based detection requires prior knowledge of malware samples [120].

In behavior (heuristic or anomaly) -based detection, malware sample behaviors are analyzed during execution in the training (learning) phase in order to label the file as malicious or benign (legitimate) during the testing phase. In contrast to signature-based detection, behavior-based detection is also able to detect the unknown type of malware in addition to malware utilizing encryption, obfuscation, or polymorphism. A significant number of false positives and considerable monitoring time requirement can be seen as the downsides of the method concerned [110]. The method incorporates a virtual machine (VM) and function call monitoring, information flow tracking, dynamic binary instrumentation, and Windows Application Programming Interface (API) call Graph. Behavior detection method benefits of utilization of traditional machine learning methods, such as Decision Tree (DT), K-Nearest Neighbor (KNN), Naïve Bayes (NB), and Support Vector Machine (SVM) to comprehend the behaviors of running files [120].

Deep learning is a subset of machine learning utilizing multiple layers of neural networks with the capability to perform better on unstructured data [69]. DL has been shown to include various advantages over traditional machine learning in areas such as speech recognition, computer vision, and natural language processing. Deep learning enables computational models to learn high-level features from original data at multiple levels. As a drawback, DL requires more computation time to train and retrain the models, which is a common phase in the malware detection process as new malware types continuously emerge. In contrast, traditional machine learning algorithms are fast but not necessarily accurate enough [24]. The deep learning model is able to learn complicated feature hierarchies and include steps of the malware detection process into one model, which can be then trained end-to-end with all of the components simultaneously [57, 58].

Deep learning has been adopted for the development of Malware Detection Systems (MDSs) due to its success when utilized in other relevant areas. In the beginning, a single deep learning model was applied to the whole dataset, which ended up causing problems as the model experience challenges in dealing with increasingly complicated data distribution of the malware samples. A Group of deep learning models has been used in conjunction (ensemble approach) in order to solve the issue, but the utilization of multiple models have ended up in similar problems. [124] presented a multi-level deep learning system for malware detection. The system can manage more complicated data distributions utilizing tree structure in order to provide means for each DL model to learn the unique data distribution for one group of a malware family. The authors demonstrated that their system improves the performance of malware detection systems compared to SVM, decision tree, the single deep learning model, and the ensemble-based approach. The system also provides more precise detection in less time to efficiently identify malware threats [124].

Kolosnjaji et al. [60] presented a hybrid Deep Learning-based neural network model for the classification of malware system call sequences. Authors combined two convolutional and one recurrent (LSTM) neural network layers into one neural network architecture in order to increase malware classification performance. The malware classification process initiates with a malware zoo, which included open source-based Cuckoo Sandbox, where acquired malware binaries can be executed in a protected environment. Results of the executions are then preprocessed to obtain numerical feature vectors, which are sent to neural networks. Neural networks act as a classifier classifying the malware into one of the predefined malware families. Malware data samples with labels were gathered from Virus Share, Maltrieve, and private collections, which provided a large and diverse number of samples. Authors utilized Tensorflow and Theano frameworks providing GPU utilization when constructing and training the neural networks. The proposed Deep Learning-based hybrid model endures simpler neural network models and, in addition, even more sophisticated and broadly used Hidden Markov Models and Support Vector machines and provided an average accuracy, precision, and recall of over 90% for most malware families.

11.7.5   Defending Against Phishing Attacks

Phishing can be counted as one of the most challenging problems in the cyber-world, causing financial worries for industries and individuals, and detecting phishing attacks accurately enough can be difficult. Phishing websites may look similar in appearance compared to equivalent legitimate websites implemented to fool users into believing they are visiting the correct and safe website [53]. Though there are several anti-phishing software and techniques for detecting potential phishing attempts in emails and detecting phishing contents on websites, phishers utilize new and hybrid techniques to circumvent the available software and techniques [16]. According to Oluwatobi et al. [86], phishing detection techniques tend to suffer relatively low detection accuracy and may induce an extensive number of false alarms, in particular, if novel and sophisticated phishing approaches have been utilized. Traditional phishing detection techniques utilized, such as the blacklist-based method, is not efficient enough countering these kinds of attacks nowadays due to easier registering of domains making blacklist databases quickly outdated.

Phishing detection techniques can be classified into the following approaches: user awareness and software detection. User awareness includes user training concerning phishing threats in order to lead users into correctly identifying phishing and non-phishing messages and mitigating the threat level. Relying on user training in the mitigating effect of phishing attacks is challenging due to human weaknesses. According to [59], end-users failed to detect 29% of phishing attacks even after training. However, phishing detection techniques are usually evaluated against so-called bulk phishing attacks, which can affect the performance with regards to targeted forms of phishing attacks. Using, e.g., proper simulated phishing platform, organization's Phish-Prone percentage (PPP) indicating how many of their employees are likely to fall for phishing or social engineering scam, could be used as a training method. User training can be an effective method, but human errors still exist, and people are prone to forget their training. Training also requires a significant amount of time, and it is not much appreciated by non-technical users.

Machine learning can be utilized as an effective tool in phishing detection due to the classification problem nature of phishing. Traditional ML classifiers, such as decision trees and random forest, can be considered as effective techniques what comes to computational time and accuracy.

Deep-learning-based methods have been recently proposed in the phishing website detection domain. Adebowale et al. [3] introduced an intelligent phishing detection system (IDPS), which uses the image, frame, and text content of a web page to detect phishing activities by utilizing deep learning methods, such as a convolutional neural network (CNN) and the long short-term memory (LSTM) to build a hybrid classification model. The proposed model was built by training the CNN and LSTM classifiers by using 1 m universal resource locators and over 10,000 images. Various types of features have been extracted from websites to predict phishing activities. The knowledge model is used to compare the extracted features to determine whether the websites are phishing, suspicious, or legitimate. Phishing websites are indicated as red, suspicious as yellow, and legitimate as green color. The experimental results showed that the model achieved an accuracy rate of 93.28% and an average detection time of 25 s.

11.8   Conclusion

In this paper, the authors reviewed the concepts of cybersecurity, cyber threats, cyber-physical systems, and artificial intelligence in critical infrastructure. The critical infrastructure field includes systems, networks, assets, services, and infrastructure essential for the continued operation of everyone from citizens to the country. Examples of these high-importance necessities include banking and business services, digital infrastructure, drinking water supply, energy, health, transport and logistics, etc. It can be argued that cyber-physical systems are the future way to guarantee the operation of these services in the modern world because they offer accessibility and ease of use in a near real-time fashion with continuous automation of tedious and arduous processes. Some of the processes can be improved utilizing artificial intelligence, for example, in the access control service of smart buildings or the energy consumption optimization of the smart grid and the local smart buildings.

The attacks towards CPSs are various, and many different attack vectors were identified, out of which the most concerning ones being adversarial attacks, false data injection attacks, malware attacks, and phishing attacks. These malicious attacks all rely on fooling humans on some level, having the capacity to harm the system itself and the human users. Especially, the malware attacks towards nuclear power plants are detesting. The DoS/DDoS attacks do not attempt to deceive human users as the other mentioned attacks; however, they too are harmful, as the case of [54] proved. The attack caused financial losses and disgruntlement in the smart building occupants in the Lappeenranta region.

In essence, the defense methods against these attacks focused on the second and fourth attribute of the cyber resilience concept, namely, "Identify and detect" and "Govern and assure." These attacks can be defended against with machine learning methods, and in the case of phishing attacks, users can be trained to detect some of the attack attempts. The authors recommend utilizing combinations of different ML models and frameworks to mitigate the risks associated with these attacks. For example, having a layered protective structure to first mitigate the DoS/DDoS attacks with trained artificial intelligence model, such as proposed by [93], and then in conjunction a more optimized ensemble structure introduced in, for example, by [124] could improve protection for the cyber-physical systems. The authors recommend that one uses defensive distillation and defense-GAN in the training of the ensemble models when applicable in order to enhance the defensive capabilities of the algorithms. Unfortunately, there exists no perfect solution to mitigate these threats. The CNN model introduced by [3] should be utilized when people governing the CI have an elevated risk of encountering phishing attacks, or those attacks are geared towards the system.

References

1. Abdallah A, Shen XS (2016) Efficient prevention technique for false data injection attack in smart grid. In: 2016 IEEE international conference on communications (ICC). IEEE, pp 1–6. https://doi.org/10.1109/ICC.2016.7510610
2. Abdullah SA, Mohd M (2019) Spear phishing simulation in critical sector: telecommunications and defense sub-sector. In: 2019 international conference on cybersecurity (ICoCSec). IEEE, pp 26–31. https://doi.org/10.1109/ICoCSec47621.2019.8970803
3. Adebowale MA, Lwin KT, Hossain MA (2020) Intelligent phishing detection scheme using deep learning algorithms. J Enterp Inf Manag. https://doi.org/10.1108/JEIM-01-2020-0036. Publishedonline
4. Ahmed M, Pathan ASK (2020) Blockchain: can it be trusted? Computer 53(4):31–35. https://doi.org/10.1109/MC.2019.2922950
5. Alazab M, Khan S, Krishnan SSR, Pham QV, Reddy MPK, Gadekallu TR (2020) A multidirectional LSTM model for predicting the stability of a smart grid. IEEE Access 8:85454–85463. https://doi.org/10.1109/ACCESS.2020.2991067

6. Albawi S, Mohammed TA, Al-Zawi S (2017) Understanding of a convolutional neural network. In: 2017 international conference on engineering and technology (ICET). IEEE, pp 1–6. https://doi.org/10.1109/ICEngTechnol.2017.8308186
7. Alelyani S, Kumar H (2018) Overview of cyberattack on Saudi organizations. J Inform Sec Cybercrimes Res 1(1):32–39. https://doi.org/10.26735/16587790.2018.004
8. Allianz (2020) Cyber attacks on critical infrastructure. Allianz Global Corporate & Specialty (AGCS), http://agcs.allianz.com/news-and-insights/expert-risk-articles/cyber-attacks-on-critical-infrastructure.html. Accessed 4 Oct 2020
9. Anderson M, Bartolo A, Tandon P (2016) Crafting adversarial attacks on recurrent neural networks. https://stanford.edu/~bartolo/assets/crafting-rnn-attacks.pdf. Accessed on 29 June 2021
10. Anton P (2020) Over 400 million malware infections detected in last 30 days, more than 10 million daily. AtlasVPN, https://atlasvpn.com/blog/nearly-404-million-malware-infections-detected-in-last-30-days-more-than-10-million-daily
11. Athiwaratkun B, Stokes JW (2017) Malware classification with LSTM and GRU language models and a character-level CNN. In: 2017 IEEE international conference on acoustics, speech and signal processing (ICASSP). IEEE, pp 2482–2486. https://doi.org/10.1109/ICASSP.2017.7952603
12. Ayad A, Farag HEZ, Youssef A, El-Saadany EF (2018) Detection of false data injection attacks in smart grids using recurrent neural networks. In: 2018 IEEE power & energy society innovative smart grid technologies conference (ISGT), pp 1–5. https://doi.org/10.1109/ISGT.2018.8403355
13. Baezner M, Robin P (2017) Hotspot analysis: Stuxnet. CSS Cyber Defense Project, Center for Security Studies, ETH Zurich, https://css.ethz.ch/content/dam/ethz/special-interest/gess/cis/center-for-securities-studies/pdfs/Cyber-Reports-2017-04.pdf
14. Barredo Arrieta A, Díaz-Rodríguez N, Del Ser J, Bennetot A, Tabik S, Barbado A, Garcia S, Gil-Lopez S, Molina D, Benjamins R, Chatila R, Herrera F (2020) Explainable artificial intelligence (XAI): concepts, taxonomies, opportunities and challenges toward responsible AI. Inform Fusion 58:82–115. https://doi.org/10.1016/j.inffus.2019.12.012
15. Barth B (2017). DDoS attacks delay trains, stymie transportation services in Sweden. SC. https://www.scmagazine.com/home/security-news/cybercrime/ddos-attacks-delay-trains-stymie-transportation-services-in-sweden
16. Basnet R, Mukkamala S, Sung AH (2008) Detection of phishing attacks: a machine learning approach. In: Prasad B (ed) Soft computing applications in industry. Studies in fuzziness and soft computing. Springer, Berlin, p 226. https://doi.org/10.1007/978-3-540-77465-5_19
17. Bencsáth B, Pék G, Buttyán L, Félegyházi M (2012) The cousins of Stuxnet: Duqu, Flame, and Gauss. Future Internet 4(4):971–1003. https://doi.org/10.3390/fi4040971
18. Berkeley AR, Wallace M (2010) A framework for establishing critical infrastructure resilience goals. Final report and recommendations by the council. National Infrastructure Advisory Council, Washington, DC, https://www.dhs.gov/xlibrary/assets/niac/niac-a-framework-for-establishing-critical-infrastructure-resilience-goals-2010-10-19.pdf
19. Biasi J (2018) Malware attacks on critical infrastructure security are growing. Burns & McDonnel. http://amplifiedperspectives.burnsmcd.com/post/malware-attacks-on-critical-infrastructure-security-are-growing
20. Biggio B, Corona I, Maiorca D, Nelson B, Srndic N, Laskov P, Giacinto G, Roli F (2017) Evasion attacks against machine learning at test time. arXiv:1708.06131v1
21. Bossetta M (2018) The weaponization of social media: spear phishing and cyberattacks on democracy. J Int Affairs 71(1.5):97–106
22. Business Finland (2016) Market opportunities in the smart grid sector in Finland 2016. Business Finland, https://www.businessfinland.fi/48cd02/globalassets/julkaisut/invest-in-finland/white-paper-smart-grid.pdf
23. CISA (2020) Critical infrastructure sectors. Cybersecurity & Infrastructure Security Agency, https://www.cisa.gov/critical-infrastructure-sectors. Accessed on 10 Sept 2020

24. Cakir B, Dogdu E (2018) Malware classification using deep learning methods. In: ACMSE '18: proceedings of the ACMSE 2018 conference, Article 10, pp 1–5. https://doi.org/10.1145/3190645.3190692
25. Cambridge (2020) Cybersecurity. Cambridge Dictionary. http://dictionary.cambridge.org/us/dictionary/english/cybersecurity. Accessed on 17 Sept 2020
26. Carlini N, Wagner D (2017) Towards evaluating the robustness of neural networks. arXiv: 1608.04644v2
27. Carnegie (2015) Computer security incident response plan. Carnegie Mellon, http://cmu.edu/iso/governance/procedures/docs/incidentresponseplan1.0.pdf
28. Chakraborty A, Alam M, Dey V, Chattopadhyay A, Mukhopadhyay D (2018) Adversarial attacks and defences: a survey. arXiv:1810.00069v1
29. Chen PY, Yang S, McCann JA, Lin J, Yang X (2015) Detection of false data injection attacks in smart-grid systems. IEEE Commun Mag 53(2):206–213. https://doi.org/10.1109/MCOM.2015.7045410
30. Cheng J, Zhang C, Tang X, Sheng VS, Dong Z, Li J (2018) Adaptive DDoS attack detection method based on multiple-kernel learning. Sec Commun Netw 2018, Article 5198685. https://doi.org/10.1155/2018/5198685
31. Christensson P (2013) SYN flood definition. TechTerms http://www.techterms.com/definition/syn_flood
32. Cisco (2020) What is cybersecurity? Cisco Systems, San Jose, CA, https://www.cisco.com/c/en/us/products/security/what-is-cybersecurity.html. Accessed on 17 Sept 2020
33. Co KT (2017) Bayesian optimization for black-box evasion of machine learning systems. Master's thesis, Imperial College London
34. Colorado (2020) Critical infrastructure protection. Planning for Hazards: Land Use Solutions for Colorado, http://planningforhazards.com/critical-infrastructure-protection. Accessed on 6 Nov 2020
35. Connecticut (2020) Critical infrastructure. Connecticut State, Division of Emergency Management and Homeland Security, https://portal.ct.gov/DEMHS/Homeland-Security/Critical-Infrastructure. Accessed on 10 Sept 2020
36. Cytomic (2019) The cybercriminal protagonists of 2019: ransomware, phishing and critical infrastructure. Cytomic, https://www.cytomic.ai/trends/protagonists-cybercrime-2019/
37. DeepAI (2019) What is defensive distillation? DeepAI, https://deepai.org/machine-learning-glossary-and-terms/defensive-distillation. Accessed on 9 Oct 2019
38. EC (2016) The Directive on security of network and information systems (NIS Directive). European Commission, https://ec.europa.eu/digital-single-market/en/news/directive-security-network-and-information-systems-nis-directive. Accessed on 10 Sept 2020
39. EC (2017) Smart building: energy efficiency application. Digital Transformation Monitor, European Commission. https://ec.europa.eu/growth/tools-databases/dem/monitor/sites/default/files/DTM_Smart%20building%20-%20energy%20efficiency%20v1.pdf. Accessed on 10 Nov 2020
40. EU (2016) Directive (EU) 2016/1148 of the European Parliament and of the Council of 6 July 2016 concerning measures for a high common level of security of network and information systems across the Union. The European Parliament and the Council of the European Union
41. Esmalifalak M, Liu L, Nguyen N, Zheng R, Han Z (2017) Detecting stealthy false data injection using machine learning in smart grid. IEEE Syst J 11(3):1644–1652. https://doi.org/10.1109/JSYST.2014.2341597
42. Ettouney MM, Alampalli S (2016) Resilience and risk management. Building innovation conference & expo. https://cdn.ymaws.com/www.nibs.org/resource/resmgr/Conference2016/BI2016_0113_ila_ettouney.pdf. Accessed on 18 Sept 2020
43. European Commission (2017) Digital transformation monitor. Smart building: energy efficiency application. https://ec.europa.eu/growth/tools-databases/dem/monitor/sites/default/files/DTM_Smart%20building%20-%20energy%20efficiency%20v1.pdf. Accessed on 10 Nov 2020

44. Farmanbar M, Parham K, Arild Ø, Rong C (2019) A widespread review of smart grids towards smart cities. Energies 12(23):4484. https://doi.org/10.3390/en12234484
45. Flores C, Flores C, Guasco T, León-Acurio J (2017) A diagnosis of threat vulnerability and risk as it related to the use of social media sites when utilized by adolescent students enrolled at the Urban Center of Canton Canar. In: Technology trends: proceedings of the third international conference, CITT 2017. Springer, Cham, pp 199–214
46. GOV-AU (2020) Critical infrastructure resilience. Australian Government, https://www.homeaffairs.gov.au/about-us/our-portfolios/national-security/security-coordination/critical-infrastructure-resilience. Accessed on 10 Sept 2020
47. Gartner (2020) Cybersecurity. Gartner Glossary, https://www.gartner.com/en/information-technology/glossary/cybersecurity. Accessed on 17 Sept 2020
48. Goodfellow I, McDaniel P, Papernot N (2018) Making machine learning robust against adversarial inputs. Commun ACM 61(7):56–66
49. Griffor ER, Greer C, Wollman DA, Burns MJ (2017) Framework for cyber-physical systems: Volume 1, overview. NIST Special Publication 1500-201, National Institute of Standards and Technology. https://doi.org/10.6028/NIST.SP.1500-201
50. Haider S, Akhunzada A, Mustafa I, Patel TB, Fernandez A, Choo KKR, Iqbal J (2020) A deep CNN ensemble framework for efficient DDoS attack detection in software defined networks. IEEE Access 8:53972–53983. https://doi.org/10.1109/ACCESS.2020.2976908
51. He Z, Zhang T, Lee RB (2017) Machine learning based DDoS attack detection from source side in cloud. In: 2017 IEEE 4th international conference on cyber security and cloud computing (CSCloud). IEEE, pp 114–120. https://doi.org/10.1109/CSCloud.2017.58
52. Ibitoye O, Shafiq O, Matrawy A (2019) Analyzing adversarial attacks against deep learning for intrusion detection in IoT networks. arXiv:1905.05137
53. Jain AK, Gupta BB (2017) Phishing detection: analysis of visual similarity based approaches. Sec Commun Netw, Article 5421046. https://doi.org/10.1155/2017/5421046
54. Janita (2016) DDoS attack halts heating in Finland amidst winter. Metropolitan.fi, http://metropolitan.fi/entry/ddos-attack-halts-heating-in-finland-amidst-winter
55. Jordan MI, Mitchell TM (2015) Machine learning: trends, perspectives, and prospects. Science 349(6245):255–260
56. Jovanovic B (2021) A not-so-common cold: malware statistics in 2021. DataProt. https://dataprot.net/statistics/malware-statistics. Accessed on 29 June 2021
57. Kaspersky (2020) Machine learning methods for malware detection. http://media.kaspersky.com/en/enterprise-security/Kaspersky-Lab-Whitepaper-Machine-Learning.pdf. Accessed on 23 Oct 2020
58. Kaspersky (2020) What is social engineering? Kaspersky, http://kaspersky.com/resource-center/definitions/what-is-social-engineering. Accessed on 7 Oct 2020
59. Khonji M, Iraqi Y, Jones A (2013) Phishing detection: a literature survey. IEEE Commun Surv Tutor 15(4):2091–2121. https://doi.org/10.1109/SURV.2013.032213.00009
60. Kolosnjaji B, Zarras A, Webster G, Eckert C (2016) Deep learning for classification of malware system call sequences. In Kang B, Bai Q (eds) AI 2016—advances in artificial intelligence: proceedings of the 29th Australasian joint conference. Lecture notes in computer science, 9992. Springer, Cham, pp 137–149. https://doi.org/10.1007/978-3-319-50127-7_11
61. Kolosnjaji B, Zarras A, Webster G, Eckert C (2016) Deep learning for classification of malware system call sequences. In: Kang B, Bai Q (eds) AI 2016: advances in artificial intelligence. AI 2016. Lecture notes in computer science, 9992. Springer, Cham. https://doi.org/10.1007/978-3-319-50127-7_11
62. LeCun Y, Bengio Y, Hinton G (2015) Deep learning. Nature 521(7553):436–444
63. Legatiuk D, Smarsly K (2018) An abstract approach towards modeling intelligent structural systems. In: 9th European workshop on structural health monitoring. NDT.net
64. Lehto M (2015) Phenomena in the cyber world. In: Lehto M, Neittaanmäki P (eds) Cyber security: analytics, technology and automation. Springer, Berlin, pp 3–29
65. Limnéll J, Majewski K, Salminen M (2014) Kyberturvallisuus. Docendo

66. Lipton ZC, Berkowitz J, Elkan C (2015) A critical review of recurrent neural networks for sequence learning. arXiv:1506.00019
67. Ma S, Liu Y, Tao G, Lee WC, Zhang X (2019) NIC: detecting adversarial samples with neural network invariant checking. In: 26th annual network and distributed system security symposium, NDSS 2019. The Internet Society. https://doi.org/10.14722/ndss.2019.23415
68. Markets (2020) Smart cities market worth $820.7 billion by 2025. Exclusive Report by MarketsandMarketsTM, https://www.marketsandmarkets.com/PressReleases/smart-cit ies.asp. Accessed on 30 Nov 2020
69. Mathew A, Amudha P, Sivakumari S (2021) Deep learning techniques: an overview. In: Advanced machine learning technologies and applications: proceedings of AMLTA 2020. Springer, pp 599–608. https://doi.org/10.1007/978-981-15-3383-9_54
70. McCreanor N (2018) Danish rail network DSB hit by cyber attack. IT governance, https://www.itgovernance.eu/blog/en/danish-rail-network-dsb-hit-by-cyber-attack. Accessed on 22 Sept 2020
71. Metropolitan (2016) DDoS Attack halts heating in Finland amidst winter. Metroplitan.fi—News from Finland in English. http://metropolitan.fi/entry/ddos-attack-halts-heating-in-fin land-amidst-winter. Accessed on 22 Sept 2020
72. Miller WB (2014) Classifying and cataloging cyber-security incidents within cyber-physical systems. Master's thesis, Brigham Young University
73. Milošević N (2013) History of malware. arXiv:1302.5392
74. Mirkovic J, Reiher P (2004) A taxonomy of DDoS attack and DDoS defense mechanisms. ACM SIGCOMM Comput Commun Rev 34(2):39–53. https://doi.org/10.1145/997 150.997156
75. Mohammad OA, Youssef T, Ibrahim A (eds) (2018) Special issue "smart grid networks and energy cyber physical systems". Issue information, MDPI, https://www.mdpi.com/journal/sensors/special_issues/smart_grid_networks
76. El Mrabet Z, Kaabouch N, El Ghazi H, El Ghazi H (2018) Cyber-security in smart grid: survey and challenges. Comput Electr Eng 67:469–482
77. Myung JW, Hong S (2019) ICS malware Triton attack and countermeasures. Int J Emer Multidiscipl Res 3(2):13–17. https://doi.org/10.22662/IJEMR.2019.3.2.0.13
78. NIST (2013) Foundations for innovation in cyber-physical systems: workshop report. National Institute of Standards and Technology, https://www.nist.gov/system/files/documents/el/CPS-WorkshopReport-1-30-13-Final.pdf
79. Nathan S (2020) What is cyber resilience? Why it is important? Teceze, https://www.teceze.com/what-is-cyber-resilience-why-it-is-important. Accessed on 11 Sept 2020
80. Niu X, Li J, Sun J, Tomsovic K (2019) Dynamic detection of false data injection attack in smart grid using deep learning. In: 2019 IEEE power & energy society innovative smart grid technologies conference (ISGT). IEEE, pp 1–6. https://doi.org/10.1109/ISGT.2019.8791598
81. NortonSantos (2016) Blackenergy APT malware. RSA Link, http://community.rsa.com/thr ead/186012. Accessed on 18 Sept 2020
82. Nweke LO (2017) Using the CIA and AAA models to explain cybersecurity activities. PM World J 6(12)
83. OSAC (2018) Ukraine 2018 crime & safety report. Overseas Security Advisory Council, U.S. Department of State, Washington, DC, http://www.osac.gov/Country/Ukraine
84. Obaid HS, Abeed EH (2020) DoS and DDoS attacks at OSI layers. Int J Multidiscipl Res Publ 2(8):1–9. https://doi.org/10.5281/zenodo.3610833
85. Oh IS, Kim SJ (2018) Cyber security policies for critical energy infrastructures in Korea focusing on cyber security for nuclear power plants. In: Gluschke G, Casin MH, Macori M (eds) Cyber security policies and critical infrastructure protection. Institute for Security and Safety, Potsdam, pp 77–95
86. Oluwatobi AA, Amiri IS, Fazeldehkordi E (2015) A machine-learning approach to phishing detection and defense. Elsevier Inc. https://doi.org/10.1016/C2014-0-03762-8
87. Paganini P (2018) Massive DDoS attack hit the Danish state rail operator DSB. Security Affairs, https://securityaffairs.co/wordpress/72530/hacking/rail-operator-dsb-ddos.html

88. Paloalto (2020) What is cybersecurity? Palo Alto Networks, https://www.paloaltonetworks.com/cyberpedia/what-is-cyber-security. Accessed on 17 Sept 2020
89. Papernot N, McDaniel P, Goodfellow I, Jha S, Celik ZB, Swami A (2016) Practical black-box attacks against machine learning. arXiv:1602.02697
90. Papernot N, McDaniel P, Wu X, Jha S, Swami A (2016) Distillation as a defense to adversarial perturbations against deep neural networks. arXiv:1511.04508
91. Papernot N, McDaniel P, Goodfellow I, Jha S, Celik ZB, Swami A (2017) Practical black-box attacks against machine learning. In: ASIA CCS '17: proceedings of the 2017 ACM on Asia conference on computer and communications security. ACM, New York, pp 506–519
92. Pawlak A (2020) Adversarial attacks for fooling deep neural networks. NeuroSYS, https://neurosys.com/article/adversarial-attacks-for-fooling-deep-neural-networks
93. Pei J, Chen Y, Ji W (2019) A DDoS attack detection method based on machine learning. J Phys Conf Ser 1237(3):032040
94. Porter E (2019) What is a DDoS attack and how to prevent one in 2020. SafetyDetectives, http://www.safetydetectives.com/blog/what-is-a-ddos-attack-and-how-to-prevent-one-in/#what. Accessed on 22 Sept 2020
95. Probst M (2015) Generative adversarial networks in estimation of distribution algorithms for combinatorial optimization. arXiv:1509.09235
96. Qureshi AS (2018) How to mitigate DDoS vulnerabilities in layers of OSI model. DZone, http://dzone.com/articles/how-to-mitigate-ddos-vulnerabilities-in-layers-of. Accessed on 22 Sept 2020
97. RSI (2019) What is cyber resilience and why is it important? RSI Security, https://blog.rsisecurity.com/what-is-cyber-resilience-and-why-is-it-important. Accessed on 11 Sept 2020
98. Rader MA, Rahman SM (2013) Exploring historical and emerging phishing techniques and mitigating the associated security risks. Int J Netw Sec Appl 5(4):23–41
99. Rajkumar R, Lee I, Sha L, Stankovic J (2010) Cyber-physical systems: the next computing revolution. 47th ACM/IEEE design automation conference (DAC). https://doi.org/10.1145/1837274.1837461
100. Rehak D, Senovsky P, Slivkova S (2018) Resilience of critical infrastructure elements and its main factors. Systems 6(2):21. https://doi.org/10.3390/systems6020021
101. Ren K, Zheng T, Qin Z, Liu X (2020) Adversarial attacks and defenses in deep learning. Engineering 6(3):346–360. https://doi.org/10.1016/j.eng.2019.12.012
102. Reo J (2018) DDoS attacks on Sweden's transit system signal a significant threat. Corero, https://www.corero.com/blog/ddos-attacks-on-swedens-transit-system-signal-a-significant-threat/
103. Riskviews (2013) Five components of resilience: robustness, redundancy, resourcefulness, response and recovery. In: Riskviews: commentary of risk and ERM. WordPress, http://riskviews.wordpress.com/2013/01/24/five-components-of-resilience-robustness-redundancy-resourcefulness-response-and-recovery. Accessed on 18 Sept 2020
104. Salmensuu C (2018) NIS directive in the Nordics: Finnkampen in the air? TietoEVRY, https://www.tietoevry.com/en/blog/2018/09/nis-directive-in-the-nordics-finnkampen-in-the-air. Accessed on 10 Sept 2020
105. Samangouei P, Kabkab M, Chellappa R (2018) Defense-GAN: protecting classifiers against adversarial attacks using generative models. arXiv:1805.06605v2
106. Sargolzaei A, Yazdani K, Abbaspour A, Crane CD, Dixon WE (2019) Detection and mitigation of false data injection attacks in networked control systems. IEEE Trans Industr Inf 16(6):4281–4292
107. Shafahi A, Najibi M, Ghiasi A, Xu Z, Dickerson J, Studer C, Davis LS, Taylor G, Goldstein T (2019) Adversarial training for free! arXiv:1904.12843
108. Shahrivari V, Darabi MM, Izadi M (2020) Phishing detection using machine learning techniques. arXiv:2009.11116
109. Short A, La Pay T, Gandhi A (2019) Defending against adversarial examples. Sandia report, SAND 2019-11748, Sandia National Laboratories, Albuquerque, NM

110. Sihwail R, Omar K, Ariffin KAZ (2018) A survey on malware analysis techniques: static, dynamic, hybrid and memory analysis. Int J Adv Sci Eng Inform Technol 8(4–2):1662–1671. https://doi.org/10.18517/ijaseit.8.4-2.6827

111. Song C, He K, Wang L, Hopcroft JE (2019) Improving the generalization of adversarial training with domain adaptation. International conference on learning representations, New Orleans, Lousiana, United States

112. Stutz D, Hein M, Schiele B (2019) Confidence-calibrated adversarial training and detection: more robust models generalizing beyond the attack used during training. arXiv:1910.062 59v2arXiv:

113. Szegedy C, Zaremba W, Sutskever I, Bruna J, Erhan D, Goodfellow I, Fergus R (2013) Intriguing properties of neural networks. arXiv:1312.6199

114. Valle C, Saravia F, Allende H, Monge R, Fernández C (2010) Parallel approach for ensemble learning with locally coupled neural networks. Neural Process Lett 32:277–291. https://doi. org/10.1007/s11063-010-9157-6

115. Valtioneuvosto (2013) Valtioneuvoston päätös huoltovarmuuden tavoitteista. Säädös 857/2013, Oikeusministeriö

116. Vorobeychik Y, Kantarcioglu M (2018) Adversarial machine learning. Morgan & Claypool

117. Wang Y, Chen D, Zhang C, Chen X, Huang B, Cheng X (2019) Wide and recurrent neural networks for detection of false data injection in smart grids. In: Biagioni E, Zheng Y, Cheng S (eds) Wireless algorithms, systems, and applications. Lecture notes in computer science, 11604. Springer, Cham, pp 335–345. https://doi.org/10.1007/978-3-030-23597-0_27

118. Westlund D, Wright A (2012) Duqu, son of Stuxnet, increases pressure for cyber security at all utilities. Newsletter of the Northeast Public Power Association, http://www.naylornet work.com/ppa-nwl/articles/index-v5.asp?aid=163517&issueID=23606

119. Wiyatno R, Xu A (2018) Maximal Jacobian-based saliency map attack. arXiv:1808.07945v1

120. Xiao F, Lin Z, Sun Y, Ma Y (2019) Malware detection based on deep learning of behavior graphs. Math Problems Eng, Article 8195395, 10 pp. https://doi.org/10.1155/2019/8195395

121. Yu X, Xue Y (2016) Smart grids: a cyber-physical systems perspective. Proc IEEE 104(5):1058–1070. https://doi.org/10.1109/JPROC.2015.2503119

122. Zhang Q, Yang Y, Ma H, Wu YN (2019) Interpreting CNNs via decision trees. In: 2019 IEEE/CVF conference on computer vision and pattern recognition (CVPR). IEEE, pp 6254–6263. https://doi.org/10.1109/CVPR.2019.00642.

123. Zhang Y, Liu Q, Song L (2018) Sentence-state LSTM for text representation. arXiv:1805. 02474

124. Zhong W, Gu F (2019) A multi-level deep learning system for malware detection. Expert Syst Appl 133:151–162

Chapter 12
Saving Lives in a Health Crisis Through the National Cyber Threat Prevention Mechanism Case COVID-19

Jussi Simola

Abstract Today's ongoing coronavirus pandemic has shown that our overall public security mechanism in Finland requires a more coherent system that combines different types of sensors with artificial intelligence-based systems. Various states may have a crucial task: creating a common early warning system with a cyber dimension. But first, the decision-making process for public safety administration must be enhanced at the national level. COVID-19 has demonstrated the difficulty of predicting the progression of a pandemic, and nearly every country on earth has faced remarkable challenges from the spread of disinformation. False information has been shared around many public health and safety-related issues—such as how the virus is spread, the usefulness of self-protection, and the side effects of vaccines. Effective early warning tools are needed to prevent the domino effect of misinformation and to ensure the vital functions of society. This research will demonstrate the need for a common emergency response model for Europe to ensure national public safety—along with a technical platform at least for the interface between the countries. Hybrid-influenced incidents require a hybrid response.

Keywords Pandemic · Emergency response · Early warning · Information sharing · Situational awareness

12.1 Introduction

In Finland, the Ministry of Social Affairs and Health (STM) and the Finnish Institute for Health and Welfare (THL) are the organizations responsible for ensuring the virus does not spread. Finland's Emergency Response Administration is responsible for the crucial administrative functions around warning and alerting the public.

It is vital to note that the ongoing COVID-19 pandemic crisis constitutes just one version of the emerging viruses that are spreading. In Finland, official reports have shown no crucial weaknesses in the national preparedness level; the society's current

J. Simola (✉)
Faculty of Information Technology, University of Jyväskylä, Jyväskylä, Finland
e-mail: jussi.hm.simola@jyu.fi

© The Author(s), under exclusive license to Springer Nature Switzerland AG 2022
M. Lehto and P. Neittaanmäki (eds.), *Cyber Security*, Computational Methods
in Applied Sciences 56, https://doi.org/10.1007/978-3-030-91293-2_12

state of vital functions is stable. Yet there is a need to enhance, for example, strategic management, political commitment, international activities, situational awareness, the protection of vital functions, legislation, and strengthening cyber security as a national competitive advantage, and as a part of overall security [32]. The vital functions of society allow it to maintain its resiliency. Meanwhile, the problems that now have emerged in central administration and middle-level administration reflect challenges around reliable information sharing and the use of evidence-based information.

Situational awareness has been lacking, for nearly the entire period of response to the COVID-19 crisis. A concise and easy-to-understand summary of the general guidelines has not been provided to citizens. This is compounded by other challenges. First, legitimate jurisdictional issues have caused political confrontation; the responsibilities of officials and politicians have been unclear for some time. Second, pandemic preparedness plans and action plans will not produce added value if they are not implemented. The political and administrative debate around separation of powers between government ministries has caused major problems in the coordination of decision-making. It is not enough merely to attempt to survive the daily challenges around the virus pandemic, while the potential for new incidents of misinformation, cybercrime incidents or public health crisis increases [50]. For example, the limited patient care capacity of hospitals makes it difficult to cope with a simultaneous accident. Yet government resources are insufficient to be distributed everywhere they are needed.

At present, Finland's social and healthcare system is overloaded. Tens of thousands of patient records were stolen from the Finnish therapy center Vastaamo [33]. The patient records of several officials and politicians have been leaked to the secret Tor network, and victims of such crimes have been subjected to blackmail [33]. Sensitive and personal data must be protected in the Finnish healthcare system and in addition at the European level. Along with grave privacy breaches like these, nearly every country has faced massive challenges due to the spread of misinformation through media and social media. Such misinformation has driven a divergence in people's perceptions and understanding of critical facts around the pandemic—as well as around the response chosen by decision-makers. False information has been shared around crucial public health and safety-related issues, including how the virus is spread, the benefits of self-protection, and vaccinations.

In this chapter, our research problem is formulated in Sect. 12.2. Section 12.3 discusses basic problems around the formation of situational awareness in a pandemic situation. Section 12.4 handles the central concepts of our review. Section 12.5 describes previous studies conducted by the researcher. Section 12.6 presents the findings and Sect. 12.7 provides discussion and conclusions.

12.2   Problem Formulation

The public debate on COVID-19 has pitted economic development and security against each other. Good economic development can help create security, because sufficient wealth provides an opportunity to create well-being and security. Lack of wealth will increase insecurity.

How can we find a balance in the flow of information? Information warfare has created barriers to forming a coherent situational picture of the COVID-19 pandemic. Figure 12.1 illustrates the formation of crisis information nationally among citizens, media (including social media), and states' decision-makers. It also shows the second crucial element: foreign influencers, including the press, scientific researchers, authorities, and politicians.

The overall formation of a situational picture has been notably difficult. Finland's government officials and members of the government have relied heavily on the World Health Organization's (WHO's) statements about the global spread of the COVID-19 pandemic. Yet is it sufficient to use one or two international organizations as sources, to support decision-making at the state level? The WHO predicted an ongoing pandemic a year ago [15]. It has been argued that WHO executives' connections with the Chinese administration would have prevented a rapid, transparent, and effective information exchange with other countries [3]. This is why we need an early warning system, at least at the European level—one that more quickly takes into account changing threat factors across the world. We need to be able to analyze raw data more quickly, we need to be able to find health abnormalities faster.

The fight against cross-border health threats requires excellent preparation and coordinated action—before, during, and after the crisis. We must be able to process and analyze scientific research more quickly. We must also be able to compile data into a sensible map of measures to be taken, and these strategic measures must be implemented quickly enough to suppress crises like pandemics on time. Solutions

Fig. 12.1 Formation of crisis information

utilizing artificial intelligence can help enormously, in such a rapidly evolving event process.

It is problematic that no separate operational "power team," or even national science adviser, has been used to advise the government of Finland. Italy was left nearly alone in its struggles against COVID-19, despite claims that the EU was acting as one front. While the European Union did not effectively work towards a common goal, it did coordinate some issues concerning all member states and placed a joint order on masks. Yet Finland was left out EC [10]. The availability of protective equipment created an almost warlike situation among different European countries.

The purpose of this publication is to look for those factors and influences that pose obstacles to our preventing the spread of a pandemic. Our focus is on a proposed hybrid model of alarm functions—as seen in Fig. 12.2—taking advantage of the scope of a cyber early warning system [53]. The study particularly emphasizes the decision-making capacity and formation of situational awareness of the Finnish government, the National Institute for Health and Welfare, and the Ministry of Social Affairs and Health. Specifically, we tackle the question of how to reduce the role of disinformation and misinformation in the state-level decision-making process. We explore how it is possible to use a hybrid emergency response model to solve multiple problems around crisis management, especially when several threats occur at the same time. For example, the combined crises of a coronavirus pandemic and cyberattacks can easily overload public safety organizations' workflow. Preventing the domino effect can become still more challenging, if separate or overlapping problem-solving methods are used in crisis management.

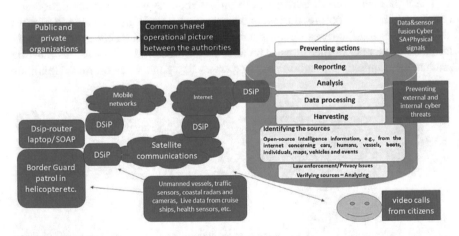

Fig. 12.2 Hybrid emergency response model (HERM)

12.3   Challenges in the Decision-Making Process with COVID-19

As the COVID-19 pandemic has shown, an international cross-border crisis can spread very quickly. It is thus crucial that decision-makers effectively share information—including around the fact that public safety organizations' preparedness levels are not sufficiently high.

12.3.1   Situation in Finland

Finland's citizens noted an enormous lack of correct information around COVID19 at the end of February 2020 [63, 66]. Ministers and responsible authorities failed to immediately offer guidelines for controlling COVID-19. In March 2020, the ministers of social affairs and health did not know how the tasks should be divided between them [20, 31]. Several countries recommended the use of protective masks. Following this, the National Emergency Supply Agency argued that it did not have enough protective masks in stock. Finland did not recommend the use of masks [28], and the masks were later reported to be out of date [35]. Eventually, the Ministry of Social Affairs and Health began to order the masks, but they did not pass the test carried out by VTT Technical Research Centre of Finland [62]. The manager of the Ministry of Social Affairs and Health and The Finnish Institute for Health and Welfare (THL) also held a different view, regarding the benefits of using masks [39, 44]. THL recommended the use of masks, but the Ministry of Social Affairs and Health doesn't.

At present, our government employs more political assistants than ever before [57]. Managing the administration is thus becoming cumbersome. State leaders need decision-making support, such as via artificial intelligence tools, to enhance administrative efficiency. External pressure has had marginal effects on the overall decision-making process, except for in the case of a few decision-makers [65]. Information about the pandemic has been made available for the decision-makers, but the response has been slow and little scientific information from abroad has been shared with the public.

In Finland, the guidelines set by the WHO have been interpreted from a national political perspective. Exceptional conditions were imposed, including a separate regional movement restriction, on the Uusimaa region. The purpose was to prevent the COVID-19 from spreading outside the metropolitan area. Despite that, it was possible to fly relatively freely between Finland and other countries for months. The classification of pandemic countries, based on disease quantity, was incomplete. Statements made by a few doctors about the development of the COVID-19 pandemic have also posed challenges to forming a coherent picture of the situation [18, 34]. They believe that by letting the coronavirus rip through the population to infect people, it is possible to achieve so-called herd immunity.

The decisions made by various Nordic countries to prevent the spread of COVID-19 have differed and continue to differ. This is also true amongst EU member countries. Sweden began to seek herd immunity for its citizens and allowed the disease to spread almost freely [21]. Finland started by following the Swedish COVID-19 strategy, but its selected strategy changed after the president intervened in the government's decision-making process [65]. After considering the situation—as well as the grounds for declaring a state of emergency by the President of the Republic and the government—the government announced a state of emergency in Finland on 16 March, 2020 [23]. The Finnish Parliament applied the Emergency Powers Act on 18 March, 2020. Regional restrictions were then put into effect, preventing needless travel among the country's regions [61].

Only one technical solution is currently in use for COVID-19 prevention. The Finnish Corona Blinker, "Koronavilkku"—an application developed by Solita and the Finnish Institute for Health and Welfare—was released in August 2020 [55]. Soon after, crucial problems were found in the app's ability to track infected people. When a person infected with coronavirus reported their infection to the app, the warning failed to reach other users of the app. Another crucial problem was the delay between a user reporting an infection and the app's recording of it. A oneweek delay slows or prevents infection chain tracing [64]. Another challenge to infection tracing is that users do not have to inform the app when they learn they have COVID-19.

There is also an online service called "omaolo". You can do an online medical check-up for COVID-19 symptoms on the internet, if you suspect you have a coronavirus infection [8]. It is free of charge and the service guides the patient to take a test or go to a hospital, if there is a need.

12.3.2 Case Vastaamo

As mentioned above, tens of thousands of patient records were stolen from the Finnish psychotherapy center Vastaamo [40]. Criminals can use stolen personal data in many ways. For example, they can try to blackmail or otherwise influence the victims. Finland's National Bureau of Investigation (KRP) has received over a thousand reports of offenses connected to the hacking and blackmailing case revolving around Vastaamo [41].

Kanta produces digital services for the social welfare and healthcare sector in Finland. According to [30], each organization associated with Kanta services has at least one Kanta-access point. Access to the service can either be carried out as an organization's activity or implemented by the organization. That means, the Kanta subscriber has an integration solution through which several systems, organizational units, or organizations are connected to the Kanta services. The purpose of the integration solution is to route messages to application servers that may be located in different organizational units or organizations. It is also possible to connect to the service via an external access point. In this model, the organization has joined the Kanta services through a Kanta access point implemented by an intermediary.

The organization may have externalized information system (e.g., a shared information system as a SaaS), messaging, and/or communications to an intermediary. There can be several access points (and server certificates) if, for example:

- the organization's units are directly connected to Kanta services from different information systems, without a centralized integration solution (messaging solution);
- the organization's reception services (for example, receipt of renewal requests) are located on a server other than that from which its systems connect to Kanta services [30].

Valvira is a national agency operating under the Ministry of Social Affairs and Health. Vastaamo is a service provider approved and supervised by Valvira. Its information system is part of the Category B systems regulated by law, for which the law does not require an external assessment of data security. Vastaamo's patient information system was developed by Vastaamo itself. It is one of 260 social and health care information systems that are monitored by the authorities only if there are particular information security-related reasons to suspect problems, or if the service provider requests it [48].

Class B patient information systems are registered with Valvira under the Customer Information Act. They may be purchased as commercial products or manufactured by the company itself. According to Valvira, their monitoring is very limited due to resource problems. It is possible that patient information from Kanta could also be stored in a private register, allowing just one healthcare professional at a time—and one who is in a care-giving role with the patient—to process patient data.

12.4   Central Concepts

This section introduces the central concepts related to the research framework and defines the meaning of the concepts, and used terminology.

12.4.1   Artificial Intelligence

Artificial intelligence (AI) is part of a system that engages in intelligent behavior by analyzing the environment and taking multiple actions—with a dimension of autonomy—to achieve specific goals [9]. AI-based systems can be software-based and act in the virtual world (e.g., image analysis software, search engines, shape and face recognition systems). AI can also be embedded in hardware devices (e.g., advanced robots, autonomous cars, unmanned vehicles, drones or Internet of Things applications) [9].

An *Intelligent Agent* (IA) is an entity that produces decisions. This allows, for example, for the performance of specific tasks for users or applications. An IA has the ability to learn during the process of performing tasks. Its two main functions are perception and action. Intelligent Agents form a hierarchical structure that comprises different levels of agents. A multi-agent system is one that consists of a number of agents interacting with one another [59] in combinations that can help solve challenging societal problems. An IA can behave in three ways: reactively, proactively, and socially [59].

12.4.2 Legislation and Regulation

Per the Emergency Powers Act, if Finland's government—in liaison with the President of the Republic—finds that exceptional circumstances exist in the country, a government decree can be issued to apply the provisions of this act (commissioning regulation). Said decree may be issued for a fixed period [13].

The ISO/IEC 27001 formally specifies an Information Security Management System (ISMS). This comprises a suite of activities concerning the management of information risks called "information security risks" in the standard ISO [27]. Information security management is an essential part of management, which should be supported by the management system. Information security ensures the confidentiality of information, as well as its availability and integrity.

ISO 27799:2016 defines guidelines for organizational information security standards and information security management practices—including the selection, implementation, and management of controls—taking into consideration the organization's information security risk environment(s). It defines guidelines to support the interpretation and implementation of the health informatics of the ISO/IEC 27002 and is a companion to the international standard ISO [26].

ISO/IEC 27032:2012 guides enhancing the state of cybersecurity, along with drawing out the unique aspects of that activity and its dependencies on other security domains—in particular: information, network, internet security, and critical information infrastructure protection (CIIP) ISO [24].

ISO/IEC 9001:2015 provides practical guidance on managing the total service produced for the customer. It also enables the healthcare organization to demonstrate that it meets customer satisfaction requirements and develops customer satisfaction by managing the risks of the operating environment International Organization for Standardization [25].

12.4.3 Situational Awareness

The Ministry of Defence of Finland [45] describes situational awareness as decision-makers' and their advisors' understanding of what has happened, the circumstances under which it has happened, the goals of the different parties, and the possible development of events. All of these are needed to make decisions on a specific issue or range of issues. A general definition of situational awareness is the perception of the elements in the environment within time and space, the comprehension of their meaning, and the projection of their status into the near future [12].

According to [14], cyber situational awareness is a subset of situational awareness—it comprises the part of situational awareness that concerns the cyber environment. Such situational awareness can be reached, for example, by the use of data from IT sensors (intrusion detection systems, etc.) that can be fed to a data fusion process or that can be interpreted directly by the decisionmaker.

12.4.3.1 Command and Control

A *command center* is any place that is used to provide a centralized command for some purpose. An *incident command center* is located at or near an incident, to provide localized on-scene command and support from the *incident commander*. *Mobile command centers* may be used to enhance emergency preparedness and back up fixed command centers. Command centers may also include Emergency Operations Centers (EOC) or Transportation Management Centers (TMC).

Supervisory Control and Data Acquisition (SCADA) systems are basically Process Control Systems (PCS) that are used for monitoring, gathering, and analyzing real-time environmental data—whether from a simple office building or a complex nuclear power plant. PCSs are designed to automate electronic systems based on a predetermined set of conditions, such as traffic control or power grid management [16].

According to [16], SCADA systems' components may involve operating equipment such as valves, pumps, and conveyors that are controlled by energizing actuators or relays. Local processors communicate with the site's instruments and operating equipment—including a Programmable Logic Controller (PLC), Remote Terminal Unit (RTU), Intelligent Electronic Device (IED), and Process Automation Controller (PAC). A single local processor may be responsible for dozens of inputs from instruments and outputs to operating equipment. SCADA also consists of instruments in the field or a facility that sense conditions such as power level, flow rate, or pressure. Short-range communications involve wireless or short cable connections between local processors, instruments, and operating equipment. Long-range communications between local processors and host computers cover a wide area—using such methods as satellites, microwaves, frame relays, and cellular packet data. The host computer acts as the central point of monitoring and control. This is where a human operator can supervise the process, as well as receive alarms, review data, and exercise control. The system may consist of automated or semi-automated processes. A Networked Control System (NCS) is a control system wherein the control loops are closed through a communication network. The defining feature of an NCS is that control and feedback signals are exchanged among the system's components, in the form of information packages, through a network CSPC [4, 49].

RIDM is a risk-based decision-making process that provides a defensible basis for making decisions. It also helps to identify the greatest risks and to prioritize efforts to minimize or eliminate them. Risk-informed decision-making (RIDM) is a deliberative process that uses a set of performance measures, together with other considerations, to "inform" decision-makers' choices [66, 36].

12.4.3.2 Management of Situational Awareness at the National Level

The Ministry of Finance of Finland is responsible for the steering and development of the state's information security [45, 51]. Government situation centers ensure that Finland's state leaders and central government authorities are kept continuously informed. Finland's government situation centre was set up in 2007. It is responsible for alerting the government, permanent secretaries, and heads of preparedness—and for calling them to councils, meetings, and negotiations at exceptional times—as required by a disruption or a crisis.

The ministries must submit the situational picture for their entire administrative branch to the government situation center and notify the center of any security incidents in their field of activity. In urgent situations, the government situation center also receives incident reports for security incidents directly from the authorities. The government situation center also follows public sources and receives situational awareness information, in its role as the national focal point for certain institutions of the European Union and other international organizations.

12.4.4 Elements of Critical Infrastructure

Very often, Critical Infrastructure is defined from the view of the public sector despite it also consist private personnel and their activities as well as public operators of assets, systems, and networks. A very common public–private partnership approach ensures cooperation and information exchange intended to protect vital functions of the society. The human, physical and cyber assets provide many critical services that are necessary for a secure society.

12.4.4.1 Classification of the Critical Infrastructure in the United States

In the United States, critical infrastructure refers to those systems and assets, whether physical or virtual, that are deemed so vital to the United States that their incapacity or destruction would have a debilitating impact on security, national economic security, national public health, or safety, or any combination of those matters [46].

The U.S. Department of Homeland Security identifies 16 different sectors for the classification of critical infrastructure [7]:

1. Chemical,
2. Commercial facilities,

3. Communications,
4. Critical manufacturing,
5. Dams,
6. Defense industrial base,
7. Emergency services,
8. Energy,
9. Financial services,
10. Food and agriculture,
11. Government facilities,
12. Healthcare and public health,
13. Information technology,
14. Nuclear reactors, materials, and waste,
15. Transportation systems, and
16. Water wastewater system.

Cyber threats—such as, for example, phishing attempts, blackmailing attempts, and hacking incidents—are an ever-changing threat to cyber systems across the sectors.

According to the National Institute of Standards and Technology NIST [38], the framework applied in U.S. is also well suited to Finland. The risk management framework consists of three elements of critical infrastructure (physical, cyber, and human), which are explicitly identified and should be integrated throughout the steps of the framework. The critical infrastructure risk management framework supports a decision-making process, which critical infrastructure actors or partners collaboratively undertake to inform their selection of risk management actions. It has been designed to provide flexibility for use in all sectors, across geographic regions and by various partners. It can be tailored to dissimilar operating environments and applies to all threats [7].

The risk management concept enables the critical infrastructure actors to focus on those threats and hazards that are likely to cause harm and to employ approaches that are designed to prevent or mitigate the effects of those incidents. It also increases security and strengthens resilience, by identifying and prioritizing actions to secure the continuity of essential functions and services and support enhanced response and restoration [7].

According to the Department of Homeland Security [7], the first point recommends setting *infrastructure goals and objectives*, which are supported by objectives and priorities developed at the sector level. To manage critical infrastructure risk effectively, actors and stakeholders must identify the assets, systems, and networks that are essential to their continued operation, considering associated dependencies and interdependencies. This dimension of the risk management process should also identify *information and communications technologies* that facilitate the provision of essential services.

The third point recommends *assessing and analyzing risks*. These risks may comprise threats, vulnerabilities, and consequences. A threat can be a natural or manmade occurrence, individual, entity, or action that has or indicates the potential to harm life, information, operations, the environment, and/or property. Vulnerability based risk may occur due to a physical feature or operational attribute that renders an entity open to exploitation or susceptible to a given hazard. A consequence can be the effect of an event, incident, or occurrence. *Implementing risk management activities* means that decision-makers prioritize activities to manage critical infrastructure risk based on the criticality of the affected infrastructure, the costs of such activities, and the potential for risk reduction. The last element, *measuring effectiveness*, implies that the critical infrastructure actors evaluate the effectiveness of risk management efforts within sectors and at national, state, local, and regional levels by developing metrics for both direct and indirect indicator measurement [7].

12.4.4.2   Smart Grid System and the Internet of Things

The Internet of Things (IoT) connects systems, sensors, and actuator instruments to the broader internet. The IoT allows things to communicate and exchange control data and other necessary information, while executing applications towards a machine goal [11].

The idea of the Internet of Things was developed in parallel to Wireless Sensor Networks (WSN). Sensors are everywhere: in our vehicles, in our smartphones, in factories controlling CO_2 emissions, and even in the ground monitoring soil conditions in vineyards. A WSN can generally be described as a network of nodes that cooperatively sense and may control the environment, enabling interaction between persons or computers and the surrounding environment. The development of WSNs was inspired by military applications—notably, for surveillance in conflict zones [2].

The Internet of Things is an emerging paradigm of internet-connected things that allows physical objects or things to connect, interact, and communicate with one another—similarly to the way humans talk via the web in today's environment. It connects systems, sensors, and actuator instruments to the broader internet [11].

The IoT allows things to communicate and exchange control data and other necessary information, while executing applications towards machine goal. The Internet of Things (IoT) is also impacted by the industrial sector, especially for industrial automation systems in which internet infrastructure makes it possible to gain extensive access to sensors, controls and actuators, with the intention of increasing efficiency [11].

Cybersecurity risks should be addressed as organizations implement and maintain their smart grid systems. According to the National Institute of Standards and Technology NIST [37], the smart grid system provides the most efficient electric network operations based on information received from consumers.

A smart grid system may involve a discrete IT system of electronic information resources organized for the collection, processing, maintenance, use, sharing, dissemination, or disposition of information. A smart grid system may also consist of operational technologies (OT) or industrial control systems (ICS), including SCADA systems, distributed control systems (DCS), and other control system configurations such as Programmable Logic Controllers (PLCs) [5, 37].

The Industrial Internet of Things (IIOT) collects data from connected devices (i.e., smart connected devices and machines) in the field or plant. It then processes this data, using sophisticated software and networking tools. The entire IIOT requires a collection of hardware, software, communications and networking technologies [11].

12.4.4.3 Intelligence Solutions for Public Safety Organizations

Open-Source Intelligence (OSINT) is any unclassified information, in any medium, that is generally available to the public—even if its distribution is limited or only available upon payment. OSINT is defined as the systematic collection, processing, analysis and production, classification, and dissemination of information derived from sources openly available to and legally accessible by the public in response to particular government requirements serving national security ATP [1, 17, 43].

Social Media Intelligence (SOCMINT) identifies social media content in particular as both a challenge and opportunity for open-source investigations [56]. Big Data is associated with OSINT and includes processes for the analysis, capture, research, sharing, storage, visualization, and safety of information. Big Data offers the ability to map standards of behavior and tendencies [47]. The availability of worldwide satellite photography, often of high resolution, on the web (e.g., Google Earth Pro) has expanded open-source capabilities into areas formerly available only to major intelligence services [14]. In the proposed hybrid emergency response model [53, 54] OSINT and SOCMINT features are integrated into the automated HERM as a part of an AI-driven decision support tool.

Threat information is any information related to a threat, which might help an organization protect itself against a threat or detect the activities of an actor [29]. Indicators are used to detect and defend against threats. These include the (IP) address of a suspected command and control server, a suspicious Domain Name System (DNS) domain name, or a Uniform Resource Locator that references malicious content. Tactics, techniques, and procedures (TTPs) describe the behavior of an actor. TTPs could describe an actor's tendency to use a specific malware variant, order of operations, attack tool, delivery mechanism, or exploitative strategy. Security alerts are vulnerability notes. Threat Intelligence reports are documents describing threat-related information that is transformed, analyzed, or enriched to provide important context for the decision-making process. Tool configurations are recommendations for using mechanisms that support the automated collection, exchange, processing, analysis, and use of threat-related information. They may comprise information on how to install and use a rootkit detection utility or on how to create, for example, router access control lists (ACLs) [29].

12.5   Previous Works

The multi-methodological approach that has been used in previous studies [53, 54] consists of four case study research strategies [42]:

1.   theory building,
2.   experimentation,
3.   observation, and
4.   systems development.

[60] identifies five components of research design for case studies:

1.   the questions of the study,
2.   its propositions, if any,
3.   its unit(s) of analysis,
4.   the logic linking the data to the propositions, and
5.   the criteria for interpreting the findings.

According to [22] (Chap. 2), information systems and organizations are complex, artificial, and purposefully designed. Such a problem-solving paradigm must lead to an artifact that solves the identified problem. This review concentrates on comparing how the proposed emergency model [53, 54] suits a pandemic situation in which information warfare is an ongoing process. Scientific publications, articles, and literary material have been comparatively reviewed with this aim. The review subject comprises the public safety organizations, procedures, and vital functions of Finland society.

The first purpose of this qualitative review was to analyze pandemic-related management and information-sharing risks, along with the formation of situational awareness, from the view of continuity management. We apply the modified risk assessment framework in this review. The second purpose was to find any hidden administrative and managerial related state-level risks that are outside the official risk classification. A simple process model helps identify those fundamental hidden management-related factors that affect to implementation process of the next-generation emergency response model proposed by [53].

12.6   Findings

In Finland, as we have seen, more than one factor influences the decision-making process at the state level. We have local and regional level administrations that form situational awareness from the view of their territorial region; decision-makers then share regional instructions and guidelines with the people. There are local corona teams that are responsible for regional security. Currently, tasks are separate from the government at the regional level and the members of the government do not give absolute commandments, such as mandatory instructions for using masks. Yet the continued lack of clarity around the workflow is a crucial barrier, when the purpose is to share relevant information with the right audience at the right time. It has been seen previously that labor movement or trade unionism can produce an agitating counterforce, by means that are not ethically valid. If the challenges to fighting the COVID-19 pandemic emerge from the nation's citizens, then the fundamental problems lie more deeply within the constructs of society.

Finland does not have an operative command and control institution for unexpected crises. The president of Finland leads foreign policy with the government, but there is no operative commander role for the president in the country's internal affairs. The ongoing COVID-19 crisis has shown that there is a lack of information exchange—both between the authorities and between the authorities and politicians. Yet citizens have likewise been kept unaware of the guidelines that should be followed. For small- or medium-sized social and healthcare companies, information security is based on self-monitoring. Public healthcare organizations also base their oversight of these operations on self-monitoring. The National Supervisory Authority for Welfare and Health (Valvira) supervises, for example, private sector licensing, healthcare, social welfare, legal protection, legal rights, and technologies [58]. A single staff member is responsible for supervising all issues like information security and privacy protection, around the Kanta-register [19].

This is not enough—especially since criminals may use private information in a variety of extremely dangerous ways. For example, criminals may try to affect the decision-making process by blackmail. A major information-sharing problem seen in the Vastaamo case was the fact that a data breach had occurred nearly two years before it was detected. There are no crucial privacy issue-related barriers to using the proposed hybrid emergency response model within a smart city infrastructure. When an alarm-based early warning procedure for data leakage is automatized, it offers possibilities to enhance privacy protection and other protective functions [52]. The proposed hybrid emergency response solution may also use sensors called flu-sensors, which can transfer data in real time from a public area—for example, from a shopping center—to the Hybrid Emergency Response Center. Data about virus particles might then indicate a need for mall closure, the early warning would allow this to be carried out immediately.

12.7 Discussion and Conclusions

By comparing different countries, crucial factors influencing the formation of information sharing can be found. For example, Finland is almost the only country in Europe that does not use scientist experts as advisors in the decision-making process at the state level. If decision-makers keep their eyes open, they can find massive amounts of research from foreign sources on how the coronavirus spreads and how its spread can be prevented.

First, there is a fundamental need to regulate new guidelines for the higher-level crisis management and command relationships for exceptional circumstances.

Temporary provisions should be made for emergency situations, which may require imposing restrictions on citizens. There must be one incident team whose leader is from the central government. This leader should take control when adjutants and instructors have too much information to share, since it is difficult to gather the correct information from a large amount of the data in a time of crisis. To date, there have been too many assistants involved in the decision-support mechanism at the state level.

In the future, it is necessary to begin using artificial intelligence solutions to support decision-making. The proposed next-generation hybrid emergency model uses artificial tools to generate information for decision-makers. Algorithm-based decision-support and decision-making mechanisms make the system effective. As Fig. 12.3 illustrates, the crucial factors in the hybrid risk management framework are risk-informed decision making (define risks and information), continuity risk management (handle risks continuously), and hybrid emergency response solutions (emergency operations). Because human beings are still decision-makers, people are responsible for the decisions they have made. Yet it is possible to combine human-based guidelines for risks and AI-driven decision-making [6].

This solution offers two possibilities to use automation. At the first level, automated protection functions are connected to semi-public spaces (e.g., shopping centers) and public open places (e.g., gardens). For example, a health sensor called "flu" may start an evacuation process if it observes several deviations from the guideline values. At the second level, an AI-aided decision support mechanism outputs analytical reports for the state level decision-makers. This level will greatly enhance the decision-making process, since the need for assisting staff will be reduced in high-level decision-making.

As mentioned above and illustrated in Fig. 12.4, the authorities' information sharing process must move towards automated functions. Still, it is an important western tradition that a parliament is democratically elected by the country's citizens.

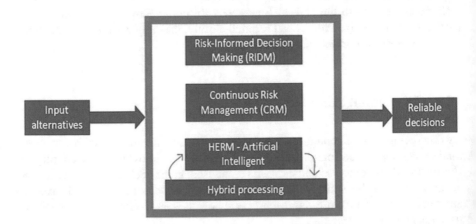

Fig. 12.3 Reliable decision-making process

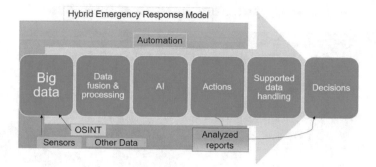

Fig. 12.4 HERM with artificial intelligence

At present, politicians' desire to maintain high levels of control over their decision-making ability may prevent the utilization and usefulness of the proposed smart hybrid emergency model. Many decision-makers want political aspects and opinions to be more represented than rational decisions. Yet Finland's politicians and other high-level decision-makers should take into consideration that cyber preparedness, operational preparedness, and reliability of decision-making are not separate parts of continuity management.

It is possible to combine operational, management, and strategic level decision support functions into a single entity. This does not mean combining all elements in one physical location. If fundamental risk factors—such as a pandemic that presents domino effects from many angles—are not recognized, then technical early warning solutions become useless. It is thus a fundamental societal requirement that a decision support mechanism be developed in jointly with the crisis management system.

It is not enough for the government of Finland to use just one international source (WHO), when they try to maintain the international level of situational awareness. Legislation around privacy issues does not cause permanent obstacles to using sensing elements (e.g., sensors) in the hybrid emergency response model. It is necessary to rationalize organizational responsibilities, for the development of overall security. A human is an individual with limited observation capability and overlapping data transmission limits the effective cooperation between politicians and authorities.

HERM's nearly tireless data handling and transmission capacity can help prevent communication problems among the authorities. Embedding preventive functions against unexpected threats in the emergency response model is an essential part of overall security, in situation awareness management and critical infrastructure protection. In particular, the analysis of global research data regarding COVID-19 can be automated. We need more detailed, standardized information systems and rules for all information systems that handle sensitive information. All that is needed is the political will to exploit intelligence solutions.

The ongoing and tremendously challenging COVID-19 crisis requires us to powerfully leverage our common will—to change the dream of digitalization into concrete actions. The proposed model for smart cities offers solutions to many problems and

HEALTH SENSORS CONNECTED TO THE HERM

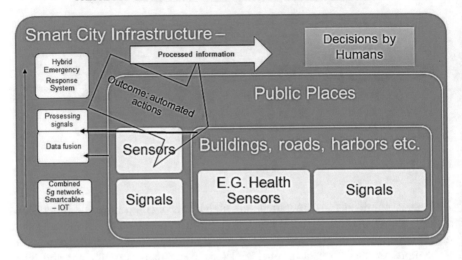

Fig. 12.5 Predictive health sensors in a smart city

questions, as Fig. 12.5 shows. The model may use health sensors, as well as traffic sensors, in a predictive way.

References

1. ATP (2012) Open-source intelligence. Army Techniques Publication No. 2–22.9, Department of the Army, Washington, DC
2. Bröring A, Echterhoff J, Jirka S, Simonis I, Everding T, Stasch C, Liang S, Lemmens R (2011) New generation sensor web enablement. Sensors 11(3):2652–2699
3. Buranyi S (2020) The WHO v coronavirus: Why it can´t handle the pandemic. The Guardian, https://www.theguardian.com/news/2020/apr/10/world-health-organization-who-v-coronavirus-why-it-cant-handle-pandemic. Accessed 10 Dec 2020
4. CSPC (2014) Securing the U.S. electrical grid. Center for the Study of the Presidency & Congress
5. Chong C, Kumar S (2003) Sensor networks: Evolution, opportunities, and challenges. Proc IEEE 91(8):1247–1256
6. Colson E (2019) What AI-driven decision making looks like. Harvard Business Review, https://hbr.org/2019/07/what-ai-driven-decision-making-looks-like. Accessed 10 Dec 2020
7. DHS (2013) NIPP 2013: Partnering for critical infrastructure security and resilience. U.S. Department of Homeland Security, https://www.cisa.gov/publication/nipp-2013-partnering-criticalinfrastructure-security-and-resilience
8. DigiFinland (2020) Welcome to take care of your health and well-being in Omaolo. DigiFinland Oy, https://www.omaolo.fi/. Accessed 10 Dec 2020
9. EC (2018) Artificial intelligence for Europe. Communication from the Commission to the European Parliament, the European Council, the Council, the European Economic and Social Committee and the Committee of the Regions, COM(2018) 237, European Commission
10. EC (2020) Coronavirus: commission delivers first batch of 1.5 million masks from 10 million purchased to support EU healthcare workers. Press release, European Commission

11. ElecTech (2016) Internet of Things (IOT) and its applications. Electrical Technology, http://www.electricaltechnology.org/2016/07/internet-of-things-iot-and-its-applications-inelectri cal-power-industry.html. Accessed 8 Nov 2016
12. Endsley MR (1988) Design and evaluation for situation awareness enhancement. Proceedings of the Human Factors and Ergonomics Society Annual Meeting 32(2):97–101
13. Finlex (2011) Emergency powers act 1552/2011. Finnish Ministry of Justice
14. Franke U, Brynielsson J (2014) Cyber situational awareness: A systematic review of the literature. Comput Secur 46:18–31
15. GPMB (2019) A world at risk: Annual report on global preparedness for health emergencies. Global Preparedness Monitoring Board
16. Gervasi O (2010) Encryption scheme for secured communication of web-based control systems. Journal of Security Engineering 7(6):609–618
17. Glassman M, Kang MJ (2012) Intelligence in the internet age: The emergence and evolution of Open Source Intelligence (OSINT). Comput Hum Behav 28(2):673–682
18. Harjumaa M (2020) HUSin Järvinen purkaisi koronarajoituksia: "Pitäisi yrittää saada sitä väestönosaa sairastamaan, jolle tauti ei todennäköisimmin ole vaarallinen". Yle, https://yle.fi/uutiset/3-11318716. Accessed 10 Dec 2020
19. Hautanen S (2020) IS: Yksi mies vastaa 260 sote-yrityksen tietoturvan valvonnasta. Verkkouutiset, https://www.verkkouutiset.fi/is-yksi-mies-vastaa-260-sote-yrityksen-tie toturvan-valvonnasta/#2fb860b4. Accessed 10 Dec 2020
20. Hemmilä I, Salminen V (2020) Oikeuskansleri moittii ministeriöiden yhteistyötä kevään suojavarustehankinnoissa—STM:ssä epäselvyyttä myös ministerien työnjaosta. Suomenmaa, https://www.suomenmaa.fi/uutiset/oikeuskansleri-moittii-ministerioiden-yhteistyota-kevaan suojavarustehankinnoissa-stmssa-epaselvyytta-myos-ministerien-tyonjaosta-2/. Accessed 4 Dec 2020
21. Henley J (2020) Sweden's Covid-19 strategist under fire over herd immunity emails. The Guardian, https://www.theguardian.com/world/2020/aug/17/swedens-covid-19-strategist-under-fireover-herd-immunity-emails. Accessed 10 Dec 2020
22. Hevner A, Chatterjee S (2010) Design research in information systems: theory and practice. Springer
23. HkiTimes (2020) Finland to close borders to non-essential travel at 12 am on Thursday. Helsinki Times, https://www.helsinkitimes.fi/finland/finland-news/domestic/17450-finland-to-closeborders-to-non-essential-travel-at-12am-on-thursday.html. Accessed 10 Dec 2020
24. ISO (2012) ISO/IEC 27032:2012: Information technology, security techniques, guidelines for cybersecurity. International Organization for Standardization (ISO), https://www.iso.org/sta ndard/44375.html
25. ISO (2015) ISO 9001:2015: Quality management systems, requirements. International Organization for Standardization (ISO), https://www.iso.org/standard/62085.html
26. ISO (2016) ISO 27799:2016 Health informatics, information security management in health using ISO/IEC 27002. International Organization for Standardization (ISO), https://www.iso.org/standard/62777.html
27. ISO (2017) ISO/IEC 27001: Information security management systems. International Organization for Standardization (ISO), https://www.iso.org/isoiec-27001-information-security.html
28. Jaskari K (2020) Ministeriö ei aio jatkossakaan suositella kangasmaskien käyttöä julkisilla paikoilla. Yle, https://yle.fi/uutiset/3-11305744. Accessed 10 Dec 2020
29. Johnson C, Badger L, Waltermire D, Snyder J, Skorupka C (2016) Guide to cyber threat information sharing. NIST Special Publication 800–150, National Institute of Standards and Technology (NIST)
30. Kela (2020) Tekniset liittymismallit Kanta-palveluihin. Ohje, Kanta-palvelut, Kela, https://www.kanta.fi/documents/20143/106828/Tekniset+liittymismallit+Kanta-palveluihin.pdf/a05 7c34a-f822-71fd-b2df-097245d582ee
31. Lakka P (2020) IS selvitti Pekosen ja Kiurun ministeriön kaaosta—ainakin nämä 5 syytä vaikuttivat taustalla: "Suksi lipsunut koko matkan". Ilta-Sanomat, https://www.is.fi/politiikka/art2000006482234.html. Accessed 10 Dec 2020

32. Lehto M, Limnéll J, Innola E, Pöyhönen J, Rusi T, Salminen M (2017) Suomen kyber-turvallisuuden nykytila, tavoitetila ja tarvittavat toimenpiteet tavoitetilan saavuttamiseksi. Valtioneuvoston selvitys- ja tutkimustoiminnan julkaisusarja 30/2017, Valtioneuvoston kanslia

33. Lyngaas S (2020) Why the extortion of Vastaamo matters far beyond Finland—and how cyber pros are responding. CyberScoop, https://www.cyberscoop.com/finland-vastaamo-hack-response/. Accessed 10 Dec 2020

34. Mediuutiset (2020) Lääkäri: Aletaan tartuttaa koronaa hallitusti hoitohenkilökuntaan immuniteetin saamiseksi. Mediuutiset, https://www.mediuutiset.fi/debatti/laakari-aletaan-tartuttaa-koronaa-hallitusti-hoitohenkilokuntaan-immuniteetin-saamiseksi/dfbc96a4-c757-44e9-988 d38cf093a7e8b. Accessed 5 May 2020

35. Mäntymaa E, Mäntymaa J (2020) Sairaalat saivat varmuusvarastoista vuosia sitten vanhentuneita hengityssuojaimia—"Ihan kuranttia ei kaikki tavara ole ollut", sanoo HUS-johtaja. Yle, https://yle.fi/uutiset/3-11286164. Accessed 10 Dec 2020

36. NASA (2010) Risk-informed decision making handbook (NASA/SP-2010-576). Technical report, NASA. https://ntrs.nasa.gov/api/citations/20100021361/downloads/20100021361.pdf. Accessed 10 Dec 2021

37. NIST (2010) Guidelines for smart grid cybersecurity. In: Privacy and the smart grid, vol 2. NISTIR 7628, National Institute of Standards and Technology (NIST)

38. NIST (2018) Framework for improving critical infrastructure cybersecurity. Version 1.1, National Institute of Standards and Technology (NIST)

39. Natri S (2020) THL:n pääjohtaja kehottaa suomalaisia pukemaan kangasmaskin julkisilla paikoilla—"Näin oireeton tartuttaja suojelee muita". Yle, https://yle.fi/uutiset/3-11305102. Accessed 10 Dec 2020

40. NewsNowFin (2020) Maria Ohisalo: Vastaamo cyber attack and blackmail demands "serious, outrageous and cowardly". News Now Finland, https://newsnowfinland.fi/crime/mariaohisalo-vastaamo-cyber-attack-and-blackmail-demands-serious-outrageous-and-cowardly. Accessed 10 Dec 2020

41. NewsNowFin (2020) Vastaamo hacking and blackmail: 25,000 police reports filed. News Now Finland, https://newsnowfinland.fi/crime/vastaamo-hacking-and-blackmail-25000-police-rep orts-filed. Accessed 10 Nov 2020

42. Nunamaker J Jr, Chen M, Purdin T (1990) Systems development in information systems research. J Manag Inf Syst 7(3):89–106

43. Nurmi P (2015) OSINT: Avointen lähteiden internet-tiedustelu. Kehitysprojektin raportti, Aaltoyliopisto

44. Ollila A (2020) Lääkintöneuvos Pälve tyrmää Kirsi Varhilan näkemykset maskien käytön esteistä. Uusi Suomi, https://puheenvuoro.uusisuomi.fi/aveollila1-2/laakintoneuvos-palve-tyr maakirsi-varhilan-nakemykset-maskien-kayton-esteista/. Accessed 10 Dec 2020

45. PM (2010) Yhteiskunnan turvallisuusstrategia: Valtioneuvoston periaatepäätös 16.12.2010. Puolustusministeriö, Helsinki

46. PPD (2013) Critical infrastructure security and resilience. Presidential Policy Directive PPD-21, U.S. White House Office

47. dos Passos DS (2016) Big Data, data science and their contributions to the development of the use of open source intelligence. Electronic Journal of Management & System 11(4):392–396

48. Ranta E (2020) Tällainen yritys on tietomurron kohteeksi joutunut Vastaamo. Ilta-Sanomat, https://www.is.fi/taloussanomat/art-2000006699437.html. Accessed 30 Nov 2020

49. Robles RJ, Kim T (2010) Communication security for SCADA in smart grid environment. In: DNCOCO'10: proceedings of the 9th WSEAS international conference on data networks, communications, computers, pp 36–40. World Scientific and Engineering Academy and Society (WSEAS), Stevens Point, WI

50. SecComm (2017) Security Strategy for Society. Government resolution, Security Committee, Helsinki

51. Simola J (2020) Privacy issues and critical infrastructure protection. In: Benson V, Mcalaney J (eds) Emerging Cyber Threats and Cognitive Vulnerabilities. Academic Press, pp 197–226

52. Simola J, Rajamäki J (2017) Hybrid emergency response model: improving cyber situational awareness. In: Scanlon M, Le-Khac N (eds) ECCWS 2017—proceedings of the 16th European conference on cyber warfare and security. Academic Conferences and Publishing International, pp 442–451
53. Simola J, Rajamäki J (2018) Improving cyber situational awareness in maritime surveillance. In: Josang A (ed) ECCWS 2018—proceedings of the 17th European conference on cyber warfare and security, pp 480–488. Academic Conferences and Publishing International
54. Solita (2020) The Finnish Covid-19 app Koronavilkku has been downloaded a million times already! Solita, https://www.solita.fi/en/the-finnish-covid-19app-koronavilkku-has-been-downloaded-million-times/. Accessed 10 Dec 2020
55. Trottier D (2015) Open source intelligence, social media and law enforcement: Visions, constraints and critiques. Eur J Cult Stud 18(4–5):530–547
56. Uosukainen R, de Fresnes T (2020) Poliitikot tulevat ja menevät, virkamiehet pysyvät—käynnissä on kamppailu siitä kenellä valta on. Yle, https://yle.fi/uutiset/3-11186910. Accessed 10 Dec 2020
57. Valvira (2020) Organizational structure. Valvira, https://www.valvira.fi/web/en/valvira/organisational_structure. Accessed 20 Oct 2020
58. Wooldridge M (2009) An introduction to multiagent systems, 2nd ed. Wiley
59. Yin RK (2017) Case study research and applications: design and methods, 6th ed. SAGE, Thousand Oaks, CA
60. Yle (2020) Daily: Gov't not planning to extend Uusimaa border closure. Yle, https://yle.fi/uutiset/osasto/news/daily_govt_not_planning_to_extend_uusimaa_border_closure/11303010. Accessed 10 Dec 2020
61. Yle (2020) Finland: Chinese face masks fail tests. Yle, https://yle.fi/uutiset/osasto/news/finland_chinese_face_masks_fail_tests/11298914. Accessed 10 Dec 2020
62. Yle (2020) Finland's first coronavirus case confirmed in Lapland. Yle, https://yle.fi/uutiset/osasto/news/finlands_first_coronavirus_case_confirmed_in_lapland/11182855. Accessed 10 Dec 2020
63. Yle (2020) Friday's paper: problem with corona alert app, more countries on restricted list, drugs in the countryside. Yle, https://yle.fi/uutiset/osasto/news/fridays_papers_problem_with_corona_alert_app_more_countries_on_restricted_list_drugs_in_the_countryside/11586606. Accessed 10 Dec 2020
64. Yle (2020) President Niinistö defends role in coronavirus crisis. Yle, https://yle.fi/uutiset/osasto/news/president_niinisto_defends_role_in_coronavirus_crisis/11303872. Accessed 12 Dec 2020
65. Yle (2020) Two possible coronavirus cases in northern Finland. Yle, https://yle.fi/uutiset/osasto/news/two_possible_coronavirus_cases_in_northern_finland/11173752. Accessed 10 Dec 2020
66. Zio E, Pedroni N (2012) Risk-informed decision-making processes: an overview. Foundation for an Industrial Safety Culture, Touloise, https://www.foncsi.org/fr/publications/cahiers-securite-industrielle/overview-of-risk-informed-decision-making-processes/CSI-RIDM.pdf. Accessed 15 Dec 2020

Chapter 13
Information Security Governance in Civil Aviation

Tomi Salmenpää

Abstract This chapter focuses mainly to proactive means in information security and more specifically governance of information security in civil aviation. The reason is that, to find sustainable, coherent and holistic way to implement information security through the complete civil aviation ecosystem, the governance plays a key role when creating sufficient framework enabling information security by design environment. The study will help aviation and other critical infrastructure sectors to consider, understand and coordinate the information security governance. The study will test how to apply information security governance with ISO27014 through such a safety critical, interconnected infrastructure sector like civil aviation.

Keywords Cybersecurity governance · Information security governance · Aviation cybersecurity · Aviation information security

13.1 Introduction

Civil aviation is a continuously evolving ecosystem in which information security plays a key role in ensuring public and societal trust and confidence in civil aviation. Other critical components, in addition to information security, are aviation safety and aviation security. Appropriate and proportionate information security measures will make sure and continuously enhance aviation safety, aviation security, and operational resilience. Information security and the governance are generally well recognized and understood at organizational level, but the role of information security governance in civil aviation at higher levels like state or international, has not yet been widely discussed. All considerations in this study represent author´s personal interpretation and expertise in this aviation cyber- security field.

T. Salmenpää (✉)
University of Jyväskylä, Jyväskylä, Finland
e-mail: tomi.salmenpaa@protonmail.com

315

13.1.1 Information Security Management

Reliably functioning critical infrastructure is necessity in the modern society. Continuously increasing complexity and connectivity of critical infrastructure systems increase the risk for information security threats and put the nation's security, economy, and public safety and health, just some most important things to mention, at risk. This chapter studies one critical infrastructure sector, civil aviation, information security governance in order to support civil aviation stakeholders' coordinated and common efforts to build holistic, standardized system of system approach to civil aviation information security. Information security must have a goal, purpose.

This chapter approaches the information security and its governance by trying to ensure the operational resilience, civil aviation security and safety. In order to have coherent, holistic balanced information security management over the complete civil aviation ecosystem, the comparison is made by using different levels in aviation with relevant industry standard for governance of information security [8]. The levels are organizational, state, regional (Europe), and international levels. Civil aviation organizations are understood to consist of all aviation domains, e.g., Air Navigation Service Providers (ANSP), Aerodrome (ADR), Airworthiness (AIR), Flight Operations (OPS), manufacturers, in other words, all aviation organizations. Other levels are state, regional (Europe) and international, for which the ISO 27014 definitions, concepts and principles are tested, because the standard itself is to improve information security management primarily within the context of the individual organization and not for higher level, like state, regional or international.

The governance is selected as the subject in the chapter, because before setting up information security actions or measures, it is fundamental to understand the objectives that one is pursuing with information security. The information security governance is discovered in the light of well-known industry standards. Because these industry standards are generally accepted and matured as the best practices by the information security industry, community and stakeholders, they provide a solid ground to apply those standards for civil aviation purposes. Then the current civil aviation existing management and governance frameworks are described, and information security governance definitions, concepts and principles are projected to those existing governance models.

In this chapter, the differences between information security and cybersecurity are not so great as to cause a problem. So the differences are therefore ignored. At the very beginning, the terms information security and management system need to be defined and understood. Information security means preservation of confidentiality, integrity and availability of information whereas the management system as for a set of interrelated or interacting elements of an organization to establish policies and objectives and processes to achieve those objectives [9]. Together they are Information Security Management System (ISMS) that consists of the policies, procedures, guidelines, and associated resources and activities, collectively managed by an organization, in the pursuit of protecting its information assets. An ISMS is a systematic

Table 13.1 Function and category unique identifiers [10]

Function unique identifier	Function	Category unique identifier	Category
ID	Identify	ID.AM	Asset management
		ID.BE	Business environment
		ID.GV	Governance
		ID.RA	Risk assessment
		ID.RM	Risk management strategy
		ID.SC	Supply chain risk management
PR	Protect	PR.AC	Identity management and access control
		PR.AT	Awareness and training
		PR.DS	Data security
		PR.IP	Information protection processes and procedures
		PR.MA	Maintenance
		PR.PT	Protective technology
DE	Detect	DE.AE	Anomalies and events
		DE.CM	Security continuous monitoring
		DE.DP	Detection processes
RS	Respond	RS.RP	Response planning
		RS.CO	Communications
		RS.AN	Analysis
		RS.MI	Mitigation
		RS.IM	Improvements
RC	Recover	RC.RP	Recovery planning
		RC.IM	Improvements
		RC.CO	Communications

approach for establishing, implementing, operating, monitoring, reviewing, maintaining and improving an organization's information security to achieve business objectives.

The National Institute of Standards and Technologies (NIST) has issued a cyber-security framework for improving critical infrastructure cybersecurity, that provides a common language for understanding, managing, and expressing cybersecurity risk both internally and externally. It can be used to help identify and prioritize actions for reducing cybersecurity risk, and it is a tool for aligning policy, business, and technological approaches to managing that risk [10].

The cybersecurity framework describes five concurrent and continuous functions: Identify, Protect, Detect, Respond, and Recover (Table 13.1). They aid an organization in expressing its management of cybersecurity risk by organizing information, enabling risk management decisions, addressing threats, and improving by learning from previous activities. When considered all together, these functions provide a high-level strategic lifecycle for an organization's management of cybersecurity risk. The framework core then identifies underlying key categories and subcategories for each function, and matches them with example informative references such as existing standards, guidelines, and practices for each subcategory [10].

The Identify function means: Develop the organizational understanding to manage cybersecurity risk to systems, assets, data, and capabilities. The activities in the Identify function are foundational for effective use of the framework. Understanding the business context, the resources that support critical functions, and the related cybersecurity risks enables an organization to focus and prioritize its efforts consistent with its risk management strategy and business needs. Examples of outcome categories within this function include: Asset management, Business environment, Governance, Risk assessment, and Risk management strategy [10]. This chapter researches the information security governance and its meaning in the whole aviation ecosystem.

13.1.2   Governance in Information Security Management

Governance means many things depending on the context or discussion. In general, governance comprises all of the processes of governing—whether undertaken by the government of a state, by a market or by a network—over a social system (family, tribe, formal or informal organization, a territory or across territories) and whether through the laws, norms, power or language of an organized society [1]. In this chapter the governance is discussed from the information security management perspective at various levels, referring to existing industry standards in information security.

In information security and standard such as ISO/IEC 27014, the governance of information security is defined as a system by which an organization's information security activities are directed and controlled [8]. NIST framework describes the governance in the following way: The policies, procedures, and processes to manage and monitor the organization's regulatory, legal, risk, environmental, and operational requirements are understood and inform the organization management of cybersecurity risk [10]. In reality, it is widely understood that for organizations it is impossible to protect everything in the cyberspace and they need to make prioritization. Therefore, information security risk management plays a key role defining organization assets that have value for organization or their stakeholders.

In order to make appropriate information security management in place, it is crucial to consider and understand the objectives of information security governance. In the standard ISO/IEC [8] it is defined:

- align the information security objectives and strategy with business objectives and strategy (strategic alignment),
- deliver value to the governing body and to stakeholders (value delivery),
- ensure that information risk is being adequately addressed (accountability).

The governance is important to understand at different levels in aviation and cybersecurity because the different levels should have converging strategies in aviation cybersecurity and governance plays a fundamental role to set right policies always down to practical solutions in aviation information security. In addition, between the aviation organizations there should not be gaps, duplication or uncoordinated areas in this chain of aviation cybersecurity governance.

In this chapter, information security governance is researched from the civil aviation ecosystem perspective at four different levels: organization, national, regional, and international. At individual organization level the information security objectives and strategy are aligned with the business objectives and strategy of that organization. At national level in civil aviation these information security objectives and strategy are different from those at organization level. It is the whole national and collectively international aviation system resiliency, safety and security, where information security objectives and strategy are aligned. Eventually at the broadest levels, regional (Europe) and international level, civil aviation information security objectives and strategy are different because the business objectives and strategy are different from those at national and organizational levels. Evaluating these differences can give an opportunity to better understand information security governance and make development according the needs, for example, when digitizing organization, society, or the whole ecosystem.

Comparing the term "system" in ISO/IEC [8], it has different meaning in information security at different levels (organizational, national, regional, and international). Also, the alignment of the information security objectives and strategy with business objectives and strategy varies because the

- information security objectives and strategy are different at every level;
- business objectives and strategy are different at every level.

In aviation, the governance at international level means all of the processes undertaken by international organization who has the necessary mandate or role to be able to govern civil aviation. The governance means all of the processes that comes to civil aviation. At state level, governance of aviation is different thing. The scope of processes is different and they are undertaken by government of a state according to a system in that society. In organization, the processes in aviation governance are again different and means the collection of mechanisms, processes and relations used by various parties to control and to operate organizations. Organization governance includes the processes through which organizations objectives are set and pursued in the context of the social, regulatory and market environment.

13.2   Information Security Management in Civil Aviation

This section focuses on efficient information security management in civil aviation in order to ensure operational resiliency, secure and safe civil aviation system. That means the information security and its management needs to be considered in the light of the overall civil aviation safety and security management.

Aviation security and safety management fundamentally complement each other. While aviation security experts often hold coordination links to threat information sources and are used to dealing with intentional threats and the respective methodologies, aviation safety experts have extensive know-how of the consequences on the safety of flight in case of system failure and know the design and set-up of systems and existing mitigation measures such as redundancies [2]. Due to existing strong, holistic and end-to-end (from the policy to practice) security and safety governance framework in civil aviation, it is strongly recommended to implement information security to the existing safety and security management frameworks.

International policies and standards in civil aviation are coordinated through the International Civil Aviation Organisation (ICAO). ICAO is specialized agency of the United Nations (UN), which is directed and endorsed by the governments. In ICAO, also industry, society groups, and other regional and international organizations participate in the exploration and development of new standards. In sector such as civil aviation where flying aircraft, commonly used processes and protocols do not recognize borders, it is paramount to have international and standardized approach in all areas of aviation safety and security.

13.2.1   Concept of Safety Management in Civil Aviation

First, it is paramount to understand the meaning of civil aviation safety management. Civil aviation safety management is commonly understood as a set of principles, framework, processes and measures to prevent accidents, injuries and other adverse consequences that may be caused by using service or product. The objective of safety management in the aviation industry is to prevent human injury or loss of life and to avoid damage to the environment and to property [11]. Safety Management System is the formal, top-down, organization-wide approach to managing safety risk and ensuring the effectiveness of safety risk controls. It includes systematic procedures, practices, and policies for the management of safety risk [3].

In aviation, there is traditional, strong, ecosystem-based and standardized safety governance concept in place. In that concept, the global chain of safety management plays a key role, where, at international level, the ICAO's Global Aviation Safety Plan (GASP) presents the strategy that supports the prioritization and continuous improvement of aviation safety. The GASP, along with the Global Air Navigation Plan (GANP), provides the framework in which regional and national aviation safety plans will be developed and implemented, thus ensuring harmonization and coordination of efforts aimed at improving international civil aviation safety, capacity, and efficiency [4, 5].

At regional level in Europe, European Plan for Aviation Safety (EPAS) constitutes the regional aviation safety plan for European Aviation Safety Agency (EASA) member states. The EPAS set out the strategic priorities, strategic enablers and main risks affecting the European aviation system and the necessary actions to mitigate those risks and further improve aviation safety.

The EASA member states have their own State Safety Programs (SSPs), which is the detailed level national description of their safety management system. They follow the ICAO's GASP and the European EPAS accordingly, but also maintain and improve them by feeding important, e.g., safety performance information to those programs.

The aviation organizations have their Safety Management Systems (SMS) to meet these safety risk management requirements and safety performance objectives. It is noteworthy that SMS requirements are not yet applicable for all aviation domains, but the most critical ones, like airlines, are mandated to have SMS.

13.2.2   Concept of Security Management in Civil Aviation

Whereas civil aviation safety focus on reducing the likelihood of accident happening, the civil aviation security focus to safeguard international civil aviation against acts of unlawful interference. The International Civil Aviation Organization (ICAO) has the leadership role at global level to develop international policies and measures at international level. In general, this contains all acts that jeopardize the safety of civil aviation. Current global level policies and measures are well implemented against the physical acts, but both digital and physical information security are underway internationally, regionally, and nationally, including aviation organizations.

Aviation security in the aviation community generally means all unlawful interference against civil aviation. Such acts or attempts jeopardize the safety of civil aviation [7]. The unlawful interference of civil aviation is safe guarded by the commonly agreed norms starting from the international level. Information security is defined as protection of information and information systems from unauthorized access, use, disclosure, disruption, modification, or destruction in order to provide confidentiality, integrity, and availability [10].

In theory, the difference between unlawful interference compared to unauthorized access, use, disclosure, disruption, modification, or destruction in order to provide confidentiality, integrity and availability is not that great. In practice, there is a grey area because it depends on the impact of information security incident in aviation security or safety. Information security incident not always impacts aviation in a way that is defined as unlawful interference, but still the incident can seriously compromise the public trust or confidence in civil aviation.

In aviation security, there is also a strong and standardized security governance concept in place. ICAO Global Aviation Security Program (GASeP) addresses the needs of states and industry in guiding all aviation security enhancement efforts. The objective of the GASeP is to help ICAO, states and aviation stakeholders enhance the effectiveness of global aviation security. The GASeP seeks to unite the international aviation security community and inspire action in this direction, taking into account that the threats and risks faced by the civil aviation community continue to evolve. It is also intended to achieve the shared and common goal of enhancing aviation security worldwide and to help states come together to fulfil the commitments set out in UNSCR 2309 (2016) and relevant ICAO assembly resolutions [7].

The states have their National Civil Aviation Security Program (NCASP) to safeguard civil aviation operations against unlawful interference. The NCASP must meet the requirements from regulations, practices and procedures, which take into account the safety, regularity and efficiency of flights [7].

13.2.3 Concept of Cybersecurity in Civil Aviation

The concept of cybersecurity is not yet in place compared to aviation safety and security and is still evolving at all levels, international, national and organizational levels. At international level there is published and agreed ICAO Aviation Cybersecurity Strategy and its Action Plan for the ICAO, states, and industry in aviation cybersecurity with a vision that civil aviation sector being resilient to cyber-attacks and remains safe and trusted globally whilst continuing innovate and growth [6]. In the ICAO Strategy and Action Plan there are eight pillars, and one is governance. In the actions it is highlighted for the states the need to develop clear national governance and accountability for civil aviation cybersecurity. Another important action is to include cybersecurity into national aviation safety and security programs. However, more specific definitions or actions about governance and its meaning are not available in the strategy or action plan.

At regional level in Europe the information security governance in aviation is discussed in some publications. The most accurate recommendation is available in the European Civil Aviation Conference (ECAC) guidance material on cybersecurity in civil aviation [2], which provides important principles on the governance that states and organizations should follow in aviation cybersecurity. The principles are about roles and accountability in the civil aviation cybersecurity, however, the meaning of governance is not defined at a detailed level. In Europe, there is aviation cybersecurity strategy by the European Strategic Coordination Platform (ESCP). The ESCP Strategy for Cybersecurity in Aviation provides a systematic approach with objectives, to build in cybersecurity into civil aviation, but does not provide direct recommendations to cybersecurity governance. For organizations, the information security industry standards provide sufficient recommendations and best practices to information security governance at organizational level, e.g., ISO 27014. For all these reasons described before, this chapter is about projecting the available standards to higher levels, such as at national, regional, and international levels.

13.3   Information Security Management Governance in Civil Aviation

The International Organisation for Standardization (ISO) provides relevant standard for Governance of Information Security ISO 27014. In addition, NIST standards, e.g., SP 800-39 was also evaluated, but the ISO 27014 was chosen because it provided from governance perspective more prescriptive model to use and meet the goals of this study. Because the ISO 27014 is applicable for all types and sizes of organizations, however primarily from the individual organization context, this encouraged to test this model at higher levels too, in order to make different needs for information security governance meaning and more tangible.

13.3.1   Definitions

To make the comparison from organization level to state or higher levels, there are some important definitions in ISO 27014 (Table 13.2) which need to be first translated and understood from that respective level. When these definitions are translated to higher levels, at state level there are sufficient ground available for cybersecurity (Table 13.3).

At regional level, e.g., in Europe, things will get more interesting. Executive management and governing body are available in the European governance model, but the system by which a regional-level information security activity is directed and controlled, does not meet the ISO 27014 recommendations (Table 13.4). In Europe,

Table 13.2 Definitions in ISO 27014

Definitions	Meaning at organization level
Executive management	Delegated responsibility from the governing body for implementation of strategies and policies to accomplish the purpose of the organization
Governing body	Accountability for the performance and conformance
Governance of information security	System by which an organization's information security activities are directed and controlled—Organization Management System
Stakeholder	Any person or organization that can affect, be affected by, or perceive themselves to be affected by an activity of the organization

Table 13.3 Definitions at state level

Definitions	Meaning at national level
Executive management	Agencies and authorities for aviation security and safety
Governing body	Ministries, agencies and authorities for aviation security and safety
Governance of information security	National civil aviation security and safety programs
Stakeholder	Any person or organization that can affect, be affected by, or perceive themselves to be affected by an activity of the agencies and authorities—Regional management system

Table 13.4 Definitions at regional level

Definitions	Regional (Europe) level
Executive management	DG MOVE, DG HOME, DG CONNECT, EASA
Governing body	European Commission
Governance of information security	European Aviation Safety (EPAS) and security (?) programs
Stakeholder	Any person or organization that can affect, be affected by, or perceive themselves to be affected by an activity of the agencies and organizations

the EPAS provide in aviation safety the needed system, but in the aviation security there is no relevant system available.

The international level has sufficient systems available, but the ICAO GASeP does not sufficiently recognize information security in its full spectrum (Table 13.5). The GASeP is more focused to unlawful interference of civil aviation recognize well, e.g., traditional terrorist threats. In the information security the threat actors are, however, very different, e.g., nation state, cybercriminals, hacktivists, terrorist groups, and insiders. The method to handle information security threat actors needs to be reviewed and coordinated in the GASeP. At international level, the GASP and GASeP can provided sufficient system for information security.

Table 13.5 Definitions at international level

Definitions	International level
Executive management	ICAO & member states
Governing body	ICAO & member states
Governance of information security	Global Aviation Safety Plan (GASP) and Global Aviation Security Plan (GASeP)
Stakeholder	Any person or organization that can affect, be affected by, or perceive themselves to be affected by an activity of the agencies and organizations

13.3.2 Concepts

Governance of information security needs to align objectives and strategies for information security with business. The governing body is ultimately accountable for an organization's decisions and the performance of the organization. In respect to information security, the key focus of the governing body is to ensure that the organization's approach to information security is efficient, effective, acceptable and in line with business objectives and strategies giving due regard to stakeholder expectations [8]. This applies to all levels, aviation organizations, state agencies and authorities, regional and international levels.

Next, the objectives and desired outcomes of the information security are projected at different levels. For the aviation organizations defining information security objectives and desired outcomes is straight forward work respect to the ISO 27014 (Table 13.6). For the state, regional and international level, there is already a solid governance model in civil aviation safety and security as previously described where information security should be implemented. From that perspective, the following objectives and desired outcomes are available in the available aviation cybersecurity strategies (ICAO and ESCP) (Table 13.7).

Table 13.6 Objectives and desired outcomes for aviation organization

Concepts	Meaning at organizational level
Governance *objectives* of information security (1) align the information security objectives and strategy with business objectives and strategy (strategic alignment) (2) deliver value to the governing body and to stakeholders (value delivery) (3) ensure that information risk is being adequately addressed (accountability)	(1) Business objectives: aviation organization specific (2) Value delivered: aviation organization specific
Desired *outcomes* from effectively implementing governance of information security include (1) governing body visibility on the information security status (2) an agile approach to decision-making about information risks (3) efficient and effective investments on information security (4) compliance with external requirements (legal, regulatory or contractual)	Indicators (need to be defined) how well the governance objectives are met: Organization management system and performance metering (indicators) from the relevant maturity metering models & information security standards [12]
Relationship with other areas of governance models (a holistic and integrated governance model with information security management usually benefits the governing body)	Governance of safety, security, legislations, information technology and business objectives

Table 13.7 Objectives and desired outcomes at state, regional and international levels

Concepts	Meaning
Governance *objectives* of information security: (1) align the information security objectives and strategy with business objectives and strategy (strategic alignment) (2) deliver value to the governing body and to stakeholders (value delivery) (3) ensure that information risk is being adequately addressed (accountability)	(1) Business objectives & strategy: Efficiently ensure public trust & confidence, operational resilience, safety and security in the digital society and aviation (2) Value delivered To governing body: timely information about industry information security (for safety and security) capability and risks → the information ensures sufficient regulatory framework, procedures, and processes in information security To stakeholders: holistic, standardized performance and risk-based legal framework, procedures and processes
Desired *outcomes* from effectively implementing governance of information security include (1) governing body visibility on the information security status (2) an agile approach to decision-making about information risks (3) efficient and effective investments on information security (4) compliance with external requirements (legal, regulatory or contractual)	Indicators (need to be defined) how well the governance objectives are met: State, regional, and international aviation safety and security programs and their indicators
Relationship with other areas of governance models (a holistic and integrated governance model with information security management usually benefits the governing body)	Coherence in aviation safety, aviation security, and information security management governance

13.3.3 Principles

Meeting the needs of stakeholders and delivering value to each of them is integral to the success of information security [8]. There are six principles in ISO 27014 to achieve the governance objective of aligning information security closely with the goals of the business and delivering value to stakeholders:

Principle 1: Establish organization-wide information security;
Principle 2: Adopt a risk-based approach;
Principle 3: Set the direction of investment decisions;
Principle 4: Ensure conformance with internal and external requirements;
Principle 5: Foster a security-positive environment;
Principle 6: Review performance in relation to business outcomes.

These principles of information security governance were the most challenging to project and compare for the perspective levels.

The principles provide a good foundation for the implementation of governance processes for information security. The statement of each principle refers to what should happen, but does not prescribe how, when, or by whom the principles would be implemented, because these aspects are dependent on the nature of the organization implementing the principles. The governing body should require that these principles be applied and appoint someone with responsibility, accountability, and authority to implement them [8].

13.3.3.1 Principles at Aviation Organizational Level

Principles at aviation organizational level can be directly transferred from the standard. The business and value delivery in this study are focused to operational resiliency, aviation security and safety. With these values the principles can be defined in the following way.

Principle 1: Establish organization-wide information security

For aviation organization, information security activities should be comprehensive and integrated with aviation safety and security. This principle emphasize the need of information security to be integrated to all aviation security and safety policies, processes, procedures and technologies. Information security responsibility and accountability should be established across the full span of organisation's activities, including aviation safety and security.

Principle 2: Adopt a risk-based approach

Governance at organizational level should be based on risk-based decisions. Determining how much security is acceptable should be based upon the risk appetite of an organization, including loss of competitive advantage, compliance and liability risks, operational disruptions, reputational harm, and financial loss [8]. In such interdependent ecosystem like civil aviation, the minimum level of information security risk appetite for aviation safety and security is based on compliance and liability through the evolving legislation. In addition in aviation, the organization and their aviation services, governance of information security should be based on consistent and integrated risk management including aviation safety, aviation security, and information security.

Principle 3: Set the direction of investment decisions

To optimize information security investments to support organizational objectives in aviation organizations, governance of information security should establish an information security investment strategy based on business outcomes achieved, resulting in harmonization between business and information security requirements and thereby meeting the current and evolving needs of stakeholders [8]. When the information security management from the operational resilience, aviation safety, and aviation security perspectives is implemented comprehensively and consistently in the aviation organizations, it enables controlled investment decisions and gives an opportunity for optimized investments, too.

Principle 4: Ensure conformance with internal and external requirements

For aviation organizations, governance of aviation information security should ensure that information security policies and practices conform to relevant mandatory legislation and regulations, as well as committed business or contractual requirements and other external or internal requirements. To address conformance and compliance issues, the governing body should obtain assurance that information security activities are satisfactorily meeting internal and external requirements by commissioning independent security audits [8]. Current legislative framework in all levels; state-, regional and international levels is strongly evolving, meaning information security management is being implemented into aviation safety and security management. Organizations should follow closely the development of this legislative framework. The business and contractual requirements also need to be emphasized, because very likely they cover more specific, but converging information security requirements with the evolving legislative framework.

Principle 5: Foster a security-positive environment

In aviation organization, governance of aviation information security should be built upon human behavior, including the evolving needs of all the stakeholders, since human behavior is one of the fundamental elements to support the appropriate level of information security. If not adequately coordinated, the objectives, roles, responsibilities, and resources may conflict with each other, resulting in the failure to meet business objectives. Therefore, harmonization and concerted orientation between the various stakeholders is very important. To establish a positive aviation information security culture, the governing body should require, promote, and support coordination of stakeholder activities to achieve a coherent direction for aviation information security. This will support the delivery of security education, training, and awareness programs [8].

Principle 6: Review performance in relation to business outcomes

In aviation organizations the governance of information security in aviation should ensure that the approach taken to protect aviation information is fit for purpose in supporting the organization, providing agreed levels of information security. Security performance should be maintained at levels required to constantly meet current and future business requirements. To review performance of information security from a governance perspective, the governing body should evaluate the performance of information security related to its business impact, not just effectiveness and efficiency of security controls. This can be done by performing mandated reviews of a performance measurement program for monitoring, audit, and improvement, and thereby link information security performance to business performance [8]. Performance measurement program is an important enabler also to make efficient investments on information security.

13.3.3.2   Principles at State Level

Principles at state level can be derived from the standard by changing the angle of view to state level. It is important to understand the differences especially in responsibilities. The state and the relevant agencies and authorities are responsible for the society, national aviation system safety and security to general public and all stakeholders.

Principle 1: Establish state-wide information security in civil aviation

At state level, civil aviation safety, aviation security, and information security agencies and authorities should co-operate closely for aviation eco-system wide information security, meaning the governance at state level ensure that information security activities are comprehensive and integrated to aviation safety and security. It is important to establish responsibility and accountability for information security for aviation safety, security and society.

Principle 2: Adopt a risk-based approach

In information security for aviation safety and aviation security, the risk appetite at state level is bound to compliance and liability with regional and international aviation legislation. Also, the societal responsibility of ensuring state aviation system is operational in all circumstances, affect risk appetite. Since the aviation information security legislation is strongly risk-based, it can be challenging to define how much information security is acceptable. Therefore it is paramount at state level to have good understanding of the overall risk picture, in order to perform state level role in civil aviation: guide, regulate and oversight aviation organizations some to mention. At state level the overall civil aviation risk picture could be part of the national civil aviation safety (State Safety Programme—SSP) and security (National Civil Aviation Security Program—NCASP) programs. That will help states and their stakeholders to better understand aviation ecosystem wide risks. Since the governance of information security should be based on risk-based decisions with an overall risk profile of the state, the state is better able to determine how much information security is acceptable in the risk-based world. This helps states consider their willingness to take risks. In addition to risk appetite, the state also need to make sure the governance of information security is based on aligned and integrated risk-based approach, meaning information security is integrated in aviation safety (SSP) and security (NCASP).

Principle 3: Set the direction of investment decisions

For the states, this principle means governance of information security could establish an aviation information security investment strategy based on aviation business, safety and security outcomes achieved, both in short and long term, thereby meeting the current and evolving needs of society and stakeholders. To enable information security investments to support national and international civil aviation objectives, the national aviation governing body should consider too that information security is integrated with existing civil aviation processes for capital and operational expenditure. This is very important aspect especially in the modern digitized society and aviation ecosystem. If information security is not coordinated through all levels, that can lead to deficiencies in the implementation of information security, in investments and high and ineffective development and operating costs in civil aviation.

Principle 4: Ensure conformance with internal and external requirements

At state level in civil aviation, there is strong and evolving framework in aviation safety and security, where information security will be implemented. Information security governance at state level should ensure the state information security policies and practices conformance with the relevant domestic, regional and international legislation and regulations, as well as with the operational or contractual requirements or other external or internal requirements. To address conformance and compliance issues, the governing body at state level should obtain assurance that information security activities are satisfactorily meeting internal and external requirements by commissioning independent security audits.

Principle 5: Foster a security-positive environment

Governance of aviation information security at state level should be built upon human behavior, including the evolving needs of all the stakeholders, since human behavior is one of the fundamental elements to support the appropriate level of information security. The human behavior is a strong asset in aviation, because there is existing strong safety and security culture, which can be leveraged into information security as well. At state level, the agencies and authorities are in the key role to stakeholders domestically or abroad, fostering the security-positive environment. If the human behavior is not adequately coordinated, the objectives, roles, responsibilities, and resources may conflict with each other, resulting in the failure to meet eventually the operational objectives. Therefore, harmonization and concerted orientation between the various stakeholders is very important. To establish a positive aviation information security culture, the governing body (relevant state agencies and authorities) should require, promote, and support coordination of stakeholder activities to achieve a coherent direction for aviation information security. This will support the delivery of security education, training, and awareness programs.

Principle 6: Review performance in relation to operational outcomes

For the states, governance of information security in aviation should ensure that the approach taken to protect aviation information at state level is fit for purpose in supporting the organizations, providing agreed levels of information security. Security performance at state level should be maintained at levels required to meet current operational and societal requirements. To review performance of information security in aviation at state level from a governance perspective, the governing body (relevant agencies and authorities) should evaluate the maturity of aviation information security related to its aviation operational resiliency, safety, and security impact, not just effectiveness and efficiency of security controls. If this principle is not identified, that can provide fallacy between theory and practice in the aviation information security. This can be done by performing mandated reviews of a performance measurement program for monitoring, audit, and improvement, and thereby link information security performance to operational performance.

13.3.3.3 Principles at Regional Level (Europe)

Principles at regional level can be defined by considering the responsibilities of European civil aviation governing body and relevant European agencies, who are responsible for the European aviation system information security, safety and security.

Principle 1: Establish Regional-wide information security in civil aviation

Regional level aviation safety and security agencies and information security agency should establish aviation eco-system wide information security. In Europe this means the governance at European level should ensure that information security activities are consistently and comprehensively integrated to aviation safety and security. To establish European level aviation information security, the responsibilities and accountabilities should be defined and established across the full span of European civil aviation activities. This is an important principle requiring all, the European civil aviation governing body and executive management in safety, security, and information security, seamlessly cooperate and coordinate information security in aviation safety and security.

Principle 2: Adopt a risk-based approach

In information security for aviation safety and security, the risk appetite at regional level could be defined similarly like at state level. It is bound to compliance and liability with international aviation legislation, but overall understanding of information security risks is paramount to assure risk-based legislation. Also at regional level the governance of aviation information security should be based on aligned and integrated risk-based approach, meaning information security is integrated in the aviation safety (European Plan for Aviation Safety) and security. It is important to however note, that currently there is no European aviation security program. Instead, there are common rules and basic standards for the states and industry on aviation security and the procedures to monitor the implementation of the common rules and standards, which are implemented by the states through the NCASP.

Principle 3: Set the direction of investment decisions

The principle at regional level is the same as at state and international levels.

Principle 4: Ensure conformance with internal and external requirements

Governance of the European level aviation information security should ensure that European civil aviation information security policies and practices conform to relevant mandatory international legislation and regulations, as well as committed business or contractual requirements and other external or internal requirements. To address conformance and compliance issues, the governing body in Europe should obtain assurance that information security activities are satisfactorily meeting internal and external requirements by commissioning independent security audits. The independent audit or relevant action would be beneficial to give an objective and comprehensive view, how well the European system currently and in the future meets the internal and external requirements.

Principle 5: Foster a security-positive environment

At regional level, governance of aviation information security at European level should be built upon human behavior, including the evolving needs of all the stakeholders, since human behavior is one of the fundamental elements to support the appropriate level of information security. The human behavior is a strong asset in aviation, because there is existing strong safety and security culture, which can be leveraged into information security as well. The European agencies and authorities are in the key role to stakeholders regionally, fostering the security-positive environment. When sufficiently coordinated, the objectives, roles, responsibilities, and resources converge with each other, resulting efficiently to meet operational objectives. Therefore, harmonization and concerted orientation between the various stakeholders, European and non-European states and stakeholders, is very important. To establish a positive aviation information security culture, the governing body (European Commission) should require, promote and support coordination of stakeholder activities to achieve a coherent direction for aviation information security. This will support the delivery of security education, training, and awareness programs. The existing aviation safety and security education-, training and awareness framework provides a good opportunity to convey information security training to aviation.

Principle 6: Review performance in relation to operational outcomes

At regional level in Europe, governance of information security in aviation should ensure that the approach taken to protect aviation information at European level is fit for purpose in supporting the organizations, providing agreed levels of information security. Security performance at European level should be maintained at levels required to constantly meet current and future operational requirements. To review performance of information security in aviation at European level from a governance perspective, the governing body (European Commission) should evaluate the performance of aviation information security related to its societal, operational resilience, safety and security impact, not just effectiveness and efficiency of security controls. This can be done by performing reviews of a performance measurement program for monitoring, audit, and improvement, and thereby link information security performance to operational performance in aviation.

13.3.3.4 Principles at International Level

Principles at international level can be defined by considering the responsibility of civil aviation governing body at international level, who is responsible for the international civil aviation information security, safety, and security governance.

Principle 1: Establish international-wide information security in civil aviation

At international level, aviation information security activities should be consistent and comprehensive and integrated with all aviation safety, security, and civil aviation activities. To establish the international level wide civil aviation information security, the responsibility and accountability for civil aviation information security for aviation safety and security, should be established across the full span of international civil aviation activities. This principle is supported by the ICAO Aviation Cybersecurity Strategy and Action Plan.

Principle 2: Adopt a risk-based approach

Similarly with the regional and state levels, the governance of information security should be based on risk-based decisions. The risk appetite determining how much information security is acceptable, should be based on the risk appetite consensus by the states and industry, including operational disruptions, reputational harm, financial loss or loss of public trust and confidence. In addition, the governance of information security should be based on aligned and integrated risk-based approach in information security, aviation safety and security. At international level, key enabler for this principle is integrated information security in the global aviation safety (GASP) and security programmes (GASeP).

Principle 3: Set the direction of investment decisions

The principle at international level is the same as at state and regional levels.

Principle 4: Ensure conformance with internal and external requirements

At the highest level, international level, it is the international aviation community, states and industry together who are developing and implementing civil aviation information security policies and practices into civil aviation. Therefore, there are no relevant mandatory legislation and regulations that information security policies and practices at international level should conform.

Principle 5: Foster a security-positive environment

The same way with regional level, the governance of aviation information security at international level should be built upon human behavior. The human behavior is a strong asset in aviation due to existing strong safety and security culture, which can be leveraged into information security as well. At international level, states and industry through ICAO, are in the key role to foster security-positive environment. Sufficiently coordinated, the objectives, roles, responsibilities, and resources converge with each other, resulting efficiently to meet operational objectives. Harmonization and concerted orientation between the various stakeholders is very important. To establish a positive aviation information security culture, the governing body (ICAO) should establish, require, promote and support coordination of stakeholder activities to achieve a coherent direction for aviation information security. This will support the delivery of security education, training, and awareness programs. The existing aviation safety and security education-, training and awareness framework provides a good opportunity to convey information security training to aviation.

Principle 6: Review performance in relation to operational outcomes

At international level, governance of information security in aviation should ensure that the approach taken to protect aviation information at international level is fit for purpose in supporting the states and aviation organizations (industry), providing agreed levels of information security. Security performance at international level should be maintained at levels required to constantly meet current and future operational requirements. To review performance of information security in aviation at international level from a governance perspective, the governing body (ICAO) should evaluate the performance of aviation information security related to its operational impact, not just effectiveness and efficiency of security controls. This can be done by performing reviews of a performance measurement program for monitoring, audit, and improvement, and thereby link information security performance to operational performance.

13.4 Conclusions

The information security management is generally well recognized in the civil aviation. However, the significance and meaning of information security governance need some attention. This study focuses on information security governance through ISO 27014 definitions, concepts, and principles at different levels in the civil aviation. In order to make sustainable and efficient aviation information security management, the meaning and objectives of information security governance should be better recognized and understood, because it is crucial for efficient performance and risk-based information security regardless of the respective level.

The objective of this study was to test and evaluate how to apply relevant standard in information security governance in civil aviation and at different levels. The study address that the ISO 27014 definitions, concepts and principles can be applied to higher levels than organizational level. No obstacles were found for the application and important information security governance objectives, desired outcomes and principles were recognized for different levels. The ISO 27014 is dedicated to ISMS in the context of organization and can be applied for all types or size of organizations. This means higher levels require special consideration with the definitions, concepts, and principles. In this study, the projected definitions, concepts, and principles for the higher levels are based on the current international, regional and state level working group work in the ICAO, ECAC and ESCP, their publications and relevant industry standards in information security. The considerations in this study represent author´s personal interpretation and expertise in this field. The author is actively involved in the relevant ICAO, ECAC and ESCP working group work, having also strong aviation safety and information security background.

The study discovered interesting aspects in information security governance which have not been properly recognized at the current international, regional, or state level aviation information security. It would be beneficial to research them and their status in more detail. It is recommended, that every organization at all levels consider their role and responsibility in the aviation ecosystem and define their information security governance at more detailed level. This study can provide one approach to help in this work.

Specific observations about the status of each principle are not presented in this study. However, it can be generally observed that there are shortages in the governance of information security in civil aviation. It is important to make sure that governing body, executive management, and governance system are available at all levels. Without those elements, it is very hard to implement sufficient information security management in civil aviation. Other observations were related to objectives and desired outcomes. At all levels, objectives and outcomes are important to define beforehand. Information security governance have also six important principles, which should be defined for all levels. This would help and ensure common effort towards consistent and coherent civil aviation information security.

References

1. Bevir M (2012) Governance: a very short introduction. Oxford University Press
2. ECAC (2020) ECAC guidance material on cyber security in civil aviation. European Civil Aviation Conference (ECAC)
3. FAA (2020) Safety management system (SMS). U.S. Department of Transportation, Federal Aviation Administration (FAA), https://www.faa.gov/about/initiatives/sms/
4. ICAO (2016) Global air navigation plan 2016–2030. Doc 9750-AN/963, 5th edn. International Civil Aviation Organization (ICAO), https://www.icao.int/publications/Documents/9750_5ed_en.pdf
5. ICAO (2019a) Global aviation safety plan. Doc 10004, 2020—2022 edn. International Civil Aviation Organization (ICAO), https://www.icao.int/safety/GASP/Documents/Doc.10004%20GASP%202020-2022%20EN.pdf
6. ICAO (2019b) Aviation cybersecurity strategy. International Civil Aviation Organization (ICAO)https://www.icao.int/cybersecurity/Documents/AVIATION%20CYBERSECURITY%20STRATEGY.EN.pdf
7. ICAO (2020) Security: safeguarding international civil aviation against unlawful interference. Annex 17, 11th edn. International Civil Aviation Organization (ICAO)
8. ISO/IEC (2013) ISO/IEC 27014:2013: Information technology, security techniques, governance of information security. International Organisation for Standardization (ISO) and International Electrotechnical Commission (IEC)

9. ISO/IEC (2017) ISO/IEC 27000:2017: Information technology, security techniques, information security management systems, overview and vocabulary (ISO/IEC 27000:2016). International Organisation for Standardization (ISO) and International Electrotechnical Commission (IEC)
10. NIST (2018) Framework for improving critical infrastructure cybersecurity. Version 1.1, National Institute of Standards and Technologies (NIST)
11. SKYbrary (2020) Safety management. SKYbrary, https://www.skybrary.aero/index.php/Safety_Management
12. Traficom (2020) Kybermittari. National Cyber Security Centre, Finnish Transport and Communications Agency Traficom, https://www.kyberturvallisuuskeskus.fi/en/our-services/situation-awareness-and-network-management/kybermittari

Chapter 14
Smart Cities and Cyber Security Ethical and Anticipated Ethical Concerns

Richard L. Wilson

Abstract Smart cities have already been developed and are beginning to change the nature of urban living. Much of the discussion on smart cities would perhaps be more of a discussion on smart technologies. The technological infrastructure needed to create smart cities is comprised of interlinking the smart technologies at the core of the smart city. This analysis will develop an anticipatory ethical analysis of the technological, social, and ethical issues of smart cities from a cyber security perspective. Anticipatory ethics concerns with the identification of ethical issues in the development of technology at an early stage in order to attempt to mitigate these issues as technology develops. Smart cities are developing, as are the technological systems that operate in them. Anticipatory ethical analysis can help provide the foundation for identifying how malevolent agents could compromise smart cities and how cyber security can be implemented to prevent this.

Keyword smart city · cyber security · urban living · smart technologies · technological infrastructure · technological development · networking of smart devices · communication technologies · anticipatory ethics

14.1 Introduction

What does it mean to be a smart city? At its broadest, it means the use of information and communication infrastructure (ICT), which enables smart devices and eventually the devices themselves to communicate (such as smartphones, automobiles, thermostats, and water meters) and connect to smart infrastructure (problem reporting, traffic signals and information, parking systems, electric grids, billing systems) to improve quality of life and productivity of those living within a smart city [1]. The name 'smart city' reflects the ways information and communication technologies (ICT) and artificial intelligence are employed within intersecting systems

R. L. Wilson (✉)

Philosophy and Religious Studies and Computer and Information Sciences Towson University Towson Md, Senior Research Scholar Hoffberger Center for Professional Ethics University of Baltimore Baltimore Maryland, Towson, USA

e-mail: wilson@towson.edu

M. Lehto and P. Neittaanmäki (eds.), *Cyber Security*, Computational Methods in Applied Sciences 56, https://doi.org/10.1007/978-3-030-91293-2_14

that are implemented and integrated into the infrastructure of smart cities. A smart city ecosystem is made up of a wide variety of systems, including smart homes, smart vehicles, and transportation systems. This ecosystem has to include a complex range of sensors, software, robots, and networks, as well as real-time surveillance. Software is also needed to integrate all the data that will enable a smart city's smart system to operate.

There are a number of cyber security, social and ethical problems, that arise due to where the data comes from and the destination to where it goes. The extensive use of ICT and connected technologies within the Internet of Things (IoT), cloud, fog and edge computing, and cyber physical systems (CPS) has led to data being generated and gathered in increasing amounts and at ever increasing speeds. This data comes from a variety of sources and may carry increasingly sensitive and private information about smart city residents. In addition to ethical issues related to the privacy of citizens there are also issues related to cyber security for a smart city itself. Cyber security refers to the body of technologies, processes, and practices designed to protect networks, devices, programs, and data from attack, damage, or unauthorized access.

This chapter will develop an anticipatory ethical analysis of the social and ethical issues confronting smart cities from the perspective of cyber security. Anticipatory ethics is concerned with identifying ethical issues with the development of technologies while they are in the early stages of development in order to attempt to mitigate these issues as the technology develops and gets implemented. Smart cities are in the process of developing as are the technological components and systems that will be operating within them. An anticipatory ethical analysis can provide the foundation for identifying areas that are important for preventing malevolent agents who could compromise smart cities. In order to identify the ethical issues that will emerge for smart cities we can construct an anticipatory ethical analysis.

The large degree of connectedness creates the conditions for a smart city, while creating the opportunity for cyber security problems and related ethical issues to arise in the smart city. From the perspective of urban development, smart cities replace earlier methods of urban development so that, smart technology produces a smart city through ICT technology. A high degree of connectedness creates alterations at a number of levels when compared with the infrastructure of traditional cities. There are a variety of problems that are directly related to planning, creating, operating and sustaining a smart city. As stated by Pelton and Singh,

> The technical challenges associated with a smart city today are enormous. It is ever more diffi-cult to prevent cyber-attacks and find safe ways to employ the power of digital communica-tions, cyber networks, information technology systems, artificial intelligence, and advanced robotics. And these are just a few of the key issues to be addressed in order to design, build and operate a smart city anywhere in the world. [11, p. 9]

While smart cities have the potential for altering the nature of urban existence, they have already begun to shift the traditional models of urban design for urban developers. There are challenges and alterations that urban planners need to address including, increased populations, housing availability and affordability, increased costs of home ownership, over stressed transportation systems, an increase in technology enabled criminal activity and behavior, and exponential technological change and development. Smart city design and planning when deployed with security in mind can work to mitigate these issues. Anticipatory ethics provides a basis for addressing a variety of important questions about smart cities and the alterations in urban design they stand to create. It can also be used to help develop a strategy for cyber security in smart cities.

This chapter will describe how smart cities will create alterations in urban design, and attempt to anticipate some of the technologies that are critical to smart city development and some of the smart city's cyber security issues. In addition, this analysis takes as critical the ethical issues that may arise by identifying the technologies that are crucial to urban planning as a result of smart city development.

14.2   Smart Cities and Cyber Security

Given the high degree of interconnectedness needed for a smart city to function one of the key problems for maintaining the function of a smart city is cyber security. What is cyber security? Kaspersky answers that

> Cyber-security is the practice of defending computers, servers, mobile devices, electronic systems, networks, and data from malicious attacks. It's also known as information technology security or electronic information security. The term applies in a variety of contexts, from business to mobile computing, and can be divided into a few common categories. [8]

These categories include

- Network security,
- Application security,
- Information security,
- Operational security,
- Disaster recovery and business continuity,
- End-user education.

Problems with the development and continued function of a smart city that create cyber security issues include:

1. Insecure legacy systems: When older and vulnerable systems are used in conjunction with newer technology, the potential attack surface increases.
2. Encryption issues: Most of the technologies are wireless, which makes them easy to hack if the technologies are not encrypted.
3. Smart transportation: Cyber-attacks can display incorrect information in public transportation systems, and it is possible to influence people's behavior by causing delays and overcrowding.
4. Cloud storage: City servers and cloud infrastructure on which smart city data is stored are exposed to common Distributed Denial of Services (DDoS) attacks, which can render these services inoperable.

5. Smart city's large and complex attack surface: Due to the complexity and interdependence of the interlocking technological systems of a smart city, it is difficult to identify what and how is exposed.
6. Lack of cyber-attack emergency plans: A smart city must properly prepare for cyber-attacks on power grids and all parts of infrastructure [8].

There are a wide variety of stakeholders who will be most affected by the development of smart cities and by the cyber security issues arising in smart cities. City planners, engineers and ICT specialists who develop the technologies used in smart cities, including sensors, robots, networks, real time surveillance, and software. Cyber security specialists *are* responsible for developing and monitoring the body of technologies, processes, and practices designed to protect networks, devices, programs, and data from attack, damage, or unauthorized access in real time. As smart cities continue to develop artificial intelligence and machine learning experts will be responsible for integrating the increasingly large amount of data into the efficiency of the operation of the smart city.

14.3   Smart Cities and Ethics

The ICT infrastructure at the center of the functioning of a smart city is directly linked to the level of technological development and related ethical issues that go into the make of smart city infrastructure. Furrow identifies the focus of ethical analysis involving a series of factors. As Furrow states, ethics is related to evaluating actions and actions are performed by those who are capable of being moral agents. Furrow says:

> When we evaluate an action, we can focus on various dimensions of the action. We can evaluate the person who is acting, the intention or motive of the person acting, the nature of the act itself, or the consequences. [5]

Two important points are made in this passage which can be applied to the smart cities. First, ethical issues related to smart cities are based upon the idea that what those who design, develop, and implement the technologies used in smart cities do, is perform actions, and second, these actions are an extension of a person's intentions, actions and goals/outcomes.

The actions related to those who design and develop the technologies used in smart cities are capable of being evaluated based upon the intentions, actions and outcomes of the actions of the person(s) engaged in the activities. If Furrow's distinctions are applied to those who design, develop, and implement the technologies used in smart cities, there are three possible levels of ethical evaluation. We can evaluate the actions of those who design, develop and implement the technologies used in smart cities, we can evaluate the intentions of those who control and direct the actions of those who design, develop and implement the technologies used in smart cities, and we can evaluate the consequences of the intentions of those who control and direct the actions of those who design, develop and implement the technologies used in smart cities. The same procedure can be performed upon those who design the cyber security measures for smart cities.

14.4   Smart Cities, Technology, and Anticipatory Ethics

At the center of anticipatory ethics is the study of a technology, the technology behind the development of artifacts, or a study of a specific technological artifact. Anticipatory ethics studies how each of these will work and projects trajectories of future technological developments, how they may work, and potential ethical issues related to how these future developments may work. We maintain that smart cities require the development of a wide variety of technologies, including technological artifacts such as sensors and, particularly robots in order to have a smart city function. Technological systems and artifacts within a smart city have different *purposes* and goals. Without an understanding of these technologies, artifacts and specific examples, it is difficult to project how a smart city currently works or such a city may work in the future. These are practicalities that are related to technological development.

According to Brey [2], as technology develops, its moves from a research and development stage to a product introduction stage, to a market saturation stage, which is also described as a power stage. It is important to recognize these stages which influence the variety of ways that smart cities, and the 'technologies' operating within them can be assessed. Following the distinctions introduced by Brey, as a set of interconnected technologies, how can smart city technologies be assessed in terms of its current stage of technological development? What can smart city technologies and related artifacts do? How can the technology be intentionally used to achieve goals projected by smart city planners and developers, and what are the end results that are expected to be achieved through the use of smart city technologies by the stakeholders? For smart cities to develop and function will require the use of innovative, emerging, and disruptive technologies. Many of these technologies are rapidly moving through a number of stages of technological development and the ethical and social problems that may emerge as this technology moves from stage to stage need to be identified. A clear statement of the stages of technological development has been presented by Brey [2].

According to Brey as technology develops there are a series of stages through which it passes. In the following diagram we have added fifth stage to Brey's fourth stage due to the rapid development of smart city technologies:

1. R & D stage of technology development,
2. Introduction stage of technology development,
3. Permeation stage,
4. Power stage,
5. Continuing innovative development.

Focusing on technologies crucial for smart cities, we will apply and extend Brey's analysis. We assume that the development of smart city technologies and variations on these technologies are the result of development of smart technology in general. Many of the specific applications of smart city technologies for humans are in the R & D and early introduction stages but the technology and variations upon the procedure the technology makes possible, and further developments with smart city technologies (the technology) will continue to occur while the current state of the technology continues to develop. Variations in smart city technologies applied to specific areas, such as robotics and Internet of Things (IoT), allow us to interpret these variations as variations in technologies and we assume these variations will in turn influence our ideas about smart cities. Thinking about smart city technologies as a technology can help us explore how the introduction of smart city technologies into urban development will introduce issues related to urban existence in the modern world.

14.5 Technologies Essential to the Development of Smart Cities

A major difference between today's cities and smart cities will be the degree to which machines and ICT will be in control of the flow of information of through the technologies that will be operating within a smart city [12]. There are a number of technologies that are at various stages of technological development and that will be critical to the future development of smart cities. Smart cities are cities where everything is interconnected and where each thing is connected to every other thing. This interconnectedness is highly dependent upon the technologies at work within a smart city. There are a number of technologies that are crucial for the implementation of smart cities. Anticipatory ethics is a good device for thinking about what technologies will be needed for smart cities because these technologies are foundational for identifying and understanding anticipated ethical issues.

Technological expertise is critical to turn a city into a smart city that is well connected, sustainable and resilient. Technology is also at the center of the cyber security needs for a smart city. Smart cities are all about providing smart services to citizens that can save their time and ease their lives. It is also about connecting them to the governance where they can give their feedback to the government as of how they want their city to be. This aim cannot be turned into reality without technology. Using technology, the city leaders and managers are able to gather city intelligence and this intelligence when integrated with the operations of a smart city, making the cities smarter and safer.

There are a number of technologies that are essential for the operation of a smart city infrastructure. Here are some of these technologies [9]:

- 5G,
- Internet of Things,
- Smart IoT devices,
- Geospatial Technology,

- Cloud Technology,
- Smart Energy,
- Information and Communication Technology,
- Smart Sensors,
- Smart Transportation,
- Smart Date,
- Smart Infrastructure,
- Smart Mobility,
- Artificial Intelligence,
- Robots,
- Virtual and Augmented Reality,
- Edge and Fog computing,
- Blockchain.

14.6 Smart Cities and Anticipated Ethical Issues

In order to understand the cyber security risks facing a smart city the risk landscape needs to be identified [10]. Smart cities face many cyber security risks which for each of the essential technologies and arise, as the essential technologies related to digital and physical infrastructure converge. To address these challenges, those managing smart cities should anticipate cyber security and privacy issues at each stage of the development of a smart city and at each stage of the development of the technologies that will be in the infrastructure of a smart city.

Smart cities are the future of urban living, combining the power of the essential technologies referred to above, while increasing the efficiency and effectiveness of city services. Three forces that will drive this transformation are carefully thought out design of systems, digital technology and the processing of immense amounts of data [10]. However, the transformation of cities based on the essential technologies, from their current state to smart cities through digital transformation also brings the threat of new cyber risks that are capable of fundamentally impact the existence of smart cities. Cyber threats have been on the rise for years, but the last few years have seen an explosion in cyberattacks that target both data and physical assets [14].

As technologies continue to develop connected devices can also be expected to increase at an accelerated rate—the number of IoT devices was expected to rise from 8.4 billion in 2018 to almost 20 billion by 2020 [4]. What this means is that cyberattacks and vulnerabilities in one area are capable of escalating and having an effect on numerous other areas within the smart city infrastructure. The consequences of this escalation could extend beyond just data loss, financial impact, and reputational damage risks to include disruption of crucial city services and infrastructure across a broad range of domains such as health care, transportation, law enforcement, power and utilities, and residential services. Such disruptions can potentially lead to loss of life and breakdown of social and economic systems.

14.7 Smart Cities and Anticipated Cyber Security Risks

The speed of the connections (the latency) related to hyperconnectivity and digitization are important factors in the acceleration of cyber threats. To address the challenge, government leaders, urban planners, and other key stakeholders should make cyber security principles an integral part of the smart city governance, design, and operations, not just an afterthought. It is *important that* key factors that influence cyber risks in a smart city ecosystem *be identified* and a broad approach that city leaders can *then* adopt to manage these risks. One way for this to occur is for a smart city to have a distinct agency that includes a chief information officer (CIO), chief technology officer (CTO), and a chief information security officer (CISO) [3].

A smart city is a matrix of essential technologies that form the infrastructure of the ecosystem of the smart city. This matrix is made up of technologies that coordinate municipal services, public and private entities, people, processes, devices, and city infrastructure that constantly interact with each other. The underlying technologies forming the infrastructure of a smart city ecosystem can be seen to be made up of three categories. These categories are: the edge, the core, and the communication channel [10]. The edge layer comprises devices such as sensors, actuators, other IoT devices, and smartphones that connect actor, things and technologies within the smart city. The core layer is made up of the essential technology platforms that process and make sense of the data and information flowing from the edge technologies within the smart city. The communication channel establishes a constant, two-way data and information exchange between the core and the edge technologies in a smart city to attempt to seamlessly integrate the various components of the infrastructure of the ecosystem of that smart city.

A smart city is characterized by having an immense amount of data and information exchanges, which requires the integration between disparate IoT devices, and dynamically changing processes which creates new cyber threats. These cyber threats are compounded by complexities in the other components of the ecosystem that are interconnected within the technology infrastructure of a smart city. For instance, data governance can be a thorny issue for cities as they need to think about whether the data is internal or external; whether it is transactional or personalized; whether the transactional data is collected via IoT devices; and how the data is stored, archived, duplicated, and destroyed. In addition, due to a lack of common standards and policies, between different agencies and organizations within a smart city many cities are experimenting with new vendors and products, which create interoperability and integration problems on the ground and increase cyber risks.

An examination of the ethical issues related to the three layers of smart cities leads to three cyber security risks that can be anticipated for smart cities. Three factors will give rise to cyber security issues for smart cities which will influence the potential cyber risk in a smart city ecosystem [10]:

1. Convergence of the cyber and physical worlds,
2. Interoperability between legacy and new systems,
3. Integration of disparate city services and enabling infrastructure.

To identify the cyber risk landscape of smart cities and to understand how it is managed, we explore these factors in more detail.

1.   **Convergence of the cyber and physical worlds**

The technology that forms the infrastructure of smart cities integrates the physical and cyber worlds. In this environment, people, processes, and places are integrated via both information technology (IT) systems used for data-centric computing and operational technology (OT) systems used to monitor events, processes, and devices and adjust city operations. At the same time this confluence of technologies allows cities to control and govern technology systems through remote cyber operations.

However, this convergence also creates risks, where many devices in the edge layer of a smart city could create an extremely large number of cyber threat vectors, giving rise to the risk of malicious actors entering the system and disrupting operations on the ground—thus exponentially expanding the cyber risk landscape [12]. With the growth of IoT and the proliferation of devices connected to IoT devices, attackers have countless entry points and attack vectors to compromise city systems and take advantage of the resulting vulnerabilities.

2.   **Interoperability between legacy and new systems**

Cities and organizations that are attempting to transform to smart technologies and that pursue digital transformation as part of the project of becoming smart cities, need to integrate new digital technologies with already existing legacy systems. Eventually these two types of systems will also need to be integrated with next generation technology. This creates significant challenges and risks. These challenges include inconsistent security policies and procedures and disparate technology platforms across different agencies, resulting in hidden security vulnerabilities within the essential technologies making up the smart city ecosystem.

3.   **Integration of disparate city services and enabling infrastructure**

Integration of disparate city services and technology infrastructure is one problem, while cyber security for these systems is another. As part of ordinary existence within cities, citizens have access to a wide range of services that were previously largely independent of each other (e.g., power, water, sewer, transportation, public works, law enforcement, firefighting, and social services). Within cities each of these services was typically provided by an agency using its own independent systems, processes, and assets. As smart cities develop these services all need to be integrated and linked through an interconnected web of digital technologies.

As cities transform to smart cities and add new technologies there will also be new services and efficiencies, this confluence of services and systems will create a unique set of challenges for the integration of these systems for cyber security. The increasing integration, interconnectedness, and data and information exchange that is at the center of a smart city ecosystem, will contribute to shared cyber security vulnerabilities. One problem that will occur is when a problem in one service area can quickly spread to and affect other areas—potentially leading to widespread and potentially catastrophic failures. In addition, cities need to rethink regulatory requirements, rationalize varied security protocols, and address data ownership and usage challenges.

14.8   Applied Anticipatory Ethics

Given these difficulties, we need to ask: What additional difficulties lie ahead for the continued development of smart cities? Projections can be made about technological development. Once technological developments have been projected then ethical issues related to these technological developments can also be projected. As stated above the center of anticipatory ethics is the study of a technology, the technology behind the development of artifacts, or a study of a specific technological artifact. Anticipatory ethics studies how each of these will work and projects trajectories of future technological developments, how they may work, and potential ethical issues related to how these future developments may work. We maintain that smart cities require the development of a wide variety of technologies, which requires technological artifacts such as sensors and, particularly robots in order to have a smart city function appropriately. Technological systems and artifacts within a smart city have different purposes and goals. Without an understanding of these technologies, artifacts and the goals of each, it is difficult to project how a smart city currently works or how such a city may work in the future.

The principles that will be employed to identify and discuss potential ethical issues will be 'the Rules for Computing Artifacts' [6]. According to the authors,

> The Rules has become a project explicitly based on Internet interactions. This aspect of The Rules is something that we often see in analyzing an ethical issue involving sociotechnical systems: there are people who are critical components in any such system, and often some of those people meet face to face. But sometimes, and increasingly often, after a computing artifact is "launched," many (and sometimes most) of the ensuing interactions and relationships are computer-mediated. [6]

14.9   Rules for Computing Applied Artifacts

How can we carry out an anticipatory ethical analysis? Here we employ five rules for computing artifacts as the basis of our anticipatory ethical analysis. Anticipatory ethics can accomplish two tasks:

1. Anticipatory ethics can be employed to identify potential ethical issues arising from the development of any one the essential technologies at work within a smart city.
2. Anticipatory ethics can be employed to analyze what technologies and issues are anticipated as developing.

What ethical issues can be anticipated as arising for smart cities? Balancing the promise of smart cities against the potential of cyber risks—and managing the cyber security risks effectively—will be critical to implementing smart cities. Cities should begin by engaging all the stakeholders and entities in the broader ecosystem. The next steps that cities should consider include the following:

All of the essential technologies at work within a smart city will need to be synchronized so that they can be brought into coordination with one another. Returning to Furrow's distinctions intention, actions and outcomes, city leaders and managers will need to apply any rule from the five rules based upon these distinctions. Here we shift the focus from computing artifacts to the essential technologies within a smart city [6]:

Rule 1. The people who design, develop, or deploy a computing artifact [any of the essential technologies in a smart city] are morally responsible for that artifact, and for the foreseeable effects of that artifact. This responsibility is shared with other people who design, develop, deploy or knowingly use the artifact as part of a sociotechnical system.

Rule 2. The shared responsibility of computing artifacts [an essential technology within a smart city] is not a zero-sum game. The responsibility of an individual is not reduced simply because more people become involved in designing, developing, deploying or using the artifact. Instead, a person's responsibility includes being answerable for the behaviors of the artifact and for the artifact's effects after deployment, to the degree to which these effects are reasonably foreseeable by that person.

Rule 3. People who knowingly use a particular computing artifact [an essential technology within a smart city] are morally responsible for that use.

Rule 4. People who knowingly design, develop, deploy, or use a computing artifact [an essential technology within a smart city] can do so responsibly only when they make a reasonable effort to take into account the sociotechnical systems in which the artifact is embedded.

Rule 5. People who design, develop, deploy, promote, or evaluate a computing artifact [an essential technology within a smart city] should not explicitly or implicitly deceive users about the artifact or its foreseeable effects, or about the sociotechnical systems in which the artifact is embedded.

Where does the application of the rules lead? Six important areas need to be addressed for cyber security in smart cities [13]:

1. **Infrastructure and inter-operability of the essential technological systems**

Smart Cities utilize sensor technology to gather and analyze information in an effort to improve the quality of life for residents. Sensors collect data on everything from rush hour stats to crime rates to overall air quality. Complicated and costly infrastructure is involved in installing and maintaining these sensors. How will they be powered?

Will it involve hard-wiring, solar energy, or battery operation? Or, in case of power failure, perhaps a combination of all three?

We can apply Rule 4 to this problem: "People who knowingly design, develop, deploy, or use a computing artifact [essential technologies within a smart city] can do so responsibly only when they make a reasonable effort to take into account the sociotechnical systems in which the artifact is embedded."

2. Greater collaboration between agencies

Due to the complex interaction of the essential technologies within a smart city the leaders and mangers of the smart city will need to make sure that an interdisciplinary base for the interaction of all of the agencies is in place so that clear communication can occur between all of the agencies. Here we can apply Rule 5 to this problem: "People who design, develop, deploy, promote, or evaluate a computing artifact [essential technologies within a smart city] should not explicitly or implicitly deceive users about the artifact or its foreseeable effects, or about the sociotechnical systems in which the artifact is embedded."

3. Privacy concerns

In any smart city, there will have to be a balance between quality of life and invasion of privacy through surveillance. While each citizen may want to enjoy a more convenient, peaceful, and healthy environment, nobody wants to feel like they are constantly being monitored through extensive surveillance technology. Cameras installed everywhere within the city may help deter crime, but they can also install fear and paranoia in law-abiding citizens. A significant concern is the amount of data being collected from all the smart sensors residents come into contact with each day. Rule 1 can be applied to privacy concerns: "The people who design, develop, or deploy a computing artifact [any of the essential technologies in a smart city] are morally responsible for that artifact, and for the foreseeable effects of that artifact. This responsibility is shared with other people who design, develop, deploy or knowingly use the artifact as part of a sociotechnical system."

4. Education, and Engaging the Community

For a smart city to truly exist, function and continue to develop, it needs "smart" citizens who are engaged with and actively taking advantage of new technologies as they are implemented in a smart city. With the implementation of any new city-wide tech project, part of the implementation process must involve city leadership helping to educate the community on benefits of the technology operating within a smart city infrastructure. It is important to develop education platforms that introduce citizens to the technology and that keeps citizens informed and engaged with smart city technology. Rule 3 can be applied to education within a smart city: "People who knowingly use a particular computing artifact [an essential technology within a smart city] are morally responsible for that use."

5.   Being socially inclusive

Smart transit programs that provide riders real-time updates are a great idea for smart city. But what if a significant percentage of the population of a smart city can't afford to take mass transit or Uber? What about a growing elderly population that does not understand how to use mobile devices or apps? How will smart technology reach and benefit these groups of people? It's important that smart city planning involves the consideration of all groups of people, not just the affluent and technologically advanced. The implementation of the essential technologies in smart cities should always be working to include the widest range of stakeholders, rather than divide them further based on age, income or education levels. Thinking of these communities, will help promote the overall success of a solution that includes the widest range of stakeholders. We can apply Rule 4 to this problem: People who knowingly design, develop, deploy, or use a computing artifact [essential technologies within a smart city] can do so responsibly only when they make a reasonable effort to take into account the sociotechnical systems in which the artifact is embedded.

6.   Cyber security and hackers

As the essential technologies within smart cities expand, in areas such as IoT and sensor technology use expands, so does the threat level to security. Recent discussion involving cyber-terror threats to vulnerable and outdated power grids has everyone a bit more concerned and skeptical about technology and cyber security.

Smart Cities have to invest more money and resources into security, while technology companies providing essential technologies for smart cities have to create solutions that include built-in cyber security mechanisms with essential technologies to protect against hacking and cyber-crimes. With technologies such as blockchain being applied to every field where they can be applied, many developers are looking for ways to incorporate these encryption techniques to increase security in new applications within smart city infrastructure. Two rules can applied to this issue, Rules 3 and 5. Rule 3 can be applied to cyber security and hackers within a smart city: "People who knowingly use a particular computing artifact [an essential technology within a smart city] are morally responsible for that use." Rule 5 can also be applied: "People who design, develop, deploy, promote, or evaluate a computing artifact [an essential technology within a smart city] should not explicitly or implicitly deceive users about the artifact or its foreseeable effects, or about the sociotechnical systems in which the artifact is embedded."

A more thorough analysis would go into the specific details of each the applications of the rules to each of these issues mentioned above.

14.10   Conclusion

The preceding analysis has attempted to carry out an anticipatory ethical analysis of smart cities from the perspectives of cyber security needs related to essential technologies for smart cities. Stakeholders concerned with cyber security include businesses, consumers, and individuals living within smart cities. The analysis proceeded by introducing Brey's view of stages of technological development and a set of ethical principles were selected as the basis of our anticipatory ethical analysis. What emerges as a result of this analysis are a number of important insights.

There is a need for interdisciplinary collaboration between different experts who interact with one another in the design and implementation of essential technologies in smart cities. Ethical issues need to be identified through emerging and disruptive technologies if they are to be incorporated into smart city infrastructure—although these technologies are still in the early stages of technological development. The complexity of the technical, social and ethical issues related to emerging technologies points to the need for this collaboration. The recent work of Johnson [7] can offer a clue for our analysis. As Johnson states, "since modern technologies involve 'many hands' both in their production and in their use, many actors may be accountable for the different aspects of the operation of the technology" [7]. This is most evident when accidents occur.

Here the insights of Johnson will be adapted to cyber attacks and cyber security within smart cities:

> The cause of the accident [cyber security failure] has to be traced back to the relevant actor/s; the cause may be in any number of places: Was the design [of cyber security measures] adequate? Did the manufactured parts [essential technologies] meet specifications? Did the instructions adequately explain how to use the [essential cyber security] technology? Did the users treat the [essential cyber security] technology as instructed? Each of the actors or actor groups is accountable for their contribution to the production of the [cyber security] technology and each may be asked to account if something unexpected happens. [7]

According to Johnson, responsibility is embedded in the relations between all of the actors involved in the development and the use of technology, which in this case would be the cyber security technology for a smart city.

It seems clear that responsibility about future ethical and social issues related to the development of technology for the development of smart cities requires that we engage in interdisciplinary discussions that involve all of the stakeholders who will be affected by the technology. According to Johnson, although the actual development of technology is contingent it is the case that the development of future technology will require an interdisciplinary negotiation between current and potential stakeholders, if design and responsibility practices are going to be established for emerging and disruptive technologies that will be used in smart cities. These ideas are important for all technologies within a smart city, which means they are important for ethical and social issues related to the large number of technological systems that will be at the center of the development and implementation of smart cities.

Finally, leaders and managers of smart cities need to have a strong team of ICT professionals who understand the wide range of technologies needed to make a smart city possible. These professionals need to be capable of understanding the confluence of technologies that go into the makeup of a smart city in order to understand the exponential range of possibilities needed to make a smart city cyber secure.

References

1. Algaze B (2016) How smart cities (will) work. ExtremeTech. https://www.extremetech.com/extreme/226739-how-smart-cities-will-work
2. Brey PAE (2012) Anticipating ethical issues in emerging IT. Ethics Inf Technol 14:305–317
3. Daswani N, Elbayadi M (2021) Big breaches: cybersecurity lessons for everyone. Apress, New York
4. Drzik JP (2018) Cyber risk is a growing challenge: so how can we prepare? World Economic Forum. https://www.weforum.org/agenda/2018/01/our-exposure-to-cyberattacks-is-growing-we-need-to-become-cyber-risk-ready
5. Furrow D (2005) Ethics: key concepts in philosophy. Continuum
6. Grodzinsky FS, Miller K, Wolf MJ (2012) Moral responsibility for computing artifacts: "The rules" and issues of trust. ACM SIGCAS Comput Soc 42(2):15–25
7. Johnson DG (2015) Technology with no human responsibility. J Bus Ethics 127:707–715
8. Kaspersky (2019) What is cyber-security? Kaspersky. https://usa.kaspersky.com/resource-center/definitions/what-is-cyber-security. Retrieved 15 July 2019
9. Maddox T (2016) Smart cities: 6 essential technologies. TechRepublic. https://www.techrepublic.com/article/smart-cities-6-essential-technologies/
10. Pandey P, Peasley S, Kelkar M (2019) Making smart cities cybersecure: ways to address distinct risks in an increasingly connected urban future. Deloitte. https://www2.deloitte.com/us/en/insights/focus/smart-city/making-smart-cities-cyber-secure.html
11. Pelton JN, Singh I (2019) Smart cities of today and tomorrow: better technology, infrastructure and security. Springer, Cham
12. Saif I, Peasley S, Perinkolam A (2015) Safeguarding the Internet of Things: being secure, vigilant, and resilient in the connected age. Deloitte Rev 17. https://www2.deloitte.com/us/en/insights/deloitte-review/issue-17/internet-of-things-data-security-and-privacy.html
13. Stone S (2018) Key challenges of cyber cities & how to overcome them. Ubidots. https://ubidots.com/blog/the-key-challenges-for-smart-cities/
14. WE-Forum (2018) The global risks report 2018, 13th edn. World Economic Forum

Chapter 15
TrulyProtect—Virtualization-Based Protection Against Reverse Engineering

Nezer Zaidenberg, Michael Kiperberg, and Amit Resh

Abstract This paper summarizes the ten-year development of virtualization-based copy protection by the TrulyProtect team. We survey the approaches used, the special problems in various operating systems and software types, and some research directions that have not been covered.

Keywords Virtualization · Reverse engineering

Abbreviations

AMD-v	AMD's technology for virtualization
ARM	A ubiquitous family of RISC processors
EPT	Extended Page Tables. Intel's implementation of a secondary-level page table for virtual machines
IOMMU	AMD's implementation for direct memory access in virtual machines
PLT	Procedure linkage table
UEFI	Unified Extended Firmware Interface. The modern implementation for booting PC systems
VMCB	Virtual Machine Control Block. A repository for all virtual machine attributes in AMD architecture

N. Zaidenberg (✉)
School of Computer Sciences, College of Management Academic Studies, Rishon LeZion, Israel

Faculty of Information Technology, University of Jyväskylä, Jyväskylä, Finland

M. Kiperberg
Department of Software Engineering, Shamoon College of Engineering, Beer-Sheva, Israel
e-mail: michaki1@sce.ac.il

A. Resh
Department of Software Engineering, Shenkar College of Engineering and Design, Ramat Gan, Israel
e-mail: amit@se.shenkar.ac.il

M. Lehto and P. Neittaanmäki (eds.), *Cyber Security*, Computational Methods in Applied Sciences 56, https://doi.org/10.1007/978-3-030-91293-2_15

VMCS   Virtual Machine Control Structure. A repository for all virtual machine
          attributes in Intel architecture
VT-x   Intel's technology for virtualization on the ×86 platform.

15.1   Introduction

A computer program is exposed to tampering and reverse engineering attacks. The
reverse engineering process's goal may be to steal trade secrets, remove copy protection or DRM [43], and change the code logic and other unwelcome attempts. The
program developer is usually interested in the prevention of such attempts and may
use several countermeasures.

The primary countermeasure used against reverse engineering is obfuscation.
Obfuscation transforms the code into a more complex program that has the same
behavior. Obfuscation is classified according to the type of transformation. Compli-
cating the code, replacing the instruction set and encryption using hidden key are all
types of obfuscation. Another countermeasure is to use modern CPU's hardware-
assisted virtualization capabilities to execute the code in a memory region not
available to the operating system.

We discuss a copy protection method based on virtualization. We published in
multiple papers as "TrulyProtect", as modern concepts such as TME and SGX
(Intel), SEV (AMD) and TrustZone (ARM), virtualization-based copy protection
and DRM solutions are phasing out. This chapter summarizes the research activities
in virtualization-based copy protection for over ten years and many variants used in
our team.

15.2   Virtualization in ×86 and ARM

Our work focused on the ×86 and ARM platform. We will describe the hardware capabilities of both platform. Modern CPUs by Intel, AMD and ARM
feature hardware-assisted virtualization. Hardware-assisted virtualization provides
new capabilities for implementing virtual machines and emulator software. These
virtualization features are the foundation of the TrulyProtect technology, so we
introduce them first.

15.2.1   ×86

Intel ×86 platform is an instruction set whose origins were in the late 1970s. Over
the years, it was the most popular CPU platform, and it is still the most widely used

CPU architecture for desktop PCs. Intel and AMD are the leading manufacturers of ×86 CPUs. Other manufacturers such as Via (or Zhaoxin), Cyrix, and Transmeta are still offering or have offered ×86 processors in the past with minimal market share. These manufacturers are usually focusing on specific limited market segments.

By design, ×86 have four permission rings: ring 0 for kernel code, ring 3 for user code and rings 1 and 2 that are unused on modern systems. Intel introduced virtualization with the VT-x instruction family in the late 2000s. AMD introduced a very similar AMD-v instruction family. Later Intel introduced Second Level Address Translation (SLAT) with the EPT instruction family and IOMMU with the VT-d instruction family.

When virtualization was released, there was no spare ring for virtualization; only two bits were available, and they were both "taken". Intel introduced a new HYP bit to specify code running in hypervisor mode (to separate from "regular" ring 0 code). The new security is typically dubbed as "ring-1" in most publications. AMD architecture has very similar (but different) virtualization concepts. Usually, translation between AMD and Intel instructions can be obtained in a one–one replacement of instructions.

15.2.2 ARM

ARM by ARM Holdings (recently acquired and merged with NVIDIA) is a famous RISC architecture that currently controls significant market share in the embedded and mobile market segments. The seventh generation of the ARM architecture introduced virtualization. Currently the eighth generation of the ARM architecture (released in 2010) is dominating the market.

The ARMv8-a platform has two execution worlds, normal and secure world. The normal world uses a standard operating system such as Linux, Android or iOS. The secure world runs on a secure operating system such as OP-TEE [26], Trusty TEE [37], OKL4 [10], Trustonic, and others. Each of the execution worlds have four exception (permission) levels:

EL0.   Normal userspace code (user processes),
EL1.   Operating system code,
EL2.   HYP mode,
EL3.   TrustZone.

EL0 is analogous to "ring 3" in the ×86 platform. All applications on a standard iPhone or Android phone run on EL0. (Kernel code) EL1 is analogous to "ring 0" in the ×86 platform. The Android or the iOS (the operating system itself) on a mobile phone runs on EL1. EL2 is analogous to "ring-1" or "hypervisor mode" in the × 86 platform. In most mobile phones and other ARM devices, nothing runs on this exception level. However, if the ARM device starts a hypervisor when it boots, some code will run on this run level. TrustZone is a special security mode that can monitor the ARM CPU as well as the operating system that it runs. TrustZone allows running a separate security real-time operating system in a secure world. There are no directly analogous modes, but similar concepts in ×86 (from security perspectives) are Intel's ME [32] and SMM [9].

Each exception level provides a set of registers and can access those registers that correspond to lower permission levels but not higher levels. We refer the reader to [41] or [39] for more details about ARM permission model.

15.3 Encrypted Code Execution

Our main work in the TrulyProtect project over the years 2011–2021 is described here. We initially considered platform independent protection against reverse engineering [2]. However, we quickly realized that an architecture means to prevent secondary program from reading the decrypted code or decryption keys is required. Our first system [3] used Intel architecture virtualization for execution protection. We augmented the system in [30], mainly improving performance. We improved the system start-up and used the N-way caching features illustrated in [17]. This paper describes the final version of the system that we focus on.

The basic flow of the system is as follows:

1. Start a hypervisor.
2. Establishing the root of trust.
3. Receive unique keys after passing some sort of attestation.
4. Performing D-E-D.

15.3.1 Starting a Hypervisor

A hypervisor is a computer software designed to support multiple operating system running in parallel on top of single system hardware. Popek and Goldberg [28] introduced the concept almost 50 years ago. Popek and Goldberg defined two types of hypervisors. A type 1 hypervisor starts when the computer starts (from BIOS) and starts operating systems as guests. A type 2 hypervisor starts after the OS starts. In both cases, the hypervisor can protect itself from subverting attacks and protect keys.

For the hypervisor to possess any protection capabilities we must be the root hypervisor i.e. the first hypervisor that runs on the system [29]. We initially tried a type 2 hypervisor in [3]. Later we moved to start a type 1 hypervisor to implement secure boot [38] as the root of trust. When starting a type 2 hypervisor our first goal was to ensure that no other hypervisor is running. We tried a modified version of the Kennell and Jamieson system [11, 12] that we modernized to current hardware [14, 15].

This method worked initially but we discovered that Intel changes their caching algorithms sometimes without notice. Intel changed caching algorithms with the third generation of core platform to improve performance and again in the eighth generation to combat Meltdown and Spectre [18, 22]. Dedicating time to reverse engineering Intel caching algorithm over and over again consumed too much time. We decided we need to move the hypervisor to UEFI. Then by ensuring that we boot first (on boot) and that the UEFI has not been modified (via secure boot), we can guarantee that we are the root hypervisor.

15.3.2 Creating the Root of Trust

As with any trusted computing system [43] virtualization-based security relies on establishing the root of trust. Virtualization-based security root of trust is the CPU itself. Modern CPUs are complex chips that consume massive amounts of energy. CPUs are manufactured on nanometer accuracy. The process of removing the ceramic casing of a CPU without ruining it, connecting external connectors and reading the CPU internal state during operation is an impossible task for most (if not all) attackers.

However, the CPU, the root of trust in the system, has to be established. There are two main ways for establishing a root of trust. The main one uses designated hardware, such as Secure Boot Method (TPM). This method is used in our later papers. Another way is to use software challenges such as the Kennell and Jamieson method [11]. The Kennell and Jamieson system was attacked and rebuffed [33] and [15] modernized it. We used this system in some early protection using some hypervisor-based copy protection systems [3].

As a result, modern TrulyProtect code encryption systems use TPM based protection. The TPM acts as a trust nexus that verify the BIOS and starts the secure TrulyProtect hypervisor [17]. The hypervisor protects itself against subverting attacks and protects the decryption keys. Resh and Zaidenberg [29] have shown that once the CPU has obtained the keys, the keys can remain protected by the hypervisor and not leak to a malicious user or kernel code.

15.3.3 Protection Against Cache Coherence Attacks

In modern Intel CPUs, the L3 cache is shared among multiple cores. Furthermore, the cache is following the N-way cache strategy. In the N-way cache, each of the memory pages competes for the same cache area. Requests to other pages that share the same cache location as the protected cache may cause eviction of the protected cache memory. If the protected data is evicted from the cache, it will travel on the bus to the RAM. A malicious user may sniff the data. Sniffing busses played critical roles in breaking the Xbox [35] or Wii security.

The initial TrulyProtect system was vulnerable to cache coherence attacks. We removed the vulnerability [17]. Pages that compete for exactly the same memory location can no longer be mapped by the OS. We have effectively created a memory region that only a hypervisor can access or evict (because the OS cannot access pages that are capable of evicting our pages). By running a hypervisor on all cores,

Fig. 15.1 TrulyProtect DED
cycle

we eliminate the risk of cross-core attacks on cache coherence. When operating this
way, no page written by the OS on any of the cores can cause cache eviction.

15.3.4   DED Cycles

We implemented a method in which encrypted code is decoded and executed. We
call it D-E-D (decode-execute-discard). After execution, the hypervisor discards the
decrypted code, so it is no longer available. The DED cycle is illustrated in Fig. 15.1.

15.4   System Implementation

We discuss various system details for implementing truly protect for various
architectures, operating systems and contents.

15.4.1   Protection of Native Code

Native code (usually compiled C/C++ binaries) is the baseline for the system
described herein. The performance of encrypted native code is very close to normal
(unencrypted) native code running on the same hardware. Figure 15.2 demonstrates
the performance penalty of using TrulyProtect system for code protection using
PCMark. Figure 15.3 demonstrates the performance penalty from using TrulyProtect
code encryption system for PassMark. (Both figures are from [17].)

Fig. 15.2 PCMark
benchmark: Performance
penalty of TrulyProtect code
encryption systems

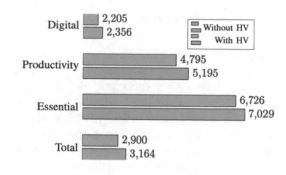

Fig. 15.3 PassMark
performance penalty of
TrulyProtect code encryption
system

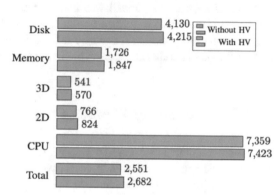

15.4.2 Protection of Managed Code

The methods described in this chapter work with native code. When running protected managed code, the VM must switch off the just-in-time compilation of managed code. If just-in-time compilation is allowed, the user can reverse engineer the compiled code (VM output) instead of the byte code.

We presented protection for managed code under TrulyProtect architecture [16]. This method incurs a very high-performance penalty as we can no longer use JIT. In practice, protecting encrypted managed code in this system was not used in any real-life system.

Turning off the just in time compilation features and embedding a basic bytecode interpreter allowed us to interpret java byte code in the hypervisor and communicate with the VM using the JVM-TI interface. The JVM-TI used in [16] is a standard interface that is available for Oracle JVM [27] and Google Dalvik VM/Android Runtime [6].

15.4.3 Protection of Linux Code

Linux code resides in ELF binaries [23]. The code resides in the .text section and is broken into functions. For practical purposes, we only allowed complete functions to be encrypted. When encrypting files in an ELF binary, we extract the code from the ELF binary and replace it with a $0 \times F4$ opcode.

We add a new section to the ELF that holds the encrypted source. The hypervisor traps the $0 \times F4$ opcode and reads the address from the ELF. The hypervisor then deciphers and performs the encrypted instructions and returns the control to the Linux OS when the application leaves the encrypted code.

Since Linux uses the same ELF format for executables, libraries and kernel module the protection method described herein works with all ELF files. It is critical that the function names in the Procedure Linkage Table (PLT) will remain unencrypted, though the actual code can be encrypted, and that the system will execute normally.

15.4.4 Protection of Windows Code

Windows executables are stored within PE-COFF files [24]. The COFF files are very similar to ELF files, and in both cases, we add a section covering the encryption code. Two differences in the windows operating system compared to Linux are.

1. In Windows, we require the Program Database (PDB) file of the "to be encrypted file" source to find the encrypted functions' actual offset.
2. Operation in "blue pill mode" requires the hypervisor to be signed to allow Windows to install the new driver.

The second limitation does not apply to type 1 hypervisor that starts from UEFI. The performance penalty between Windows and Linux under Intel is nearly identical.

15.4.5 Protection Under Intel

The system described so far in this chapter is designed on the Intel platform. In our Intel implementation, we have used Intel AES- NI instructions were available to randomize keys and decompress code efficiently.

When running under Intel architecture, we ported the Kennell test to $\times 86$ only to discover that Intel has changed their caching algorithm twice, once for performance reasons (on their 3rd generation CPU) and once to combat speculative execution attacks such as Meltdown and Spectre. We decided that reversing the caching algorithms is too unstable, and we have migrated to TPM based approach. The system is currently operational in Linux and Windows, kernel code, user code, applications, and libraries with an overhead of less than 5%.

15.4.6 Protection Under AMD

The system under AMD works in a similar way as the Intel-based protection. All the Intel-based virtualization instruction technologies and concepts are replaced with their AMD counterparts. For example, VMLAUNCH is replaced with VMRUN, EPT is replaced with RVI, and VMCS is replaced with VMCB.

Mutatis-mutandis, the Truly Protect system, works on AMD using the same methods as under Intel. The performance penalty under AMD is almost identical to Intel. The performance penalty between Intel and AMD under Intel is nearly identical.

15.4.7 Protection Under ARM

The ARM architecture does not offer protection rings like the Intel platform. Instead, ARM has four exception levels in a secure and insecure world. The insecure world is running the standard operating system (for example, Android or iOS) and has.

- user mode: EL0, analogous to ring 3 under ×86 platform,
- kernel-mode: EL1, similar to ring 0 under ×86 platform,
- a hypervisor (if present) running on EL2 analogous to ring-1 under ×86 platform.

Furthermore, ARM also has a "secure world". It is a concept in which another secure operating system runs alongside an insecure world, monitors it, and provides security services.

TrustZone model is similar to virtualization (as the two operating systems co-exist on the same hardware) but has one significant difference. The virtualization model calls for separate operating systems. On the ARM TrustZone model, there is separation protecting the secure world from the insecure world. The opposite does not occur, and the ARM platform allows the secure world to inspect the insecure world, read the memory etc. Separation on memory level also exists, and the secure world may have memory space and addresses that are utterly inaccessible from the insecure world.

Like the insecure world, the secure world may also have user code (EL0), kernel code (EL1), and hypervisor code (EL2). The monitor mode or TrustZone (EL3) exists in both worlds. Thus, ARM has seven exception levels: three for the secure world, three for the insecure world, and a shared monitor mode. ARM is a RISC processor with seven sets of registers (one for each exception level). On the ARM architecture, higher exception levels have access to lower exception levels registers.

There is less requirement for establishing the root of trust on the ARM architecture since only the hardware vendor can write the boot code and TrustZone code. If we make the necessary assumption that we can trust the hardware vendor, then we can trust TrustZone to deliver us trustworthy keys. Furthermore, if TrustZone has not been present, the vendor signed the boot code. Therefore, we can trust that a trustworthy hypervisor is booted and acts as a root of trust, booting only a trustworthy OS (which boots only trustworthy apps).

Ben Yehuda and Zaidenberg [5] implemented hypervisor-based protection of code in ARM using the hyplet concept [4]. The ARM model does not rely on buffered execution [17]. It is similar to the in-place execution described in [30], as context switch to hypervisor mode is not very expensive and we did not want to lose some of the memory and cache. Also, a context switch in ARM is much cheaper compared to ×86. Performance of encrypted code in ARM64 architecture bears a performance penalty of 6%. Therefore, we did not see the need to implement a buffered execution, which would remove a part of the memory and cache.

15.4.8 Protection for Rich Media

Protection is possible for rich media and other types of digital contents [42]. We improved the system by encrypting only the sign of the motion vectors [8]. David and Zaidenberg [8] propose a way to protect video delivery (such as pay-per-view) using similar methods.

The system provided low latency encryption without noticeable loss of performance or increased latency. The performance required to run the hypervisor and decrypt the motion vector resulted in a zero increase in latency and slightly higher CPU load, which was not noticeable.

15.4.9 Future Work

There are very few unexplored issues in this field. We are now examining virtualization-based security of GPGPU code (to be released soon). However, code encryption protection using virtualization has become an inherent part of modern CPUs.

Intel SGX (used in Intel core CPUs between the 7th and 10th generations) allows code to run in an encrypted enclave so that it is no longer accessible unencrypted to external software. AMD SEV, Intel TME (used in Intel core CPUs since the 11th generation) and MK-TME enable the creation of entirely encrypted memory regions. MK-TME also provides for remote attestation and the establishment of a root of trust. ARM TrustZone and separate memory regions also make hypervisor-based code protection less necessary for protection against reverse engineering.

15.5 Related Work

There are many other code protection systems. Game consoles used various types of obfuscation and hypervisors, usually with limited success, PS3 being a notable exception [40]. Similarly, many systems were used to protect PC games and video contents. Because a significant portion of the success of these systems is a company secret, not much has been published about them. We explore protection based on hardware features (such as TrulyProtect) and protection through process virtual machines (managed core).

15.5.1 Protection Based on CPU Features

A modern CPU contains many features that others used to create a secure execution environment. HARES [36] is a system that uses the Translation Lookaside Buffer (TLB). However, system keys can always be exposed to a malicious hypervisor if the system does not create root of trust and maintain trust using a hypervisor.

CAFE [13] uses virtualization to protect sensitive data in cloud environment. If the system's trust hypothesis is valid, the system uses similar principals to protect the data. ARM Trustzone provide a similar secure area for code execution. Provided ARM security is not broken (e.g., [34]), similar results can be obtained as with TrulyProtect in ARM.

15.5.2 Process-Virtualization Based Obfuscators

Software developers may use special process virtual machines (like JVM or CLR) to run obfuscated and modified code. These machines have proprietary undocumented, often obfuscated "instructions" (equivalent to CIL or bytecode).

Such machines may use obscure machine code, which can be difficult to decipher. [31] describes means to defeat such obfuscated code, while [2] describe means to defeat Rolles' automatic deobfuscation. Since the virtual machine itself runs on the client's physical device, it is always possible to reverse the generated code if the virtual machine generates the code ("just in time"). Also, even if the machine byte code is encrypted, the keys can be extracted from the local machine process (by debuggers, kernel, or otherwise), reducing the problem to simple reverse engineering of the native code.

Kuang et al. [19] and Lim et al. [21] propose such a system, but if the operating system captures performing instructions, they can always be reversed. Therefore, such systems are weak in practice. Thus, obfuscation may be time consuming, but does not prevent reverse engineering. In practice, hacker groups find time to defeat obfuscation-based security.

15.5.3   Hypervisor Based Protection of Other Contents

We described how only by the hypervisor can encrypt binary code and accessed it. However, the system does not rely on that protected content is code in any way. Protected content may also be data, a docking enclave [25], etc.

SCONE [1] is a container-based approach that enhances docking security using the virtualization-like system SGX. Costa et al. [7] offer a similar approach to [8], using SGX to protect digital content with newer decoders.

We used hypervisor memory (unmapped by the operating system) to secure the software signatures allowed to run at the endpoint [20].

15.6   Conclusions

We believe that we will still see virtualization-based copy protection in the short term. However, over time, as CPUs supporting memory encryption gain popularity, we think virtualization-based copy protection to fade and remain only on legacy systems.

References

1. Arnautov S, Trach B, Gregor F, Knauth T, Martin A, Priebe C, Lind J, Muthukumaran D, O'Keeffe D, Stillwell ML, Goltzsche D, Eyers D, Kapitza R, Pietzuch P, & Fetzer C (2016) SCONE: secure Linux containers with Intel SGX. In: 12th USENIX symposium on operating systems design and implementation (OSDI 16). USENIX Association, pp 689–703
2. Averbuch A, Kiperberg M, Zaidenberg N (2011) An efficient VM-based software protection. In: 2011 5th International conference on network and system security. IEEE, pp 121–128
3. Averbuch A, Kiperberg M, Zaidenberg N (2013) Truly-protect: an efficient VM-based software protection. IEEE Syst J 7(3):455–466
4. Ben Yehuda R, Zaidenberg N (2020a) The hyplet: Joining a program and a nanovisor for real-time and performance. In: 2020 International symposium on performance evaluation of computer and telecommunication systems, SPECTS 2020. IEEE, pp 1–8
5. Ben Yehuda R, Zaidenberg N (2020b) Protection against reverse engineering in ARM. Int J Inf Secur 19(1):39–51
6. Chang CW, Lin CY, King CT, Chung YF, Tseng SY (2010) Implementation of JVM tool interface on Dalvik virtual machine. In: Proceedings of 2010 international symposium on VLSI design, automation and test. IEEE, pp 143–146
7. Costa RS, Pigatto DF, Fonseca KVO, Rosa MO (2018) Securing video on demand content with SGX: a decryption performance evaluation in client-side. In: Simpósio Brasileiro em Segurança da Informação e de Sistemas Computacionais (SBSeg). Sociedade Brasileira de Computação, pp 127–140
8. David A, Zaidenberg N (2014) Maintaining streaming video DRM. In: 2nd International conference on cloud security management (ICCSM 2014). Academic Conferences, pp 36–41
9. Embleton S, Sparks S, Zou CC (2013) SMM rootkit: a new breed of OS independent malware. Secur Commun Netw 6(12):1590–1605

10. Heiser G, Leslie B (2010) The OKL4 Microvisor: Convergence point of microkernels and hypervisors. In: First ACM Asia-Pacific workshop on systems (APSys2010). USENIX Association, pp 19–23

11. Kennell R, Jamieson LH (2003) Establishing the genuinity of remote computer systems. In: Proceedings of the 12th USENIX security symposium. USENIX Association, pp 295–310

12. Kennell R, Jamieson LH (2004) An analysis of proposed attacks against genuinity tests. CERIAS Tech Report, 2004-27, Purdue University, West Lafayette, IN

13. Kim CH, Park S, Rhee J, Won JJ, Han T, Xu D (2015) CAFE: a virtualization-based approach to protecting sensitive cloud application logic confidentiality. In: ASIA CCS '15: Proceedings of the 10th ACM symposium on information, computer and communications security, pp 651–656

14. Kiperberg M, Zaidenberg N (2013) Efficient remote authentication. In: Proceedings of the 12th European conference on information warfare and security ECIW 2013. Academic Conferences, pp 144–148

15. Kiperberg M, Resh A, Zaidenberg N (2015) Remote attestation of software and execution-environment in modern machines. In: 2015 IEEE 2nd international conference on cyber security and cloud computing. IEEE, pp 335–341

16. Kiperberg M, Resh A, Algawi A, Zaidenberg N (2017) System for executing encrypted Java programs. In: Proceedings of the 3rd international conference on information systems security and privacy (ICISSP 2017). Scitepress, pp 245–252

17. Kiperberg M, Leon R, Resh A, Algawi A, Zaidenberg N (2019) Hypervisor-based protection of code. IEEE Trans Inf Forensics Secur 14(8):2203–2216

18. Kocher P, Genkin D, Gruss D, Haas W, Hamburg M, Lipp M, Mangard S, Prescher T, Schwarz M, Yarom Y (2018) Spectre attacks: exploiting speculative execution. arXiv:1801.01203

19. Kuang K, Tang Z, Gong X, Fang D, Chen X, Wang Z (2018) Enhance virtual-machine-based code obfuscation security through dynamic bytecode scheduling. Comput Secur 74:202–220

20. Leon RS, Kiperberg M, Zabag AA, Resh A, Algawi A, Zaidenberg N (2019) Hypervisor-based white listing of executables. IEEE Secur Priv 17(5):58–67

21. Lim K, Jeong J, Cho SJ, Choi J, Park M, Han S, Jhang S (2017) An anti-reverse engineering technique using native code and obfuscator-LLVM for android applications. In: RACS '17: Proceedings of the international conference on research in adaptive and convergent systems. ACM, pp 217–221

22. Lipp M, Schwarz M, Gruss D, Prescher T, Haas W, Mangard S, Kocher P, Genkin D, Yarom Y, Hamburg M (2018) Meltdown. arXiv:1801.01207

23. Lu H (1995) Elf: From the programmer's perspective. NYNEX Science & Technology Inc

24. Margosis A, Russinovich ME (2011) Windows sysinternals administrator's reference. Microsoft Press

25. Merkel D (2017) Docker: lightweight Linux containers for consistent development and deployment. Linux J 2014(239):2

26. Optee (2021) OP-TEE. GitHub. https://github.com/OP-TEE/. Accessed in May 2021

27. Oracle (2010) JVM^{TM} tool interface, version 1.0. Oracle. http://docs.oracle.com/javase/1.5.0/docs/guide/jvmti/jvmti.html

28. Popek GJ, Goldberg RP (1974) Formal requirements for virtualizable third generation architectures. Commun ACM 17(7):412–421

29. Resh A, Zaidenberg N (2013) Can keys be hidden inside the CPU on modern windows host. In: Proceedings of the 12th European conference on information warfare and security: ECIW 2013. Academic Conferences, Reading, pp 231–235

30. Resh A, Kiperberg M, Leon R, Zaidenberg N (2017) Preventing execution of unauthorized native-code software. Int J Digit Content Technol Appl 11(3):72–90

31. Rolles R (2009). Unpacking virtualization obfuscators. In: WOOT'09: proceedings of the 3rd USENIX conference on offensive technologies. USENIX Association

32. Skochinsky I (2014) Intel ME secrets: hidden code in your chipset and how to discover what exactly it does. Presentation at Hex-Rays Code Blue, Tokyo

33. Shankar U, Chew M, Tygar JD (2004) Side effects are not sufficient to authenticate software. In: Proceedings of the 13th USENIX security symposium. USENIX Association

34. Stajnrod R, Ben Yehuda R, Zaidenberg N (2021) TrustZone attack. In press
35. Steil M (2005) 17 mistakes Microsoft made in the Xbox security system. In: Proceedings of the 22nd chaos communication congress 2005. Verlag Art d'Ameublement, pp 378–390
36. Torrey J (2015) HARES: Hardened anti-reverse engineering system. In: Proceedings of the symposium on security for Asia network
37. Trustytee (2021) Trusty TEE. Android Open Source Project. https://source.android.com/security/trusty. Accessed 23 May 2021
38. Wilkins R, Richardson B (2013) UEFI secure boot in modern computer security solutions. UEFI Forum
39. Zaidenberg N (2018) Hardware rooted security in Industry 4.0 systems. In: Cyber defence in industry 4.0 systems and related logistics and IT infrastructures. IOS Press, pp 135–151
40. Zaidenberg N (2020) Game console protection and breaking it. In: Khosrow-Pour M (ed) Encyclopedia of criminal activities and the deep web. IGI Global, pp 449–461
41. Zaidenberg N, Ben Yehuda R, Leon RS (2020) Arm hypervisor and TrustZone alternatives. In: Khosrow-Pour M (ed) Encyclopedia of criminal activities and the deep web. IGI Global, pp 1150–1162
42. Zaidenberg N, David A (2013) Truly protect video delivery. In: Proceedings of the 12th European conference on information warfare and security. Academic Conferences, Reading, pp. 405–406
43. Zaidenberg N, Neittaanmäki P, Kiperberg M, Resh A (2015) Trusted computing and DRM. In: Cyber security: analytics, technology and automation. Springer, Cham, pp 205–212

Chapter 16
Refining Mosca's Theorem: Risk Management Model for the Quantum Threat Applied to IoT Protocol Security

Mikko Kiviharju

Abstract Large-scale quantum computation (QC) presents a serious threat to many modern cryptographic primitives. This has profound implications to the critical information infrastructure (CII) as well. The main mitigation techniques involve migrating to cryptographic schemes that are postulated to be "quantum resilient", i.e. there are no known quantum algorithms for them. The solution requires that all the known instances of the vulnerable cryptographic schemes are replaced with quantum resistant schemes. In a lecture at the NIST premises in 2015, Dr. Mosca gave his famous theorem of when different stakeholders should start upgrading their systems. This theorem is, however, very general and assumes the worst possible scenario. Does it apply equally well for all or even the majority of use cases of cryptography? Is the answer same across all domains of CII or all types of networks? Currently the most challenging environments are those with very little computational and networking capabilities, i.e. Internet-of-Things (IoT). IoT is today used in many CI subcategories, such as power grid (including SCADA), water supply, logistics, agriculture and dangerous goods handling. In this text, we develop a more detailed risk management model to prepare for QC and apply this model to survey the status of Internet-of-Things (IoT) protocols. We cover in total 17 different IoT protocols or protocol families.

Keywords IoT · Critical Information Infrastructure · Mosca's Theorem · Risk Assessment · Quantum Resilient · IoT Protocols · LoRaWAN · ZigBee · EnOcean · NB-IoT · SigFox · Cryptography · Security · PQC · Quantum Threat

16.1 Introduction

The Critical Information Infrastructure (CII) is an essential component of many critical national infrastructures (CNI). As CII is essentially a set of information systems

M. Kiviharju (✉)
Finnish Defence Research Agency, FDRA, Riihimäki, Finland
e-mail: mikko.kiviharju@mil.fi

© The Author(s), under exclusive license to Springer Nature Switzerland AG 2022
M. Lehto and P. Neittaanmäki (eds.), *Cyber Security*, Computational Methods
in Applied Sciences 56, https://doi.org/10.1007/978-3-030-91293-2_16

369

with technical security controls, it also follows that CII uses cryptography for a multitude of security functions. Today, many essential modern cryptographic primitives (most prominently elliptic curve cryptography (ECC) and the RSA-cryptosystem used for, e.g., certificates and key exchange) are considered to be vulnerable in the framework of large-scale, general-purpose quantum computation. There are multiple approaches to counter this threat, but many of them lack foundation other than the necessity to move away from old systems.

Cryptology as a science offers many exact security models to counter the QC threat, but it appears to be rather case-dependent and system-specific when and how to upgrade the critical systems to be quantum resistant (QR). This is especially relevant for CII, since they are easily picked as a target by such operators in this field, which are likely to implement a large QC early.

In 2015, Dr. Mosca posed a simple, but thought-provoking idea on how and when system owners will need to upgrade to QR-systems [51]. The idea basically pitted different time-frames against each other: the lifetime of the assets to be protected, the time to build or upgrade to new QR-systems, and the time it actually takes to build a large-scale, general-purpose quantum computer. Then, with the most optimistic scenario for the "quantum event" and with very long-lived and highly sensitive data (e.g., diplomatic communication), it would already be too late.

However, the application areas and use cases of the potentially vulnerable systems cover an extraordinarily large spectrum. This may mean that "secrets revealed" is not actually a secret at all, but an integrity marker; or that the data or the cryptographic key to be protected is outdated in days. Additionally, there are multiple definitions and estimates for the "quantum event", as well as definitions on what it actually means to be quantum resistant. Natural questions are then: How badly is an upgrade needed? How soon does it need to be installed? What type of mitigation is the most effective given a fixed budget and a certain type of system? Currently there is no methodology to answer this question, other than the necessity to move away from old systems.

Our main contribution is to suggest a more refined risk management model to evaluate the effects of QC to systems using the different vulnerable cryptographic schemes. To construct the model, we also apply more exact definitions and information of the quantum threat and quantum resilience. Our model will classify systems in ten different threat levels in three different points in time, and case-by-case risk reduction techniques. We stress the aim of the model: it does not represent cryptologic threat (although it plays a vital role in the model), but rather the need to upgrade the system.

To evaluate the viability of risk management model, we also apply it to a vital technology area within CI, namely the IoT-domain. We selected Internet-of-Things (IoT) as our focus, as it is (as a technology) very pervasive within the CI, has a very versatile set of use cases and it is developed with a quick pace. More specifically, we evaluated 17 different IoT communication protocols or protocol families with respect to the quantum threat.

The IoT protocol evaluation shows that protocols operating in the lower ISO OSI layers are generally transmitting only short-lived data and are very resource-conscious, thus employing only symmetric cryptography, which makes their vulnerability to QC relatively low. Higher OSI-layer protocols, however, do not need to resort to pure symmetric cryptography schemes, and thus they are also more vulnerable to QC. However, we found out that there are also many QR schemes planned or at least already in research for these cases.

16.2 Quantum Computation and C(I)IP

Critical Infrastructure Protection (CIP) refers to critical infrastructure (CI), which is a concept tied to the adequate functioning of a modern society. In short, if any of the areas of CI fail, the society using them will experience severe discontinuities in operation. A related concept is Critical Information Infrastructure (CII).

However, the exact definitions of CI and CII differ somewhat between nations [33], which has resulted in the more accurate term Critical National Infrastructure (CNI). In Finland there is a slight difference between CII and CI information systems: CII is defined as any information infrastructure critical to the vital functions of the society [33] (seven functions [62]), but CI information systems are limited to the specifically identified nine CI subcategories. In this study, we are concerned with the CII, and CII protection (CIIP). In case there are differences between the definitions and subcategories of CII, we refer to the Finnish CNI.

Quantum computation (QC) as a general concept refers to all models of computation and implementations, which strive to gain significant advantage in computational power with respect to classical (conventional) computation or implementations. The main promise of quantum computation is an exponential speedup for some algorithms, due to the capability of the quantum computational units (qubits) to preserve multiple states within them simultaneously, and allowing arbitrary combinations. This leads to the fact that N qubits can encode 2^N states simultaneously. This exponential speedup for algorithms has also drawbacks for security systems, which rely on the intractability of some problems, most notably cryptography.

The impact of large-scale QC would be immense: modern cryptographic algorithms, such as AES, ECC, and RSA, are used ubiquitously in nearly everywhere within the cyberspace infrastructure, including CII. Cryptography underlies almost every technical security control from smartcards to password checking, from device authentication to secure communications and from power grid controls to banking applications.

There are some immediate and simple approaches to counter the QC threat to cryptography, such as hiding the quantum algorithm-specific cryptographic scheme parameters needed above all to make the specific quantum algorithm work, mixing vulnerable schemes with less vulnerable schemes, extending the scheme parameter sizes sufficiently that any foreseeable QC implementation cannot scale such a number during the lifetime of the asset, etc. However, all these types of "solutions" are more like patches and do not represent any long-term solution.

Two known long-term solutions include switching from vulnerable classical cryptographic algorithms to such algorithms that are not (known to be) vulnerable, even though they still represent the classical computation paradigm; and replacing classical cryptography with quantum cryptography. The former is captured with the umbrella-concept "Post-Quantum Cryptography" (PQC), while the latter contains multiple cryptographic schemes. In quantum cryptography, the most prominent scheme is the quantum key-distribution scheme, or QKD, for which there are already commercial products. On the other hand, PQC is still under standardization by NIST (and other organizations).

QKD has more sound security premises than PQC, but Implementing QKD is currently thought to be expensive and integration to other security controls difficult. QKD-enabled networks are called "the quantum internet", but due to the difficulties of implementation, they can currently only be recommended for the core networks and high-security niche applications. PQC is likely to be cheaper, more versatile and easier to install by just replacing older ECC-/RSA-implementations with a suitable PQC implementation.

16.3   QC Threat Model

16.3.1   General

The term QC covers all types of implementations. However, not all types are suitable for the construction of universal quantum computer, which would be able to work with a sufficiently large set of quantum operations ("gates"), to allow using them to solve arbitrary problems. Furthermore, the applicability of even universal QC implementations to classical (non-quantum) problems very much depends on the actual problems and its size. Therefore, it is prudent to differentiate between any QC and such computation that forms a threat to modern cryptographic systems. These are called "cryptographically relevant QC" or CRQC [16]. It has not been standardized exactly how mature or large QC implementations would count as CRQC, but the cryptographic community often cites the capability to crack the 2048-bit RSA-scheme as the milestone.

The model we are presenting here is inspired by Mosca's theorem: "How soon do we need to worry". This theorem essentially tries to answer the question of when it would be prudent to start migrating to quantum-safe systems (PQC, in Mosca's context). Mosca's theorem uses a very simplistic risk scale: it includes levels from zero to one, considers confidentiality only, and represents the worst-case scenario and independence from any differences in strength of the different solution types. However, the question "when" was addressed for QC to some degree.

Our risk management model for quantum resiliency (RMQR) has three aspects:

1. Estimation of the resources needed by QC implementations, both the time to develop them and their scalability,
2. Qualitative classification of QR method strengths by type (note that the strength measures are not commensurate over all classes),
3. Qualitative classification of risk levels defined by cryptography and asset use cases (including nine, not necessarily equally high, risk levels).

16.3.2   Timescales and Resources

There are many estimates on when CRQC will become reality. The cryptography after the "quantum event" is called "post-quantum", referring to the goal that post-quantum cryptography (PQC) schemes should be secure even after the CRQC event (here: breaking 2k RSA). In 2018, the cryptologic community predicted CRQC to happen in 2032 [11], but, based on simple meta-research with the Google search engine, the prediction varies between 2025 and 2040. Based on an exponential model[1] of the size of evidenced QC implementations, the CRQC event would be between 2024 and 2050 (the current speed of development is faster than long-term). In any case, if large scale QC is possible at all, the consensus is that it should be realistic by 2050 with a very high probability, the peak probability being somewhere in the 2030s.

The quantum computation paradigm has the largest impact on certain asymmetric cryptographic primitives, namely RSA and ECC, where there is expected exponential speedup in the algorithms intended to break them, which renders them completely unusable in cryptography. Also symmetric primitives, notably AES, suffer some weakening, but this affects the security level only. There are special cases in symmetric cryptography as well that may suffer a more drastic impact [40], but the security model is considered to be unrealistic. Table 16.1 presents the most important classical cryptographic schemes and their most recent resource estimates for quantum computing to successfully "break" them (per instance).

The number of qubits refers to logical qubits used by the logical-level quantum algorithm operating in a universal quantum computer. This may differ significantly from the number of physical qubits in the implementation. The number of quantum gates can roughly be interpreted as the number of (sequential) qubit-operations in the quantum computer. This can be translated into time units by considering the "clock frequency" of each type of implementation. For the IBM quantum computer, 5–7 GHz is expected [8], so, e.g., RSA-2048 would be broken in minutes or even seconds,[2] given a sufficiently large implementation.

It should be noted that even though, e.g., AES-128 requires implementations sizes on the same scale as the "easy" ECC-256, it requires a far larger number of operations, lasting nearly a billion (10^9) years, or requiring a billion parallel implementations

---

[1] FDRA internal report.

[2] For very short "run times", the effect of measurement setups should also be considered, which may be, for practical reasons, much longer than the actual run time.

Table 16.1 Number of qubits and quantum gates required for different current cryptographic schemes

Scheme	Qubits		Quantum gates	
	Formula	Number	Formula	Number
RSA-1024	$2n + 2$	2050	$O(\log(n)n^3)$	3.2×10^9
RSA-2048		4098		2.8×10^{10}
RSA-3072		6146		1.0×10^{11}
RSA-4096		8194		2.4×10^{11}
ECC-256	$9n + 2\text{lb}(n) + 10$	2330	$448n^3\text{lb}(n) + 4090n^3$	1.3×10^{11}
SHA2-256	N/A	6200	$2^{O(n)}$	2.0×10^{46}
SHA3-256		1.0×10^6		1.3×10^{44}
AES-128	$O(n^{1+\epsilon})$	2953	$2^{O(n)}$	2.1×10^{26}
AES-192		4449		1.4×10^{36}
AES-256		6681		9.2×10^{45}

to manage that within a year. Thus, even though the security level of symmetric cryptography drops, it is still infeasible to crack them in practice. The NIST security limit of 2^{80} "operations" [50] reflects this as well.

The main outcomes from this analysis are as follows. Firstly, implementing large scale universal QC has still very many unknown elements and it may become available in five or thirty years, if at all. Majority of cryptographic development effort assumes naturally the lower limits and expects an approximate decade to be the likely timeframe. Additionally, when (if) large scale QC becomes feasible, RSA and ECC will be completely broken. Some parameter sizes will last a few years longer than the smaller instantiations, but for all practical sizes the scheme is broken. Lastly, symmetric primitives will, as a rule of thumb, lose half of their key lengths. In practice, this seems to be the asymptotic lower limit and, e.g., AES-128 reduces to more like 87 bits than 64 bits [32]. Special cases should always be checked, but for symmetric primitives this implies a drop (of usually 1–2 levels) in security level.

16.3.3 Strength of Quantum Resilient Solutions

In cryptography, security can be measured in a number of ways, not all of them quantitative [34], and many cryptographic measures tend to be security-model-specific, and thus use-case specific. Furthermore, not much can be said about the measures to estimate vulnerability to quantum computation.

Theoretical computer science has a sub-branch of complexity theory, which can answer to theoretical questions also in the world of quantum computational complexity.[3] However, current responses to the quantum threat are more heuristic in

---

[3] E.g., the complexity class BQP (Bounded-Error Quantum Polynomial-time) [12].

Table 16.2 Levels of quantum resilience

Level	Description
C_0	No large enough QC implementations are at the moment known for a classical problem with a given size. Note that due to the current small scale of QC implementations, even the RSA- and ECC-schemes with the smallest parameter sizes fit into this category
C_1	The problem size of the classical problem is notably larger than commonly used parameters, giving a few extra years of protection after the "canary" problem has been cracked. An example is the current NSA CNSA[a] (former Suite B) cryptography, which still endorses 3k RSA [20]
C_2	There is a quantum algorithm for a classical problem, which for all problem sizes provides an advantage over classical brute-force search that is at most sub-exponential in terms of speedup. Symmetric cryptography under Grover's algorithm falls into this category
C_3	There are no known quantum algorithms for a classical problem that would perform asymptotically better than the best-known classical algorithm
Q_1	There are no known "efficient" (polynomial, in the number of quantum gates) quantum algorithms for a quantum problem
Q_2	The quantum problem is provably "hard" for a quantum computer (exponential number of gates, or at least as hard as the hardest in the class BQP)

[a] Commercial National Security Algorithm suite

nature, and most of them not capable of reaching quantum-computational security guarantees; those that reach these levels, are often impractical to realize.

In order to capture the current selection of solutions offered to provide quantum resiliency (QR), we use a six-level scale from "ignoring the problem" to quantum-complexity-theoretic guarantees (see Table 16.2). Table 16.2 shows four levels of QR strength that can be achieved by classical means (conventional cryptography). Of these, C_2 corresponds to using pure symmetric cryptography solutions due to their relative strength. The PQC-schemes fall in category C_3.

The categories Q_1 and Q_2 refer to schemes that are only achievable by harnessing the capabilities of some quantum phenomena, such as entanglement or superposition in quantum key distribution (QKD). These are further divided in two categories, depending on whether there are any security guarantees for the scheme in some practical security model, or whether the security is based just on the absence of attacks. For example, "plain" QKD would be in Q_1, but, e.g., device-independent QKD in Q_2, as well as some quantum cryptographic primitives.

Mixtures of QR-techniques should be evaluated case-by-case. For example, the so-called hybrid QR schemes may employ techniques from level C_0 and C_3 such that if either one fails, the remaining will still hold. In this case, the combined level is obviously the maximum of the two (C_3). It should be noted that as the QR techniques mature, there is likely to arise a need to revise this categorization, especially the Q-levels.

16.3.4 QC Threat Levels

The quantum computation threat to cryptography is directed mainly towards primitives (higher-level structures in special cases only). These primitives, in turn, are combined to applications in different ways, keyed differently and provide different security services to a variety of assets. In order to estimate risk management level threat, it is therefore more suitable to include more aspects to the model than just the cryptographic primitive and the related technical security model.

In this study, we are interested in the following fundamental aspects affecting cryptographic protection, in general:

1. How long-lived is the protected asset? If the assets have a long lifespan (years or even decades) they are more likely to be relevant during the CRQC-event, and also more prone to eavesdropping or exfiltration.

2. Does the asset sensitivity need large security margins (high security levels) for the security controls? If the assets are not very sensitive, and only "unfeasible at the time of evaluation" is a sufficient criteria for any attack, many such schemes that suffer only a drop in the level of their security (as opposed to becoming completely useless) may not need to be changed at all.

3. Does cryptography protect the asset confidentiality or integrity? It is difficult to know when confidentiality is broken, since the cryptographic scheme used for protection does not report it directly. Integrity-related goals, on the other hand, are by their very nature checked separately, before acting on the trust claim. If an encrypted message is captured, there is no way of knowing if and when it is decrypted. But if a signature is forged, it is always possible to evaluate its trustworthiness at the time of use. For this reason, integrity (including authenticity, non-repudiation, etc.) is not under immediate threat until the CRQC event.

4. Which algorithm is used? Some of the algorithms are more vulnerable than others, for some there are no known QC algorithms at all (see Table 16.1).

5. How readily is the asset available to the attacker? Especially Mosca's theorem assumes that the assets are transferred encrypted over public channels, and possibly stored for years in case they could be decrypted someday. Not all assets are communicated at all and some only with a certain probability during their lifetime. Additionally, even the data-in-transit may contain several layers of encryption, some of them more resistant to QC than others.

6. What kind of key lifecycles are in use? For confidentiality-related services and the current QC threat model, the lifetime of cryptographic the key is not that relevant[4]: once the cipher text is available to the attacker, the key for that encryption becomes fixed for the adversary.

7. How long does it take to upgrade the cryptographic implementation to new schemes? The time of evaluation is highly relevant to the threat level we are

---

[4] It could be argued that re-encrypting data-in-rest would alleviate the threat, since the owner has more control over all the data in the databases. However, in this model we factor the availability of the data into one variable, and in this case we would consider events only after-the-fact.

considering. Obviously, the threat level jumps significantly at the CRQC event. Also, if the upgrade cycle of a particular system is long enough, care needs to be taken to be prepared sufficiently ahead of time. However, system upgrades are very much use-case specific, and can be easily just subtracted from the estimated CRQC-event to factor them into the threat level time-of-assessment.

We present the threat levels as a combination of different cryptographic scheme and asset parameters, as well as the time of assessment (ToA). We do not factor the cryptographic upgrade time to this model: if this is known, the reader is advised to subtract that estimate from the CRQC-event estimate and use the result as the new CRQC-event ToA. The levels range from zero to nine, nine being the highest. We note that zero does not mean "no threat at all", merely the lowest threat level. Furthermore, the threat levels are not quantitative or even consistent over summation: it is possible to obtain level X via different combinations, but the combinations on the same level may or may not be exactly the same. We do postulate, however, that arriving at level X is still a higher threat than $X-1$ and lower than $X+1$.

The threat level is calculated in such a way that a certain value in each scheme/asset parameter column adds the level by one, and the ToA advances the value 0–2 levels (see Table 16.3). The ToA CRQC refers to the selected CRQC-event (here: the breaking of RSA-2048), and ToA K_{long} refers to a time before the CRQC-ToA subtracted with the lifespan of the longest lived key material used for that particular algorithm in the system for integrity purposes.

The calculation method presented in Table 16.3 is linear, whereas the threat model itself is not. We address this by adding additional rules to the calculation method. There are three types of additional rules:

Table 16.3 QC threat level base calculation

Parameter type	Parameter value	Score (levels)
Scheme type	Asymmetric (RSA & ECC)	1
Key length (Conf.)	Short[a]	1
Tag length (Int.)	Long[b]	1
Key lifespan	Long	1
Security goal	Confidentiality	1
Asset lifespan	Long	1
Availability of the material	Easy	1
Security margin	Large	1
ToA	Between K_{long} and CRQC	1
ToA	After CRQC	2

[a] 128 bits or less
[b] 128 bits or more

1.   general rules,
2.   rules applied to cases using vulnerable asymmetric cryptography, and
3.   rules applied to symmetric cryptography schemes.

The *first general rule* is that if a parameter is deemed irrelevant, it is also necessary to define, whether the combination of other parameters in general increases the threat level or not. Also, due to the security model, whenever something becomes available for the attacker, the secrecy is considered lost. Thus for confidentiality, the key lifetime does not matter (we consider here direct attacks only, not secondary attacks to uncover further secrets, so e.g. forward-secrecy is outside the scope here). Key lifetime score "1" is used. This also makes the integrity-related ToA irrelevant, and for the K_{long} column, ToA-score "0" is used.

Long-lived integrity/authentication keys may provide a false sense of trust, if their lifetime exceeds the CRQC event. Here, if either the data or key lifetime is "short", it doesn't matter if the other element lifespan is longer, as the checkpoint time is forced by the element with the shorter lifespan. If one of these is "short", the other will also receive a score of "0". If the security target is confidentiality, and the protected assets have a sufficiently long lifespan, the time of assessment is irrelevant, and the highest ToA-score is used throughout. Also, for short-lived keys, the K_{long} column ToA is irrelevant, and for this, ToA-score "0" is used.

In the case of *vulnerable asymmetric cryptography* schemes (ECC; RSA), the additional rules consider that the break with QC is considered to be total, irrespective of the security margin needs or used key lengths. We then do not consider them explicitly but set the minimum threat level to "3". Since after the CRQC the vulnerable schemes break in minutes, this type of cryptography does not provide any protection; only additional, non-cryptographic protection applies. For this reason, the threat level in all of the cases is maximal. Only the availability of the ciphertext ("Hard") will decrease this by one level.

In the case of *symmetric cryptography schemes*, as the Grover's algorithm scales linearly according to the key length, not the block/tag length, many symmetric integrity-related schemes with short tag lengths are far more vulnerable to classical attacks, and do not suffer from QC. This is especially true for tag lengths <87 bits (considering that breaking AES-128 with Grover is roughly equivalent to 87 bits of security classically). This means that for tag lengths of 96 bits or less, the quantum "speedup" would be small enough to fit inside a single security level (considering the prevalent steps of 64 bits in security parameter [41]). Thus, for integrity-related protection (e.g., authentication), the QC threat is deemed to be irrelevant, if it is protected with symmetric key mechanisms (such as MACs), and the tag length is sufficiently small.[5] Instead of marking the threat level to "0" (there is, after all, some threat from QC), we express this with a specific marker.

The maximal threat to symmetric schemes is very different from that to asymmetric schemes. Comparable levels would arise if.

---

[5] Note that short tag lengths are by no means a desirable property, due to the increased conventional threats.

- the symmetric scheme use case included the use of very long-lived keys to protect high-security-margin assets for decades, with the shortest key-lengths (so that there would be incentive to invest large amounts of QC power to analyze the encrypted asset for years), and
- the asymmetric scheme use case was any easily available asset, in pre-CRQC-level (since post-quantum levels would be the maximum in any case).

As the minimum threat level in the asymmetric case for easily available assets, pre-quantum, is "4", and maximum possible threat is "9", this gives us a scaling factor of 0.5 for the symmetric cases, compared to the asymmetric ones.

The final procedure to build the threat table is then as follows:

1. Compile a list of all possible combinations of the parameters ($2^7 = 128$ entries).
2. Apply the scoring to the combinations in Table 16.3.
3. Apply the general and asymmetric scheme additional rules.
4. Apply the symmetric scheme exceptions, except the scaling.
5. Apply the scaling factor to symmetric scheme levels as follows:

$$L_1 = L_0 \text{DIV } 2.$$

6. Here, L_0 is the threat level calculated above, and L_1 is the scaled level. "DIV" is a truncating division operator, i.e. the decimal part of the result is stripped.
7. Prune the list for duplicates.

The threat levels are depicted in Table 16.4. The table originally contained 128 entries, but due to the exceptions and dependencies mentioned above, many of them collapse together. Markings "0" and "1" imply that they are given a constant score, independent of what the actual value would be. By default, the value is indicated by the first letter of the description (e.g., "L" = "Long"). In some cases a certain ToA-point is not relevant, but as the threat level should be continuous over ToA-points, the value is copied from adjacent cells (earlier ToA). This is indicated by shading the irrelevant cell with diagonal lines. The column "Key/Tag length" refers to key length for confidentiality schemes and to tag length for integrity (/authenticity) schemes.

16.4 IoT Protocols

The whole concept of Internet-of-Things (IoT) is not a scientifically exact definition. According to Wikipedia, IoT is defined as "the network of physical objects that are embedded with sensors, software, and other technologies for the purpose of connecting and exchanging data with other devices and systems over the Internet". The whole concept overlaps similar concepts from different ages and areas, e.g., Machine-to-Machine (M2M), industrial control systems (ICS), and smart home [13], and exact definition may depend on the background of the speaker. The Wikipedia definition binds the concept to the connectivity over Internet, but the technical

Table 16.4 QC threat levels for different system attribute combinations

Symmetric

Key/tag length (short/long)	Long-/short-lived key	Integrity / Confidentiality	Long-/short-lived asset	Easy/hard availability	Small/large security margin	Now	K_{long} - CRQC	After CRQC
S	0	I	0	0	0			
	1	C	L	E	L	4	4	4
	1	C	L	H	L	3	3	3
	1	C	S	E	L	2	2	3
	1	C	S	H	L	2	2	3
	L	I	L	E	L	2	2	3
	L	I	L	H	S	1	1	2
	L	I	L	H	L	1	2	2
	0	I	0	0	S	0	0	1
	0	I	0	0	L	0	0	1
	1	C	L	E	S	3	3	3
	1	C	L	E	L	3	3	3
	1	C	L	H	S	2	2	2
	1	C	L	H	L	3	3	3
	0	0	S	0	0	0	0	1
	S	I	0	0	0	0	0	1

Asymmetric

Key/tag length (short/long)	Long-/short-lived key	Integrity / Confidentiality	Long-/short-lived asset	Easy/hard availability	Small/large security margin	Now	K_{long} - CRQC	After CRQC
1	L	I	L	E	1	6	7	9
	0	I	S	E		4	4	9
	1	C	L	E		9	9	9
	1	C	S	E		6	6	9
	S	I	0	E		4	4	9

Shaded cells indicate that the threat level is inherited from other same level cells

defining characteristics of IoT are not particular to the fixed Internet; rather instead to the low-power, low-bandwidth, low-connectivity and small processing-power edge of the Internet. Furthermore, in order to bring as much functionality as possible to the edge, layered solutions[6] are often not sufficiently capable, but integrated silo-like implementations are needed.

To make things even more complicated, definitions of IoT security also form a loose, complex and case-dependent whole [78, 84]. A report by ENISA [10] makes a survey of IoT security in CI via asset and threat taxonomies, listing 54 different asset types and 61 individual threats, yielding nearly 3300 possible combinations.

---

[6] As in the ISO OSI-model.

Luckily, many security goals (e.g., confidentiality, privacy) can be used to cover most of the threats, and cryptography is an important mechanism used to implement many of the goals.

Following the categorization of IoT security requirements in [78], which lists 21 requirements in five categories (network security, identity management, privacy, trust and resilience), cryptography in general can be used to fulfill at least 12 of the requirements. This in turn speaks for the fact that IoT cryptography needs to be trustworthy and securely implemented.

In this study, we are interested mostly in the "Network security"-part of the security requirements in [78]. This is justified by the fact that over-the-air communication is easily available to the attacker, and the protocols are one of the few somewhat common properties shared by the IoT implementations, not being vendor or implementation specific. Communication protocols categorized under the IoT umbrella appear in all of the ISO OSI[7] model layers, covering a varying degree of layers, or ranges of layers. Then, also the security requirements and foundations differ per protocol.

According to an ENISA report on IoT security within CII [10], IoT technologies are pervasive in the core of plethora of both critical and non-critical infrastructures, pertinent to almost all aspects of daily life, including sectors such as industry, energy, transport, health, retail, etc. [10, Chap. 2]. Thus it is not a question of *whether* IoT is involved in CII or not, but how *deep* it is.

When CII systems are discussed, the most prominent example is usually drawn from industrial control systems, namely SCADA.[8] However, SCADA is an integrated system, with the data acquisition and some of the control mechanisms implemented with IoT devices and protocols [65]. In some cases the whole SCADA system may comprise of standard IoT components. Furthermore, the SCADA developments are taking it to directions indistinguishable from the IoT concept, even towards the standard IoT protocols [58]. We can then safely argue that SCADA is also included in the IoT-concept itself, and that IoT reaches to the very core of many CNI information systems and also CII.

Since there are literally tens of IoT communication protocols (nearly 30 in a listing given in [71]), it is then justifiable to question the relevance of each protocol to the focus of this study in the totality of IoT. From various sources it can be seen that certain protocols surface time and again. We selected 17 protocols or groups of them to closer study, meaning that some items represent a single specification while the widest groups represent multiple standards from various organizations. In the latter case it is infeasible to survey the totality of the group, rather some instances only. We list the protocols in the next chapter, where we survey their cryptographic security.

---

[7] Open Systems Interconnection.

[8] SCADA = Supervisory Control and Data Acquisition.

16.5   Cryptographic Primitives in IoT Protocols

IoT security and IoT protocol security is a well-studied area. Some survey-level research of general IoT protocol security includes, e.g., [5, 24, 42, 48, 73]. These form a good basis for this work, but they do not assess the quantum security per sé.

Most of the work towards securing IoT in the quantum sense involves general resource estimations on whether the PQC algorithms allow sufficiently fast implementations or how much bandwidth they require, e.g., in [17]. A good PQC algorithm resource survey in IoT environments can be found in [26]. However, protocol-specific plans or even research to integrate PQC to the most popular IoT protocols seems to be rare, considering the large variety of protocols. Category-specific surveys can be found (code-based authentication protocols for RFID), and also some protocol-specific schemes, such as [57] where the authors integrate PQC both in plain and hybrid mode to the handshake in OPC Unified Architecture [75–77]. Other such research includes for instance EU Commission SAFEcrypto-project's smart tag case, where the TLS RLWE key exchange was demonstrated for CoAP [80], or a general PQC key exchange and attestation scheme for CoAP over LoRa-networks [47] and an RLWE-based Bluetooth SSP [29].

Some protocols rely on lower layer security protocols, such as TLS. The most common security protocols and protocol libraries have already plans to include quantum resiliency on the strength level C_3 (PQC):

- TLS has proposals for hybrid[9] quantum protocols: [14, 69] as well as research on many levels to investigate how the integration to TLS can be done and what are its effects (e.g., [38, 43])
- IPSec IKEv2: a hybrid framework standards draft exists [72]
- OpenSSL: Does not support PQC directly (as of Oct/2020, [74]), but this is possible via the Open Quantum Safe—project plugins [70].

We list the IoT-protocols with their main purpose and properties in Table 16.5. The relevant protocol standards (if clearly defined), OSI-layer, and main uses are given in the table. In Table 16.5, standard(/s, or equivalent) is (are) given, if there is a clearly defined set of specifications for the protocol. The reach is relevant, if the protocol is designed or tied particularly to certain physical layer standards. For higher-layer independent protocols, we have signified this with "*Inf*".

Some of the entries in Table 16.5 are not actually even protocol families, but rather a loosely defined collection of protocols and standards considered to form a coherent concept, such as RFID and cellular. In these cases, we handle them via their main representative standards/protocols, such as 5G standardization work for cellular and NFC, and ISO 18000 security standards for RFID. There are also categories,

---

[9] Hybrid quantum protocol means that conventional, well-studied but QC-vulnerable cryptography is somehow "mixed" with more quantum-resilient mechanisms, such as PQC or symmetric cryptography.

Table 16.5 IoT protocols under considerations, with their main properties

Protocol name	Standard	OSI	Reach (m)	Uses
MQTT	ISO/IEC 20922	6/7	*Inf*	Brokering, pub/sub
CoAP	IETF (RFC 7252)	7	*Inf*	Low-power lossy networks
AMQP	OASIS	6/7	*Inf*	Brokering, pub/sub
DDS	OMG [22]	3–5	*Inf*	Brokering, pub/sub
OPC UA	OPC Foundation	5–7	*Inf*	Industrial applications
ZigBee	E.g., IEEE 802.15.4	1–7	10–20	Home, WSN, ICS, medical, power
Thread/6LowPAN	Thread Group Licensed	3–4	10–20[a]	Home automation, connect 802.15.4. to IP networks
EnOcean	ISO/IEC 14543-3-10	1–7	30–300	Smart Homes, buildings
LoRaWAN	LoRa Alliance [1]	2	10k+	Long-range IoT
NB-IoT	3GPP	1–3	<10k	WSN, smart meters
SigFox	None (SigFox propr.), PHY-layer opened	1–7	40 k+	Smart meters, remote monitoring, transportation
Z-Wave	None (Silicon Labs propr.)	1–7	10	Home automation
Bluetooth	IEEE 802.15.1 (Bluetooth SIG)	1–7	10	General purpose PAN
NFC	Several (ISO: 14443, 18092, ECMA-340, NFC Forum: NDEF, SNEP, TNEP, CH)	1–7	0.04	Mobile payment, identity cards, keycards
WiFi	WiFi Alliance, IEEE	1–2	10–30k	General purpose LAN—MAN
RFID	Several (ISO, EPC Global)	1–2	0.1–50	Identification
Cellular	Multiple	1–5	80k	General purpose WAN

[a] *Note* 6LowPAN is an L3 adaptation layer, and meant for Zigbee L2, thus the Zigbee range

which although well-defined, have a large number of sub-standards and specifications collected with them. These include WiFi and Bluetooth, which we handle via representative examples.

Our approach includes charting the "cryptographic landscape" of IoT protocols, involving the main cryptographic algorithms, their parameters and key management principles. The results of this study are presented in Table 16.6. The columns in the table are divided as follows:

- Confidentiality-only modes of operation for symmetric cryptography:

  – CTR: Counter

Table 16.6 Cryptographic properties of different IoT protocols

Protocol	Secur. Docum. Ref.	Conf.				Integr.											Key mgmt																		Libs used		
		CTR	OFB	CBC	XCBC	CCM	GCM	CMAC	XMAC	Propr.	DES	3DES	AES128	AES256	Other block	Other stream	PSK	OOB	Hierarchy	I-Key life	C-key life	DSA	Schnorr-	2k RSA	ECC-163	ECC-192	ECC-256	ECC-283	ECC-384	DH	EC-	Other	I-Key life	C-key life	TLS	DTLS	SASL
MQTT	[9]	x		D			3						D	D						S	S	D		D			D			D			L	L	X		
CoAP	[66]	x		X		1-3							X	X			X			S	S	X		X			X			X			L	L		X	
AMQP	[53]	x		X		1-3	3						X	X						D	D	X		X			X			X			D	D			X
DDS	[22]	x					3						X	X						S	S	X		X			X			X			L	L			
OPC UA	[54]			X										X	X					-	S											X	L	L			
Zigbee	[85]	x				1-3							X				X			D	D				D			D			D		D	D			
Thread	[24]	x				1							X				X	X		D	D		X				X			X	X		L	L/S		X	
EnOcean	[61]							1		X							X			L	L												-	-			
LoRaWAN	[46]	X						1					X				X		X	-	L/S												-	-			
NB-IoT	[15]	X						1					X			X	X		X	L/S	L/S												-	-			
SigFox	[27]	X						1					X				X			L	L												-	-			
Z-Wave	[28]		X			1							X				X			L	L/S												-	-			
Bluetooth	[55]	x				1							X			X	X	X	X	L	L/S					X	X			X	X		L/S	L/S			
NFC	[52]				X				2		X	X	X	X	X	X	X	X		L/S	L/S									X		X	L	L			
RFID	[37]			X			3	2			X	X	X	X	X	X	X			L/S	L/S	X	X		X					X		X	L/S	L/S			
WiFi	[83]	x				1	3						X	X	X	X	X	X	X	L/S	L/S	X	X				X		X	X	X	X	L	L/S	X		

Legend
X = yes
x = implied
D = Impl.dep.

L = Long
S = Short
L/S

Tag length
1 = 64 bits or less
2 = 96 bits
3 = 128 bits

Blue marking — Informational (not used for assessment)

- OFB: Output-Feedback
- CBC: Cipher text block chaining
- XCBC: a CBC-variant specified in [52]

- Modes of operation with integrated authentication properties for symmetric cryptography:

  - CCM: Counter with Cipher Block Chaining-Message Authentication Code [25]
  - GCM: Galois Counter Mode
  - CMAC: Cipher-Based Message Authentication Code (NIST SP-80038B)
  - XMAC: XCBC-MAC: specified in IETF RFC3566
  - Propr.: Proprietary, such as "Variable AES" in EnOcean

- Symmetric primitives:

  - DES: Data Encryption Standard
  - 3DES: Triple-DES
  - AES128: Advances Encryption Standard, with 128 bits of key-length
  - AES256: AES with 256 bits of key-length
  - "Other block": Other block cipher primitives, such as PRESENT in RFID-standard ISO-29167-11
  - "Other stream": Other stream cipher primitives, such as ZUC in LTE security by 3GPP

- Symmetric cryptography-only key management:

  - PSK: Pre-shared key (assumes the existence of some pre-shared secret between parties)
  - OOB: Out-of-Band (standard declares key exchange to be OOB, usually amounts to PSK)
  - Hierarchy: a symmetric key-hierarchy with master keys and key-encryption keys

- Symmetric key lifecycle:

  - "I-key life": the key period of integrity-related (incl. authentication) keys. "L" for "Long", meaning years, or requiring manual procedures (=product lifetime); "S" for "Short", otherwise, both may coexist.
  - "C-key life": as "I-key life", but for confidentiality purposes

- Asymmetric primitives and their use:

  - DSA: Digital Signature Algorithm, and variants
  - Schnorr-type: Schnorr signatures [64] with variants (e.g., ISO-29167-17)

- Asymmetric primitives:

  - 2 k RSA: RSA-scheme, with 2048bit key-length
  - ECC-X: Elliptic curve cryptography scheme, using a curve with a key length of X bits

- Asymmetric key management:
  - DH: Diffie-Hellman
  - EC-JPAKE: Elliptic-Curve PAKE (Password-Authenticated Key Exchange)
  - "Other": Other types, including Dragonfly key exchange for WPA3 Private mode and proprietary ones (OPC UA)
  - Asymmetric key lifecycle: as in the symmetric case
- Libraries used: security may depend also on the following standards / implementations of them:
  - TLS: Transport-Layer Security
  - DTLS: Datagram TLS
  - SASL: Simple Authentication and Security Layer

The entries in Table 16.6 describe the cryptographic attributes and properties of each protocol as follows:

- "X": the protocol recommends or mandates the method in question;
- "x": method is implied by another method (e.g., the GCM mode also implies the CTR-mode);
- "D": implementation dependent—the protocol does not define any recommendation, direct or implied as whether to use the method. Entries taken from popular implementations;
- "L": long key lifecycle, e.g., a manufacturer key or certificate permanently in the device;
- "S": Short key lifecycle, e.g., a session key.
- Integrity tag length types: "1" means a "short" tag length (64 bits or less), "2" refers to "medium" (96 bits) and "3" to "long" tag length. Length is taken to be in relation to the Grover's algorithm's run time for AES-128.
- Blue markings are indicative, and are not used for actual risk evaluation due to the fact that they are only examples and do not represent the whole of the group in question (e.g., RFID).

RFID cryptographic security is very much implementation-dependent. Parallel standards abound: ISO/IEC 29167 [37] intended for ISO/IEC-18000 air interface cryptography, defines two block ciphers (PRESENT and AES in different modes), one stream cipher (Grain128a) and four different asymmetric mechanisms starting from conventional ECC and ending up in Rabin-like schemes [37], used both for key exchange and digital signatures (DSA- and Schnorr-type constructions alike, e.g., in ISO-29167-17).

Another widely used standard within RFID is ISO/IEC 29192, which covers the use of two lightweight block ciphers (PRESENT and CLEFIA) and two stream ciphers (Trivium and Enocoro). Some of these symmetric ciphers are, already based on their key length (80-bit PRESENT), barely secure in the classical sense, not to mention the classical vulnerability of many of the stream ciphers (e.g., Grain [3]).

Some of the RFID symmetric schemes are thus irrelevant for QC (due to their weak classical security). Furthermore, the usage and scenarios for both symmetric and asymmetric cryptography already within these two standards cover the whole IoT spectrum. This implies that the possible threat levels would also cover the whole spectrum in our model, which makes handling the RFID protocols as one consistent category impossible, and we decided to exclude them from the threat assessment.

WiFi mostly refers to the IEEE 802.11 set of standards. However, the *security* used in and for these standards is a very versatile concept. Most prominent security standards include 802.11i, WPA, and EAP via 802.1X, all of them having several versions and multiple fielded devices using older versions. Data encryption varies from RC4 in WEP with a 32-bit CRC [36] to AES-GCM with 384 bit SHA-MICs [83], and asymmetric operations may or may not be used at all. When they are used, several key exchange mechanisms may be at use. ECC-curves may be selected from the NIST recommendations [83] or Brainpool [82]. WiFi implementations in IoT also represent such a wide variety of techniques that we disqualify it from the threat assessment.

Cellular technologies are a third category, which by nature refers to several standards and implementations of different ages. Cellular wireless standards are separated by generations, ("G's") starting from analog 1G and going up to digital 6G or even further in research. Standards from 3G onwards have been under the custody of the 3GPP industry consortium. Currently the largest development effort concerns 5G. The 3GPP and ITU-T have defined three major use case categories for 5G technologies, one of which is called mMTC, or massive Machine-Type Communications, such as Low-Power Wide-Area (LPWA) techniques. The mMTC and LPWA address the IoT needs [49]. 3GPP also presented that NB-IoT and LTE would be the main IoT technologies to enhance, in order to address the LPWA use cases for the foreseeable future [49, 79].

While we cannot address cellular technologies as one group, we have already considered the main 5G technologies for IoT, the NB-IoT. Furthermore, since NB-IoT security is based on LTE security, the security considerations for 5G within the IoT context can be reduced to LTE security. We note also that some NFC main standards (ISO-14443 and 18092) do not specify encryption, this is left for the application layer. The entries in Table 16.6 reflect the Mifare and ECMA-386 [52] standards. DDS specifies an API but not the communication protocol. We used the OpenDDS implementation [45] as an example.

Protocols residing solely on the OSI-model session layer (5) or higher, rarely need to specify security features independently. MQTT, CoAP, and AMQP all rely on TLS and variants. OPC UA is an exception to this. In these cases, we have referred to the recommended cipher suites to be used in each protocol, or a popular implementation of it. For example, MQTT cryptography suites are extracted from HiveMQ specification default cipher suites [60]. NB-IoT security is based on LTE security [15], and the entries in Table 16.6 reflect LTE choices. Bluetooth is an extensive collection of specifications,[10] but the most relevant for IoT concern the Bluetooth Low-Energy (BLE) specifications, and the security parameters reflect BLE. OPC UA specifies a

---

[10] 16 actual specifications, nearly a hundred profiles/parameter sets and over a hundred testing conformance documents.

possibility to use SHA-256 MICs, i.e. symmetric hash-based integrity checks in the payload (the current table lists this under "other block" column).

The key-management in each protocol is often loosely defined, or including a multitude of options. Many of the protocols studied here use a symmetric cryptography key hierarchy. In these cases, the lower-level keys are usually short-lived, whereas the master-keys may be hard-coded (do not allow remote rekey), and last for the lifetime of the node, for example, 10 years in LoRaWAN [19]. This principle extends to other protocols as well (Sigfox and NB-IoT [19]). We present an overview of the most relevant key-management solutions here.

The OPC UA uses a conventional Diffie-Hellman procedure with 2 k RSA for the key exchange, but uses the same private key both for signing and encryption [54]. In ZigBee, the 802.15.4 layer security does not specify key management [24], but higher layers may use several types of keys, whose life-cycle is left application-dependent. Also pre-configured (during the device lifetime) keys are possible [59]. A separate certificate-based key-exchange (CBKE)-profile is possible, enabling a J-PAKE-based asymmetric key establishment [6, 81].

According to [24], in Thread, the network authentication and key agreement is based on an elliptic curve variant of J-PAKE (EC-JPAKE), using NIST P256 elliptic curve. It uses ECDHE (Elliptic-Curve Diffie-Hellman Exchange) for key agreement and Schnorr signatures as NIZK proof mechanism for the authentication. Thread integrates EC-JPAKE with DTLS in order to provide security. The EC-JPAKE-procedure forms ephemeral asymmetric keys for confidentiality, based on symmetric long-term key [35]. The same procedure also delivers proofs of knowledge of the long-term key, so this same pre-shared key is used also for integrity. EnOcean security has high-security profiles [2], but they all use pre-placed keys (which have a lifetime as long as the master node's lifetime), which are copied by a *teach-in*-procedure to possible slave nodes.

Many IoT protocols do not use the full authenticated encryption tag length, but instead truncate them. These include, for example:

- BLE: the RFC 4493 for AES CMAC [67] recommends tag length of 128 bits, but BLE uses 64 bits [39]
- Thread: NIST AES CCM [25] recommends tag lengths of 64 bits or more, but Thread MIC is 32 bits [23]
- EnOcean: CMAC tag length in the message is 3–4 bytes [39], so at most 32 bits
- DDS: The specification [22, clause 9.5.2.5] specifies 128 of tag length
- CoAP: the underlying DTLS recommendation also accepts tag lengths of exactly 64 bits for CCM.

16.6 QC Threat Assessment for IoT Protocols

After cataloguing the cryptographic parameters of each IoT-protocol (-family), we apply our QC-threat model to the protocols. Since the QC threat model is attached to case-dependent risk analysis but the protocols themselves are (at least intended to be) case-independent, some of the model parameters are more difficult to extract from protocols only. In these cases, we estimate the parameters based on the typical uses of the protocol (as per Table 16.5), protocol range or just a more conservative security approach.

A symmetric key-hierarchy represents a non-trivial case for the evaluation. On one hand, only the higher-level keys have a long lifecycle and it is difficult to find known plaintext pairs for cryptanalysis from KEK-transfers only. On the other hand, the adversary is assumed to be able to monitor all the traffic, and can start lower-level key extraction from the actual payload, "climbing" up the hierarchy. Also, in the QC security model, e.g., with the Grover's algorithm for symmetric ciphers, only few known-plaintext pairs are required. The master keys are, in any case, used to encode lower-level secrets, and a QC adversary would need to monitor only few such key establishments encrypted with the same master key. In most cases, this is quite feasible.

Analogous to the situation in typical asymmetric cryptography schemes the asymmetric key acts as a key encryption key for the symmetric key of the actual payload. The threat level for symmetric key hierarchies is then established by considering the lifetime of the highest level key ("master key") and assessing, how readily ciphertext-plaintext-pairs are available for it.

This approach for key hierarchy also means that if there are both long- and short-lived (symmetric) keys and they form a hierarchy the keys with the longer lifetime determine the threat level, as the keys higher up in the hierarchy have a longer lifespan. The asset lifetime is estimated from the main use cases of the protocol: e.g., sensor readings are assumed to have a short lifetime, whereas brokering may transmit long-lived industrial control system configuration information.

By default, we adopt mostly a conservative approach in estimating the main parameters for use-case-specific or ambiguous cases:

- Communication protocols imply easy availability of the material for the adversary, including certificates and long term keys, since keys need to be exchanged or transported anyway. We then assume the availability to be Easy in our QC threat model.
- The protocols and applications are generally unaware of the security needs of the use case. We thus assume that security margin is Large.
- When estimating the asset lifetime for general-purpose protocols, we assume that the short range protocols are used mostly to transmit short-lived sensor data, unless otherwise specified
- For implementation-dependent cases (marked with "D" in Table 16.6) we again assume the worst case.

When we estimate the total effect of different cryptographic elements to the protocol, we need to take into account the fact that in some cases there are dependencies between different levels of keys. In many cases, symmetric keys for integrity and confidentiality are both derived from an asymmetric key-exchange. In this case the threat profile of the asymmetric case overrides the one used for symmetric schemes only.

Some protocols also use the same certificate for key exchange, authentication and signatures. Even though the exposure of the certificate affects the same way to both confidentiality and integrity *post-quantum*, the effect is not the same *pre-quantum*: we still assume we are able to make the decision of the validity of the data at the time of data use. Thus confidentiality threat-profiles do not override integrity threat profiles, pre-quantum. Higher-level integrity-profiles are still assumed to override lower-level integrity profiles (also pre-quantum), if such dependencies exist. If the protocol offers a possibility to use the schemes independently, the complete threat profile is more complicated.

We start with the detailed threat model, which is evaluated independently of the possible key management dependencies (Table 16.7) and then prune those parts of the model which are dependent on higher levels of keys or invalidate the use of more secure techniques in the protocol. Table 16.7 shows the threat profile for the protocols in terms of four sub-profiles (only, the dependencies are accounted for later on): asymmetric and symmetric cryptography integrity and confidentiality profiles, A_I, A_C, S_I and S_C, respectively.

Table 16.7 Independent QC-threat-profiles to IoT protocols

Protocol	Symmetric					Asymmetric					Dependencies
	S_C		S_I			A_C		A_I			
	Now	After CRQC	Now	K_{long} - CRQC	After CRQC	Now	After CRQC	Now	K_{long} - CRQC	After CRQC	
MQTT	4	4	0	0	1	9	9	6	7	9	Default TLS suite: symm.keys are derived
CoAP	4	4	0	0	1	6	9	4	4	9	Mandat. DTLS suite: symm.keys are derived
AMQP	4	4	2	2	3	9	9	6	7	9	Symm.keys derived from DH
DDS	4	4	0	0	1	6	9	4	4	9	Symm.keys derived from DH
OPC UA	3	3	2	2	3	9	9	6	7	9	RSA-key exchange provides symm.keys
Zigbee	4	4	2	2	3	6	9	4	4	9	Ind.use of PSK possible
Thread	2	3				6	9	4	4	9	DTLS provides 802.15.4 keys
EnOcean	2	3									
LoRaWAN	4	4									
NB-IoT	2	3									
SigFox	2	3									
Z-Wave	2	3									
Bluetooth	4	4				6	9	4	4	9	OOB use of symm.keys possible
NFC	4	4	2	2	3	6	9				Ind.use of SKH possible

From Table 16.7 it can be seen that there are almost exactly two types of threat profiles for each type. This stems from the facts that the actual IoT protocols are use-case- independent, and especially the modern protocols follow the same cryptographic standards. Only the older protocols, such as those based on cellular 3G or RFID still present more variety. The differences between the profiles arise from:

- Differences in the expected lifetime of the assets, based on the expected use cases (especially CoAP and DDS low A_I- and A_C-profiles)
- Long tag length (DDS low S_I-profile)
- Short I-key lifetime (CoAP and MQTT low S_I-profile).

We used a different marking to indicate that the QC threat does not apply. This has two reasons: no schemes are specified for the category (pure symmetric, or no asymmetric integrity protection), and the classical threat for the type is greater than the threat caused by QC (MIC tag length is shorter then Grover's expected runtime for AES-128).

The possible dependencies between the asymmetric and symmetric keying are:

- Thread: The DTLS (asymmetric) layer provides the keys to the (symmetric) 802.15.4 layer
- Bluetooth (Low Energy): Independent use of Out-of-Band keys is possible
- NFC: There are several standards and some of them allow the use of symmetric key hierarchies. However, ECMA-386 does not specify authentication in the key exchange.
- ZigBee: independent use of pre-shared keys is possible
- OPC UA: the RSA-based key-exchange is used to derive the symmetric keys. Furthermore, the asymmetric integrity- and confidentiality key are the same [54]. This means, that the A_C threat profile represents the whole of OPC UA profile.
- DDS and AMQP: the symmetric keys are derived from the ECDHE in most cases
- CoAP: some security modes in RFC7252 allow the use of PSK, and decoupling the asymmetric layer from the symmetric layer. However, these are not mandated, and we base the coupling to the mandated "RawPublicKey" security mode.
- MQTT: the sample implementation default TLS suite uses asymmetric key exchange to derive all symmetric keys.

If we implement the dependencies between the S- and A-profiles, the resulting profiles change somewhat (see Table 16.8). The dependent profiles inherit the respective parent profiles (presented without the number, but with the same color). In Table 16.8, the protocols on the higher OSI-level are more vulnerable to QC than lower levels. This is due to the use of vulnerable asymmetric schemes on the top of the key hierarchy. Protocols using pure symmetric-key schemes meant for short-lived data are relatively unaffected by QC.

Table 16.8 The QC-threat-profiles to IoT protocols with key management dependencies

Protocol	Symmetric					Asymmetric				
	S_C		S_I			A_C		A_I		
	Now	After CRQC	Now	K_{long}-CRQC	After CRQC	Now	After CRQC	Now	K_{long}-CRQC	After CRQC
MQTT						9	9	6	7	9
CoAP						6	9	4	4	9
AMQP						9	9	6	7	9
DDS						6	9	4	4	9
OPC UA						9	9	6	7	9
Zigbee	4	4	2	2	3	6	9	4	4	9
Thread						6	9	4	4	9
EnOcean	2	3								
LoRaWAN	4	4								
NB-IoT	2	3								
SigFox	2	3								
Z-Wave	2	3								
Bluetooth	4	4				6	9	4	4	9
NFC	4	4	2	2	3	6	9			

16.7 Quantum Resilience in IoT

In order to estimate the actual risk to the protocols (pre-quantum as well as post-quantum) the current research and standardization activities need to be mapped (see Table 16.9). From each protocol it is stated, whether there are implied or direct plans to upgrade the standard or implementations to a quantum-resilient (PQC or QKD, in practice) form; or if there is any research to suggest something like this is possible.

For cases that to do not directly support security, but instead rely on security on other layers, we distinguish to cases:

1. *"Implied"*: The security layer has announced plans to support PQC and the "customer" IoT-protocol standard has indicated it will be taken into use. As it turns out, this option was not to be found at all.
2. *"Possible"*: As no. 1, but the "customer"-protocol standard has so far ignored this possibility, e.g., in the recommended TLS-suites.

Table 16.9 The quantum-resiliency possibilities for IoT protocols

Protocol	PQC plans?	PQC research?
MQTT	Possible, via TLS	No
CoAP	Possible, via DTLS	Attestation [47], RLWE-KEX [80]
AMQP	Possible, via TLS	No
DDS	No	No
OPC UA	No	Both hybrid and direct [57]
Zigbee	No	No
Thread	Possible, via DTLS	No
EnOcean	No	No
LoRaWAN	No	PQC-KEX [47]
NB-IoT	No	Some (QKD in the backbone [7])
SigFox	No	No
Z-Wave	No	No
Bluetooth	No	R-LWE for SSP [29]
NFC	No	Auth.prot. [56]; CH use of BLE SSP [29]
WiFi	Partly via TLS	Both QKD [31] and PQC ([21, 68] and via TLS)
Cellular (5G)	Crypto agility	Also QKD is relevant in the backbone, e.g., [7]
RFID	No	Auth.prot. [18], sig. schemes [44]
Libraries		
DTLS	Partly via TLS	NTRU-KEX [63], RLWE-PAKE [30]
TLS	Yes [14, 69]	
IPSec	Yes [72]	
OpenSSL	Yes [70]	

We note that to protect symmetric schemes from analysis with Grover's algorithm, it is not sufficient to generate merely the symmetric scheme key with a quantum-resistant manner, such as QKD. The Grover's algorithm does not care for the origin of the symmetric key: thus, if QKD is used to generate just a small number of (classical) bits, instead of full keystream, they do not contribute to the QR-properties of symmetric schemes. Thus QKD is useful in two scenarios:

1. It replaces an asymmetric key-exchange
2. It replaces the symmetric key scheme as a whole, turning it to one-time pad (OTP).

In this part of the assessment, we have also included the relevant libraries or security standards and the protocol families that are not part of the actual numerical evaluation due to their diverse nature. This is done for informational purposes and for completeness. Some protocol-specific particulars here include:

- CoAP: Remote attestation refers to a type of device integrity-check remotely, and RLWE-KEX refers to Ring-Learning With Errors Key-Exchange, a PQC implemented with a certain type of lattice scheme.
- OPC UA: A "hybrid" (PQC-) scheme refers to a scheme, were conventional (e.g., ECC) schemes are combined with post-quantum schemes in a secure way. "Direct" refers to pure PQC only.
- Bluetooth: The research refers to Bluetooth Low Energy (BLE) Secure Simple Pairing protocol (SSP).
- NFC is a protocol family, and the cases refer to subsets of them: authentication protocols for payment applications, and the possibility to use lattices in BLE SSP in the NFC Forum's Connection Handover (CH) scenario.
- WiFi: There is research to integrate QKD to 802.11i [31]. Cisco purports to keep 802.1X as symmetric-key only, until they can migrate to suitable PQC TLS suites in EAP and WPA [21]. There is also some research to replace the main 802.11 standard's PAKE with SIDH (an isogeny-based PQC key-exchange protocol) [68].
- In cellular technologies, 5G security has a built-in capability for crypto agility, i.e. the ability to change cryptographic primitives and modules easily. This does not necessarily carry over to derivative standards (such as NB-IoT) so readily.
- TLS has already solid plans for migration to PQC schemes. These can be inherited to DTLS and other "customer" protocols if so desired.
- DTLS: The referenced research means research directed solely for DTLS, not those inherited from TLS.

16.8   IoT Protocol Risk Assessment

In this chapter, we pull together the threat profiles for IoT and their projected future for quantum resilience into a protocol-specific risk assessment. Here, again, we exclude protocol families, which are deemed to be too diverse in their standards, i.e. we use Table 16.9 as our basis for the assessment.

The main ideas in the risk assessment are as presented below:

- The post-quantum threat level for profiles A_C and A_I is reset to zero, if (a) there is at least existing research to replace the asymmetric schemes with PQC schemes or (b) there is a good possibility the protocol integrates post-quantum security from some other security layer.
- If there are ready-made plans for the standard itself, also the threat level "K_{long} – CRQC" in the profile A_I is reset to zero. (Such were not identified though.)
- PQC solutions do not affect symmetric schemes (S_C- and S_I-profiles are unaffected when considered independently)
- QKD solutions reset also those symmetric schemes threat to zero, which are affected by it directly. QKD does not have effect on schemes on the lower levels of the key-hierarchy.

We note that the PQC-algorithms themselves are not by default invulnerable to other quantum attacks, e.g. the Grover's algorithm in quantum lattice sieving methods. However, it appears that at least for the most important cases, the best-known quantum algorithms are only a little more efficient than classical algorithms [4]. This is especially the case with the finite parameter sizes of actual algorithms, when the asymptotic behavior is not that noticeable yet.

In contrast to threat modelling, where exposing the highest-level key in a key-hierarchy extends the threat to the lowest levels, protecting the highest level only protects that particular level. Thus for PQC, which is suggested to replace the highest level asymmetric keys, the A_I and A_C profiles would be affected, but the S_I and S_C-profiles would still be "exposed" (since they could be attacked independently also). NFC is a special case in this assessment, because only some cases are suggested to be PQC-protected. We then assume a 50% chance that all relevant protocol parts are updated to be quantum resistant in time. This reduces the threat level by half, or to the maximum threat level attainable with pure symmetric-key schemes.

QKD resets those profiles to zero, for which keys are directly affected. The case-specific considerations are presented as follows:

- Arul et al. [7]: QKD is suggested to be used in the LTE backbone to provide the master key to the top of the key hierarchy. This would protect the possible highest level asymmetric key exchanges, but the protocols evaluated here do not use those services. This scheme does not reduce the threat in the IoT context.
- Ghilen et al. [31]: QKD is proposed to replace the highest level keys in a symmetric key hierarchy. As explained in the previous chapter this does not provide any additional security, as the main threat (Grover's algorithm) is agnostic to the key generation method.

The results of the combination can be seen in Table 16.10. For most of the higher level IoT-protocols, it is likely that the vulnerable asymmetric schemes will be patched in due time. Exceptions include DDS and ZigBee asymmetric key exchange use cases. There are no such patches expected, which would affect the integrity threat profiles mid-term. Symmetric threat profiles are reset to those profiles that are independent of the asymmetric profiles, post-quantum, unless there is no evidence for such PQC-migration present. Symmetric-key-only protocols do not benefit from planned QR-solutions, not from the PQC nor from QKD.

16.9 Conclusions

In this study, we have presented a risk management model to estimate the risk level to different systems, when subjected to the power of large scale quantum computation. We found out that the current hype only concerns the worst-case scenario: the most sensitive and long-lived information that is protected with the most (quantum-) vulnerable cryptography, that is easily captured and out of control of the data owner. There are, however, other scenarios and other decision points, when the question is

Table 16.10 The QC risk assessment in IoT protocols, QR-possibilities factored

Protocol	Symmetric					Asymmetric				
	S_C		S_I			A_C		A_I		
	Now	After CRQC	Now	Klong-CRQC	After CRQC	Now	After CRQC	Now	Klong-CRQC	After CRQC
MQTT		4			1	9	0	6	7	0
CoAP		4			1	6	0	4	4	0
AMQP		4			3	9	0	6	7	0
DDS		4				6	9	4	4	9
OPC UA		3			3	9	0	6	7	0
Zigbee	4	4	2	2	3	6	9	4	4	9
Thread		3				6	0	4	4	0
EnOcean	2	3								
LoRaWAN	4	4								
NB-IoT	2	3								
SigFox	2	3								
Z-Wave	2	3								
Bluetooth	4	4				6	0	4	4	0
NFC	4	4	2	2	3	6	4			

if and when to be prepared for the "quantum event". It is not even universally agreed, what the "quantum event" actually is.

The risk management model includes a threat model, which can be updated according to use-case specific plans for quantum-safe migration. The threat model involves many non-linear elements and many dependencies, and may require further refinement based on new applications. We performed an extensive study on the cryptographic status of the most important IoT protocols, and applied the threat model for them. Within the IoT protocols, it turns out that they fall mainly into two categories: symmetric-key only, with a relatively low threat level, and asymmetric-key schemes, with some plans to migrate to PQC. Some exceptions exist as well.

An interesting discovery was that quantum resiliency for IoT is also offered by presenting quantum key-distribution (QKD) schemes to replace already rather secure symmetric key-hierarchy schemes, in the highest levels of the symmetric key hierarchy. QKD does offer quantum security to those bits it delivers, but if QKD is used only to agree on 128-bit symmetric keys, then those keys are equally vulnerable to quantum algorithms (Grover's) as classically generated keys. QKD with fixed number of key bits offers any additional security only when compared to vulnerable asymmetric schemes; and unconditional security only, if used as a one-time-pad (OTP, or Vernam's cipher).

When we take a look at the QC threat specifically within the IoT framework, it turns out that they do not present the most urgent targets to be upgraded to be quantum resilient:

- Many of the IoT applications transmit very short-lived data: currently the data lifetime needs to be of the order or a decade, for it to be in an immediate threat
- Short-range wireless IoT objects are generally connected to more powerful devices, for which post-quantum solutions are expected to be available anyway in the future
- Due to the low processing power of IoT nodes, symmetric cryptography is used more than asymmetric. Symmetric cryptography in most cases is not as vulnerable to QC as the main asymmetric cryptography schemes.
- Many important IoT protocols reside on the OSI Layer 7, the application layer, and are able to use common lower layer security technologies, such as TLS, for which there are plans and demonstrations to incorporate PQC as soon as they are standardized.

However, there is a global drive to modernize and standardize encryption systems throughout the CNI. Furthermore, IoT protocols are not used solely to transmit such information that is critical to keep hidden from adversaries, but also information that guarantees the authenticity or integrity of devices, other information or services. The availability and integrity of many critical services in CNI depends on multiple chains of trust, which are composed of arbitrary sequences of integrity and confidentiality services. Leaving some of the links poorly protected invalidates the enhancements on the other parts of the chain. We thus still see the QR-upgrade to be relevant in all of the CNI IoT protocols as well.

References

1. About the LoRaWAN® specification. LoRa Alliance, 2020. https://lora-alliance.org/lorawan-for-developers
2. Adding security to EnOcean receivers: Notes for security in different receiver application. Application Note 510, EnOcean, 2013. http://advanceddevices.com/sites/default/files/documents/AN510_Adding_Security_to_EnOcean_Receivers.pdf
3. Ågren M, Hell M, Johansson T, Meier W (2011) A new version of Grain-128 with authentication. Paper presented at Symmetric Key Encryption Workshop 2011, Lyngby
4. Albrecht M, Gheorghiu V, Postlethwaite E, Schanck J (2020) Estimating quantum speedups for lattice sieves. In: Advances in cryptology—ASIACRYPT 2020: 26th International conference on the theory and application of cryptology and information security (Daejeon, 2020), Proceedings, Part II. Springer, Cham, pp 583–613
5. Ammar M, Russello G, Crispo B (2018) Internet of things: a survey on the security of IoT frameworks. J Inf Secur Appl 38:8–27. https://doi.org/10.1016/j.jisa.2017.11.002

6. AN708: Setting smart energy certificates for Zigbee devices. Silicon Labs, Austin, TX. https://www.silabs.com/documents/public/application-notes/an708-setting-manufacturing-certificates.pdf

7. Arul R, Raja G, Almagrabi AO, Alkatheiri MS, Hussain CS, Bashir AK (2019) A quantum-safe key hierarchy and dynamic security association for LTE/SAE in 5G scenario. ResearchGate. https://www.researchgate.net/publication/337019593_A_Quantum_Safe_Key_Hierarchy_and_Dynamic_Security_Association_for_LTESAE_in_5G_Scenario

8. Asfaw A, Alexander T, Nation P, Gambetta J (2019) Get to the heart of real quantum hardware. IBM, 2019. https://www.ibm.com/blogs/research/2019/12/qiskit-openpulse/

9. Banks A, Briggs E, Borgendale K, Gupta R (eds) (2019) MQTT Version 5.0: OASIS Standard. OASIS. https://docs.oasis-open.org/mqtt/mqtt/v5.0/os/mqtt-v5.0-os.pdf. 7 Mar 2019

10. Baseline security recommendations for IoT in the context of critical information infrastructures. European Union Agency for Network and Information Security (ENISA) (2017)

11. Bernstein D (2018) Interview in public key cryptography conference (PKC'2018), Rio de Janeiro, 25–29 March 2018

12. Bernstein E, Vazirani U (1997) Quantum complexity theory. SIAM J Comput 26(5):1411–1473

13. Berte D (2018) Defining the IoT. In: Proceedings of the international conference on business excellence, vol 12(1), pp 118–128.https://doi.org/10.2478/picbe-2018-0013

14. Campagna M, Crockett E (2020) Hybrid post-quantum key encapsulation methods (PQ KEM) for transport layer security 1.2 (TLS). Working document, IETF Datatracker, 2019, https://datatracker.ietf.org/doc/html/draft-campagna-tls-bike-sike-hybrid/. Last updated 10 Sept 2020

15. Cavo L, Fuhrmann S (2017) Implementation and benchmarking of a crypto processor for a NB-IoT SoC platform. M.Sc. thesis, Lund University

16. Chen L, Jordan S, Liu Y, Moody D, Peralta R, Perlner R, Smith-Tone D (2016) Report on post-quantum cryptography. NISTIR 8105, National Institute of Standards and Technology (NIST). https://doi.org/10.6028/NIST.IR.8105

17. Cheng C, Lu R, Petzoldt A, Takagi T (2017) Securing internet of things in a quantum world. IEEE Commun Mag 55(2):116–120. https://doi.org/10.1109/MCOM.2017.1600522CM

18. Chikouche N, Cherif F, Cayrel PL, Benmohammed M (2017) RFID authentication protocols based on error-correcting codes: a survey. Wireless Pers Commun 96(1):509–527. https://doi.org/10.1007/s11277-017-4181-8

19. Coman FL, Malarski KM, Petersen MN, Ruepp S (2019) Security issues in internet of things: vulnerability analysis of LoRaWAN, Sigfox and NB-IoT. In: 2019 Global IoT Summit (GIoTS). IEEE, New York, pp 1–6. https://doi.org/10.1109/GIOTS.2019.8766430

20. Commercial National Security Algorithm (CNSA) Suite. National Security Agency (NSA) (2016) https://apps.nsa.gov/iaarchive/library/ia-guidance/ia-solutions-for-classified/algorithm-guidance/commercial-national-security-algorithm-suite-factsheet.cfm

21. Configuring post-quantum MACsec in Cisco switches. Cisco, 2020, https://www.cisco.com/c/dam/en_us/about/doing_business/trust-center/docs/configuring-post-quantum-macsec-in-cisco-switches.pdf

22. Data distribution service category—specifications associated. Object Management Group (OMG) (2020) https://www.omg.org/spec/category/data-distribution-service/

23. Dinu D, Kizhvatov I (2018) EM Analysis in the IoT context: lessons learned from an attack on thread. IACR Trans on Cryptograph Hardware Embedded Syst 2018(1):73–97

24. Dragomir D, Gheorghe L, Costea S, Radovici A (2016) A survey on secure communication protocols for IoT systems. In: 2016 International workshop on secure internet of things. IEEE, New York, pp 47–62. https://www.computer.org/csdl/pds/api/csdl/proceedings/download-article/12OmNzd7bCZ/pdf

25. Dworkin M (2007) Recommendation for block cipher modes of operation: The CCM mode for authentication and confidentiality. NIST Special Publication 800–38C, National Institute of Standards and Technology (NIST), Gaithersburg, MD. https://nvlpubs.nist.gov/nistpubs/Legacy/SP/nistspecialpublication800-38c.pdf

26. Fernández-Caramés TM (2020) From pre-quantum to post-quantum IoT Security: a survey on quantum-resistant cryptosystems for the Internet of Things. IEEE Internet Things J 7(7):6457–6480. https://doi.org/10.1109/JIOT.2019.2958788

27. Ferreira L (2021) (In)security of the radio interface in Sigfox. Paper presented at Financial Cryptography and Data Security 2021. https://eprint.iacr.org/2020/1575.pdf
28. Fouladi B, Ghanoun S (2013) Security evaluation of the Z-Wave wireless protocol. Paper presented at BlackHat USA 2013 (Las Vegas). https://sensepost.com/cms/resources/confer ences/2013/bh_zwave/Security%20Evaluation%20of%20Z-Wave_WP.pdf
29. Gajbhiye S, Karmakar S, Sharma M, Sharma S (2017) Design and analysis of pairing protocol for bluetooth enabled devices using R-LWE lattice-based cryptography. J Inf Secur Appl 35:44–50. https://doi.org/10.1016/j.jisa.2017.05.003
30. Gao X, Ding J, Li L, Saraswathy RV, Liu J (2017) Efficient implementation of password-based authenticated key exchange from RLWE and post-quantum TLS. Cryptology ePrint Archive, Report 2017/1192. https://eprint.iacr.org/2017/1192
31. Ghilen A, Azizi M, Bouallegue R (2015) Integration of a quantum protocol for mutual authen-tication and secret key distribution within 802.11i standard. In: 2015 IEEE/ACS 12th Interna-tional conference of computer systems and applications (AICCSA). IEEE, New York, pp. 1–7. https://doi.org/10.1109/AICCSA.2015.7507110
32. Grassl M, Langenberg B, Roetteler M, Steinwandt R (2016) Applying Grover's algorithm to AES: quantum resource estimates. In: Tsuyoshi T (ed) Post-quantum cryptography: 7th International workshop, PQCrypto 2016 (Fukuoka, 2016). Springer, Berlin, pp 29–43
33. Hagelstam A (2015) CIP—kriittisen infrastruktuurin turvaaminen: Käsiteanalyysi ja kansain-välinen vertailu. Julkaisuja 1/2005, Huoltovarmuuskeskus
34. Halunen K, Kiviharju M, Suomalainen J, Vallivaara V, Kylänpää M, Latvala O-M (2020) A taxonomy of metrics for cryptographic systems. In: Thirteenth international conference on emerging security information, systems and technologies (SECURWARE 2019). International Academy, Research, and Industry Association (IARIA), pp 69–77
35. Hao F (ed) (2016) J-PAKE: password authenticated key exchange by juggling. Internet-draft, IETF Networking Group. https://tools.ietf.org/html/draft-hao-jpake-04
36. IEEE standard for information technology—Telecommunications and information exchange between systems-local and metropolitan area networks-specific requirements—Part 11: Wire-less LAN Medium Access Control (MAC) and Physical Layer (PHY) specifications. IEEE STD 802.11-1997, IEEE; 1997. https://doi.org/10.1109/IEEESTD.1997.85951
37. ISO/IEC 29167-16:2015(en) Information technology—automatic identification and data capture techniques—part 16: Crypto suite ECDSA-ECDH security services for air interface communications. International Organization for Standardization (ISO) and International Elec-trotechnical Commission (IEC), 2015, https://www.iso.org/obp/ui/#iso:std:iso-iec:29167:-16: ed-1:v1:en
38. Jao R, Azarderakhsh R, Campagna M, Costello C, De Feo L, Hess B, Jalali A, Koziel B, LaMacchia B, Longa P, Naehrig M, Pereira G, Renes J, Soukharev V, Urbanik D (2019) Supersingular isogeny key encapsulation. Technical report, ResearchGate. https://doi.org/10. 13140/RG.2.2.26543.07847
39. Kambourakis G, Kolias C, Geneiatakis D, Karopoulos G, Makrakis G, Kounelis I (2020) A state-of-the-art review on the security of mainstream IoT wireless PAN protocol stacks. Symmetry 12(4):579. https://doi.org/10.3390/sym12040579
40. Kaplan M, Leurent G, Leverrier A, Naya-Plasencia M (2016) Breaking symmetric cryptosys-tems using quantum period finding. arXiv:1602.05973 [quant-ph]
41. Kiviharju M (2017) On the fog of RSA key lengths: verifying public key cryptography strength recommendations. In: 2017 International conference on military communications and infor-mation systems (ICMCIS). IEEE, New York, pp 1–8. https://doi.org/10.1109/ICMCIS.2017. 7956481
42. Krejčí R, Hujňák O, Švepeš M (2017) Security survey of the IoT wireless protocols. In: 2017 25th Telecommunication forum (TELFOR). IEEE, New York, pp 1–4. https://doi.org/10.1109/ TELFOR.2017.8249286
43. Kwiatkowski K, Valenta L (2019) The TLS post-quantum experiment. The Cloudflare blog, Cloudflare. https://blog.cloudflare.com/the-tls-post-quantumexperiment/

44. Li D, Chen H, Zhong C, Li T, Wang F (2017) A new self-certified signature scheme based on NTRUSign for smart mobile communications. Wireless Pers Commun 96:4263–4278. https://doi.org/10.1007/s11277-017-4385-y
45. License: OpenDDS (2005) Object Computing Incorporated (OCI). St. Louis, MO. https://opendds.org/about/license.html
46. LoRaWAN™ security: Full end-to-end encryption for IoT application providers (2017) A white paper prepared for the LoRa Alliance™ by Gemalto, Actility and Semtech, LoRa Alliance. https://lora-alliance.org/sites/default/files/2019-05/lorawan_security_whitepaper.pdf
47. M, Niranjan M, Kenchaiah N (2020) Secure and optimized method of providing trustworthiness for IoT sensors in low-power WAN deployments. Technical Disclosure Commons. https://www.tdcommons.org/dpubs_series/3692
48. Marksteiner S, Expósito Jiménez VJ, Valiant H, Zeiner H (2018) An overview of wireless IoT protocol security in the smart home domain. In: 2017 Internet of things business models, users, and networks. IEEE, New York, pp 1–8. https://doi.org/10.1109/CTTE.2017.8260940
49. Mobile IoT in the 5G future: NB-IoT and LTE-M in the context of 5G. GSM Association (2018) https://www.gsma.com/iot/wp-content/uploads/2018/05/GSMAIoT_MobileIoT_5G_Future_May2018.pdf
50. Moody D (2017) The ship has sailed: the NIST post-quantum crypto "competition". Invited talk at ASIACRYPT 2017, National Institute of Standards and Technology (NIST). https://csrc.nist.gov/CSRC/media/Projects/Post-Quantum-Cryptography/documents/asiacrypt-2017-moody-pqc.pdf
51. Mosca M (2015) Cybersecurity in a quantum world: will we be ready? Invited talk at NIST workshop on Cyber Security in a Post-Quantum World (Gaithersburg, MD, 2015), National Institute of Standards and Technology (NIST). https://csrc.nist.gov/csrc/media/events/workshop-on-cybersecurity-in-a-post-quantum-world/documents/presentations/session8-mosca-michele.pdf
52. NFC-SEC-01: NFC-SEC cryptography standard using ECDH and AES (2015) Standard ECMA-386, Ecma International. https://www.ecma-international.org/wp-content/uploads/ECMA-386_3rd_edition_june_2015.pdf
53. OASIS Advanced Message Queuing Protocol (AMQP), version 1.0, Part 5: Security. OASIS Standard, OASIS (2012) http://docs.oasis-open.org/amqp/core/v1.0/amqp-core-security-v1.0.html
54. OPC UA security analysis. Federal Office for Information Security (BSI), Bonn (2017) https://www.bsi.bund.de/SharedDocs/Downloads/EN/BSI/Publications/Studies/OPCUA/OPCUA.html
55. Padgette J, Bahr J, Batra M, Holtmann M, Smithbey R, Chen L, Scarfone K (2017) Guide to bluetooth security. NIST Special Publication 800-121, rev. 2, National Institute of Standards and Technology (NIST), Gaithersburg, MD. https://nvlpubs.nist.gov/nistpubs/SpecialPublications/NIST.SP.800-121r2.pdf
56. Park S, Lee I (2016) Mutual authentication scheme based on lattice for NFC-PCM payment service environment. Int J Distrib Sens Netw 12(7):2016. https://doi.org/10.1177/1550147794 71539
57. Paul S, Scheible P (2020) Towards post-quantum security for cyber-physical systems: Integrating PQC into industrial M2M communication. In: Computer security—ESORICS 2020. Lecture Notes in Computer Science, vol 12309. Springer, Cham, pp 295–316
58. Rajeswar K (2019) Industry 4.0 wave: relevance of SCADA in an IoT world and journey towards a true digital enterprise. IEEE India Info 14(3):78–88. http://site.ieee.org/indiacouncil/files/2019/10/p78-p88.pdf
59. Rudresh V (2017) ZigBee security: basics (Part 2). Kudelski Security Research. https://research.kudelskisecurity.com/2017/11/08/zigbee-security-basics-part-2/
60. Security (2021) Chapter in HiveMQ User Guide, version 4.5. HiveMQ. https://www.hivemq.com/docs/hivemq/4.5/user-guide/security.html
61. Security of EnOcean radio networks v2.3. System Specification, EnOcean Alliance, San Ramon, CA (2018) https://www.enocean-alliance.org/wp-content/uploads/2018/06/Security_of_EnOcean_Radio_Networks_v2.3.pdf

62. Security Strategy for Society (2017) Government resolution, Finnish Security Committee
63. Sepúlveda J, Liu S, Mera J (2019) Post-quantum enabled cyber physical systems. IEEE Embed Syst Lett 11(4):106–110. https://doi.org/10.1109/LES.2019.2895392
64. Seurin Y (2012) On the exact security of Schnorr-type signatures in the Random Oracle Model. Cryptology ePrint Archive: Report 2012/029, International Association for Cryptologic Research (IACR). https://eprint.iacr.org/2012/029.pdf
65. Shawn (2018) What is the difference between SCADA and IoT?. Digital Connect Mag. https://www.digitalconnectmag.com/what-is-the-difference-between-scada-and-iot/
66. Shelby, Z., Hartke, K., Bormann, C. The Constrained Application Protocol (CoAP). RFC 7252, Internet Engineering Task Force (IETF), 2014. https://tools.ietf.org/html/rfc7252
67. Song JH, Poovendran R, Lee J, Iwata T (2006) The AES CMAC algorithm. RFC 4493, The Internet Society. https://tools.ietf.org/html/rfc4493
68. Soukharev V, Hess B (2019) PQDH: a quantum-safe replacement for Diffie-Hellman based on SIDH. Cryptology ePrint Archive, Report 2019/730, International Association for Cryptologic Research (IACR). https://eprint.iacr.org/2019/730
69. Stebila D, Fluhrer S, Gueron S (2019) Design issues for hybrid key exchange in TLS 1.3. Internet Engineering Task Force (IETF). https://datatracker.ietf.org/doc/html/draft-stebila-tls-hybrid-design-01
70. Stebila D, Mosca M (2017) Post-quantum key exchange for the Internet and the Open Quantum Safe project. In: Avanzi R, Heys H (eds) Selected areas in cryptography—SAC 2016. Lecture Notes in Computer Science, vol 10532. Springer, Berlin, pp 14–37
71. The complete list of wireless IoT network protocols (2016) Link Labs, Annapolis, MD. https://www.link-labs.com/blog/complete-list-iot-network-protocols
72. Tjhai C, Tomlinson M, Bartlett G, Fluhrer S, Van Geest D, Garcia-Morchon O, Smyslov V (2019) Framework to integrate post-quantum key exchanges into Internet Key Exchange Protocol version 2 (IKEv2). Internet Engineering Task Force (IETF). https://tools.ietf.org/html/draft-tjhai-ipsecme-hybrid-qske-ikev2-04
73. Tournier J, Lesueur F, Le Mouël F, Guyon L, Ben-Hassine H, A survey of IoT protocols and their security issues through the lens of a generic IoT stack. Internet of Things (in press). https://doi.org/10.1016/j.iot.2020.100264
74. Tuveri N (2020) OpenSSL 3.0 Alpha7 Release. OpenSSL Blog. https://www.openssl.org/blog/blog/2020/10/20/OpenSSL3.0Alpha7/
75. Unified Architecture, Part 1: Overview and concepts. Developer tool specification, OPC Foundation, 2017.
76. Unified Architecture (2017) Part 4: services. Developer tool specification, OPC Foundation
77. Unified Architecture (2017) Part 6: mappings. Developer tool specification, OPC Foundation
78. Vasilomanolakis E, Daubert J, Luthra M, Gazis V, Wiesmaier A, Kikiras P (2015) On the security and privacy of Internet of Things architectures and systems. In: 2015 International workshop on secure internet of things (SIoT). IEEE, New York, pp 49–57. https://doi.org/10.1109/SIOT, 9/2015
79. Vos G (2018) With LTE-M and NB-IoT you're already on the path to 5G. Sierra Wireless. https://www.sierrawireless.com/iot-blog/lte-m-nb-iot-5g-networks/
80. Ward N, Barnett A, Byrne A, Catterall N (2018) Secure architectures of future emerging cryptography. SAFEcrypto: D9.3– Testbed Design Report, European Commission
81. Wireless Gecko Series 2: xG21 and xG22 families (2020) Document of Webinar Silicon Labs. Sfera. https://sfera.mo.it/wp-content/uploads/2020/04/Melchioni_xG21_22_Webinar.pdf
82. WPA3^{TM} security considerations (2019) Wi-Fi Alliance. https://www.wi-fi.org/file/wpa3-security-considerations
83. WPA3^{TM} specification, version 3.0 (2020) Wi-Fi Alliance. https://www.wi-fi.org/download.php?file=/sites/default/files/private/WPA3_Specification_v3.0.pdf
84. Yin L, Fang B, Guo Y, Sun Z, Tian Z (2020) Hierarchically defining internet of things security: from CIA to CACA. Int J Distrib Sensor Networks 16(1)
85. ZigBee specification (2015) ZigBee Document 05-3474-21, ZigBee Alliance, Davis, CA. https://zigbeealliance.org/wp-content/uploads/2019/11/docs-05-3474-21-0csg-zigbee-specification.pdf

Chapter 17
Intelligent Solutions for Attack Mitigation in Zero-Trust Environments

Mikhail Zolotukhin, Timo Hämäläinen, and Pyry Kotilainen

Abstract Many of today's smart devices are rushed to market with little considera-tion for basic security and privacy protection, making them easy targets for various attacks. Therefore, IoT will benefit from adapting a zero-trust networking model which requires strict identity verification for every person and device trying to access resources on a private network, regardless of whether they are located within or out-side of the network perimeter. Implementing such model can, however, become challenging, as the access policies have to be updated dynamically in the context of constantly changing network environment. In this research project, we are aim-ing to implement a prototype of intelligent defense framework relying on advanced technologies that have recently emerged in the area of software-defined networking and network function virtualization. The intelligent core of the system proposed is planned to employ several reinforcement machine learning agents which process current network state and mitigate both external attacker intrusions and stealthy advanced persistent threats acting from inside of the network environment.

Keywords Network security · Deep learning · Reinforcement learning · Software-defined networking

17.1 Introduction

Increasing computing and connectivity capabilities of smart devices in conjunction with users and organizations prioritizing access convenience over security makes such devices valuable asset for cyber criminals. The intrusion detection in IoT is

M. Zolotukhin (✉) · T. Hämäläinen · P. Kotilainen
Faculty of Information Technology, University of Jyväskylä, Jyväskylä, Finland
e-mail: mikhail.m.zolotukhin@jyu.fi

T. Hämäläinen
e-mail: timo.t.hamalainen@jyu.fi

P. Kotilainen
e-mail: pyry.kotilainen@jyu.fi

403

limited due to lack of efficient malware signatures caused by diversity of processor architectures employed by different vendors [1]. In addition to that, owners use mostly manual workflows to address malware-related incidents and therefore they are able to prevent neither attack damage nor potential attacks in the future. Furthermore, since not all devices support over-the-air security updates, or updates without downtime, they might need to be physically accessed or temporarily pulled from production. Thus, many connected smart devices may remain vulnerable and potentially infected for long time resulting in a material loss of revenue and significant costs incurred by not only device owners, but also users and organizations targeted by the attackers as well as network operators and service providers. A potential solution to these and other emerging challenges in IoT is employing zero-trust networking model, that implies that all data traffic generated must be untrusted, no matter if it has been generated from the internal or external network [8].

In this research, we aim to design and implement an intelligent zero-trust networking solution capable of detecting attacks initiated by both external attackers and smart devices from the inside, adapt detection models under constantly changing network context caused by adding new applications and services or discovering new vulnerabilities and attack vectors, make an optimal set of real-time crisis-action decisions on how the network security policy should be modified in order to reduce the ongoing attack surface and minimize the risk of subsequent attacks in the future. These decisions that may include permitting, denying, logging, redirecting, or instantiating certain traffic between end-points under consideration, are based on behavioral patterns observed in the network and log data obtained from multiple intrusion and anomaly detectors and deployed on the fly with the help of cutting-edge cloud computing technologies such as software-defined networking and network function virtualization. Our implementation of the decision making mechanism in the system proposed is planned to rely on recent advances in reinforcement learning (RL), machine learning paradigm in which software agents automatically determine the ideal behavior within a specific context by continually making value judgments to select good actions over bad. RL algorithms can be used to solve very complex problems that cannot be solved by conventional techniques as they aim to achieve long-term results correcting the errors occurred during the training process.

Recent advent of cutting-edge technologies such as cloud computing, mobile edge computing, network virtualization, software-defined networking (SDN) and network function virtualization (NFV) have changed the way in which network functions and devices are implemented, and also changed the way in which the network architectures are constructed. More specifically, the network equipment or device is now changing from closed, vendor specific to open and generic with SDN technology, which enables the separation of control and data planes, and allows networks to be programmed by using open interfaces. With NFV, network functions previously placed in costly hardware platforms are now implemented as software appliances located on low-cost commodity hardware or running in the cloud computing environment. In this context, the network security service provision has shifted toward replacing traditional proprietary middle-boxes by virtualized and cloud-based network functions in order to enable automatic security service provision.

Software-defined perimeter (SDP) is an architecture for zero trust that borrows concepts from SDN and NFV. An SDP controller functions as a broker of trust between a client and a gateway, which can flexibly establish a transport layer security tunnel terminating on the gateway inside the network perimeter, allowing access to applications. Each device establishes a unique VPN tunnel with the service that is requested, and the origin is cloaked from public view. Each device establishes a unique VPN tunnel with the service that is requested, while the origin is cloaked from public view. SDP relies on the concepts of network access control in an attempt to minimize the impact of existing and emerging network threats by adding authentication of the hosts. Similar to micro-segmentation, SDP enforces the principle of only providing access to the services that are required. Besides this authentication function the SDP controller can enforce authorization policies that may include host type, malware checks, time of day access, and other parameters. The data plane will typically rely on an overlay network to connect hosts via VPN tunnels. However, SDP approach has several drawbacks. For example, an SDP implementation usually requires usage of specific hardware and software gateways and controller appliances. Gateways may be needed at each site where applications are located making the deployment, management, and maintenance of this infrastructure challenging, especially in large globally distributed, high availability environments. In addition, security appliances are supposed to be configured to accept connections and allow traffic from the SDP gateways. Intrusion detection system and firewall rules introduce complexity, holes in the perimeter, and added IT maintenance. In this research project, we focus on solving these drawbacks with the help of state-of-art machine learning techniques.

The purpose of this study is to highlight the implementation process of the defense framework proposed. The rest of the document is organized as follows. In Sect. 17.2, we evaluate various deep learning algorithms needed for implementation. Reinforcement learning algorithms are discussed in Sect. 17.3. Traffic generation problem is addressed in Sect. 17.4. Section 17.5 outlines implementation of SDN flows and security VNFs. The resulting system prototype is discussed in Sect. 17.6. Section 17.7 concludes the report and outlines future work.

17.2   Intrusion Detection with Deep Learning

Artificial intelligence and deep learning are revolutionizing almost every industry with a seemingly endless list of applications ranging from object recognition for systems in autonomous vehicles to helping doctors detect and diagnose cancer. This list includes multiple branches of the field of cyber-security that include intrusion detection, malware classification, network traffic analysis and many others. A deep neural network consists of multiple layers of nonlinear processing units. The main idea behind deep learning is using the first layers to find compact low-dimensional representations of high-dimensional data whereas later layers are responsible for achievement of the task given, e.g., regression or categorical classification. All the neurons of the layers are activated through weighted connections. In order the network being capable to approximate a nonlinear transformation, a non-linear activation function is applied to the neuron output. The learning is conducted by calculating error in the output layer and backpropagating gradients towards the input layer. In regular deep neural network layer, each neuron in a hidden or output layer is fully connected to all neurons of the previous layer with the output being calculated by applying the activation function to the weighted sum of the previous layer outputs. Such layers have few trainable parameters and therefore learn fast compared to more complicated architectures.

To evaluate deep learning model capabilities to detect intrusions we use network packet captures from CICIDS2018 [17] dataset. It contains 560 Gb of traffic generated during 10 days by 470 machines. The dataset in addition to benign samples includes following attacks: infiltration of the network from inside, HTTP denial of service, web, SSH and FTP brute force attacks, attacks based on known vulnerabilities. We concentrate on the intrusion detection based on the analysis of network traffic flows. A flow is a group of IP packets with some common properties passing a monitoring point in a specified time interval: IP address and port of the source and IP address and port of the destination. Resulting flow measurements provide us an aggregated view of traffic information and drastically reduce the amount of data to be analyzed. After that, two flows such as the source socket of one of these flows is equal to the destination socket of another flow and vice versa are found and combined together. This combination is considered as one conversation between a client and the server. A conversation can be characterized by following four parameters: source IP address, source port, destination IP address and destination port.

For each such conversation, at each time window, or when a new packet arrives, we extract the most essential features including flow duration, total number of packets in forward and backward direction, total size of the packets in forward direction, minimum, mean, maximum, and standard deviation of packet size in forward and backward direction and overall in the flow, number of packets and bytes per second, minimum, mean, maximum and standard deviation of packet inter-arrival time in forward and backward direction and overall in the flow, total number of bytes in packet headers in forward and backward direction, number of packets per second in forward and backward direction, number of packets with different TCP flags, backward-to-forward number of bytes ratio, average number of packets and bytes transferred in bulk in the forward and backward direction, the average number of packets in a sub flow in the forward and backward direction, number of bytes sent in initial window in the forward and backward direction, minimum, mean, maximum and standard deviation of time the flow is active, minimum, mean, maximum and standard deviation of time the flow is idle [17]. All the features can have different scale and therefore they are supposed to be standardized.

In our numerical experiments, we process raw packet capture files. First, we extract necessary packet features, then combine separate packets into conversations and, after that, we extract conversation features. It is worth noticing, that every time a new packet is transferred during the conversation or a certain time period (one second in our case) passes, we recalculate the conversation features and add a new

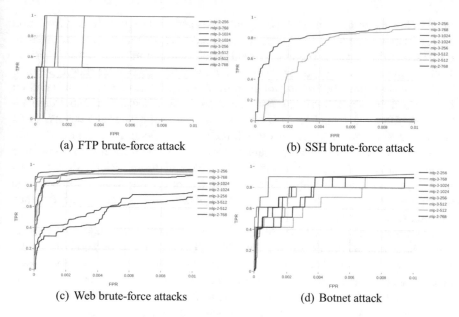

(a) FTP brute-force attack   (b) SSH brute-force attack

(c) Web brute-force attacks   (d) Botnet attack

Fig. 17.1 Dependence of TPR on FPR for different intrusion detection with deep learning models applied to flow features

data sample for the updated conversation. The idea behind that is that we attempt to evaluate how well the deep learning methods can detect intrusions in real time not when the conversation is over. Some results are presented in Fig. 17.1.

As one can notice from the figures, basic neural networks allow us to detect malicious connections without many false alarms. Results for the classification models slightly vary in terms of true and false positive rates depending on the architecture. It is also worth noticing that increasing the number of trainable parameters does not improve accuracy of the models significantly. In the sense of efficiency, simple MLPs look the most promising solution. It is worth noticing that we also experimented with more complicated neural network layers, e.g., residual [7] and self-attention [19], but for our classification task those do not provide any increase in the detection accuracy. We also tested unsupervised models such as autoencoders, however we did not manage to obtain good results using those.

17.3 Deep Reinforcement Learning

Reinforcement learning is a machine learning paradigm in which software agents and machines automatically determine the ideal behavior within a specific context by continually making value judgments to select good actions over bad. A reinforcement learning problem can be modeled as Markov Decision Process (MDP) that includes three following components: a set of agent states and a set of its actions, a transition probability function which evaluates the probability of making a transition from an initial state to the next state taking a certain action, and an immediate reward function which represents the reward obtained by the agent for a particular state transition. If the transition probability function is known, the agent can compute the solution before executing any action in the environment. However, in real-world environment, the agent often knows neither how the environment will change in response to its action nor what immediate reward it will receive for executing the action. It is not enough to only account the immediate reward of the current state, the far-reaching rewards should also be taken into consideration. Most of the time RL algorithms focus on the optimization of infinite-horizon discounted model, implying that the rewards that come sooner are more probable to happen, since they are more predictable than the long term future reward.

There are three main approaches for the reinforcement learning: value-based, policy-based and model-based. In value-based RL, the goal is to maximize the value function which is essentially a function that evaluates the total amount of the reward an agent can expect to accumulate over the future, starting at a particular state. The agent then uses this function by picking an action at each step that is believed to maximize the value function. On the other hand, policy-based RL agent attempts to optimize the policy function directly without using the value function. The policy function in this case is the function that defines the action the agent selects at the given state. Finally, model-based approach focuses on sampling and learning the probabilistic model of the environment which is then used to determine the best actions the agent can take. Assuming the model of the environment has been properly learned, model-based algorithms are much more efficient than model-free ones, however, since the agent only learns the specific environment model, it becomes useless in a new environment and requires time to learn another model.

As a rule, a neural network is used to estimate an RL agent's policy with loss function being estimated based on the probability of the action taken multiplied by the cumulative reward obtained from the environment. Updating the policy network parameters by taking random samples may introduce high variability in probabilities and cumulative reward values, because trajectories during training can deviate from each other at great degrees. This results in unstable learning and the policy distribution skewing to a non-optimal direction. One way to reduce variance and increase stability is subtracting the value function from the cumulative reward. This allows one to estimate how much better the action taken is compared to the return of an average action. The value function can be estimated by constructing the second neural network, which estimates the environment's state value in the manner similar to DQN. The resulting architecture is called advantageous actor-critic (A2C), where the critic estimates the value function, while the actor updates the policy distribution in the direction suggested by the critic [11].

To improve stability of the learning even further, trust region policy optimization (TRPO) relies on minimizing a certain surrogate objective function that guarantees policy improvement with non-trivial step sizes [15]. TRPO uses average KL divergence between the old policy and updated policy as a measurement for a region around

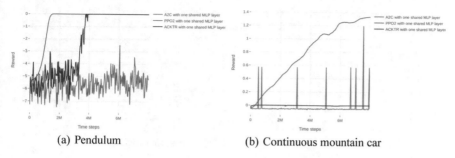

(a) Pendulum   (b) Continuous mountain car

Fig. 17.2 Performance of three different RL algorithms in two basic OpenAI gym environments

the current policy parameters within which they trust the model to be an adequate representation of the objective function, and then chooses the step to be the approximate minimizer of the model in this region. Although TRPO has achieved great and consistent high performance, the computation and implementation of it is extremely complicated. The current state-of-art algorithm policy optimization (PPO) attempts to reduce the complexity of TRPO implementation and computation by tracing the impact of the actions with a ratio between the probability of action under current policy divided by the probability of the action under previous policy and artificially clipping this value in order to avoid having too large policy update [16]. Another option to reduce the complexity closer to a first-order optimization is proposed in [20]. Actor-critic using Kronecker-Factored trust region (ACKTR) speeds up the optimization by reducing the complexity using the Kronecker-factored approximation.

To evaluate performance of different RL algorithms, we use OpenAI gym that has emerged recently as a standardization effort [3]. We run multiple copies of the environment in parallel. The training process is divided into episodes. Each episode lasts a certain fixed amount of time steps, during which one of the tasks is performed by an agent implemented using OpenAI baselines [4]. The tasks include swinging an inverted pendulum up from a random position and moving a two-dimensional car. We test three state-of-art RL algorithms A2C, ACKTR and PPO in those environments. We concentrate on these algorithms as they can be applied for both discrete and continuous environments. In our experiments, PPO consistently provides good results in terms of both average reward and convergence speed (see Fig. 17.2). We run several experiments with different network architectures. The results in Fig. 17.3 show that the network with one shared layer followed by two separate streams for policy and value function looks the most promising architecture variant. We also experimented with using shared LSTM layer for both policy and value function, but the results showed that much more steps is required for the algorithm convergence in this case, which can be critical in case of more complicated environment that requires more time per iteration.

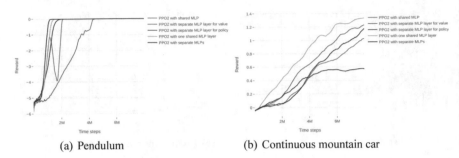

(a) Pendulum   (b) Continuous mountain car

Fig. 17.3 Performance of PPO2 with different policy and value network architectures in several OpenAI gym environments. Architectures: shared MLP (blue), one separate layer for value (green), one separate MLP layer for policy (red), one shared MLP layer (yellow), separate MLPs (purple)

17.4 Traffic Generation

In order to train the reinforcement learning agents a simulation environment is supposed to be constructed since we cannot deploy, train and test those agents in a real production network. Traffic in such environment can be attempted to be generated with the help of conditional generative adversarial networks (GANs) [6]. In GANs, the discriminator generates an estimate of the probability that a given sample is real or generated. The discriminator is supplied with a set of samples which include both real and generated ones and it would generate an estimate for each of these inputs. The error between the discriminator output and the actual labels would then be measured by cross-entropy loss. GAN can be extended to a conditional model if both the generator and discriminator are conditioned on some extra information [10]. We can perform the conditioning by feeding this information into the both the discriminator and generator as additional input layer. In our case, this extra information includes several packets sent in a network traffic flow, and the GAN is trained to generate features of the next packet, i.e. its payload size, TCP window size, TCP flags, inter-arrival time, etc. However, cross-entropy loss fails in some cases and not point in the right direction in other cases. This may lead to mode collapse when the generator only learns a small subset of the possible realistic samples which the discriminator cannot recognize. One potential solution for this problem is using Wasserstein distance metric [2]. The Wasserstein metric looks at the distribution of each variable in the real and generated samples, and determines how far apart the distributions are for real and generated data. The Wasserstein metric looks at how much effort, in terms of mass times distance, it would take to push the generated distribution into the shape of the real distribution.

We use conditional Wasserstein GANs to generate inter-arrival time between two consecutive packets, payload size and TCP window size, the second generates n-gram distribution for the payload of the packet. Features for the condition include direction (request or reply) and TCP flags. In the generator network, a random noise vector is concatenated with the condition and the result is fed to an MLP, output of which is a feature vector for the next packet. The discriminator network also takes

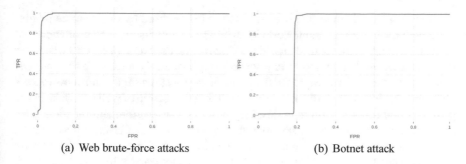

(a) Web brute-force attacks   (b) Botnet attack

Fig. 17.4 Dependence of TPR on FPR for different intrusion detection with deep learning models applied to flow features extracted from fake traffic generated with Wasserstein GANs

features extracted from the previous packets of the flow as an input. The second input is the feature vector generated by the generator. The generator produces packets that are closer to the real ones extracted from the datasets while the discriminator network tries to determine the differences between real and fake packets. The goal is to have a generative network which can produce traffic flows whose features resemble the ones extracted from the real flows.

We trained such GANs separately for different attacks presented in the dataset and the normal HTTP traffic. Figure 17.4 shows the results of applying the classifiers trained in the previous stage to the traffic generated with GANs. As one can see, the results for models which are trained with flow features are more or less inline with the ones obtained using the real data.

We implement an application for traffic generation in form of a Docker container. Docker allows users to package an application with all of its dependencies into a standardized unit for software development. Unlike virtual machines, containers do not have high overhead and hence enable more efficient usage of the underlying system and resources. The container includes client and server application implemented with Scapy module which is able to forge or decode packets of a wide number of protocols. We set the first ECN bit of each packet generated with the generator trained using malicious traffic to one in order to be able to provide ground truth labels for the AI in order to calculate the reward. Generator models of the trained GANs are first converted into a compressed .tflite format and added to the application. This has been done in order to deploy the trained model without installing the entire TensorFlow library.

17.5   Software-Defined Networking and Network Function Virtualization

The main purpose of the SDN controller in our defense framework, which is to transform the security intent of the AI core to SDN flows and push them to the switches, can be implemented as an internal module of an existing SDN controller or an external application that uses RESTful APIs exposed by one or more plugins existing in the controller framework. There are many open-source controller options currently available, that can be modified in order to be used to redirect traffic between devices under protection and virtual security appliances. According to several SDN controller surveys, OpenDayLight [9] is one of the most featured controllers that are able to run on different platforms. Being under the partnership of well-known network providers and research communities, they have a clear development plan and good documentation. Even though Java-based OpenDayLight is inferior in performance compared to the controllers implemented in C in terms of throughput, they perform on similar level in terms of latency [14], which is alongside with high modularity and proper documentation makes it the most optimal choice to serve as an SDN controller in the defense system proposed.

Once all the necessary features have been installed on the OpenDaylight controller, we implement a simple application for receiving information from operational data store of the controller and manipulating its data stores, which include pushing a flow into a switch table, finding existing tables on a switch, finding existing flows on a switch, deleting flow from a switch table and deleting entire table from a switch. These functions allow us to setup basic network configurations by resubmitting a packet with certain Ethernet protocol to another table, replying to an ARP request with a MAC address, redirecting an ARP packet with certain target protocol address to a port, redirecting an IP packet with certain destination to a port, outputting an IP packet with certain source to a port and resubmit to another table, modifying ECN of an IP packet with certain destination to a port and resubmit to another table. The purpose of the last action is to change the second ECN bit of a packet to one when it arrives at the first switch and change it back to zero when it sent from the last switch. The idea is to account for packets which are dropped in the environment in order to calculate the impact of the defense framework.

Concerning the virtual security functions, there are many open-source intrusion detection and packet inspection software available that can be implemented as security middle boxes for timely attack detection and mitigation. We implement our own security middle box in order to use deep learning models trained. For that purpose, we first install OpenVSwitch on an Ubuntu virtual machine and connect it to our Open-Daylight controller. It can then be connected to other network switches via VXLAN tunnels. For intercepting and analyzing network traffic we use Libnetfilter_queue [12] and Iptables firewall rules to gain access to network packets and the ability to reject or accept these packets for forwarding. Python library Netfilterqueue is used to interface with Libnetfilter_queue [13] from a python program. The interceptor program receives a network packet and extracts relevant features from it and uses pre-trained classifier to determine whether it is malicious or not. The malicious flows are flagged by setting a desired bit in the TCP-protocol DSCP field (upper 6 bits of the TOS field) facilitating detection and further actions by downstream devices. All packets are then forwarded regardless of the analysis result.

Similarly to the traffic generation containers we use TensorFlow Lite interpreter in order to avoid installing the entire TensorFlow library. We select the best classifier in terms of the metric selected (e.g., accuracy or AUC) for each attack class tested

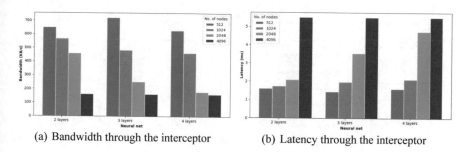

(a) Bandwidth through the interceptor   (b) Latency through the interceptor

Fig. 17.5 Network performance of the interceptor with varying number of layers and nodes in each layer

and copy those models to the VNF. We finally implement a simple API using Flask which allows to manipulate two parameters: classifier model to use for the analysis and the threshold according to which we differentiate normal traffic from malicious one. This is basically a very straight forward implementation of transfer learning approach, i.e. a model developed for a task is reused as the starting point for a model on a second task. We train a model using traffic contained in the dataset and then use all its layers except the last one as a foundation for the classifier inside our VNF. The last layer is essentially one number since we only classify traffic as either normal or malicious. Thus, an intelligent agent can manipulate VNfs implemented by selecting the most optimal combination of the classifier and the value of the threshold.

The effect on network performance can be measured by setting up a three machine virtual network: two virtual machines in separate subnets and a virtual machine running the interceptor program acting as a router between the subnets. The software tool Qperf [5] is used to analyze network performance. A Qperf server is started on one of the test machines and a client running a test against the server was started on the other. All traffic passed through the interceptor machine and through the analysis. The recorded metrics are bandwidth and latency of TCP traffic. The test is repeated several times with the interceptor analyzing packets with several different classifiers. The tested configurations have 2, 3 or 4 layers of 512, 1024, 2048 or 4096 nodes. The results are shown in Fig. 17.5. These results accompanied with the ROC curves obtained previously allow us to conclude that it is reasonable to use MLP classifiers with less trainable parameters, since increasing the number of trainable parameters does not improve accuracy of the models significantly, however it negatively affects the network performance.

17.6 Prototype Environment

The biggest drawback of the reinforcement learning approach is its hunger for data: RL methods require to interact with the environment at each new training iteration. In order to train an intelligent agent in a reasonable amount of time, the training process

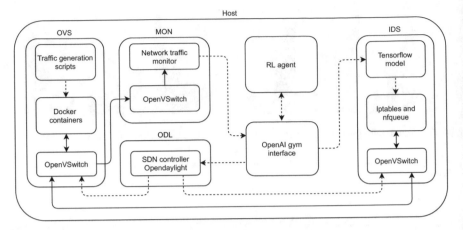

Fig. 17.6 The environment for training RL agent implemented using Vagrant. Network traffic flows and commands are shown by solid and dashed lines respectively

can be carried out in several environments in parallel. For this reason, we build our training environment as a network of several virtual machines using Vagrant. Vagrant [18] is an open-source program which allows for automatic building and managing virtual machines. Vagrant uses existing hypervisors, in our case Qemu/KVM through Libvirt, to deploy and run the machines. Vagrant manages these machines through SSH connections and can provide access to files for the virtual machines through NFS shares. We use Vagrant to configure the VMs required for the system implementation which include SDN controller, several VMs with Docker containers, several VMs with TensorFlow network traffic flow classifiers, and one VM for traffic monitoring. All VMs except for the controller have Openflow-enabled OpenVSwitch pre-installed. Switches are connected between each other with VXLAN tunnels. It is worth noticing that switch of the traffic monitor is not controlled by OpenDaylight, it simply acts as a "sink" for the network traffic in order to provide information on the network state for an RL agent. We use OpenAI gym [3] to implement the frontend for the virtualized environment. The RL agent is implemented using OpenAI baselines [4]. The resulting environment is shown in Fig. 17.6.

The RL agent observes packet and byte counts sent from one host to another for each pair of hosts in the environment, one-hot encoded indexes of the classifiers deployed in the security boxes, threshold values used in the classifiers. Action set of the RL agent consists of changing the classifier model index for a certain VNF, changing the threshold for a certain VNF, redirecting traffic between a certain pair of subnets having certain DSCP label to a certain middle box, and blocking traffic between a certain pair of subnets having certain DSCP label. For the reward function calculation, we utilize our flow monitor VM. Since all malicious traffic packets generated have the first ECN bit equal to 1 and all packet which are received on the first switch have the second ECN bit equal to one, we can calculate percentages of the normal and malicious traffic which are blocked by the environment. The reward

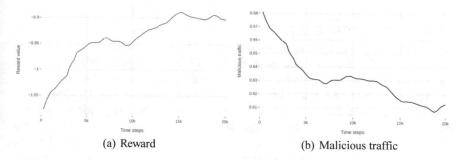

(a) Reward   (b) Malicious traffic

Fig. 17.7 Reward function during the training and percentage of the malicious traffic received by the attack targets

function is then calculated as a sum of these two components. We initialize flow tables of each SDN switch with basic flows to forward each packet to its destination. SDN flows to drop packets or redirect them to a particular security appliance are pushed to the dedicated flow tables with priority higher than default forwarding rules.

In order to evaluate the framework proposed, we consider the following attack scenario. Eight devices are connected to the internal network. These devices can be accessed via SSH and HTTP by both internal and external hosts. To generate malicious traffic, we generate three types of the attacks: SSH password brute-force, web application password brute-force and communication between C&C and one of the devices which is considered infected. The training process is divided into episodes. Each episode lasts for one minute, during which both benign and malicious traffic flows are generated. The RL agent is implemented using OpenAI baselines. The agent selects one of the actions for one of the flows that are sent to the environment back-end where they are transformed to SDN rules. We train the RL agent using PPO algorithm with multi-layer perceptron (MLP) as both policy and value function to detect and mitigate the attacks mentioned. Figure 17.7 shows the evolution of the reward function throughout few training episodes in the attack scenario mentioned. As one can notice, the agent starts to identify and block malicious connections reducing the number of malicious flows and subsequently increasing the reward value.

17.7 Conclusion and Future Work

The main contribution of this research is developing a proof-of-concept of an intelligent network defense system which relies on SDN and NFV technologies and allows customers to detect and mitigate attacks performed against their smart devices by letting an artificial intelligent agent control network security policy. On infrastructure level, the defense framework proposed includes cloud compute servers in order to emulate elements of real infrastructure as well as launch security appliances. In order to forward traffic from the network under protection to these appliances as well as connect the appliances to each other, the system relies on SDN capabilities that include global visibility of the network state and run-time manipulation of traffic forwarding rules. The key component of the defense system proposed is a reinforcement learning agent that resides on top of the SDN and NFV controllers and is responsible for manipulating security policies depending on the current network state. In particular, the agents processes traffic flowing through edge switches as well as log reports from security appliances deployed and manipulates the network traffic by instructing the SDN controller to pass, forward or block certain connections. We used the resulting prototype to evaluate two state-of-art reinforcement learning algorithms for mitigating three basic network attacks against a small virtual network environment.

There are, however, still numerous issues which have to be addressed. Those are mostly related to the traffic generation procedure. We managed to implement simple traffic generation application, it, however, does not really represent the realistic traffic and therefore its usage is limited in a real world scenario. A potential solution would be to use real devices and applications and generate traffic using those. In the future, we are planning to continue this research by conducting experiments in the environment prototype, as well as testing various reinforcement learning algorithms for different attack scenarios. We are also aiming to improve the scalability of the framework proposed and evaluate the system performance for bigger network environments. We are also going to implement adversarial module for the traffic generators which would allow for spoofing a neural-network-based intrusion detection system by manipulating flow parameters. Finally, we are going to test the working prototype of the network defense system developed during the project in a non-SDN enterprise network environment.

References

1. Alhanahnah M, Lin Q, Yan Q, Zhang N, Chen Z (2018) Efficient signature generation for classifying cross-architecture IoT malware. In: 2018 IEEE conference on communications and network security (CNS), pp 1–9
2. Arjovsky M, Chintala S, Bottou L (2017) Wasserstein generative adversarial networks. In: ICML'17: Proceedings of the 34th International conference on machine learning, vol 70, pp 214–223
3. Brockman G, Cheung V, Pettersson L, Schneider J, Schulman J, Tang J, Zaremba W (2016) OpenAI Gym. arXiv:1606.01540
4. Dhariwal P, Hesse C, Klimov O, Nichol A, Plappert M, Radford A, Schulman J, Sidor S, Wu Y, Zhokhov P (2017) OpenAI baselines. GitHub repository. https://github.com/openai/baselines
5. GitHub (2018) GitHub—linux-rdma/qperf. Retrieved 1 Dec 2020, from https://github.com/linux-rdma/qperf
6. Goodfellow IJ, Pouget-Abadie J, Mirza M, Xu B, Warde-Farley D, Ozair S, Courville A, Bengio Y (2014) Generative adversarial nets. In: NIPS'14: Proceedings of the 27th International conference on neural information processing systems, vol 2, pp 2672–2680
7. He K, Zhang X, Ren S, Sun J (2016) Deep residual learning for image recognition. In: 2016 IEEE conference on computer vision and pattern recognition (CVPR), pp 770–778

8. Kindervag J (2010) No more chewy centers: introducing the zero trust model of information security. Forrester Research. https://media.paloaltonetworks.com/documents/Forrester-No-More-Chewy-Centers.pdf

9. Medved J, Varga R, Tkacik A, Gray K (2014) OpenDaylight: towards a model-driven SDN controller architecture. In: Proceeding of IEEE international symposium on a world of wireless, mobile and multimedia networks 2014, pp 1–6

10. Mirza M, Osindero S (2014) Conditional generative adversarial nets. arXiv:1411.1784

11. Mnih V, Badia AP, Mirza M, Graves A, Lillicrap TP, Harley T, Silver D, Kavukcuoglu K (2016) Asynchronous methods for deep reinforcement learning. arXiv:1602.01783

12. Netfilter (2020) The netfilter.org "libnetfilter_queue" project. Retrieved 1 Dec 2020, from https://www.netfilter.org/projects/libnetfilter_queue/index.html

13. PyPI (2017) NetfilterQueue—PyPI. Retrieved 1 Dec 2020, from https://pypi.org/project/NetfilterQueue/

14. Salman O, Elhajj IH, Kayssi A, Chehab A (2016) SDN controllers: a comparative study. In: Proceedings of the 18th Mediterranean electrotechnical conference (MELECON 2016), pp 1–6

15. Schulman J, Levine S, Moritz P, Jordan MI, Abbeel P (2015) Trust region policy optimization. arXiv:1502.05477

16. Schulman J, Wolski F, Dhariwal P, Radford A, Klimov O (2017) Proximal policy optimization algorithms. arXiv:1707.06347

17. Sharafaldin I, Lashkari A, Ghorbani A (2018) Toward generating a new intrusion detection dataset and intrusion traffic characterization. In: Proceedings of the 4th International conference on information systems security and privacy (ICISSP 2018), pp 108–116

18. Vagrant (2020) Vagrant by HashiCorp. Retrieved 19 Nov 2020, from https://www.vagrantup.com

19. Vaswani A, Shazeer N, Parmar N, Uszkoreit J, Jones L, Gomez AN, Kaiser Ł, Polosukhin I (2017) Attention is all you need. In: NIPS'17: Proceedings of the 31st International conference on neural information processing systems, pp 6000–6010

20. Wu Y, Mansimov E, Liao S, Grosse R, Ba J (2017) Scalable trust-region method for deep reinforcement learning using Kronecker-factored approximation. arXiv:1708.05144

Chapter 18
Insecure Firmware and Wireless Technologies as "Achilles' Heel" in Cybersecurity of Cyber-Physical Systems

Andrei Costin

Abstract In this chapter, we analyze cybersecurity weaknesses in three use-cases of real-world cyber-physical systems: transportation (aviation), remote explosives and robotic weapons (fireworks pyrotechnics), and physical security (CCTV). The digitalization, interconnection, and IoT-nature of cyber-physical systems make them attractive targets. It is crucial to ensure that such systems are protected from cyber attacks, and therefore it is equally important to study and understand their major weaknesses.

Keywords Cybersecurity · Firmware · Binare · Vulnerabilities · Exploits · Reverse engineering · Protocols · Cyber-physical systems · Critical infrastructure · Aviation · ADS-B · Video surveillance · CCTV · Remote firing systems · Wireless pyrotechnics · RF · Zigbee

18.1 Introduction

The exponential growth of the (industrial) Internet of Things and the increasing digitalization of everything in modern life will inherently lead to an increase in the number of cyber-physical system applications. Such systems, which have the most direct interaction with the physical world, are mainly in the medical, industry/energy, transportation, safety/surveillance, entertainment, and military sectors. This means that they have a somewhat direct connection or control over the (quality of) lives of many people. Often, however, such systems are not (completely) secure. In addition, as we demonstrate, in most cases, the main cause of insecurity lies in firmware or wireless communications.

This chapter is organized as follows. In Sect. 18.2, we analyze the cyber-physical system (CPS) related to the aviation transport. We will focus in particular on the next-generation ADS-B system, which will be used for radars, situational awareness, air-traffic control, and air-traffic management. In Sect. 18.3, we perform cybersecurity

A. Costin (✉)
University of Jyväskylä, Jyv äskyl ä, Finland
e-mail: ancostin@jyu.fi

© The Author(s), under exclusive license to Springer Nature Switzerland AG 2022   419
M. Lehto and P. Neittaanmäki (eds.), *Cyber Security*, Computational Methods
in Applied Sciences 56, https://doi.org/10.1007/978-3-030-91293-2_18

analysis and attack implementations for a wireless firing system used in fireworks pyrotechnics. In Sect. 18.4, we survey vulnerabilities in and attacks against CCTV and Video Surveillance Systems (VSS).

18.2   ADS-B in Air Transport

In this section, our main focus is to demonstrate the ease, feasibility, and practicality of ADS-B compared to previous works that covered the theoretical aspects of ADS-B insecurity. To this end, we set up a practical, cost-effective, and moderately sophis- ticated attack against the next-generation ADS-B technologies, which are expensive and safety-critical. Although the use of a manual validation procedure [3] can partial mitigate attacks, conducting attacks on air traffic controllers and aircraft in continu- ous and/or decentralized manner greatly increases the potential for human error. For example, repeated erroneous messages on the air traffic control display and critical response time requirements affect the security of the entire system.

This section is based on the author's original work in [23].

18.2.1   ADS-B in General

Automatic Dependent Surveillance-Broadcast (ADS-B) is an Air Traffic Manage- ment and Control (ATM/ATC) surveillance system intended to replace traditional radar-based systems. It is expected to be an essential part of the next generation (NextGen) air transport system. The concept behind ATM, ATC and ADS-B is quite simple and can be summarized as follows. ADS-B avionics transmits plain text, unencrypted error-protected messages over radio transmission links approximately once per second. These messages include aircraft location, velocity, identification, and other air traffic control information.

ADS-B can be used for many purposes. It is useful in

- Improving air-traffic management and control security,
- Improving the detection and resolution of air-traffic conflicts,
- Optimizing and condensing air traffic.

ADS-B aims to dramatically improve pilots' situational awareness by providing them with access to real-time air-traffic information similar to that of air traffic controllers. For example, they receive information about other aircrafts as well as weather and terrain. ADS-B lets pilots know the position of the aircraft they are flying in relation to other aircraft without recourse to the infrastructure.

A traditional passive radar system has a relatively low resolution. Moreover, with traditional radars, the accuracy of the position depends on the distance to the plane. Radars are also usually unable to provide altitude information. ADS-B has much better coordinate accuracy and an effective range of 100–200 nautical miles [36].

Therefore, it is expected that ADS-B will allow for much better use of airspace by allowing the distance between planes to be reduced, especially near busy airports.

18.2.2   ADS-B in Detail

ADS-B operates on the following radio frequencies:

- 1030 MHz for active interrogation, for example from ATC towers, radars or other aircraft, and
- 978MHz/1090 MHz for active response or normal transmissions, for example from aircraft or, less frequently, from airport vehicles.

For interoperability, regulation, and tradition, ADS-B is supported by two different data connections, specifically 1090 MHz Mode-S Extended Squitter (1090ES) and 978 MHz Universal Access Transceiver (UAT). As part of the next generation ATM system, ADS-B will be developed and deployed in conjunction with Flight Information Services-Broadcast (FIS-B) and Traffic Information Service-Broadcast (TIS-B). Both FIS-B and TIS-B can be susceptible to attacks similar to those described in this section. However, such protocols are used for less critical data processing, so we have not investigated actual feasibility of the attacks but left it to be done in future work.

In terms of active response and normal broadcast, the role of the unit in the ADS-B architecture can be either a transmitter called ADS-B OUT or a transceiver called ADS-B IN. Currently, most aircraft are designated as transmitters and equipped with ADS-B OUT technology. Therefore, their role in ADS-B is to broadcast their location information for further analysis and compilation at ATC towers and ATM stations. ADS-B IN technology is currently used mainly in ATC towers. As one of the most advertised benefits of the ADS-B is the superior situational awareness it provides to the pilot of an aircraft, testing of the ADS-B IN has begun on aircraft. According to [35], SWISS is a pioneer in the use of ADS-B IN in Europe and one of only five airlines in the world to participate in the Airborne Traffic Situational Awareness (ATSAW) project. ADS-B IN is intended to enable ATSAW, spacing, separation and self-separation applications. However, from a security point of view, ADS-B IN technology in aircraft brings new challenges. Examples of challenges include verifying Online 2 reliably and real-time validation of identity, location, and flight route from a received broadcast. While the situation is well controlled at an ATC station on the ground where high-speed connection is not a problem, control is more difficult on an aircraft.

The ADS-B protocol is encapsulated in Mode-S frames. The ADS-B uses Pulse-Position Modulation (PPM) and the responses/transmissions are encoded in a certain number of pulses, each pulse being 1.0 μs long. Therefore, the data rate of the ADS-B is 1.0 Mbit/s. The response/transmission frames consist of a preamble and a data-block. A preamble of length 8.0 μs is used to synchronize transmitters and receivers. It consists of four pulses, each of 0.5 μs in length, with intermediate space (relative to the first pulse) of 1.0, 3.5, and 4.5 μs. The ADS-B protocol does not specify whether Collision Detection (CD) or Collision Avoidance (CA) is used on medium radio frequency (especially considering that the transmission is plain text and digitally unsigned wireless transmission channel is used as the channel). The data blocks are either 56-bit or 112-bit and are used to encode various Downlink Format (DF) messages. DF packets are used by a receiver, which is usually an airplane, Unmanned Aerial Vehicle (UAV), or Unmanned Aircraft System (UAS). Uplink Format (UF) messages are usually are sent by a ground station (e.g., air traffic control tower, UAT tower), but can also be sent by another airplane UAV, or UAS [e.g., Traffic Collision Avoidance System (TCAS), Airborne Collision Avoidance System (ACAS)]. Related to this study, the most interesting DFs are DF11 (Mode S Only All-Call Reply) and DF17 (1090 Extended Squitter or 1090ES).

The secure Mode-S/ADS-B mode, used in the military, is encoded in DF19 Military Extended Squitter, DF22 for military use only (discussed in [44, 71]), and cryptographically coded Mode-5, which uses enhanced cryptography based on time of day and direct sequence spread spectrum modulation as specified in NATO STANAG 4193 [76] and ICAO's Annex 10 [4]. To our knowledge, the exact specifications of DF19, DF22, and Mode-5 are not public at the time of writing.

As the ADS-B is intended to support mission-critical automated and human decision-making and has a direct impact on overall air safety, it is imperative that the technology behind the ADS-B meets operational performance and security requirements. However, the main problem with ADS-B is the lack of security mechanisms, specifically lack of

- entity authentication to protect against messages send by unauthorized entities,
- message integrity checking [e.g., digital signatures, Message Authentication Codes (MAC)] to protect against message forgery or aircraft impersonation,
- message encryption to protect against eavesdropping,
- challenge-response mechanisms to protect against recurrence attacks,
- ephemeral identifiers to protect against privacy tracking attacks,
- prevention of jamming, although we did not include Denial of Service (DoS) (e.g., by jamming with radio signals) because it affects RF communication in general and is not specific to ADS-B alone.

Surprisingly, despite years of standardization [89–93], development, and thorough testing, the ADS-B protocol used in commercial air-traffic does not specify mechanisms to ensure that protocol messages are authentic and non-replayed, or that they comply with other security requirements.

18.2.3   ADS-B Attacker and Threat Models

Building the right attacker model is essential when assessing their potential actions in the system. In the ADS-B system, an attacker can be classified using several factors, such as his/her place in the system, physical position, and goals.

The attacker's place in the system can be *external* or *internal*. An external attacker is more likely. As an outsider, he/she does not require authentication or authorization, so he/she can easily execute low-cost attacks. This type of attacker can virtually belong to any group of the Classification III-A3. An internal attacker (insider) is a person the system trusts. For example, he/she could be a pilot, an air traffic controller, an airport technician, etc. This type of attacker is encountered less frequently. He/she is mostly observed in intentional or unintentional prankster group, e.g., [29].

An attacker can be physically located on the ground or in the air. Ground attackers are most commonly analyzed. Various detection and mitigation techniques can be used against their attacks. Airborne attackers are still ignored, and such attacks may not be well understood and modeled. However, taking advantage of technological advances, they can use drones, UAVs, automatically activated luggage check-in, or passenger miniature devices capable of performing attacks.

The attacker's motivation/goal may be a prank, abuse, crime, or military intelligence. Pranks are usually considered the least offensive. However, the impact on safety can be significantly greater than expected. Attackers can include, for example, unaware pilots, "curious" and unaware technical experimenters. Abusers can have a variety of motivations, such as money, fame, message conveying. They can be invasive of privacy (e.g., paparazzi) or even pilots who intentionally abuse their access to ADS-B technology (e.g., by sending obscenities [29], drawing obscene trajectories [2]). Criminals usually have two main motivations: money and/or terror. Attackers conducting military intelligence may have state-level motivations, such as espionage, sabotage, etc. Attacks may target military intelligence agencies as well as nation states.

During the development and deployment of ADS-B, both academia and industry sought to create threat and vulnerability models to develop mitigation techniques and solutions. A wide range of identified and described threats can be found throughout the literature:

- jamming, denial of service,
- eavesdropping,
- spoofing, impersonation,
- message injection/replay
- message manipulation.

18.2.4   Implementation of a Wireless Attack

A similar hardware and software environment is required to trigger and demonstrate a potential attack. Next, we present the hardware and hardware settings, as well as the software modules we have used to carry out the attacks and exploits.

As the main hardware support, we used a radio device defined by USRP1 software [110]. The USRP was combined with an SBX transceiver daughter board [94] covering the frequency range of 400 MHz–4.4 GHz. This was a good enough combination for 1030 MHz interrogation and 1090 MHz response frequencies. In addition, the transmission and reception chains could be controlled separately to provide greater flexibility for the scenarios being tested. To assess the correctness of our implementation and the effectiveness of the attacks, we used the PlaneGadget ADS-B virtual radar [85]. It is an enthusiast-level ADS-B receiver chosen for its good price-quality ratio. However, a large number of similar ADS-B receivers are currently available and any of them could be used in such an experimental setup.

We used the open GNU Radio software package [1] as the main software base. GNU Radio is a FOSS implementation of several basic radio technologies that are useful for higher-level SDR design and applications. In particular, it provides very good software support for USRP1 and USRP2. We used USRP hardware in Universal Hardware Driver (UHD) mode, which is recommended because it supersedes the original hardware mode. In addition to the PlageGadget, we also used our USRP1 as a secondary ADS-B receiver as well as a backup device. Using USRP1 as an ADS-B IN device requires demodulation and decoder support. Fortunately, there are two public implementations of the Mode-S/ADS-B receiver module for GNU Radio. Eric Cottrell performed the historic first implementation of the Mode-S/ADS-B demodulator and decoder for pre-UHD-mode. The latest implementation for UHD-mode was done by Nick Foster [39]. Since the USRP1 was in UHD-mode, we used the gr-air-modes software module [39].

For reproducible attacks, we used the out-of-box functions of USRP1 and GNU Radio. Thus, our approach at the frame level is as follows:

- Capture ADS-B using uhd_rx_cfile at 1090 MHz;
- In UHD-mode, use TX samples to transmit reproducible captured data via GNU Radio;
- Or in pre-UHD mode, use usrp_replay_file.py to transmit reproducible captured data via GNU Radio.

(Please define TX samples!)

For message impersonation attacks, i.e. spoofing, it is necessary to implement ADS-B for PPM encoding and PPM modules. As usual, there are several ways to accomplish this. One of them is writing the original C/C++-based GNU Radio modulator and encoder [87]. Another approach we used is to perform most of the encoding and modulation in MatLab. In outline, we follow these steps:

1. Encode the detailed ADS-B data into a MatLab array as a bitstream;
2. Modulate it using PPM's modulate() function with a ppm argument;
3. Or read I/Q formatted data into MatLab (or Octave) using read_float_binary.m and modify the downloaded data;
4. Write the modulated data to I/Q format using write_float_binary.m;
5. In UHD-mode, use TX samples to transmit the modulated data via GNU Radio;
6. Or in pre-UHD-mode, use usrp_replay_file.py to transmit the modulated data via GNU Radio.

18.2.5 Key Results

Section 18.2 clearly verifies the inherent insecurity in the design of the commercial ADS-B protocol. Despite the fact that security vulnerabilities in ADS-B technology have been widely covered in previous academic studies and more recently in the hacking community, fundamental problems in the architecture and design of ADS-B have never been addressed and fixed. Given the time and money invested so far and still to be invested, it is unclear why such a mission-critical safety protocol does not address safety at all and there is not even a security chapter in the main requirements specification document [93].

In conclusion, the most important and intended contribution of this study is to raise awareness among academia, industry, and policy makers that critical infrastructure technologies, such as ADS-B, require real security to operate safely and in accordance with requirements. We can do this by showing that a low-cost hardware setup combined with moderate software in multi-million dollar technology is enough to expose the system to dangerous security and operational failures while failing to take advantage of basic security mechanisms such as message authentication.

18.3 Wireless Firing Systems for Remote Explosives and Robotic Weapons

In this section, we examine the risks of the firing system. We describe our experience in discovering and exploiting a wireless firing system in a short period of time without prior knowledge of such systems. We demonstrate our methodology starting from firmware analysis to discovering vulnerabilities. Our static analysis helped us acquire a system suitable for the purpose, which we then analyzed in depth. This allowed us to confirm the presence of exploitable vulnerabilities in the actual hardware. Finally, we stress the security of hardware and software, as well as the need to monitor the safety of the use of pyrotechnic firing systems.

This section is based on the author's original work in [24].

18.3.1 Main Motivations

Fireworks are mainly explosives for entertainment purposes. A fireworks event, also called a pyrotechnic show or fireworks show, is a demonstration of the effects produced by fireworks devices. Fireworks devices are designed to produce, among other things, noise, light, smoke, and floating materials (e.g., confetti). Fireworks events and fireworks devices are controlled by fireworks firing systems. In addition to fireworks, firing systems often serve other primary industries as well. These include special effects production and military training or simulation.

Despite the fact that fireworks are intended for celebrations, their usage is often associated with a high risk of destruction, injury, and even death. Many recent news and studies show the dangers of fireworks [30, 83]. Sometimes fireworks are even used as real weapons in street clashes [108]. Fireworks accidents are often the result of improper handling of equipment, non-compliance with safety regulations, or poor quality fireworks. Fatal consequences can also occur when CPS-style systems with software defects are connected to ammunition/explosive firing systems [14]. Another risk factor is that fireworks are generally intended to be displayed in densely populated areas. Accidents continue to occur despite the strict control of the distribution of fireworks and the mandatory professional license of a fireworks shooter.

Classically, fireworks firing systems consist of mechanical or electrical switches and electrical wires (often called shooting wires). This type of setup is simple, efficient and relatively safe [38]. However, it dramatically limits the effects, complexity, and implementation of fireworks systems and events. The development of software, embedded and wireless technologies can be fully utilized in fireworks systems. A modern (wireless) firing system is at the same time a complete Embedded Cyber-Physical System (ECPS) and a combination of Wireless Sensor/Actuator Network (WSAN). As fireworks firing systems increasingly rely on wireless, embedded, and software technologies, they are exposed to the same risks as other ECPS, WSAN, or computer systems. Recent research has shown that both critical and embedded systems have acquired a poor security reputation. For example, airplanes can be fooled by new radar systems [23], car control can be taken over [18, 65], car driving can be compromised by failure [55], an implanted insulin pump can be made to malfunction [86], or nuclear plant PLCs can be rendered inoperative [37, 68].

18.3.2 Overview of Fireworks and Pyrotechnics Systems

The pyrotechnics of fireworks is typically composed of:

- Remote control modules,
- Firing modules,
- Wired connections,
- Wireless transceivers,
- Igniter clips,
- Mortars,
- Pyrotechnic devices.

Remote control modules (sometimes also called main controls) control the entire show, which includes sequencing cues and the transmission of fire commands. They connect to the firing modules via wired or wireless connections. In simple systems, one remote control module is connected to all firing modules, while in more complex shows, there are several remote control modules, each of which is connected to a specific set of firing modules depending on the show. All remote control modules work independently. These devices rely on a micro-controller embedded in its own firmware.

The *firing modules* receive fire commands from the remote control modules and activate the minimum ignition current to the igniter clips. The firing modules are based on micro-controllers and have their own firmware. *Wired connections* are described here for completeness, but in our case, all remote control and firing modules were wireless. Classic fireworks firing systems consist of electrical wiring between the remote control and the firing modules [38]. Simple connection cables with End-Of-Line (EOL) resistors are used for secure termination of wire loops. EOL resistors allow the remote control to detect wiring problems or tampering in short circuit situations while monitoring field wiring.

Wireless transceivers enable wireless connections between remote control modules and firing modules. These connections are usually implemented with 433.92 MHz modules (often capable of using rolling codes [10]), or 2.4 GHz ZigBee-compliant (IEEE 802.15.4) modules that support AES according to the standard.

The *igniter clips* connect the firing modules to the pyrotechnic devices inside the mortar. They ignite a fire when the firing module activates the minimum current. *Mortars* contain pyrotechnic devices. They also ensure the safe launching and firing of the pyrotechnic device into the sky. *Pyrotechnic devices* are actual pyrotechnic compositions that produce visual and sound effects in the sky after a firing.

18.3.3 Preliminary Analysis

First, we performed a large-scale firmware analysis by gathering firmware images from the Internet, reaching 172,000 firmware candidates [26]. Once the firmware images were unpacked, we processed each image with simple static analysis, correlation, and reporting tools, leading us to discover 38 previously unknown vulnerabilities. In the process, we accidentally discovered firmware images for the wireless firing system. We omit the name of the vendor and the system for safety and ethical reasons. Analysis of the firmware images for that system revealed components (strings, binary codes, configurations) that appeared insecure. The findings were convincing enough, so we acquired the devices for a detailed analysis. Another motivating factor for the acquisition was that, according to the vendor, this system is used by "over 1000 customers in over 60 countries". These systems seem to be particularly popular in fireworks companies.

18.3.3.1 Firmware Analysis

Our crawlers collected, among others, several Intel Hexadecimal Object File (Intel Hex) firmware images dedicated to the wireless firing system from the Internet. After unpacking, we used several heuristics, including keyword matching. Keyword matching searches for specific keywords such as backdoor, telnet, UART, shell, which often allows to find multiple vulnerabilities. The firmware images matched with the string Shell.

Based on this, we isolated those firmware images and further analyzed them using automated and manual approaches. We detected several security issues from the analyzed images. First, the Intel Hex format alone does not provide encryption or authentication, so the functionality is openly explorable by an attacker and thus likely to be open to malware. In addition, the Intel Hex format provides attackers with mechanisms to insert code or data to memory regions that may not be designed to be accessed.

18.3.3.2   Wireless Communication Analysis

Wireless communication systems, like many others from other vendors, include a 2.4 GHz ZigBee (IEEE 802.15.4) CEL MeshConnect transceiver. Discovering, configuring, installing and pairing these units, as well as updating the firmware, is done through Synapse Portal [101]. We installed Synapse Portal and then ran a discovery and configuration query.

The wireless chipsets for the remote control, firing, and firmware reprogramming modules include AES-128-compatible firmware. However, encryption is not enabled, the encryption key is not present, and the AES-128 appears to be unused. In addition, the system documentation does not appear to support AES-128-secured configuration steps. Surprisingly, even if those devices conform to the standards and have AES-128 capabilities, message authentication or encryption is not used. This is likely due to difficulties in properly configuring key management and distribution. Thus, when used in this way, AES-128 carries the risk of functional failure to the fireworks rather than acts as a safety mechanism.

Further analysis revealed that it is possible to load Python application code into wireless remote chipsets. These scripts are executed in a Python interpreter on a wireless chipset microcontroller (MCU), see [100]. The provided interpreter framework is a subset of Python. Before downloading to target nodes, Synapse Portal compiles these Python scripts in binary format and stores them as SNAPpy files (with extension .spy), see [102]. The binary format is assigned to a specific MCU that drives a particular wireless chipset. These scripts expose the entry-points (functions) that other wireless nodes can call remotely (via RPC). Scripts can interact with the wireless chipset MCU or General-Purpose Input/Output (GPIO) ports. Usually, these GPIO ports are connected to the main MCU of the remote control or firing module. This allows interaction with main MCUs as well as IO peripherals such as buttons, displays and igniter clips. The typical use of script entry-points is as follows. The remote control module processes CSV orchestration scripts. When it decides that a fire command is required, it sends a ZigBee packet containing a higher-level message to a specific entry-point of a particular remote module.

The usual standard firing procedure is as follows:

1. Each firing module is connected to a specific remote control module.
2. The physical keys of the firing modules are turned to standby mode.

3. Staff move to the statutory safety distance to fire cues.
4. The keys for the remote controls are turned on.
5. After making sure everything is safe and ready, the staff presses the power button on the remote control. The remote, in turn, sends a wireless digital command to the firing module, which enters standby mode for incoming fire commands.
6. Staff begin the show by sending commands to each shooting module, either manually or according to the script.

Each firing module accepts arming, disarming, and firing commands only from its paired remote control. The pairing is forced by checking the remote control's 802.15.4 short address (similar to MAC address filtering).

18.3.4 Wireless Threats

The lack of encryption and mutual unit authentication opens up the system to multiple attacks, particularly sniffing, spoofing, and replaying. We describe a simple attack, however, that we consider the most dangerous to the fireworks show staff.

The attacker would proceed as follows. He/she eavesdrops on packets (broadcasts, multicasts, node-to-node) by learning from them the 802.15.4 addresses of each remote control and firing modules and the corresponding pairing. For each pair learned, the attacker spoofs the 802.15.4 addresses on the remote control and the digital arm command sent to the paired firing module, and immediately sends a fire command from all cues when the digital arm confirmation comes from the firing module. As a result of such an attack, when the show operator turns the physical key of a firing module to the arming position, that firing module immediately receives a series of digital arming and firing commands from all cues. This fires all pyrotechnic loads and in the worst case does not give the staff enough time to move to a safe distance. Thus, it overrides the security of the physical key and the separation of functions. We successfully implemented this attack on the systems we acquired using the components described in this section.

Alternatively, an attacker could easily replace the default Python functions responsible for firing cues with arbitrary malicious Python functions. For example, each malicious firing cue function could fire a firing module from all of the cues at once instead of its own cue, which could cause a massive chain explosion. Or it cannot fire from the cues at all or fires randomly, leaving the fireworks show below expectations. Last but not least, an attacker can remotely set random encryption keys on remote nodes. This would mean a denial of service to the legitimate user, as his/her devices would no longer be able to communicate with other devices used for fireworks. This can definitely ruin a holiday party or harm competitors in professional fireworks competitions.

18.3.5   Implementing a Wireless Attack

We implemented simple attacks, such as *message replay* and *unauthorized message injection* (e.g., the command "fire all"). However, it is obvious and trivial to extend the implementation to automatically and continuously sniffing out new firing modules and subsequently spoofing remote control sequences. Next, we present details of the software and hardware we used to carry out the attacks.

SNAP Stick SS200 The SNAP Stick SS200 [103] is firmware software primarily for remote control and firing modules and is based on Atmel's well-known ATmega128RFA1 chipset. Using the SNAP Portal utilities and its own firmware (Synapse ATmega128RFA1 Sniffer), the SNAP Stick SS200 can be converted to a SNAP-specific 802.15.4 sniffer that sniffs and decodes 802.15.4 packets based on Synapse's higher-level protocol semantics (e.g., multicast, broadcast, peer-to-peer or multicast RPC calls). We used it to sniff and record packets between the remote control and firing modules during their normal operation. Finally, we also used it to validate packet injection and replay attacks. If this sniffer received them, the remote control and firing modules would see our rogue packets. Otherwise, we had to fix our injector (regardless of whether our lower-level raw packet sniffer saw them) and then re-test the sniffed packets and the actual behavior of the devices.

GoodFET GoodFET [43] is an embedded bus adapter for various microcontrollers and radios, while also providing great open source support for advanced attacks. Its TelosB-compatible firmware allows sniffing, among other functionalities. We tested our attack with GoodFET firmware running on TelosB.

KillerBee KillerBee [64] is a framework and tools for exploiting ZigBee and 802.15.4 networks. It provides convenient pre-compiled GoodFET firmware for extra attack functionality. We tested our attack with such GoodFET firmware running on TelosB.

TelosB An sniffer based on SS200 is useful for SNAP protocols and visualization, but it filters and strips down the packets, which is largely limiting. We needed a lower level raw packet sniffer. We also needed an cheap and open source supported approach. Crossbow's TelosB [104] hardware and GoodFET firmware fit perfectly, so we used them as an additional, much more verbose and raw sniffer. After learning the SS200 higher-level packets for critical commands, we correlated them with the raw packets recorded by TelosB (running GoodFET firmware). Alternatively, Zigduino [118] could have been used for this task.

Econotag Redwire Econotag is an inexpensive and convenient open source platform for 802.15.4 networks. We assembled sequences of packets that sent commands to arm and fire from the remote control to the firing module. Finally, we encoded an infinite loop of these sequences in custom firmware. Once plugged, Econotag performs an attack on the firing module when its key is turned to the physical arm position. Alternatively, Zigduino [118] could have been used for this task as well.

18.3.6 Main Outcomes

We were able to quickly and automatically isolate the firmware of critical remote firing systems and identify several potential vulnerabilities using both automatic and manual static analysis. These vulnerabilities include unauthorized firmware updates, unauthenticated wireless communications, wireless communications sniffing and spoofing, arbitrary code injection, functionality trigger, and temporary denial of service. We have successfully implemented and tested an unsophisticated attack that can have devastating consequences. Our conclusion is that, given the risk posed by use, the security of wireless firing systems should be taken very seriously. We also conclude that such systems need to be more rigorously certified and regulated.

We stress the need and urgency to introduce software and hardware compliance verification similar to that of the DO-178B and DO-254 respectively. We strongly believe that these small improvements, along with the suggested solutions, can definitely help improve the security and safety of wireless embedded systems. Last but not least, we discussed the issues with the vendor. The firmware update now deployed fixes most security issues. Unfortunately, due to more than 20 vendors, wireless firing systems may be vulnerable to similar attacks, especially those for which a firmware update is not available.

18.4 CCTV for Physical Security

Video surveillance, Closed-Circuit Television (CCTV), Digital/Network Video Recorder (DVR/NVR), and IP-camera (IPcam) systems[1] have become very common all over the world. Currently, the use of VSSs is central to most, if not all, areas of life in modern society. They are used very widely, from law enforcement and crime prevention to transport safety, traffic monitoring, and industrial process and retail control. Unfortunately, their unauthorized [109, 116], illegal [117], and even criminal [63] use is also common. Their number is incredibly large; in some reports it is estimated at 245 million cameras/systems [56]. It is expected that by 2021, there will be more than a billion CCTV cameras worldwide [20].

This section is based on the author's original work in [21, 22].

18.4.1 CCTV in General

Most of the concerns about video surveillance systems are related to privacy protection for obvious reasons. Improving the privacy of VSSs is particularly important in the light of global surveillance revelations, and specifically video surveillance scandals [31]. However, in addition to privacy issues, an insecure or compromised VSS can raise a myriad of other non-privacy issues. For example, data breaches

---

[1] We call such a system a Video Surveillance System (VSS).

were shown to endanger prison security [63], pose theft risks to money-based institutions such as banks [6] and casinos [117], emotionally affect people (especially children) [54], and interfere with police and law enforcement [80].

At a time when embedded devices are increasingly being analyzed on a large scale for security vulnerabilities [26, 27], it is no surprise that security researchers have dramatically increased their focus on VSSs [21, 52, 70, 81, 96]. These and similar studies found more than a handful of vulnerabilities [7, 8, 16, 17, 33, 34, 53, 59, 60, 73, 77, 111] that have a large-scale impact in real life [61, 114]. The number of vendors and the variety of vulnerabilities revealed in the investigations clearly indicate the unhealthy state of cyber-security in video surveillance systems.

18.4.2 Visual-Layer Attacks

Compared to other embedded systems, video surveillance systems have an additional level of abstraction, i.e. *visual layer*. Therefore, it is possible to (ab)use this layer to carry out novel attacks on video surveillance systems that take advantage of imaging semantics and image recognition. Costin [21] first presented such an attack on CCTV cameras as the back door of the visual layer. Mowery et al. [75] carried out a similar attack on a full-body scanner as a secret knock-on image.

This attack is multi-stage and works on the visual layer as follows. In the first stage, the VSS is infected with a malicious component (e.g., hardware, firmware). In some scenarios, this can be achieved locally via a malicious firmware update over a USB port and remotely via a command injection or a malicious firmware update over a web interface. In other scenarios, a VSS or CCTV system with pre-installed malware could be sold through legitimate sales channel [78, 113]. In the second stage, the malicious component is triggered and controlled through the input of a malicious image that is "visualized" by the cameras and video sensors.

In the most general case, the trigger command can be coded in any arbitrary data-to-image encoding scheme.[2] First, a malicious component could be pre-programmed to blur an attacker's face or the license plates of an attacker's car, or to disable certain functions of a surveillance system (e.g., video recording functions or scanning a prohibited object such as a gun in a full-body scan [75]). Such malicious functions could be used for theft and other crime. Second, the malicious component could read commands from QR-like codes. Malicious images [62] could be printed on t-shirts, cars, or any accessory that is sufficiently visible to cameras. The command could be "stop recording", "blur the face of an attacker with a malicious image/QR-code", "contact the command and control center" or "update malicious components". A variation of such an attack was carried out in the hacking of Google Glass [45]. It used a specially crafted QR code as malicious image input to control (unauthorized and unattended) Google Glass and force it to visit a malicious URL.

---

[2] QR-codes are a popular implementation of such data-to-image encoding schemes.

Optical covert channel techniques could be used to hide the visual layer attack and the resulting load from human operators. Taking advantage of the camera's sensitivity to the infrared and near-infrared spectra, an attacker could send "invisible" information. An attacker could also use techniques similar to VisiSploit [46], except that the channel would be used to inject data and commands and not to exfiltrate data.

Finally, visual-layer attacks are certainly not far-fetched. Because visual layer information is processed at a certain point (e.g., image compression, face recognition, Optical Character Recognition (OCR)), both intentional and unintentional errors can occur. An infamous example of an unintentional error is Xerox scanners and copiers that randomly altered document numbers and data [66]. Because incredibly complex processing (e.g., image compression, face recognition, Automatic License Plate Reading (ALPR)) is built into modern video surveillance systems, it is reasonable to assume that similar (both intentional and unintentional) problems in the visual processing layer can allow an attack against them as well.

18.4.3   Covert-Channel Attacks

In recent years, covert channels and data exfiltration (especially in air-gap environments) have been the subject of productive research. The channel used can be electromagnetic [47, 48, 67, 112], acoustic [50, 51, 79], thermal [49, 72], or optical [46, 69, 88, 95]. With regard to VSS and CCTV systems, we will introduce one novel covert channel and look more broadly at the use of several existing covert channels. Although the channels we present can mainly be used to exfiltrate data using the compromised VSS and CCTV component [78, 113], they can also be used for autonomous and distributed command-and-control functions.

18.4.3.1   Normal and Infrared LEDs

In modern electronic equipment, such as device status indicators, LED lights have been used repeatedly in covert channels and data exfiltration [19, 69, 95]. Smart LED bulbs have also recently been shown to pose similar threats [88]. Although LEDs are sometimes physically connected to hardware and cannot be controlled from software/firmware, recent attacks show that manipulating LEDs from software/firmware is becoming increasingly practical and feasible [13]. VSS and CCTV systems usually have plenty of status LEDs on both core equipment and outdoor CCTV cameras. Therefore, LEDs in VSS and CCTV systems could also be used in data exfiltration attacks.

There is one major drawback to (ab)using normal LEDs in such attacks. If the LEDs are handled in an eye-catching way (e.g., abnormal blinking frequencies, unusual luminosity levels), they are distinguishable to the human eye, allowing the covert channel to be exposed. Therefore, we propose the use of InfraRed (IR) LEDs in optical covert channels. IR-LED arrays are installed in almost any modern CCTV camera. IR LEDs are used for illumination and provide IR night vision for cameras and VSSs. One important characteristic of IR LEDs is that when they operate, they are often invisible.[3] For example, another camera without IR cut-off filters (e.g., another IR-compatible CCTV camera) must be used to detect the operation of the IR LEDs. Therefore, IR-compatible CCTV cameras can use the intensity of IR LEDs (or their on and off mode) to modulate and exfiltrate data. Such exfiltration would be invisible to the human eye.

Ambient lighting can affect the success of an attack. When it is dark, changing the intensity/status of the IR LEDs is immediately reflected in the surveillance camera image, so operating personnel may notice that something is wrong. When the environment is lighted, changes to the IR LEDs would not be very visible in the surveillance camera image, but an attacker could still intercept the exfiltrated data remotely.

18.4.3.2   Covert Channels

Recently, Guri et al. [46] presented VisiSploit, a new type of optical covert channel that, taking advantage of the limitations of human visual perception, leaks data imperceptibly through the LCD display of a standard computer. Most VSS and CCTV systems are connected to screens that are fully or partially visible to the public. These screens display real-time images from one or more cameras in the system. For example, this is especially popular in supermarkets to deter shoplifting and to help staff early detect potential illegal or unethical activity. VSS and CCTV systems can also be seen to be used in this way in the operational centers of large car parks, in the reception lobbies of organizations (e.g., companies, hotels, elite residences) and in many other places. Therefore, the compromised VSS and CCTV component could use screens installed in this way in conjunction with VisiSploit techniques to exfiltrate the data.

Steganography is the art of hiding information inside other information (e.g., images, documents, media streams, or network protocols). Although many different "carrier media" can be used for this purpose, digital images are the most popular due to their prevalence on the Internet and their concealment efficiency. A comprehensive overview of image steganography is presented in [74, 82]. A special feature of VSS and CCTV systems is that virtually all systems provide both video and image streams [32]. Image streams can be either motion images (e.g., MJPEG) or still snapshots and can usually be accessed in URLs such as http://CAM-IP/now.jpg, http://CAM-IP/shot.jpg, or http://CAM-IP/img/snapshot.cgi?size=2.

---

[3] Almost always invisible, but it also depends on characteristics of the IR LEDs used. Here, we assume that it is difficult, if not impossible, for the human eye to easily distinguish between normal and abnormal use of IR LEDs.

Therefore, the compromised VSS component (e.g., CCTV camera, DVR, NVR) can exfiltrate the data employing steganography when generating the above-mentioned images/image streams. The attacker then only needs to capture digital snapshots of well-known URLs and recover the exfiltrated data. Whether an attacker has access to image streams and how he/she can access is not the purpose of this section. However, recent projects such as TRENDnet Exposed [106], Insecam [57], Shodan images [97], corroborate studies such as [28], which demonstrate that it is feasible, even very easily in VSS and CCTV systems protected by current cyber security practices. To prevent exfiltration of data in steganography, as discussed above, automated methods could be used to detect steganography [12, 41].

18.4.3.3 Mechanical Movement and Position of the CTTV Camera

Many modern CCTV cameras have so-called Pan-Tilt-Zoom (PTZ) functionality. With PTZ, a CCTV camera can move or stay fixed in almost any direction in 3D (e.g., with pan and tilt functions) and also zoom in and out with multiple zoom factors (e.g., using a high-precision lens). Such functionality is usually implemented with stepper motors built into specific camera models and is generally controlled by PTZ data protocols. PTZ data protocols are byte sequences of commands sent over a communication channel to control pan, tilt, and zoom. PTZ commands are classically sent over RS-422 or RS-485 links, but can also be sent over classical Ethernet and WiFi channels. PTZ commands can be sent to PTZ-compatible cameras from custom PTZ-controllers (e.g., a special joystick keyboard for surveillance personnel) or from software (e.g., OS-specific heavyweight clients or browser-based lightweight clients).

In this context, a compromised CCTV camera can exfiltrate data to an external attacker by encoding data about its position or changes in movement. For example, it could change its normal fixed position to another specific fixed position that would encode a certain value. Assume that a compromised camera on the wall in its normal position "looks" *down-and-right*. To exfiltrate data, the compromised camera would then encode:

- bits 00, moving itself to "look" *up-and-right*;
- bits 01, moving itself to "look" *up-and-left*;
- bits 10, moving itself to "look" *down-and-left*.

Adding bits to the data resolution (which increases exfiltration data rate) would increase the number of abnormal positions – just like in Phase-Shift-Keying (PSK) modulation – which would require an attacker to observe the compromised camera more closely from the outside.

Many VSS and CCTV systems are audio-capable, which allows them to record and process one or more audio channels coming from external microphones or microphones built into CCTV cameras. Therefore, a compromised VSS component (e.g., CCTV camera, DVR, NVR) can use the audio layer as a command-and-control channel, for example, by means of *hidden voice command* techniques [15].

18.4.4 Denial-of-Service and Jamming Attacks

We would like to emphasize the importance of Denial-of-Service (DoS) and jamming attacks on video surveillance systems. In this case, the emphasis is on the VSSs as the *final target* of the attack. In cases where VSSs are infected and used in botnets to carry out DDoS attacks on other systems as final targets, VSS plays a role as the source of the attack [116].

In most cases, uninterrupted and untampered operation is critical to video surveillance systems, for example because they are used to monitor and record crimes or other important activities. Producing a DoS attack on a CCTV system for just a minute could cause it to miss an important event, such as an extremely fast bank robbery [6, 105] or a worse crime [63]. While a DoS attack on a home router could be a minor nuisance, DoS attacks on video surveillance systems have a critical impact that needs to be considered in design, evaluation, and testing. However, this in itself is non-trivial, as explained in detail in [42].

18.4.5 Online Network Attacks

The most useful and used feature of a modern video surveillance system is the *plug-and-play* feature for ease of installation and deployment, as well as for *remote access control* and video monitoring. As a result, many video surveillance systems are connected to the Internet [57]. Thus, they are directly exposed to the Internet, often even with default settings and credentials [28]. Therefore, we tried to estimate the number of video surveillance systems on the Internet in order to estimate the magnitude of potential exposure.

For this purpose, we compiled an extensive list of queries about video surveillance systems and then ran the queries in both online services and existing Internet scanning databases. Using the Shodan [98] online service, these queries revealed an incredible amount of over 2.2 million video surveillance systems produced by more than 20 vendors. Using the Internet Census 2012 database [58], these queries returned more than 400,000 video surveillance systems produced by more than 10 suppliers. At the same time, according to some reports [56], in 2014, there were nearly 245 million video surveillance cameras installed in the world. Unsurprisingly, finding, tracking, and publishing[4] online video surveillance systems that are vulnerable, compromised or poorly protect the privacy of their owners has always been an interesting topic of discussion. Projects such as TRENDnet Exposed [106], Insecam [57], Shodan images [97] and EFF ALPR [84] are examples of such initiatives. As a result, these projects received an incredible amount of media attention, public scrutiny, and outrage, again raising the issue of the lack of security and privacy in modern video surveillance systems.

---

[4] Many times along with their screen shots and video feeds.

Cui and Stolfo [28] reported that the inferior 39.72% of the cameras and surveillance systems they analyzed on the Internet in 2010 used default credentials. This basically means that they are completely vulnerable to all kinds of attacks, such as video feed eavesdropping,[5] malicious firmware updates, and DNS hijackings. As a further example, we analyzed a set of firmware images from the DVR system and discovered a full admin back door. We then correlated the identification information extracted from the firmware images with the results of the above queries. The result was more than 130,000 affected devices using an online connection.

Even though some of these systems (i.e., their IP addresses) and vendors may overlap (or cannot be accurately calculated), these results give a lower limit on the vulnerability of video surveillance systems to cybersecurity threats. Running Internet queries and using vulnerability estimations of previous works [28] proved to be a very effective method for estimating the number of potentially exposed and vulnerable video surveillance systems.

18.4.6 Key Takeaways

Section 18.4 provides a systematic review of the security of video surveillance systems, detailing threats, vulnerabilities, attacks, and mitigation. The review is based on publicly available data as well as existing classifications and taxonomies. It provides comprehensive information on how video surveillance systems can be attacked and protected at different levels. This structured information can then be used to better understand and identify the security and privacy risks associated with the development, deployment, and use of these systems.

18.5 Conclusions

In this chapter, we looked in more detail at several CPS use-cases. In Sect. 18.2, we analyzed CPS related to the aviation transport sector and focused in particular on the next-generation ADS-B system used in radars, situational awareness, air-traffic control, and air-traffic management. We have demonstrated through real lab attacks that the wireless communication on which the entire ADS-B system is based is inherently insecure and vulnerable to most wireless attacks (e.g., jamming, eavesdropping, spoofing, impersonation).

In Sect. 18.3 we performed a cybersecurity analysis and attack implementations for a wireless firing system used in fireworks pyrotechnics. These types of CPS are particularly troublesome because they deal directly with explosives, thus threatening human lives and the physical world. We demonstrated that by starting with insecure

---

[5] Practically extensively demonstrated in projects such as TRENDnet Exposed [106], Insecam [57], and Shodan images [97].

firmware, we were able to quickly find cybersecurity issues in wireless communication related to the triggering the explosives. We also demonstrated that carrying out dangerous attacks is relatively easy and feasible even by incompetent attackers.

In Sect. 18.4 we surveyed the vulnerabilities in and attacks on CCTV and video surveillance systems. More than a billion CCTV cameras by 2021 [20] will represent perhaps the largest IoT and CPS attack surface in terms of number of devices. The Mirai botnet fully demonstrated the devastating power of just a relatively tiny fraction of compromised CCTV/DVR/VSS systems [5]. As a CPS, the risks for and from CCTV comes from the direct interaction with the physical world in terms of privacy, face recognition, and (un)lawful surveillance. As discussed, mainly CCTV firmware vulnerabilities (but also wireless attacks against it) are the main cybersecurity risk factors for such systems.

In summary, we have found that CPSs in several critical sectors are prone to security vulnerabilities and attacks. All of the attacks presented can be executed with limited knowledge and affordable hardware/software setups. However, most of the vulnerabilities are in the firmware of the devices or in the wireless communications used in the system.

Last but not least, we invite the interested reader to take a deeper look at the following related works [9, 11, 14, 25–27, 40, 99, 107, 115].

Acknowledgements The author of this chapter would like to acknowledge the contributions of the author's collaborators, editors, editing assistants, and everyone involved in the production of this book. In particular, the author would like to acknowledge the contributions of Prof. Aurélien Francillon (EURECOM) as part of co-authoring the original papers related to Sects. 18.2 and 18.3.

References

1. About GNU Radio. GNU Radio. https://gnuradio.org/about
2. Adjei-Darko K (2020) Pilots draw penis in sky over Russia: investigation over flight path. Nationwide News Pty Limited. https://www.news.com.au/travel/travel-updates/travel-stories/pilots-draw-penis-in-sky-over-russia-investigation-over-flight-path/news-story/4d436870952e06eec36ec70eb8d79298
3. ADS-B radar-like services: preliminary hazard analysis. Capstone Safety Engineering Report #1 1, Federal Aviation Administration (FAA). https://www.faa.gov/nextgen/programs/adsb/Archival/media/SERVOL1.PDF
4. Aeronautical telecommunications. Volume I: Radio navigational aids. Annex 10. International Civil Aviation Organization (ICAO) (2006)
5. Antonakakis M, April T, Bailey M, Bernhard M, Bursztein E, Cochran J, Durumeric Z, Halderman JA, Invernizzi L, Kallitsis M, Kumar D, Lever C, Ma Z, Mason J, Menscher D, Seaman C, Sullivan N, Thomas K, Zhou Y (2017) Understanding the Mirai botnet. In: SEC'17: proceedings of the 26th USENIX conference on security symposium. Berkeley, CA, pp 1093–1110 USENIX Association
6. Aron J (2013) Want to rob a bank? Hack your way in. NewScientist 220(2937)
7. Austin B (2012) Trendnet cameras: i always feel like somebody's watching me. Console Cowboys. http://goo.gl/sYkUAF
8. Austin B (2013) Swann song: DVR insecurity. Console Cowboys. http://goo.gl/oY3z3w
9. Avoine G, Hernandez-Castro J (eds) Security of ubiquitous computing systems. Springer

10. AVR411: secure rolling code algorithm for wireless link. Atmel Application Note 2600E-AVR-07/15, Atmel, 2015
11. Baheti RS, Gill H (2011) Cyber-physical systems. In: Samad T, Annaswamy A (eds) The impact of control technology: overview, success stories, and research challenges. IEEE Control Systems Society, pp 161–166
12. Berg G, Davidson I, Duan MY, Paul G (2003) Searching for hidden messages: automatic detection of steganography. In: Proceedings of the fifteenth conference on innovative applications of artificial intelligence. AAAI, pp 51–56
13. Brocker M, Checkoway S (2014) iSeeYou: disabling the MacBook webcam indicator LED. In: SEC'14: proceedings of the 23rd USENIX conference on security symposium. USENIX Association, Berkeley, CA, pp 337–352
14. Cardenas A, Cruz S (2019) Cyber-physical systems security knowledge area. Issue 1.0, The Cyber Security Body Of Knowledge (CyBOK)
15. Carlini N, Mishra P, Vaidya T, Zhang Y, Sherr M, Shields C, Wagner D, Zhou W (2016) Hidden voice commands. In: SEC'16: proceedings of the 25th USENIX conference on security symposium. USENIX Association, Berkeley, CA, pp 513–530
16. CCTV systems. CVE Details. http://goo.gl/IB1Hk7
17. CCTV systems. InsecureOrg. http://insecure.org/search.html?q=cctv
18. Checkoway S, McCoy D, Kantor B, Anderson D, Shacham H, Savage S, Koscher K, Czeskis A, Roesner F, Kohno T (2011) Comprehensive experimental analyses of automotive attack surfaces. In: SEC'11: proceedings of the 20th USENIX conference on security, vol 4. USENIX Association, Berkeley, CA, pp 447–462
19. Clark J, Leblanc S, Knight S (2009) Hardware Trojan horse device based on unintended USB channels. In: 2009 third international conference on network and system security. IEEE, pp 1–8
20. Cosgrove E (2019) One billion surveillance cameras will be watching around the world in 2021, a new study says. https://www.cnbc.com/2019/12/06/one-billion-surveillance-cameras-will-be-watching-globally-in-2021.html
21. Costin A (2013) Poor man's panopticon: mass CCTV surveillance for the masses. Presentation at POC 2013
22. Costin A (2016) Security of CCTV and video surveillance systems: threats, vulnerabilities, attacks, and mitigations. In: TrustED'16: proceedings of the 6th international workshop on trustworthy embedded devices. ACM, New York, pp 45–54
23. Costin A, Francillon A (2012) Ghost in the air(traffic): on insecurity of ADS-B protocol and practical attacks on ADS-B devices. Presented at Black Hat USA 2012. http://lib.21h.io/library/Y8STRIX5
24. Costin A, Francillon A (2014) A dangerous 'pyrotechnic composition': fireworks, embedded wireless and insecurity-by-design. In: WiSec'14: proceedings of the 2014 ACM conference on security and privacy in wireless and mobile networks. ACM. Short Paper, New York, pp 57–62
25. Costin A, Zaddach J (2018) IoT malware: comprehensive survey, analysis framework and case studies. Presentation at Black Hat USA 2018
26. Costin A, Zaddach J, Francillon A, Balzarotti D (2014) A large-scale analysis of the security of embedded firmwares. In: SEC'14: proceedings of the 23rd USENIX conference on security symposium. USENIX Association, Berkeley, CA, pp 95–110
27. Costin A, Zarras A, Francillon A (2016) Automated dynamic firmware analysis at scale: a case study on embedded web interfaces. In: ASIA CCS '16: proceedings of the 11th ACM on Asia conference on computer and communications security. ACM, New York, pp 437–448
28. Cui A, Stolfo SJ (2010) A quantitative analysis of the insecurity of embedded network devices: results of a wide-area scan. In: ACSAC'10: proceedings of the 26th annual computer security applications conference. ACM, New York, pp 97–106
29. Dodgy callsigns from SkyWest flights. RadioReference.com. https://forums.radioreference.com/threads/dodgy-callsigns-from-skywest-flights.216125/

30. Dolak K, Shaw A (2013) Fireworks mishaps, parade fatalities Mar Fourth of July displays: July 4 celebrations across the country befell accidents and fatalities. ABC News. https://abcnews.go.com/US/fourth-july-accidents-mar-parades-fireworks-displays/story?id=19583533
31. Domain Awareness Center. Oakland Wiki. http://oaklandwiki.org/Domain_Awareness_Center
32. Download the iSpy source code. iSpyConnect.com. https://www.ispyconnect.com/source.aspx. Accessed July 26, 2016
33. DVR systems. CVE Details. http://goo.gl/Xmv1jN
34. DVR systems. InsecureOrg. http://insecure.org/search.html?q=dvr
35. Eco-care reaches new (flight) levels. SWISS Magazine:104–106 (2012)
36. Fact sheet: automatic dependent surveillance-broadcast (ADS-B). Federal Aviation Administration (FAA). https://www.faa.gov/news/fact_sheets/news_story.cfm?newsKey=4172
37. Falliere N, Murchu LO, Chien E (2010) W32.Stuxnet dossier. White paper, Symantec Corporation
38. Fireworks electric firing systems. Skylighter Inc. https://www.skylighter.com/fireworks/how-to/setup-electric-firing-systems.asp (2018)
39. Foster N (2012) gr-air-modes. GitHub. https://github.com/bistromath/gr-air-modes
40. Francillon A, Thomas SL, Costin A (2021) Finding software bugs in embedded devices. In: Avoine G, Hernandez-Castro J (eds) Security of ubiquitous computing systems. Springer, Cham, pp 183–197
41. Fridrich J, Goljan M, Du R (2001) Reliable detection of LSB steganography in color and grayscale images. In: MM&Sec'01: proceedings of the 2001 workshop on multimedia and security: new challenges. ACM, New York, pp 27–30
42. Gasser M (1988) Building a secure computer system. Van Nostrand Reinhold
43. GoodFET. GitHub. https://github.com/travisgoodspeed/goodfet
44. Grappel RD, Wiken RT (2007) Guidance material for Mode S-specific Protocol application avionics. Project Report ATC-334, Massachusetts Institute of Technology
45. Greenberg A (2013) Google Glass hacked with QR code photobombs. Forbes. https://www.forbes.com/sites/andygreenberg/2013/07/17/google-glass-hacked-with-qr-code-photobombs/?sh=62bc8ef57e49
46. Guri M, Hasson O, Kedma G, Elovici Y (2016) VisiSploit: an optical covert-channel to leak data through an air-gap. arXiv:1607.03946
47. Guri M, Kachlon A, Hasson O, Kedma G, Mirsky Y, Elovici Y (2015) GSMem: data exfiltration from air-gapped computers over GSM frequencies. In: SEC'15: proceedings of the 24th USENIX conference on security symposium. USENIX Association, Berkeley, CA, pp 849–864
48. Guri M, Kedma G, Kachlon A, Elovici Y (2014) AirHopper: bridging the air-gap between isolated networks and mobile phones using radio frequencies. In: 2014 9th international conference on malicious and unwanted software: the Americas (MALWARE). IEEE, pp 58–67
49. Guri M, Monitz M, Mirski Y, Elovici Y (2015) BitWhisper: covert signaling channel between air-gapped computers using thermal manipulations. In: 2015 IEEE 28th computer security foundations symposium. IEEE, pp 276–289
50. Guri M, Solewicz Y, Daidakulov A, Elovici Y (2016) Fansmitter: acoustic data exfiltration from (speakerless) air-gapped computers. arXiv:1606.05915
51. Hanspach M, Goetz M (2014) On covert acoustical mesh networks in air. arXiv:1406.1213
52. Heffner C (2013) Exploiting surveillance cameras: like a Hollywood hacker. Presentation at Black Hat USA 2013
53. Hill K (2013) 'Baby monitor hack' could happen to 40,000 other Foscam users. Forbes. http://goo.gl/2cdYy0
54. Hill K (2013) How a creep hacked a baby monitor to say lewd things to a 2-year-old. Forbes. http://goo.gl/92yg9G
55. Hirsch J, Bensinger K (2013) Toyota settles acceleration lawsuit after $3-million verdict. Los Angeles Times. https://www.latimes.com/business/la-xpm-2013-oct-25-la-fi-hy-toyota-damages-20131026-story.html

56. IHS: 245 million surveillance cameras installed globally in 2014. SecurityIn-foWatch.com. https://www.securityinfowatch.com/video-surveillance/news/12082966/245-million-surveillance-cameras-installed-globally-in-2014-ihs-says (2015)
57. Insecam: live cameras directory. http://insecam.org
58. Internet Census 2012: port scanning /0 using insecure embedded devices. http://census2012.sourceforge.net/paper.html (2012)
59. IP cameras. CVE Details. http://goo.gl/ObpWCg
60. IP cameras. InsecureOrg. https://insecure.org/search.html?q=IP%20camera
61. Israeli road control system hacked, caused traffic jam on Haifa highway. The Hacker News. https://thehackernews.com/2013/10/israeli-road-control-system-hacked.html (2013)
62. Kharraz A, Kirda E, Robertson W, Balzarotti D, Francillon A (2014) Optical delusions: a study of malicious QR codes in the wild. In: 2014 44th annual IEEE/IFIP international conference on dependable systems and networks. IEEE, pp 192–203
63. Kidman A (2012) How a prison had its CCTV hacked. http://goo.gl/sKombD
64. KillerBee. Google. http://code.google.com/p/killerbee/
65. Koscher K, Czeskis A, Roesner F, Patel S, Kohno T, Checkoway S, McCoy D, Kantor B, Anderson D, Shacham H, Savage S (2010) Experimental security analysis of a modern automobile. In: 2010 IEEE symposium on security and privacy. IEEE, pp 447–462
66. Kriesel D (2014) Xerox scanners/photocopiers randomly alter numbers in scanned documents. https://www.dkriesel.com/en/blog/2013/0802_xerox-workcentres_are_switching_written_numbers_when_scanning
67. Kuhn MG, Anderson RJ (1988) Soft tempest: hidden data transmission using electromagnetic emanations. In: International workshop on information hiding. Springer, Berlin, pp 124–142
68. Langner R (2011) Stuxnet: dissecting a cyberwarfare weapon. IEEE Secur Privacy 9(3), 49–51
69. Loughry J, Umphress DA (2002) Information leakage from optical emanations. ACM Trans Inf Syst Secur 5(3), 262–289
70. Marpet J (2010) Physical security in a networked world: video analytics, video surveillance, and you. Presentation at Black Hat DC 2010
71. McMath J (2007) Automated dependent surveillance-broadcast military (ADS-M). US Air Force Slides
72. Mirsky Y, Guri M, Elovici Y (2015) HVACKer: bridging the air-gap by manipulating the environment temperature. DeepSec. https://deepsec.net/docs/Slides/2015/Bridging_the_Air-Gap_Data_Exfiltration_from_Air-Gap_%20Networks_-_Yisroel_Mirsky.pdf
73. Moore HD (2013) Ray sharp CCTV DVR password retrieval and remote root. Rapid7. http://goo.gl/Hnp3TO
74. Morkel T, Eloff JHP, Olivier MS (2005) An overview of image steganography. In: Proceedings of the fifth annual information security South Africa conference (ISSA2005)
75. Mowery K, Wustrow E, Wypych T, Singleton C, Comfort C, Rescorla E, Checkoway S, Halderman JA, Shacham H (2014) Security analysis of a full-body scanner. In: SEC'14: proceedings of the 23rd USENIX conference on security symposium. USENIX Association, Berkeley, CA, pp 369–384
76. NATO—STANAG 4193 PT I: technical characteristics of the IFF Mk XIIA system. https://standards.globalspec.com/std/14346734/STANAG%204193%20PT%20I (2016)
77. Obermaier J, Hutle M (2016) Analyzing the security and privacy of cloud-based video surveillance systems. In: IoTPTS'16: proceedings of the 2nd ACM international workshop on IoT privacy, trust, and security. ACM, New York, pp 22–28
78. Olson M (2016) Beware, even things on Amazon come with embedded malware. http://artfulhacker.com/post/142519805054/beware-even-things-on-amazon-come, Apr 2016. Accessed July 25, 2016
79. O'Malley S, Choo KKR (2014) Bridging the air gap: inaudible data exfiltration by insiders. In: Proceedings of the 20th Americas conference on information systems (AMCIS 2014)
80. Owning a cop car. Digitalmunition. http://www.digitalmunition.com/OwningCopCar.pdf
81. Owning big brother (or how to crack into Axis IP cameras). Purple paper, ProCheckUp, London. https://www.procheckup.com/media/1k0fv4mf/vulnerability_axis_2100_research.pdf

82. Provos N, Honeyman P (2003) Hide and seek: an introduction to steganography. IEEE Secur Privacy 1(3), 32–44
83. Puri V, Mahendru S, Rana R, Deshpande M (2009) Firework injuries: a ten-year study. J Plast Reconstr Aesthetic Surg 62(9), 1103–1111
84. Quintin C, Maass D (2015) License plate readers exposed! How public safety agencies responded to major vulnerabilities in vehicle surveillance tech. Electronic Frontier Foundation (EFF). https://www.eff.org/deeplinks/2015/10/license-plate-readers-exposed-how-public-safety-agencies-responded-massive
85. Radar Gadgets Planegadget Radar. RadioPics. http://www.radiopics.com/Flight%20(Air %20Band)/SS-Radar/Planegadget/Planegadget_Radar.htm
86. Radcliffe J (2011) Hacking medical devices for fun and insulin: breaking the human SCADA system. Presentation at Black Hat USA 2011
87. Rondeau T (2012) Re: [discuss-gnuradio] a chunks to symbols related question. Free Software Foundation (FSF). https://lists.gnu.org/archive/html/discuss-gnuradio/2012-01/msg00144.html
88. Ronen E, Shamir A (2016) Extended functionality attacks on IoT devices: the case of smart lights. In: 2016 IEEE European symposium on security and privacy (EuroS&P). IEEE, pp 3–12
89. RTCA DO-242A (2002) Minimum aviation system performance standards for automatic dependent surveillance broadcast (ADS-B). RTCA
90. RTCA DO-249 (1999) Development and implementation planning guide for automatic dependent surveillance broadcast (ADS-B) applications. RTCA
91. RTCA DO-260A (2003) Minimum operational performance standards for 1090 MHz automatic dependent surveillance broadcast (ADS-B) and traffic information services (TIS-B). RTCA
92. RTCA DO-263 (2000) Application of airborne conflict management: detection, prevention, and resolution. RTCA
93. RTCA DO-282B (2009) Minimum operational performance standards for universal access transceiver (UAT) automatic dependent surveillance-broadcast (ADS-B). RTCA
94. SBX 400-4400 MHz Rx/Tx (40 MHz). Ettus Research. https://www.ettus.com/all-products/SBX/
95. Sepetnitsky V, Guri M, Elovici Y (2014) Exfiltration of information from air-gapped machines using monitor's LED indicator. In: 2014 IEEE joint intelligence and security informatics conference. IEEE, pp 264–267
96. Shekyan S, Harutyunyan A (2013) To watch or to be watched: turning your surveillance camera against you. Presentation at HITB security conference 2013, Amsterdam
97. Shodan images. https://images.shodan.io/
98. Shodan search engine. http://www.shodan.io
99. Sun J (2020) The 1090 megahertz riddle: a guide to decoding mode S and ADS-B signals, 2nd edn. TU Delft OPEN Publishing
100. Synapse module comparison chart. Solarbotics. https://solarbotics.com/wp-content/uploads/synapse_comparison_table.pdf
101. Synapse Wireless. Portal: reference manual for version 2.6.6. http://help.synapse-wireless.com/Portal/Portal-Reference-Manual.pdf
102. Synapse Wireless. SNAP network operating system: reference manual for version 2.4. https://cdn.sparkfun.com/datasheets/Wireless/General/SNAP%20Reference%20Manual.pdf
103. Synapse Wireless. SNAP stick user guide (2011). https://usermanual.wiki/Synapse-Wireless/SS200/html
104. TelosB. Crossbow. https://www.willow.co.uk/TelosB_Datasheet.pdf
105. The fastest robbery—1 min in bank. YouTube. http://youtu.be/LFArxqcP4MI, 2012
106. TRENDnet Exposed. https://twitter.com/trendnetexposed. This account doesn't exist!
107. Turtiainen H, Costin A, Hämäläinen T, Lahtinen T (2020) Towards large-scale, automated, accurate detection of CCTV camera objects using computer vision: applications and implications for privacy, safety, and cybersecurity. arXiv:2006.03870

108. Ukraine protests: Kiev fireworks 'rain on police'. BBC News. https://www.bbc.com/news/world-europe-25820899

109. Ullrich JB (2014) This is why your DVR attacked my synology disk station (and now with bitcoin miner!). SANS ISC InfoSec Forums. https://isc.sans.edu/forums/diary/More+Device+Malware+This+is+why+your+DVR+attacked+my+Synology+Disk+Station+and+now+with+Bitcoin+Miner/17879

110. USRP1 (Universal Software Radio Peripheral). Ettus Research. https://www.ettus.com/products/

111. van Berkum M (2011) ABUS TVIP 11550/21550 multiple vulnerabilities (and possibly other ABUS cams). SecurityFocus. http://www.securityfocus.com/archive/1/520045

112. Vuagnoux M, Pasini S (2009) Compromising electromagnetic emanations of wired and wireless keyboards. In: SSYM'09: proceedings of the 18th conference on USENIX security symposium. USENIX Association, Berkeley, CA, pp 1–16

113. Walsh C (2015) Police body cameras infected with conficker worm. https://www.carmelowalsh.com/tag/martel/

114. Welch C (2013) FTC settles with Trendnet after 'hundreds' of home security cameras were hacked. http://goo.gl/94Ibmv

115. Zaddach J, Costin A (2013) Embedded devices security and firmware reverse engineering. Presentation at Black Hat USA 2013

116. Zeifman I, Gayer O, Wilder O. CCTV DDoS botnet in our own back yard. Imperva, Inc. https://www.incapsula.com/blog/cctv-ddos-botnet-back-yard.html

117. Zetter K (2013) Crooks spy on casino card games with hacked security cameras, win $33m. Wired. http://goo.gl/zmxVXe

118. Zigduino r2. Logos Electromechanical. https://www.logos-electro.com/zigduino/

Chapter 19
Physical Weaponization of a Smartphone by a Third Party

Juhani Rauhala

Abstract In the literature and media, the treatment of the dangers and exposures posed by smartphones has generally focused on information security or privacy concerns. There have also been reports of fires, explosions, electric shocks, or loss of phone functionality due to faulty design or manufacture. This article provides an overview of acute physical and physiological dangers of smartphones that could be induced or triggered by a third party. It proposes a categorical discussion framework to describe and define the dangers in terms of attack vectors, effects on the smartphone, harms, and potential culprits/instigators. Counterfeit smartphones are themselves a significant potential threat in this context. Finally, some possible solutions and mitigation are suggested as preventive measures. Some templates for threat assessment forms are also proposed.

Keywords Technology acceptance · Smartphone dangers · Technology abuse · Unorthodox weaponization

19.1 Introduction

It may soon be possible to remotely "self-destruct" a smartphone [19]. Previous reports have shown that ISPs and mobile operators may soon be able to disable smartphones remotely [14]. Smartphone self-destruction differs from remote disablement in that consumers are not only able to disable their device (similar to PIN locking) but also destroy device data and even components at the hardware level [20]. Self-destruction would make the device unusable for a thief, even if a sophisticated thief could override a disabled state to reactivate the device. User data cannot be physically restored.

A common signal-initiated (or software-based) disablement that can be activated by a user or operator is different from self-destruction. With software-based disabling, a smartphone's memory cards and chips remain intact, so data may be recoverable.

J. Rauhala (✉)
University of Jyväskylä, Jyväskylä, Finland
e-mail: jussi@ieee.org

In the self-destruction method described in [20], the system data or hardware of the device would be destroyed, making reactivation, data recovery, and use of the device impossible.

The problems and threats related to malicious software and hardware hacking are well known in the cybersecurity community. Connected devices such as computers and even automobiles have been hacked remotely. Such hacking has been done for eavesdropping, remote control of functions, or other purposes. Recent WikiLeaks revelations show that remote hacking is possible, at least on Android and iPhone devices [40]. It was revealed that it is possible for an intelligence agency to override smartphone firmware in the supply chain [11]. Android and Apple smartphones have also been subject to malware attacks by actors such as individual hackers who are not affiliated with any government [5, 12]. In addition, there are software methods that allow complete remote control of some iPhone and Android phones by a third party [30, 40].

This chapter deals with hypothetical actions that are intended to impact the owner of a given smartphone, or more precisely, the primary user (either as an actual or misidentified target, either by design or coincidence). For the hypothetical scenarios, we use a research approach that is abductive and formally semi-escapist. The use of the smartphone by the primary user is assumed to be typical, i.e., users use their devices in ordinary ways. The literature seems to lack an overview of potential third-party induced acute direct manipulations of smartphone hardware that result in physical or psychological threats and dangers. Our intention is to draw attention to the issue to help catalyze the development of preventive and mitigating measures by stakeholders. We attempt to present a discussion framework outlined by a profile of potential threats. Profiling is done by characterizing potential threat vectors, potential third-party actors or culprits, and estimated consequences for the user.

In this work, we do not address certain non-physical dangers posed by weaponized smartphones, such as fraud, privacy threats, security threats, financial loss, or identity theft. Nor do we deal with the weaponization of information, such as an attack on a user by notifications, messages, or alerts designed to manipulate the user. The misuse of smartphones to trigger the detonation of externally connected explosives (e.g., a roadside bomb to which the phone is connected) is also excluded. We do not treat the abuse of smartphones as blunt force instruments or projectiles. We do not deal with technical details.

The terms "smartphone", "phone" and "device" are used interchangeably.

19.2   Remote Destruction of the Smartphone

Researchers have developed a method to remotely trigger the destruction of a smartphone by directing power from the smartphone battery to heat and expand the phone material. The material expands to physically destroy some critical hardware, rendering device data physically unrecoverable and the phone useless [20]. While the remote destruction capability of a smartphone is legal and useful under the intended

Fig. 19.1 M-80 firecracker [43]

use scenario, it may lead to more severe and damaging results that can extend beyond the small integrated circuits and components of the target device. Every smartphone has a battery, a lithium cell, designed to store enough energy to run the device for as long as possible. With the development of battery technology, it has been possible to design and manufacture more efficient batteries. Lithium-ion batteries commonly used in smartphones have a very high energy density [9] and are around 90% efficient [44]. A typical smartphone battery contains about 5 Wh of energy, which is equivalent to 18,000–20,000 J. Utilizing information from [17, 43], this can be calculated to be roughly equivalent to the energy of five grams of TNT or about two M-80 firecrackers (Fig. 19.1).

These small and efficient batteries are not always harmless. Problems with the design or manufacture of the battery can cause malfunctions that result in fires or explosions. Some battery issues can be caused by smartphone design, user operations, or software errors. Explosions in a smartphone battery have been sufficient to cause a short-term shock, injury, or fire [6, 23]. In cases where the user does not suffer physical harm, many users consider the loss of a smartphone alone to cause almost as much stress as the threat of terrorism [31].

A smartphone is typically owned and used by a single individual. Most people carry their smartphones with them or keep them close all day. Once a person and their smartphone are identified, it is reasonably sure that most of the day the person will carry the smartphone with them, the person will handle it, or it will be close to them. It is conceivable that techniques similar to those described by [20] (which trigger a rapid rise in the internal temperature of the device with battery electrodes) could be applied to rapidly cause an uncontrolled thermal reaction of the battery. This in turn can result in a fire or explosion. Thus, it may be possible for a remote hacker to attack a device, causing physical harm to the user. For example, unauthorized tampering with the device firmware or operating system could cause a fire in the device or an explosion of the battery. Hacking could also cause the device to malfunction, which drains the battery very quickly. Indeed, there are smartphone apps freely available that are designed to cause rapid but safe battery discharge [24].

High ambient temperature is one factor known to cause battery fires [8]. Over-charging, abnormally rapid discharge, short circuiting of the battery can produce heat. Storage of the smartphone in a hot enclosed area can cause the smartphone and battery to heat, which in turn could cause an explosion or fire. Alternatively, firmware hacking can result in activity that could cause the battery to explode or catch fire. Explosive destruction of the phone battery can even result in the death of the user [3, 10, 21, 32, 37, 46]. At least one death has been reported due to electric shock when the phone was connected to a charger [2]. It should be noted that some of the reported deaths or injuries due to smartphone explosions appear to be hoaxes [33, 45].

Battery-powered devices that are frequently used with smartphones may also pose threats. Smartphone accessories, such as headphones, are on rare occasion known to overheat or explode, causing burns to the user's face, see Fig. 19.2 [15, 28]. Even if smartphone batteries are designed to withstand hacking (e.g., with robust short-circuit protection), hacking into any of the user's battery-powered accessories could still pose a danger. Examples of such accessories include wireless headphones [28] or a Bluetooth earpiece that is used very close to the ear. There has been at least one report of a Bluetooth speaker bursting into flames [38].

Hackers or culprits who produce and distribute malware or commit cyberattacks can be individuals or organizations. Recent WikiLeaks documents have revealed the extensive hacking capabilities of a national intelligence agency [40]. Hacking against smart TVs was developed in cooperation with intelligence agencies in different nations [41]. Some governments around the world are certainly able to develop and implement such hacking or install backdoor capabilities on after-market devices. This ability could give powerful bad actors a personal level "kill switch" to an affected smartphone or accessory. The device could be disabled or destroyed by causing a fire or explosion in the battery. Bad actors could also develop a program or hack that causes the device to emit radiofrequency (RF) radiation at high levels. If the user

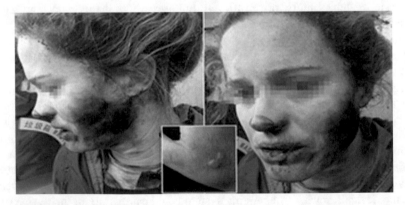

Fig. 19.2 Battery-operated headphones exploded while the passenger was listening to music [1]

becomes aware of such an attack, they may feel psychological distress. The distress would depend on their concern about possible radiation exposure and where they usually keep the device relative to their body.

19.3 Categorical Framework for Smartphone Dangers

Various threat modeling techniques and frameworks exist, but many of them are intended to model threats to large organizations or entities that have high economic value and strict security requirements or other high-stakes targets. Examples of such models are listed by Shevchenko [35]. Some of these techniques can be applied, perhaps in awkward ways, to model the threats to individual smartphone users. Based on the author's literature review, there are currently no threat modeling techniques designed to model the specific threats that this chapter focuses on.

19.3.1 Characteristics of Attack Effect

In this and the following sections, we propose the following parameters and corresponding descriptions to help assess the potential harm caused by a third-party attack. Some characterstics of attack effects are:

- Acute versus chronic,
- Sudden versus long-term,
- Obvious/salient versus hidden/obscured,
- Catastrophic versus undetectable,
- Maintained functionality versus compromised functionality vs. eliminated functionality.

Is the effect sudden or long-term? This applies to the first two parameters. For example, a battery explosion will have sudden consequences while increased radio frequency emissions can have a long-term effect. The effect is obvious to the user, for example, when the phone overheats or ignites. An example of a non-obvious effect would be intensified radio frequency emissions. The catastrophic effect significantly impairs the functionality of the smartphone and threatens the user's well-being. Otherwise, the user will not detect any inconvenience or danger during normal use.

An example of the effect of maintaining functionality (excluding battery life) is the increase in radio frequency emissions. Compromised functionality is a scenario in which some functions, such as an Internet connection or a camera/gallery or other function, are forced off, but other important functions, such as the ability to make a call, remain. Eliminated functionality means a case where the smartphone is completely disabled or "bricked."

19.3.2 Attack Vectors

Different attack vectors can be used to carry out a smartphone attack:

- Implanted software,
- Voluntarily downloaded software,
- Hijacked default or hijacked downloaded software,
- Implanted firmware,
- Update with malicious firmware,
- Rogue or fake cell towers,
- Using a counterfeit smartphone.

Implanted software is malware or other software that is designed to cause a particular effect through an embedded payload. Voluntarily downloaded software is malware that a user has intentionally downloaded from the Internet. Hijacked default or hijacked downloaded software is firmware or apparently legitimate software that has been infected with a payload of malware. Implanted firmware is firmware that has malware embedded on it when it comes from the factory. Update with malicious firmware occurs when a user updates his/her device with malware-embedded firmware. The user has obtained it from a malicious website or elsewhere.

Rogue or fake cell towers spoof an authentic operator tower. This vector enables communication monitoring of connected devices and the sending of spoofed text messages to these devices [25]. Thus, it is possible to organize SMS-based hacking from a fake tower to the victim, such as receiving an image as a text message as described by [30]. When using a counterfeit smartphone, the user is using an unauthorized copy of the branded smartphone product. The device manufacturer has not been authorized to manufacture this device and may not be known.

19.3.3 Attack Perpetrators

The culprit/perpetrator/source of the attack may be

- Single hacker,
- Hacker group,
- Nation state actor,
- Private company,
- Criminal gang/organization.

The perpetrator of an attack may be an individual using one of the attack vectors. In the case of a group of hackers, the attack is carried out in cooperation by several hackers. A national state actor is any entity with the resources and operational support of a national government. A private company refers to a criminal company or part of a private company that makes an attack. A criminal gang/organization is an organized criminal group that carries out an attack, perhaps as part of a turf war or through proxies.

19.3.4   Weaponizable Components

Weaponizable components can be the following:

- RF transmitter,
- Battery,
- User interface (UI) function.

An RF transmitter is a (radio frequency) hardware module that could transmit electromagnetic signals abnormally. The battery inside the smartphone may be damaged. The interactive UI components of the device may start to malfunction.

19.3.5   Attack Effects

Effects of an attack on a smartphone could be

- Device heating/overheating,
- Battery swelling,
- Battery fire,
- Battery explosion,
- Abnormal radiation from the device,
- Disabling the device,
- Destruction of the device.

As a result of the attack, the device may become hot or overheated. The battery generates enough heat to cause injury to the user and damage the smartphone. Swelling of the battery will damage the operation of the smartphone due to physical damage to the device. When a battery catches fire, it causes (typically) a hot and rapid fire in the smartphone. Explosive energy from the battery can cause injury to the user but may not necessarily destroy data on the device or its functions.

An attack may cause excessive abnormal radiation from the device. In this case, the device's RF modules and antennas emit abnormally high levels of electromagnetic radiation. The required power for such emissions would cause a faster drain of the battery. An awareness of the effect or of the associated battery drain can cause distress to the user. A direct or indirect (timed or user-triggered) disablement of the device by a remote/third party will cause some or all of the device's functions to stop. The functions that are disabled may be critical for a particular user. The remote/third party may cause the device to be destroyed so that no operations can be performed and all data is destroyed. This could be accomplished by a battery explosion or by less visible means, e.g., expansion of a polymer layer that destroys essential components, as described by [20].

The harm caused to the user by an attack can be physical. For example, the user suffers from a burn or physiological shock. Psychological consequences can include distress, anxiety, or emotional shock.

In addition to the acute effects, the realization of an attack may have significant secondary effects. Consider a passenger flight. Nearly every passenger carries a battery-powered device. If the battery of the passenger's device burns or explodes during a flight, the flight may be disrupted. Secondary social impacts may include decreased user confidence in smartphone technology and willingness to use smartphones. The public's confidence in the safety of air travel may also be affected.

A hypothetical assessment of weaponizable smartphone components can be found in Table 19.1 in the Appendix. Using Tables 19.2, 19.3 and 19.4 in the Appendix, a researcher or threat analyst can cross-reference the above parameters against each other to analyze threats. The cells in the tables can be filled with a suitable scale parameter, such as a number ranging from zero to ten. For example, 0 means no threat is detected, and 10 means that the combination has a certain or current manifestation.

19.4 Nation State as a Bad Actor

Advances in technology have made it possible for various entities to abuse technology. Such entities include nation-states with significant sovereign authority and access to substantial resources. Because of the scale of the influence of nation states, an abuse of technology by them may be a grievous a threat to human rights. The discovery and public awareness of the threats of such abuse often follow only after the new technology has been massively adapted.

WikiLeaks' Vault 7 revelations have revealed state-sponsored hacking and malware used on smartphones. NightSkies 1.2, designed to enable complete remote control and management of iPhones, has apparently been implanted in devices during the product supply chain [11]. With RoidRage software, a third party can monitor the device's RF functions and SMS messages [29]. The Vault 7 revelations were released in 2008 and comprised only 1% of the leaks [42]. Thus, there is no doubt that more sophisticated device hijacking and surveillance tools exist today.

Apps such as TikTok and at least one private technology company that manufactures smartphones have been accused of being channels for international espionage [22, 34]. The benefits and risks of remotely activated self-destruction of a smartphone should be thoroughly considered for possible abuse. The damaging effects of unethical or illegal hacking on a smartphone battery could be prevented by physical protection measures during design and manufacture. However, manufacturers of counterfeit smartphones, batteries, and accessories may not implement all of the safety features of copied products.

19.5 Counterfeit Smartphones

A significant risk factor for the threats described in this chapter is the widespread availability of counterfeit smartphones. The counterfeit electronics industry as a whole is in the order of US$100 billion and it is estimated that 10% of the world's electronics are counterfeit [36]. Counterfeit smartphones are relatively cheap to buy, widely available online, and compose a US$48 billion market [16]. Authorities have fought against such trafficking [18, 39]. A carefully manufactured counterfeit smartphone may appear nearly identical to authentic ones [13]. Thus, some consumers may not be able to distinguish counterfeit smartphones. Consumers may also knowingly use a counterfeit without much concern for the risks involved. A study by [26] found that consumers agreed with the perceived risks of buying counterfeit (or "grey-market") smartphones. However, they only slightly disagreed with the idea or intention of purchasing them: the mean user response was 2.78 on the Likert scale (from 1 = strongly disagree to 5 = strongly agree).

It can be extremely difficult for a consumer to discover or begin to suspect hidden functionalities or backdoors that can be designed for any smartphone. Counterfeit smartphones pose additional risks [13]. Detecting malicious or exploitable features that can be embedded in tiny integrated circuits used in smartphones can require considerable technical expertise and expensive sophisticated equipment. At the technology level, counteracting the use of counterfeit smartphones, batteries, and accessories can be difficult. It requires a great deal of involvement from the original manufacturers. One measure to prevent the use of counterfeit batteries has required advanced cryptographic security-based technology [7]. Counterfeit devices are often designed and manufactured in areas where government quality control, regulations, and policies are questionable.

In addition to counterfeit smartphones, counterfeit batteries and chargers are widely available. The varying quality of these devices poses its own danger [4]. With modern technology, it is possible to embed concealed electronics or functionality in a counterfeit product housing, including smartphone accessories. As the Vault 7 revelations suggest, very sophisticated concealed functionality can be embedded in legal and authentic devices. Hidden functionalities could also be embedded in authentic batteries or accessories. One possible scenario is a counterfeit battery installed in an authentic smartphone (or an authentic battery in a counterfeit smartphone) that, together with a malware app, could cause unexpected or dangerous damage. In other words, a malware app or firmware could perform as [19] suggests but in a malicious way, weaponizing the smartphone by causing an explosive reaction in the battery. Alternatively, the malware app or firmware may act as a malicious variation of the battery drainage app [24], causing a rapid drainage and (assuming the battery has sufficient charge) a significant temperature rise inside the device. This could also pose a danger to the device and the user.

The use of smartphones is widespread. Globally, about 6.4 billion people use smartphones [27]. Entities that can control remote connections to such devices generally have, figuratively speaking, the vicinity of each smartphone user on a wireless tether. The vicinity is either the user's pocket, hand, handbag, nightstand and so on.

19.6 Discussion

When considering a potential threat posed by a remote-weaponized smartphone, the cybersecurity officer should take security measures as appropriate. For example, for high-profile or VIP personnel gatherings or meetings, a protocol can be implemented that requires attendees to hand over their smartphones to a separate and secure location. Alternatively, guests could be asked to remove the batteries from their phones (though such batteries are not designed to be user-accessible in most modern smartphones). Another possible security measure would be to prevent potential wireless signal triggers by creating an RF interference field around the secured area. RF jamming can also block connections from fake cell towers. During the jamming, smartphones are also rendered incapable of normal wireless communication. A similar effect could be achieved with a Faraday-shielded storage or meeting area.

Prevention of the described hypothetical threats can be promoted by advising smartphone users to avoid downloading unknown or unauthorized apps and opening suspicious messages from unknown senders. However, compliance with the advice is not effective against modified firmware embedded in a supply chain or against text message hacking that is activated merely upon delivery. If a bad actor has significant technology resources and expertise at its disposal, threat prevention can be difficult or impossible. Such actors may include a manufacturer of counterfeit phones under the control of a criminal organization or an arm of an authoritarian regime.

Designers could choose materials and configuration models for the smartphone chassis so that the smartphone body would withstand a catastrophic battery fire or explosion. This would provide the user with some protection from injury. This mitigation is problematic in the case of counterfeit phones—not to mention phones specifically designed to be weaponized.

Further research could focus on analyzing suspected counterfeit smartphones and batteries for malicious or dangerous functions. The analyses should include studies of whether such functions are designed or coincidental, whether they are in the smartphone ICs or battery, and whether they are pre-programmed into software or firmware. If physically harmful functions are found, the analyses should try to determine their triggering mechanisms.

19.7 Conclusion

The pervasive use of smartphones creates a potentially highly vulnerable target for those malicious parties with sufficient technical means. The technology developed to enable remote-triggered self-destruction of a smartphone could, hypothetically and in combination with malicious technology, be abused by a third party to cause catastrophic battery fires and explosions. For the victim, severe heating or explosion of their device can cause distress (about the destruction of the device and the data contained in it and possible thermal damage to property), injury or, at worst, death. The widespread availability of counterfeit devices makes it more difficult to combat such threats. Simply disabling the smartphone can cause significant stress to the victim. A third party guilty of physical weaponization of a smartphone could be any actor, including a nation state-sponsored actor, organization, mafia, company, criminal gang, hacker group, or individual hacker. Regardless of possible culprits, authorities and security analysts should consider the interests of citizens and fundamental human rights, the role of regulators, and the interests of operators and the high-tech industry when proactively assessing the potential threats and preventive measures.

By no means does the author imply or suggest that any individual or organization was or will be involved as a perpetrator or culprit for any of the hypothetical malicious attack scenarios described. The author is also not aware of any realizations of the attack scenarios that are the focus of this chapter.

19.8 Appendix: Threat Analysis

Threat assessment matrices are presented in Tables 19.1, 19.2, 19.3 and 19.4.

Table 19.1 Threat analysis of third-party induced weaponization of a smartphone, a hypothetical example

Component/module	Potential result	Attack vector/trigger
RF transmitter	Exposure to abnormal levels of RF radiation Rapid battery drain Heating	Firmware programming (call to certain number, opening of certain website [malicious code in the site, firmware sniffing for opening of the site, …] Firmware trigger for permanent abnormally excessive transmission strength with every activity that requires a transmission Firmware trigger for maximum transmission power during mundane background transmission activity and/or disabling of OLPC (open-loop power control)
Battery	Swelling Fire Explosion	Remote activation Firmware programming (Timer, push-button sequence, phone call, download, malicious app [malware, …]
UI functionality	Stress and distress to users via partial or full disabling of functionality	Firmware (implanted during manufacture, or malicious update) Malware/virus Fake cell tower (via malicious or rogue (hacked) base station) Physical damage (via "self-destruct" or battery damage hack) Rogue operator employee

Table 19.2 Threat assessment table: threat versus potential culprit

		Culprit				
		Hacker	Nation-state actor(s)	Private corporation	Criminal gang/organization	Hacker group
Threat	Device emits excessive heat/overheats					
	Battery swelling					
	Battery fire					
	Battery explosion					
	Abnormal RF emissions					
	Remotely induced disablement of device					
	Remotely induced destruction of device					

Table 19.3 Threat assessment table: threat versus potential trigger/attack vector

	Potential trigger/attack vector						
	Implanted software	Voluntarily downloaded software	Hijacked default or hijacked downloaded software	Implanted firmware	Updated with malicious firmware	Rogue or fake cell towers	Using a counterfeit smartphone
Threat							
Device emits excessive heat/overheats							
Battery swelling							
Battery fire							
Battery explosion							
Abnormal RF emissions							
Remotely induced disablement of device							
Remotely induced destruction of device							

Table 19.4 Threat assessment table: potential trigger/attack vector versus potential culprit

		Potential culprit				
		Hacker	Nation-state actor(s)	Private corporation	Criminal gang/organization	Hacker group
Potential trigger/attack vector	Implanted software					
	Voluntarily downloaded software					
	Hijacked default or hijacked downloaded software					
	Implanted firmware					
	Updated with malicious firmware					
	Rogue or fake cell towers					
	User is using a counterfeit smartphone					
	User is using a counterfeit battery/accessory					

References

1. ATSB (2017) Battery explosion mid-flight prompts passenger warning. Australian Transport Safety Bureau. https://www.atsb.gov.au/media/news-items/2017/battery-explosion-mid-flight/
2. Azman KK (2019) Man dies of electrocution after his counterfeit phone charger caused an explosion. Says. https://says.com/my/news/man-dies-of-electrocution-after-his-counterfeit-charger-caused-an-explosion
3. Beschizza R (2007) Man killed by exploding cell phone. Wired. https://www.wired.com/2007/07/man-killed-by-e/
4. Best S (2017). Use an iPhone? Check your charger NOW: study finds 98% of fake Apple power leads risk causing fatal 'electric shocks or house fires'. MailOnline. https://www.dailymail.co.uk/sciencetech/article-5155765/98-fake-iPhone-chargers-users-risk-DEATH.html
5. Brewster T (2015) Stagefright: It only takes one text to hack 950 million Android phones. Forbes. https://www.forbes.com/sites/thomasbrewster/2015/07/27/android-text-attacks/
6. Brown H (2013) Student's cell phone battery explodes, starts a fire. CBS Minnesota. http://minnesota.cbslocal.com/2013/02/21/students-cell-phone-battery-explodes-starts-a-fire/
7. Bush T (2014) Fighting the fakes: Algorithmic security combats counterfeit batteries. Medical Design Technology
8. Chen A, Goode L (2016) The science behind exploding phone batteries. The Verge. http://www.theverge.com/2016/9/8/12841342/why-do-phone-batteries-explode-samsung-galaxy-note-7
9. CEI (2021) Lithium-ion battery. Clean Energy Institute, University of Washington. https://www.cei.washington.edu/education/science-of-solar/battery-technology/

10. DailyMail (2009) Man killed after his mobile phone explodes, severing an artery in his neck. Daily Mail. http://www.dailymail.co.uk/news/article-1134838/Man-killed-mobile-phone-explodes-severing-artery-neck.html
11. Durden T (2017) Wikileaks releases "NightSkies 1.2": Proof CIA bugs "factory fresh" iPhones. The Liberty Beacon. https://www.thelibertybeacon.com/wikileaks-releases-nightskies-1-2-proof-cia-bugs-factory-fresh-iphones/
12. Eadicicco L (2017) Watch out for this iPhone-crashing text message. Time. https://time.com/4637574/iphone-crash-text-2017/
13. Evans C (2019) From the depths of counterfeit smartphones. Trail of Bits. https://blog.trailofbits.com/2019/08/07/from-the-depths-of-counterfeit-smartphones/
14. FoxNews (2012) Wireless providers to disable stolen phones. Fox News. http://www.foxnews.com/politics/2012/04/10/wireless-providers-to-disable-stolen-phones.html
15. FoxNews (2016) Cell phone battery catches fire aboard Delta Air Lines flight to Atlanta. Fox News. http://www.foxnews.com/travel/2016/09/16/cell-phone-battery-catches-fire-aboard-delta-air-lines-flight-to-atlanta.html
16. Gilchrist K (2017) Fake smartphone sales cost global industry $48 billion. CNBC. https://www.cnbc.com/2017/02/28/fake-smartphone-sales-cost-global-industry-48-billion.html
17. Herskowitch J (1963) The combustion of a granular mixture of potassium perchlorate and aluminum considered as either a deflagration or a detonation. Technical report, 3063. Picatinny Arsenal, Dover, NJ. https://apps.dtic.mil/sti/pdfs/AD0296417.pdf
18. HK-CED (2018) Hong Kong customs combats sale of suspected counterfeit smartphones and accessories. Press release, Customs and Excise Department, Government of the Hong Kong Special Administrative Region of the People's Republic of China. https://www.customs.gov.hk/en/publication_press/press/index_id_2372.html
19. Hsu J (2017) Self-destructing gadgets made not so mission impossible. IEEE Spectrum. https://spectrum.ieee.org/tech-talk/consumer-electronics/gadgets/selfdestructing-gadgets-made-not-so-mission-impossible
20. Hughes O (2017) Mission possible: self-destructing phones are now a reality. International Business Times. http://www.ibtimes.co.uk/mission-possible-self-destructing-phones-are-now-reality-1605897
21. India (2019) 22-year-old man dies as mobile phone explodes while charging. India News. https://www.india.com/technology/22-year-old-man-dies-as-mobile-phone-explodes-while-charging-3840866/amp/
22. Kaska K, Beckvard H, Minárik T (2019) Huawei, 5G and China as a security threat. NATO Cooperative Cyber Defence Centre of Excellence (CCDCOE)
23. Kerr D (2013) Samsung cell phone battery explodes in man's pocket. CNET. https://www.cnet.com/news/samsung-cell-phone-battery-explodes-in-mans-pocket/
24. Kushwaha N (2020) 6 best free battery drain apps for Android. List of Freeware. https://listoffreeware.com/free-battery-drain-apps-for-android/
25. Leiva-Gomez M (2014) Everything you need to know about fake cell towers. Make Tech Easier. https://www.maketecheasier.com/fake-cell-towers/
26. Liao C-H, Hsieh I-Y (2013) Determinants of consumer's willingness to purchase gray-market smartphones. J Bus Ethics 114(3):409–424. https://doi.org/10.1007/s10551-012-1358-7
27. O'Dea S (2021) Number of smartphone users worldwide from 2016 to 2026 (in billions). Statista. https://www.statista.com/statistics/330695/number-of-smartphone-users-worldwide/
28. Olding R (2017) Safety warning after passenger's headphones explode on Beijing to Melbourne flight. The Sydney Morning Herald. https://www.smh.com.au/technology/safety-warning-after-passengers-headphones-explode-on-beijing-to-melbourne-flight-20170315-guy6va.html
29. Paganini P (2017) WikiLeaks Vault 7 data leak: Another earthquake in the intelligence community. Infosec Resources. https://resources.infosecinstitute.com/topic/wikileaks-vault-7-data-leak-another-earthquake-intelligence-community/
30. Pagliery J (2015) Android phones can be hacked with a simple text. CNN Business. https://money.cnn.com/2015/07/27/technology/android-text-hack/index.html

31. PhySoc (2017). Stress in modern Britain. The Physiological Society. https://static.physoc.org/app/uploads/2020/02/20131612/Stress-in-modern-Britain.pdf
32. Prabhu A (2018). Cradle Fund CEO killed by smartphone explosion. Gizbot. https://www.gizbot.com/mobile/news/smartphone-explosion-kills-ceo-cradle-fund-051647.html
33. Ram S (2014) This FB post about a boy getting killed due to an exploding phone is a hoax. Says. https://says.com/my/tech/explosion-of-exploding-phone-that-killed-10-year-old-boy-is-a-hoax
34. Ryan F, Fritz A, Impiombato D (2020) TikTok and WeChat: curating and controlling global information flows. Australian Strategic Policy Institute. https://www.aspi.org.au/report/tiktok-wechat
35. Shevchenko N (2018) Threat modeling: 12 available methods. Carnegie Mellon University. https://insights.sei.cmu.edu/sei_blog/2018/12/threat-modeling-12-available-methods.html
36. Spiegel R (2009) Counterfeit components remains a huge electronics supply chain problem. Engineering Design News. https://www.edn.com/counterfeit-components-remains-a-huge-electronics-supply-chain-problem/
37. Stewart W (2019) Girl, 14, killed in her sleep 'by exploding phone' after going to bed listening to music while device was charging. The Sun. https://www.thesun.co.uk/news/10032279/schoolgirl-14-killed-sleep-exploding-smartphone-listening-music-device-charging/?utm_campaign=sunmainfacebook300919&utm_medium=Social&utm_source=Facebook#comments
38. Strahan T, Novini R (2017) Bluetooth speaker starts smoking on bed, bursts into flames. NBC New York. http://www.nbcnewyork.com/news/local/Bluetooth-Speaker-Bursts-into-Flames-Seen-Smoking-on-Bed-Sources-417596643.html
39. US-CBP (2019) Philadelphia CBP seizes nearly $1 million in counterfeit smartphones from China. United States Customs and Border Protection. https://www.cbp.gov/newsroom/local-media-release/philadelphia-cbp-seizes-nearly-1-million-counterfeit-smartphones-china
40. WikiLeaks (2017) Vault 7: CIA hacking tools revealed. WikiLeaks. https://wikileaks.com/ciav7p1/
41. Wikileaks (2017) Weeping angel (extending) engineering notes. In Vault 7: CIA Hacking Tools Revealed. WikiLeaks. https://wikileaks.org/ciav7p1/cms/page_12353643.html
42. WikiLeaks (2017) WikiLeaks has released less than 1% of its #Vault7 series in its part one publication yesterday 'Year Zero'. Twitter. https://twitter.com/wikileaks/status/839475557721116672
43. Wikipedia (2020) M-80 (explosive). Wikipedia, Retrieved September 2, 2020 from https://en.wikipedia.org/wiki/M-80_(explosive)
44. Xiong S (2019) A study of the factors that affect lithium ion battery degradation. M.Sc. thesis, University of Missouri-Columbia. https://mospace.umsystem.edu/xmlui/bitstream/handle/10355/73777/Xiong-Shihui-Research.pdf?sequence=1&isAllowed=y
45. Yarow J (2010) The Droid phone that exploded and blew up a guy's ear? It was just dropped, says Motorola source. Business Insider. https://www.businessinsider.com/droid-phone-explosion-motorola-2010-12?international=true&r=US&IR=T
46. Zamfir G (2018) Girl, 18, killed when mobile phone explodes while she is chatting to relative. Mirror. https://www.mirror.co.uk/news/world-news/girl-18-killed-mobile-phone-12215521

Chapter 20
Practical Evasion of Red Pill in Modern Computers

Amit Resh, Nezer Zaidenberg, and Michael Kiperberg

Abstract The blue pill is a malicious stealthy hypervisor-based rootkit. The red pill is a software package designed to detect blue pills or hypervisors in general. Ever since the blue pill was originally proposed, there has been an ongoing arms race between developers trying to develop stealthy hypervisors and developers trying to detect such stealthy hypervisors. Hypervisors can also be used for monitoring and forensic purposes, while malicious software may include a red pill component to discover such a hypervisor in order to evade it. This chapter discusses a practical approach to counter such malicious software by evading the red pill components.

Keywords Virtualization · Forensics · Information security · Red pill

20.1 Introduction

The blue pill was introduced by Johanna Rutkowska at Blackhat 2006 [16]. There were others who used hypervisors for security. However, with the introduction of the blue pill and red pill concept, the virtualization concept became so closely related to cyber security.

The blue pill is a rootkit that takes control of a victim host computer. Unlike other rootkits the blue pill is actually a malicious hypervisor. The original blue pill starts after the OS has already booted and through a series of hardware instruction, the blue pill gains control of the victim host. In fact, after the blue pill is deployed it has

A. Resh (✉)
Department of Software Engineering, Shenkar College of Engineering and Design, Ramat Gan, Israel
e-mail: amit@se.shenkar.ac.il

N. Zaidenberg
School of Computer Sciences, College of Management Academic Studies, Rishon LeZion, Israel

Faculty of Information Technology, University of Jyväskylä, Jyväskylä, Finland

M. Kiperberg
Department of Software Engineering, Shamoon College of Engineering, Beer-Sheva, Israel
e-mail: michaki1@sce.ac.il

M. Lehto and P. Neittaanmäki (eds.), *Cyber Security*, Computational Methods in Applied Sciences 56, https://doi.org/10.1007/978-3-030-91293-2_20

461

gained higher privileges then the OS that started it. Of course, like all rootkits the blue pill must camouflage its existence, or it will be removed by the user. Using a hypervisor assists in camouflage because the blue pill can now start tasks outside the OS scope. In order to counter the blue pill the red pill was invented. The red pill is a hardware or software tool designed to detect such malicious blue pill rootkit.

The red pill is a special case of the related "trusted computing" and the attestation concept [21]. In trusted computing attestation of a remote third party or even local software tries to ensure the integrity of the local machine in terms of software (mainly) and hardware (sometimes).

When hypervisors are concerned the attestation of lack of hypervisor (or blue pill) was first researched by [3] who discussed methods to establish the "genuinity of the host" (i.e. ensure that the host is a physical machine running the correct software as opposed to an emulator or a virtual machine or a physical machine running non-genuine software). Kennell and Jamieson proposed running a series of tests that will pass only if the inspected system is genuine. The test will fail if a hypervisor is running due to side effects involved with the running of ೮ hypervisor, mainly more expensive memory traversal.

Today the original Kennell's hypothesis is questionable due to the availability of specific hardware instructions and features designed to remove memory access side effects (Second Level Address Translation instructions such as EPT™ in Intel's case and RVI™ in AMD's case). However, the core idea of using side effects is still in use in many modern red pills.

Since Rutkowska introduced the blue pill malware concept, multiple attempts to create red pills that detect such blue pills have also been proposed. However, more advanced blue pills have been designed to avoid detection. Which led to more advanced red pills and so on and so forth. So, the goals of the blue and red pills are conflicting and naturally technology advances in one front requires advancement by the other front in order to keep up.

Nowadays, modern CPUs by Intel, AMD and ARM feature hardware-assisted virtualization. Hardware-assisted virtualization provides new capabilities for implementing virtual machines and emulator software. Thus, hardware-assisted virtualization can be used to evade some "red pill" attempts, for example by eliminating several virtualization side effects [20]. However, modern hardware platforms are much more complex than before, hosting multiple processors and many more side effects, which give rise to new openings with which to create new red pills.

This chapter describes a practical approach geared at the evasion of red pills on Intel virtualization architecture platforms.

20.2 Background

Recent advances in ×86 hardware-assisted virtualization allow for seamless introspection of the OS operation in order to verify that the system is operating in a secure environment in addition to supporting multiple OSs on a single hardware platform.

These advents are backed by new instruction families called VT-x, VT-d and EPT on the Intel architecture. AMD-v, IOMMU and RVI are the corresponding names on AMD architectures.

20.2.1   Hypervisors and Thin-Hypervisors

A hypervisor is a computer software concept designed to run multiple operating systems on the same hardware. As its name implies, a hypervisor has higher privileges (hyper = "above") than the operating system (i.e. the operating system = the supervisor). The operating system supervises memory and hardware resources for the processes that run on top of it. Likewise, the hypervisor supervises the hardware resources for each operating system that runs on top of it.

Hypervisors research started with [14] who classify hypervisors into two main categories:

Type 1   Boot hypervisors,
Type 2   Hosted hypervisors.

Type 1 hypervisors are started by the machine at startup. The machine starts the guest operating system after booting the hypervisor. VMWare ESXi is an example of a modern Type 1 hypervisor. Type 2 hypervisors start only after the operating system has started. A modern example of a Type 2 hypervisor is VMWare Desktop or Oracle Virtual Box hypervisors. The original blue pill described by Rutkowska was a Type 2 hypervisor, but Type 1 "blue pill" hypervisors (boot kits) are also possible.

Hypervisors reside logically between the hardware and the supervisor (OS) layers. The hypervisors can intercept a variety of instructions and catch interrupts and deliver them to the correct operating system and control memory addresses. The hypervisor uses its own translation tables for deciding which operating system owns each memory address and which operating system should handle each hardware interrupt. This is analogues to the MMU in OS environment where each memory address is assigned to a different process.

However, there also exists a special case of hypervisors that do not attempt to run multiple operating systems. Instead, these hypervisors, called "thin hypervisors", support running only one operating system on the target hardware. All interrupts and memory accesses are either blocked or transferred to the operating systems for handling. Indeed, thin hypervisors act as a microkernel that provides certain services to the underlying operating system. The thin hypervisor includes very little memory management and relies on the single guest OS system for both memory management and interrupt handling.

Table 20.1 ×86 virtualization instructions

	Intel architecture Name	AMD architecture Name	Usages
Virtualization instructions	VT-x	AMD-v	This family is required for starting a hypervisor
SLAT (second-level address translation)	EPT (Extended page tables)	RVI (Rapid virtualization indexing)	Allows the HV to run Multiple MMUs for multiple guest operating system
IO MMU	VT-d	IOMMU	Allow the HV to assign IO memory to specific guest operating systems
VM data structure	VMCS	VMCB	Holds all the VM information (one per VM)

20.2.2   ×86 Virtualization

In Intel's ×86 CPU architecture, hardware-assisted virtualization is provided by unique instruction families. Intel architecture and AMD architecture each provide three such instructions families for handling hypervisors. These instructions are further optimized with each new processor generation.

Newer generations include new hardware assisted virtualization capabilities such as shadow VMCS. Furthermore, virtualization instructions take less CPU cycles to complete in newer CPU generations. The ×86 instruction families are presented in Table 20.1.

20.2.3   Rootkits and Bootkits

The recommended and common protocol for a system administrator to respond to cyber incidents once an attack is detected on any web server is to format and reinstall the operating system on the target server. If the operating system is reinstalled (and has been completely patched with security patches), the hacker may find herself locked outside of the compromised system. It follows that hackers want to hide their tracks. Thus, if the hack was not detected, the hacker can maintain a persistent access to the hacked servers.

Therefore, hackers frequently install software packages known as a "*rootkit*." A rootkit is a software package that allows the hacker access to the victim system resources. Furthermore, the rootkit hides itself and all the processes that the hacker runs on the infected system in an effort to mask its existence. Therefore, the rootkit has two goals. First, the rootkit provides the hacker ease of access to victim computer resources. Second, the rootkit provides measures to hide the hack and its own existence on the victim system.

There are many ways to build rootkits from hijacking system calls and library functions and installing *setuid* programs to replacing innocent-looking binaries. The blue pill is a special type of rootkit. Unlike normal rootkits that modify the operating system in order to hide files and gain access to the system, the blue pill starts a hypervisor. Thus, the blue pill gains more permissions than the operating system. The blue pill can run processes in the hypervisor address space. Thus, the processes or their address spaces are not visible by the operating system.

One of the ways that the rootkit can be instantiated is as a "*bootkit*". A bootkit is a special type of rootkit that boots (from the hard drive master boot record, UEFI, PXE, or other means) before the operating system starts. The bootkit runs its own software (i.e. a hypervisor in the blue pill case) before the OS starts and later boots the OS (by calling the OS boot). The bootkit may also patch the OS system calls in order to hide its processes and files.

20.2.4   Hypervisors, Forensics and Cyber Security

Some cyber-security technologies rely on hardware-assisted virtualization ("hypervisor instructions") and hardware-assisted virtualization technology to monitor computer systems against malicious software attacks. Zaidenberg [18] summarized hypervisor usage in cyber security. Hypervisors can be used to inspect the target system. Such forensic efforts can be used to assist developers [5], to profile the code [6], to directly obtain the memory of the inspected system [9] or to detect malware [19].

Furthermore, thin-hypervisors may provide security services to the guest system. Microsoft's Deviceguard, TrulyProtect hypervisor for protection against reverse engineering [1, 10] and Execution Whitelisting [15] are examples of thin hypervisors utilized for cyber-security purposes. Other thin hypervisors monitor Video DRM [2], provide forensics data or provide end point security [10]. Thin-hypervisors, which monitor the operation of a system, are used not only to verify its secure operation, but also as a means to assist in tracking down and examining methods of operation used by suspected malicious software.

However, some modern malicious software now incorporate a red pill component in order to inspect a hypervisor presence. If it detects such a hypervisor, it can refrain from part (or all) of its malicious activity that is prone to hypervisor detection. This will give the malicious software an upper hand by delaying its discovery and analysis, allowing it to propagate unnoticed to additional systems. Thus, it is clear that modern cyber-security hypervisors must evade red pill detection, if they are to be successfully used to monitor, detect and analyze malicious software behaviors.

20.2.5 Kennell's Timing Method and Derived Attacks

Kennell and Jamieson [3] proposed a method to perform a remote hypervisor red pill based on computation side effects on the inspected system. Such side effects include Translation Lookaside Buffer (TLB) and cache hits and misses as well as real world time that preforming the computation consumes. According to Kennell's method, the attested computer receives a "challenge" (computation request) from the trustworthy remote server.

The computation of the challenge causes several computational side effects such as TLB hits, TLB misses, cache hits, cache evictions, etc. The challenge contains several stages in each of these stages the side effects of the prior stage are added to the computation result. Thereby the prior stages result, and side effects affect the result of the next phase computation. To pass the test, the tested computer must not only produce the same results for the computation itself but also compute accurate results for the side effects as well. Furthermore, the entire computation (of all stages) must be completed in a short time (the time it would take a non-virtual machine to calculate).

Kennell's test relies on the fact that if a blue pill or an emulator is running then the computation side effects are bound to be different. Thus calculation of side effects must be done separately and consume more time. Therefore, it is impossible for the response to the challenge on an emulated machine or virtual machine to be right and arrive on time. Furthermore, since the challenge is constructed from many stages that must be computed in the correct order Kennell's test cannot be emulated on parallel machines. Kennell's test will declare the machine is not genuine if the answer is wrong or arrives too late.

Kennell's method came under direct attack the following year. Shankar et al. [17] claimed that performance side effects are not sufficient as a method for software detection. Kennell and Jamieson [4] has answered these claims and the matter rested until virtualization became commonly available in modern PCs. Kennell's method cannot be emulated directly on modern system as modern systems are more complex than the model assumed by Kennell and Jamieson with multiple caches. Also EPT provides much faster memory traversal even if an hyper visor is present.

Kiperberg et al. [7, 8] claimed that Kennell's method can be replicated on modern PCs with hardware virtualization. This result was short-lived as Intel changed their caching algorithm the following year (between the second and third generation of core processor). Furthermore, Intel has not shared their caching algorithms.

However, Kennell's tests rely on the availability of certain algorithms, such as CPU caching algorithms, which are not commonly available. These algorithms are considered trade secrets. Furthermore, Intel has changed the caching algorithms of their core platform between the second and third generations and changed them again to combat the "meltdown" [12] and "specter" [11] weaknesses. Thus supporting Kennell's tests on modern hardware require reversing the architecture of the caching algorithm. It is difficult and extremely time consuming. Supporting Kennell's tests on all recent Intel/AMD architectures can be a menial task that will require further research.

20.3   Local Red Pills

Local red pills are tests performed by the tested machine and contained within the tested machine. These tests cannot be considered reliable as computation is performed on an untrustworthy machine (the very machine that is being inspected). A red pill component running as part of a malicious software that gains a foothold on a computer systems is such a case of a local red pill. Since no trustworthy root of trust is available, the malicious software can only attempt to provide attestation in a best effort attempt to detect a hypervisor on the inspected machine.

20.3.1   *Paranoid Fish and Other Modern Red Pills*

Paranoid fish (Pafish) [13] is currently the de facto standard in "red pill for hypervisor detection" software. Pafish includes multiple tests capable of detecting most known hypervisors when running under Linux or Windows. Many of these hypervisor detection tests assume the hypervisor is not attempting to hide, and simply lookup specific values in memory. For example, VMWare routinely reports 440BX (1990s era hardware) on all machines.

   However, some Pafish tests are specifically geared toward hypervisors that do try to hide. These methods rely on timing and side effects [20] of running hypervisors. Local timing methods are employed to try to flush out these hypervisors. The local timing tests perform the following steps:

Step 1   Take local time (for example, RDTSC instruction).
Step 2   Execute an operation that must be intercepted by a hypervisor (for example, CPUID instruction).
Step 3   Take local time again to obtain elapsed time.

   The underlying assumption is that Step 2 takes significantly longer to execute when a hypervisor is active.

20.3.2   *Paranoid Fish Timing Tests*

Paranoid Fish (Pafish) timing tests include the following:

Test 1   RDTSC, RDTSC < 750 cycles,
Test 2   RDTSC, CPUID, RDTSC < 1000 cycles.

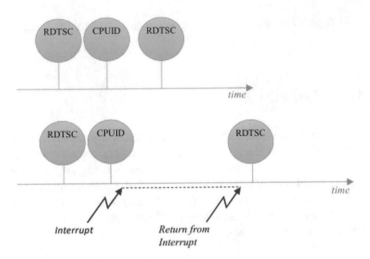

Fig. 20.1 RDTSC timing and interrupt intercepts

Test 1 involves a sequence of two consecutively called RDTSC instructions and measuring response times. It is designed to ensure that the hypervisor does not intercept the RDTSC instruction. Test 2 sequence measures the number of execution cycles of the CPUID instruction that the hypervisor must intercept. Executing the required context switch from guest to host (and back) under hypervisor control adds significant cycles, which is used as a tell-tale sign of hypervisor presence. In both tests, an irregular high result may be obtained due to a random interruption between two RDTSC instructions. To avoid false positives due to irregular results, Pafish performs an average of over 10 runs (Fig. 20.1).

The likelihood of such events is extremely low, since these instruction sequences complete within a few dozen cycles. However, it is not zero probability. Therefore, the Pafish algorithm tests results against rather high thresholds (750 and 1000 cycles) to account for this. The spacious threshold will cover, in most practical cases, a single interrupt event during 10 runs. However, rare cases of false positives have been viewed during extensive Pafish test runs, as explained below. It therefore stands to reason that malicious software incorporating such a red pill component will activate several such detection attempts to rule out false-positives, see Fig. 20.2.

Such false positive cases are perpetuated until the intercept interrupt acts as an OS round-robin timer, causing task to be rescheduled. In this case, the result of the cycle length after the task has been rescheduled to run is enormously larger than the first reading, so the test fails.

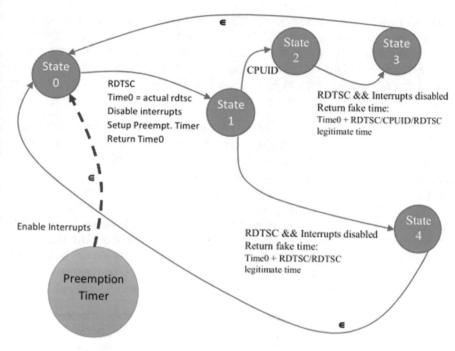

Fig. 20.2 Evading user-mode Pafish states

20.3.3 *Paranoid Fish Timing Tests in User-Mode*

Pafish timing tests are used as a model that represents a larger category of timing tests that red pills utilized to attempt to flush-out the existence of hypervisors. This category being characterized by local timing of instruction sequences that contain instructions which hypervisors *must* intercept. The following method assumes that the Pafish test is executing from user mode and is therefore prone to interrupt interceptions, as described above.

The approach taken by the hypervisor to evade the Pafish timing test is based on interception of the RDTSC (and RDTSCP) Intel command, detecting a sequence of closely spaced RDTSC >> RDTSC or RDTSC >> CPUID >> RDTSC commands and faking the results of the second RDTSC reading. The hypervisor is set to intercept the RDTSC (and RDTSCP) instruction. The CPUID instruction is intercepted as well, as it is a must-intercept instruction. The hypervisor maintains a state-machine that detects the two possible instruction sequences. On receiving the first RDTSC instruction, it stores the actual RDTSC result in temporary storage. When receiving the second RDTSC, it responds with a fake and "legitimate" calculation depending on which sequence was detected.

With this setup in place, the hypervisor must deal with several problems that arise:

Problem 1 Hypervisor intercepts in RDTSC and CPUID instructions cause a context switch from guest mode to host mode. The time span accumulated during the intercept includes the context switch to the host, the instructions carried out in the host's intercept routine, and the context switch back to the guest. This time span is an order of magnitude larger than for non-intercepted instructions. Therefore, they are much more prone to interrupting interceptions.

Guest interrupts are not serviced in host mode. If interrupts occur while the processor is *not* in guest mode, the interrupts are pending and occur after the context switch back to the guest. When this chain of events occurs during the intercept of the first RDTSC instruction in the sequence, the hypervisor sets it as the reference point and position of the state-machine and returns to guest. However, a pending interrupt is triggered before the guest Pafish task regains control. In most cases, a pending interrupt acts as an interrupt for the OS round-robin timer, causing task rescheduling. The guest will regain control of the processor and, with significant likelihood, will be scheduled on a different processor, where a different instance of the hypervisor runs as the host. This completely breaks the solution approach, as the state machine transitions have spread over different contexts.

Problem 2 RDTSC instructions are called by the OS's user mode components. They may be called by other user tasks in the systems as well. Therefore, not every RDTSC instruction intercepted is part of a Pafish timing test sequence. These sporadic RDTSC instructions tend to throw the state machine out of sync.

The solution proposed for Problem 1 suggests that when intercepting an RDTSC instruction the hypervisor disables interrupts for a predefined period. This is achieved by resetting the Interrupt bit in the guest's flag register and setting up a hypervisor preemption timer for that predefined period. When the preemption timer expires, interrupts are re-enabled. This assures that when an RDTSC instruction is intercepted, it is regarded as the first RDTSC in the Pafish sequence. If so, this sequence will have ample interrupt-less time to complete. This aligns well with the non-hypervisor situation, where Pafish sequences run with negligent probability of intercepting interrupts. RDTSC instruction that are not part of a Pafish sequence shall also "suffer" interrupt disabling, however, the predefined period is kept relatively short, therefore interrupts will be quickly restored. During rigorous tests of this methodology, no negative effects were observed.

This approach also plays well in solving Problem 2. Since every RDTSC instruction now creates a short *interrupts-disabled* period, this period can be utilized as a marker delineating a time-period during which a Pafish sequence must conclude. In other words, once the preemption timer is activated and re-enables interrupts, it will also reset the hypervisors detection state machine to "*state-zero*", see Fig. 20.2.

20.3.4   *Paranoid Fish Timing Tests in Kernel-Mode*

Pafish timing tests may, in theory, be activated from kernel mode. Granted there is a much smaller probability of malicious software gaining kernel mode access by utilizing OS weaknesses and zero-days. However, the growing complexity of OSes contains a handful of such weaknesses and it is relevant to consider malicious software both gaining kernel mode access and having a red pill component to detect monitoring hypervisors.

The problems associated with evading Pafish timing tests activated in kernel mode are much more complex than those that exist when timing tests are activated from user mode. Here we highlight only a few.

Problem 3 Interrupts cannot be disabled by the hypervisor in response to kernel mode RDTSC intercepts. Interrupts and interrupt priorities are used extensively by the OS. Any mucking around with interrupt settings by a hypervisor host during an intercepted guest instruction that executes in kernel mode will cause OS chaos and must be avoided at all cost.

Problem 4 As a consequence of Problem 3, this approach also loses its important sequence-time-limit marker, absolutely required to force a return to "State 0".

Problem 5 The (Windows) OS makes extensive use of the RDTSC and CPUID instructions in kernel mode. Therefore, a hypervisor that intercepts RDTSC instructions to locate Pafish timing sequences will also get an abundance of RDTSC/CPUID intercepts from a variety of kernel threads executing during the execution of the timing sequence. Therefore, the hypervisor must detect timing sequences on a per-thread basis.

Problem 6 It is possible for a non-timing-sequence RDTSC instruction to appear before, and very close by, to the first RDTSC instruction of a timing sequence. In this case, the hypervisor cannot distinguish this situation from two consecutive RDTSC instructions of a timing sequence. Therefore, it must respond to the first RDTSC (the second one in this case) with a fake time. Now, when the actual second RDTSC instruction is intercepted, the hypervisor must also fake its result. However, to do that, the hypervisor must have remembered its previous fake result. This chain of events may continue for some time and cause a time-lapse avalanche between real and fake time results. Therefore, provisions must be made to break this chain at the correct point and return to real-time.

The approach to solve these (as well as some other) problems involves creating data structures in the hypervisor that can continuously monitor RDTSC and CPUID intercepts along with their kernel-mode coordinates (for example, kernel thread ID, real RDTSC time, fake RDTSC reported). Every intercept first stores the intercepted instruction and its coordinates in the appropriate slot in this data structure followed by a backtrack analysis of the intercepts for this slot to determine if a fake response is required and if so, what value needs to be reported.

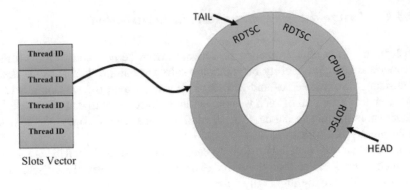

Fig. 20.3 Evading Pafish from kernel-mode data structure

One possible approach to defining this data structure is creating a vector of slots, where each slot is defined by a unique kernel thread ID. The vector size reflects the number of concurrent kernel threads anticipated. When a kernel thread is inactive for a predefined extent of time, it can be discarded from the data structure, freeing up a slot for new kernel threads. Each slot will point to a cyclic-list of intercepted instruction attributes. A cyclic-list will allow an ongoing process of registering RDTSC and CPUID instructions. The list depth will reflect the longest instruction sequence expected to cover a practical Pafish timing sequence. The cyclic-list is a very good candidate to store a continuous stream of information, while offering backtracking memory that supports the analysis stage, see Fig. 20.3.

The authors have successfully coded a workable solution, along the lines of this approach, and demonstrated evasion of Pafish timing-tests activated from kernel mode on a Windows 10 system. Full details of implementation are beyond the scope of this chapter. They are the subject of a future follow-up paper.

20.4   Conclusion

Recent use of hypervisors as a security tool for monitoring and introspection of computer systems may be hindered by red pill components included in modern malicious software. However, this chapter demonstrates that it is feasible for a hypervisor to evade red pill detection. It should be noted that this chapter was focused on the Pafish timing tests to model red pill detection. Other models may exist in the wild and will most likely require modified evasion techniques when they are revealed.

References

1. Averbuch A, Kiperberg M, Zaidenberg NJ (2013) Truly-Protect: an efficient VM-based software protection. IEEE Syst J 7(3):455–466
2. David A, Zaidenberg N (2014) Maintaining streaming video DRM. In: 2nd international conference on cloud security management (ICCSM 2014). Academic Conferences, pp 36–41
3. Kennell R, Jamieson LH (2003) Establishing the genuinity of remote computer systems. In: Proceedings of the 12th USENIX security symposium. USENIX Association, pp 295–310
4. Kennell R, Jamieson LH (2004) An analysis of proposed attacks against genuinity tests. CERIAS Tech Report, 2004–27, Purdue University, West Lafayette, IN
5. Khen E, Zaidenberg NJ, Averbuch A (2011) Using virtualization for online kernel profiling, code coverage and instrumentation. In: 2011 international symposium on performance evaluation of computer & telecommunication systems. IEEE, pp 104–110
6. Khen E, Zaidenberg NJ, Averbuch A, Fraimovitch E (2013) LgDb 2.0: Using Lguest for kernel profiling, code coverage and simulation. In: 2013 international symposium on performance evaluation of computer and telecommunication systems (SPECTS). IEEE, pp 78–85
7. Kiperberg M, Zaidenberg N (2013) Efficient remote authentication. In: Proceedings of the 12th European conference on information warfare and security ECIW 2013. Academic Conferences, pp 144–148
8. Kiperberg M, Resh A, Zaidenberg NJ (2015) Remote attestation of software and execution-environment in modern machines. In 2015 IEEE 2nd international conference on cyber security and cloud computing. IEEE, pp 335–341
9. Kiperberg M, Leon R, Resh A, Algawi A, Zaidenberg NJ (2019) Hypervisor-assisted atomic memory acquisition in modern systems. In: Proceedings of the 5th international conference on information system security and privacy ICISSP. Scitepress, pp 155–162
10. Kiperberg M, Leon R, Resh A, Algawi A, Zaidenberg NJ (2019) Hypervisor-based protection of code. IEEE Trans Inf Forensics Secur 14(8):2203–2216
11. Kocher P, Genkin D, Gruss D, Haas W, Hamburg M, Lipp M, Mangard S, Prescher T, Schwarz M, Yarom Y (2018) Spectre attacks: exploiting speculative execution. arXiv:1801.01203
12. Lipp M, Schwarz M, Gruss D, Prescher T, Haas W, Mangard S, Kocher P, Genkin D, Yarom Y, Hamburg M (2018) Meltdown. arXiv:1801.01207
13. Ortega A (2016) Pafish. https://github.com/a0rtega/pafish
14. Popek GJ, Goldberg RP (1974) Formal requirements for virtualizable third generation architectures. Commun ACM 17(7):412–421
15. Resh A, Kiperberg M, Leon R, Zaidenberg NJ (2017) Preventing execution of unauthorized native-code software. Int J Dig Content Technol Appl 11(3):72–90
16. Rutkowska J (2006) Subverting Vista^{TM} kernel for fun and profit. Presentation at Black Hat Briefings, Las Vegas
17. Shankar U, Chew M, Tygar JD (2004) Side effects are not sufficient to authenticate software. In Proceedings of the 13th USENIX security symposium. USENIX Association
18. Zaidenberg NJ (2018) Hardware rooted security in Industry 4.0 systems. In: Cyber defence in industry 4.0 systems and related logistics and IT infrastructures. IOS Press, pp 135–151
19. Zaidenberg NJ, Khen E (2015) Detecting kernel vulnerabilities during the development phase. In: 2015 IEEE 2nd international conference on cyber security and cloud computing. IEEE, pp 224–230
20. Zaidenberg NJ, Resh A (2015) Timing and side channel attacks. In Cyber security: analytics, technology and automation. Springer, Cham, pp 183–194
21. Zaidenberg N, Neittaanmäki P, Kiperberg M, Resh A (2015) Trusted computing and DRM. In: Cyber security: analytics, technology and automation. Springer, Cham, pp 205–212

Chapter 21
Malware Analysis

Michael Kiperberg, Amit Resh, and Nezer Zaidenberg

Abstract The number of malware is constantly growing. Better detection techniques need to be developed to keep detection times as short as possible and to reduce the cost of malware attacks. Malicious programs use a variety of techniques to avoid detection. Each technique has been developed as a countermeasure to a particular method of detection. This chapter introduces analytical techniques and their countermeasures so that the reader can understand the evolution of armaments between the two research camps.

Keywords Malware · Hypervisor · Evasion

21.1 Introduction

The number of registered malware increases each year. The AV-TEST Institute registered 1139 million malicious programs in 2020 [22], a 14% increase over 2019. According to IBM [7], the average time to identify a breach was 206 days in 2019—a 5% increase over the identification time in 2018. The costs associated with a breach increase with its identification time. Therefore, better identification techniques are required to shorten the identification time and lower the associated costs.

Malicious programs vary by their goal, their replication method, and their abilities [37]. Spyware [5, 8, 18] records the victim's activity [15], e.g., keystrokes,

M. Kiperberg (✉)
Department of Software Engineering, Shamoon College of Engineering, Beer-Sheva, Israel
e-mail: michaki1@sce.ac.il

A. Resh
Department of Software Engineering,, Shenkar College of Engineering and Design, Ramat Gan, Israel
e-mail: amit@se.shenkar.ac.il

N. Zaidenberg
School of Computer Sciences, College of Management, Rishon LeZion, Israel

Faculty of Information Technology, University of Jyväskylä, Jyväskylä, Finland

© The Author(s), under exclusive license to Springer Nature Switzerland AG 2022
M. Lehto and P. Neittaanmäki (eds.), *Cyber Security*, Computational Methods
in Applied Sciences 56, https://doi.org/10.1007/978-3-030-91293-2_21

mouse movement, active window title, screen snapshots, camera's image, and microphone's sound. Spyware then sends the recorded data to the attacker. Ransomware [11] encrypts the victim's information and sends the decryption key to the attacker. Then, the attacker asks the victim for a ransom to receive the decryption key and restore the encrypted information. Adware [5] is a malicious program that constantly displays unwanted advertisements. A botnet [10] is a form of a malicious program that awaits a command from the attacker. When the command is received, the botnet uses the computer hardware to attack other computers or networks. For example, distributed denial-of-service attacks are typically implemented using botnets. Cryptojacking software [14] uses the victim's computing resources for the mining of cryptocurrencies.

Malware has three main methods of propagation between computers:

1. Virus,
2. Worm,
3. Trojan horse.

Viruses propagate by inserting their copies into other programs or documents. Worms are standalone program the replicate themselves to another location. Worms do not use other programs and documents as carrying containers. Trojan horses are programs that have both legitimate and malicious functionality. Users typically install Trojan horses at will to utilize their legitimate functionality without knowing about their malicious functionality.

A malicious program can execute in user-mode, thus allowing them to steal and corrupt user's programs and files. More sophisticated malicious programs attack the kernel itself [34], allowing them to steal information from programs' memory and files belonging to other users. Besides, such malware is more resistant to detection by antivirus software. Hypervisor-based malware [16] improves the detection resistance even further by moving the malicious program to an isolated environment. An even better-isolated environment is the devices' firmware [9]. The CPU cannot directly access the firmware of devices, and therefore, antivirus software executing on the CPU cannot analyze the devices' firmware.

Malicious programs employ various techniques to avoid detection. Each technique is a countermeasure to a specific detection method. For example, polymorphism and metamorphism [36] are countermeasures for static analysis. Debugger and hypervisor detection techniques [1] are countermeasures for dynamic analysis. In this chapter, we present the analysis techniques and their countermeasures, allowing the reader to understand the evolution of the arms race between the two camps of researchers.

21.2 Static Analysis

The most naive approach for malware detection is checking program samples against a predefined repository of patterns. A pattern is a regular expression over instructions [27]. The patterns should be sufficiently general to describe slight malware variations but not too general to capture benign programs. YARA [28] is an example of a tool for malware classification. YARA defines malware using regular expressions, which are called YARA rules. ClamAV is an antivirus capable of using YARA rules for malware definition. Commercial antiviruses use similar repositories of malware definitions.

Each regular expression captures a narrow spectrum of malware samples. By inserting slight modifications to existing malware samples, attackers produce new malware variants resistant to previous definitions. In response, commercial antiviruses update their repositories with definitions of the new samples. This arms race between the attackers and the antivirus vendors requires the end-users to respond quickly to every new threat by updating the malware definitions.

Polymorphic malware [36], a new approach devised by the attackers, simplified the production of malware variants. Polymorphic malware consists of two parts:

1. Encrypted body: contains the malicious business-logic,
2. Decryption routine: decrypts the body before its actual execution.

When the malware replicates itself, it encrypts the body using a different key and mutates the decryption routine. The mutation engine can vary in its complexity from register re-assigning to total rewrites of the original decryption routine. The mutation engine itself remains unchanged in different copies of the malware.

Unlike polymorphic malware, metamorphic [36] malware mutates not only its body but also the mutation engine itself. The replication operation of a metamorphic malware consists of three steps:

1. Disassembly, in which the binary representation is translated to an intermediate form,
2. Random mutation, in which the intermediate form is altered,
3. Assembly, in which the altered intermediate form is translated back to the binary representation.

This replication operation allows metamorphic malware to achieve a high degree of variance between two copies.

A new class of semantic-aware detectors was proposed [6] to detect polymorphic and metamorphic malware. In these detectors, malware definitions describe the behavior of the malware rather than malware's structure. Determining whether a sample corresponds to a definition is undecidable in the general case. However, a limited number of transformations can be handled, thus making this approach efficient in practice against various variants of the Netsky and Beagle malware. Later, a general undecidable transformation was presented [27], thus outlining the limit to which malware semantic can be analyzed for classification.

Static analysis is a fast and reliable method capable of detecting malware based on malware's syntactic and semantic properties. Although the effectiveness of static analysis in modern malware is questionable, the low overhead and simplicity of this approach make it a default malware countermeasure in many security system bundles.

21.3 Dynamic Analysis

Metamorphic malware evades static analysis by obfuscating its semantics and altering its syntactic structure on each replication. However, malware's behavior remains intact between replications. This property forms the foundation for malware's dynamic analysis. Unlike static analysis, dynamic analysis [31] is immune to evasions based on obfuscation, making it the preferred choice for analyzing unknown, zero-day malware.

There are multiple perspectives for the classification of the dynamic analysis techniques. First, the analysis can be performed on-line, while the potential malware executes on the system. Alternatively, the memory can be recorded for later off-line analysis. Second, the analysis techniques vary by their privilege level, from user-mode and kernel-mode to external devices. Third, the analysis techniques vary by the information being analyzed, from system calls and functions to data movement. We address these perspectives in the following sections.

21.3.1 Memory Acquisition

Memory acquisition [19] is a process of obtaining a reliable image of the memory under analysis. In various circumstances, it may be desirable to obtain the memory image of a single process, a group of processes, or the entire RAM. After obtaining the memory image, a security specialist can analyze them using various tools, e.g., Rekall [39] and Volatility [42]. The analysis can be automated to some extent.

The most straightforward memory acquisition technique can be realized as a kernel-mode driver, which maps and dumps the entire RAM. The driver can perform the acquisition autonomously or use the operating system's services to accomplish its task. Examples of such realizations include Pmem [41], LiME [13], FTK [4] and DumpIt [26]. Despite their simplicity, kernel-mode memory acquisition techniques suffer from deficiencies. Malware that attacks the kernel itself can detect them and disrupt their operation.

Naturally, to protect themselves from malware, memory acquisition techniques moved one privilege level higher to the virtual machine monitor. LibVMI [32] provides API for the extraction of virtual and physical memory snapshots. Another solution, which is based on a thin hypervisor, is HyperSleuth [25]. HyperSleuth provides atomic memory acquisition of a running system. We note that kernel-mode memory acquisition techniques never demonstrated atomicity. Vis [43], a HyperSleuth successor, provides a more efficient implementation using Intel's EPT mechanism. A similar approach was demonstrated by Kiperberg et al. [17].

Hypervisors are a much more challenging target for evasion. However, multiple works demonstrated the possibility of hypervisors' detection, thus allowing malware to disable its malicious behavior in the presence of a hypervisor. Oleksiuk [30] demonstrated memory acquisition from the System Management Mode (SMM), which is usually active. SMM executes in a higher privilege mode, thus protecting the memory acquisition process from malware. Unfortunately, the code that executes in SMM is part of the system vendor's signed firmware. Hence, the deployment of an SMM-based system requires cooperation with the system vendor.

Finally, external devices can use DMA to acquire the system's memory. PCILeech [33] and Inception [21] are examples of such external devices. The main benefit of memory acquisition via external devices is their undetectability and immunity to malware. However, external devices suffer from two problems. The first is their inability to make an atomic memory snapshot. The second is their inability to operate when the Input/Output Memory Management Unit (IOMMU) [3] is active. IOMMU is a security mechanism, which Intel recently introduced to allow operating systems and hypervisors to construct a page-table between the IO-devices and the main memory. In particular, IOMMU can prevent external devices from accessing certain memory regions. Malware can alter IOMMU's configuration, thus preventing reliable memory acquisition by PCILeech and Inception.

The memory snapshot obtained using one of the described methods must be analyzed. The snapshot often lacks critical information stored in CPU registers, e.g., the CR3, GS_BASE, and LSTAR registers' values. Without this information, it is impossible to reconstruct the actual system state reliably. Moreover, in most cases, the snapshot provides a non-atomic view of the actual memory. For example, linked-lists, trees, and any other compound object may appear corrupted in the snapshot. Finally, the memory acquisition approach's most significant drawback is its inability to react to malware promptly.

21.3.2 Behavioral Analysis

The goal of behavioral analysis [31] is to classify a program as malicious based on its behavior. Behavioral analysis systems can monitor a single process or the entire operating system using a wide range of monitoring techniques. The observed behavior is then classified as malicious according to a predefined policy [38] or using machine-learning techniques [12]. Therefore, conceptually, the behavioral analysis system consists of two components:

1. process or system monitor,
2. analyzer.

In this chapter, we discuss only the monitoring component.

While policies provide deterministic guarantees regarding the classification, writing a policy requires a deep understanding of the overall system operation and the security risks. Erroneous policies can allow malicious programs to carry out their attacks. Therefore, machine-learning techniques become more favorable for securing systems from known and unknown malware. The purpose of the monitor is to log the events that occur during the system's execution. Monitors vary by: (a) the granularity of the logged events, (b) the additional information collected for each event, and (c) their evasions resistance.

The logged events' granularity can be high-level, e.g., process or file creation, sending a document to a printer, opening a socket. The granularity can be mid-level, which corresponds to a system call interface, e.g., on Windows, the socket opening function performs the DeviceIoControl system call. Finally, the granularity can be low-level, which corresponds to an invocation of a program's or an operating system's inner function, e.g., an invocation of a function that performs RSA encryption.

For each event, the monitor logs its identification, e.g., the system call number and possibly additional information. The additional information can include the arguments passed to the system call, the return value, and the context in which the system call was made. Gathering this information is not an easy task since arguments often include direct and indirect pointers to compound objects.

Evasion resistance refers to two properties of the monitor:

1.   Ability to protect its functionality from a direct attack by malware,
2.   Ability to hide its presence.

These prevent malware from behaving maliciously and avoiding detection. Evasion resistance is usually achieved using emulation, virtualization, or an external device.

Monitoring high-level API is a widespread technique that is not limited to malware analysis. Process Monitor from Windows Sysinternals [24] can record process, file, and registry activities. Process Monitor collects the arguments of each operation and the context of its invocation. Unfortunately, monitoring malware from within the system itself is prone to detection and evasion by malware.

In an attempt to make them less detectable, monitors were moved outside the monitored system to virtual machine monitors and emulators. The emulation-based methods construct the simulated environment in software without any assistance from the hardware. Unfortunately, perfect emulation is hard to achieve due to the complexity of the underlying hardware. Therefore, emulation-based methods, such as Anubis [23] and Bitblaze [40], can be detected by malware, as was described by Lindorfer et al. [20].

Virtualization solutions use two types of hypervisors [31]:

Type 1   Referred to as hypervisor-based solutions,
Type 2   Referred to as virtual machine-based solutions.

Type 1 hypervisors can operate directly over the hardware, whereas Type 2 hypervisors require an operating system to mediate between the hardware and the hypervisor itself. Both cases use a full hypervisor, like KVM or Xen, which runs at least one operating system acting as the malware's execution environment. Such system suffer from a high performance penalty and an increased detectability, due to the emulated devices that they provide. MAVMM [29] is another solution, based on a thin hypervisor, thus reducing its detectability.

Virtualization-based malware analysis techniques detect invocation of system calls and record their arguments. These techniques are efficient in terms of performance and cannot be easily evaded by malware. However, all these techniques classify a program as malicious based on the system calls it issues. As such, these techniques cannot adequately handle zero-day attacks because these attacks may use a previously unused system call or use a standard system call in a previously unknown way.

21.3.3 Evasion

Malicious software attempts to avoid analysis by employing a variety of evasion techniques [1]. The purpose of these techniques is to detect the presence of a monitor and disable its malicious behavior. The evasion techniques can be divided into two categories:

1. General,
2. Monitor-specific.

General evasion techniques can evade any monitor, while monitor-specific techniques evade a specific monitor or a group of similar monitors.

The two general and widespread evasion techniques are based on the assumption that the analysis is not continuous but rather terminates after a predefined amount of time. Therefore, to prevent the malicious behavior analysis, it is sufficient to postpone it by this amount of time. Alternatively, the malicious behavior can be triggered by some event—receiving a network packet, for example—thus hiding the malicious behavior during normal execution. In both cases, under the assumption that the analysis is limited in time, the malware can successfully prevent its detection.

Another general evasion technique called the "reverse Turing test" verifies the underlying operating system's genuineness. It is important to note that malware analysis is usually performed on a pre-generated synthetic operating system image. An operating system restored from this image may not have artifacts typically present in an actively used operating system. For example, an actively used Windows stores a list of recently opened documents. The reverse Turing test technique can use this observation to determine whether the underlying operating system is synthetic.

Targeted malware demonstrates its malicious activity only in a predefined environment. The environment can be defined by a specific network topology, a particular external device, or a presence of a special application. For example, Stuxnet activated its malicious behavior only in the presence of a specific industrial controller.

Monitor-specific techniques attempt to detect the monitor itself. Commercial hypervisors reveal themselves through the devices that they emulate and additional applications that they install. For example, VirtualBox installs the "vboxservice.exe" application. VMWare installs the "vmmouse.sys" driver for its emulated device. Thin hypervisors do not emulate devices; they do not require installing additional applications or device drivers. Therefore, monitors based on thin hypervisors are not vulnerable to this evasion technique.

General hypervisor detection techniques are based on timing attacks [2, 35]. These attacks execute a particular set of instructions in a tight loop and measure the average execution time. The set of instructions include at least one instruction that causes the hypervisor, if present, to handle it, thus requiring additional handling time. The time discrepancy can reveal the presence of a hypervisor. These techniques use various time-sources ranging from the CPU's tick counter to an external time server.

Emulators can be detected similarly to hypervisors using timing attacks. However, another method exists that does not require a time source. Due to the complexity of the emulated instruction-set architecture and imperfections of the emulator, some instructions may behave differently on physical and emulated processors. These discrepancies can be used by malware to detect an emulator.

21.4 Conclusion

Malware analysis and evasion techniques have evolved over the last decade to high levels of complexity. In this chapter, we outlined this evolution and the current state-of-the-art to encourage security systems vendors to introduce academic novelties to their commercial products.

References

1. Afianian A, Niksefat S, Sadeghiyan B, Baptiste D (2019) Malware dynamic analysis evasion techniques: A survey. ACM Comput Surv (CSUR) 52(6):1–28
2. Algawi A, Kiperberg M, Leon R, Resh A, Zaidenberg N (2019) Creating modern blue pills and red pills. In: ECCWS 2019 18th European conference on cyber warfare and security. Academic Conferences, pp 6–14
3. Amit N, Ben-Yehuda M, Yassour BA (2010) IOMMU: Strategies for mitigating the IOTLB bottleneck. In: Computer architecture: ISCA 2010 international workshops A4MMC, AMAS-BT, EAMA, WEED, WIOSCA. Springer, Berlin, pp 256–274
4. Carbone F (2014) Computer forensics with FTK. Packt Publishing
5. Chien E (2005) Techniques of adware and spyware. In: Proceedings of the fifteenth virus bulletin conference, Citeseer
6. Christodorescu M, Jha S, Seshia SA, Song D, Bryant RE (2005) Semantics-aware malware detection. In: 2005 IEEE symposium on security and privacy (S&P'05). IEEE, pp 32–46
7. Cost of a data breach report 2019. IBM Security, 2019
8. Egele M, Kruegel C, Kirda E, Yin H, Song D (2007) Dynamic spyware analysis. In ATC'07: 2007 USENIX annual technical conference on proceedings of the USENIX annual technical conference, Article 18. USENIX Association, pp 1–14
9. Embleton S, Sparks S, Zou CC (2013) SMM rootkit: a new breed of OS independent malware. Secur Commun Netw 6(12):1590–1605

10. Feily M, Shahrestani A, Ramadass S (2009) A survey of botnet and botnet detection. In: 2009 third international conference on emerging security information, systems and technologies. IEEE, pp 268–273
11. Gazet A (2010) Comparative analysis of various ransomware virii. J Comput Virol 6(1):77–90
12. Gibert D, Mateu C, Planes J (2020) The rise of machine learning for detection and classification of malware: research developments, trends and challenges. J Netw Comput Appl 153:102526
13. Heriyanto A, Valli C, Hannay P (2015) Comparison of Live Response, Linux Memory Extractor (LiME) and Mem tool for acquiring Android's volatile memory in the malware incident. In: 13th Australian digital forensics conference. Edith Cowan University, pp 5–14
14. Hong G, Yang Z, Yang S, Zhang L, Nan Y, Zhang Z, Yang M, Zhang Y, Qian Z, Duan H (2018) How you get shot in the back: a systematical study about cryptojacking in the real world. In: CCS '18: Proceedings of the 2018 ACM SIGSAC conference on computer and communications security. ACM, pp 1701–1713
15. Hussain M, Al-Haiqi A, Zaidan AA, Zaidan BB, Mat Kiah ML, Anuar NB, Abdulnabi M (2016) The rise of keyloggers on smartphones: a survey and insight into motion-based tap inference attacks. Pervasive Mob Comput 25:1–25
16. King ST, Chen PM, Wang YM, Verbowski C, Wang HJ, Lorch JR (2006) SubVirt: implementing malware with virtual machines. In: 2006 IEEE symposium on security and privacy (S&P'06). IEEE, 14 pp
17. Kiperberg M, Leon R, Resh A, Algawi A, Zaidenberg N (2019) Hypervisor-assisted atomic memory acquisition in modern systems. In: Proceedings of the 5th international conference on information systems security and privacy (ICISSP 2019). Scitepress, pp 155–162
18. Kirda E, Kruegel C, Banks G, Vigna G, Kemmerer R (2006) Behavior-based spyware detection. In: Security '06: 15th USENIX security symposium. USENIX Association, pp 273–288
19. Latzo T, Palutke R, Freiling F (2019) A universal taxonomy and survey of forensic memory acquisition techniques. Digit Investig 28:56–69
20. Lindorfer M, Kolbitsch C, Comparetti PM (2011) Detecting environment-sensitive malware. In: Recent advances in intrusion detection: proceedings of the 14th international symposium, RAID 2011. Springer, pp 338–357
21. Maartmann-Moe C Inception. GitHub, https://github.com/carmaa/inception
22. Malware statistics and trends report. AV-TEST Institute. https://www.av-test.org/en/statistics/malware/
23. Mandl T, Bayer U, Nentwich F (2009) ANUBIS: ANalyzing unknown BInarieS the automatic way. Presentation at Virus Bulletin Conference 2009, Geneva
24. Margosis A, Russinovich ME (2011) Windows sysinternals administrator's reference. Microsoft Press
25. Martignoni L, Fattori A, Paleari R, Cavallaro L (2010) Live and trustworthy forensic analysis of commodity production systems. In: Recent advances in intrusion detection: proceedings of the 13th international symposium, RAID 2010. Springer, Berlin, pp 297–316
26. McRee R (2011) Memory analysis with DumpIt and volatility. ISSA J 35–38
27. Moser A, Kruegel C, Kirda E (2007) Limits of static analysis for malware detection. In: Twenty-third annual computer security applications conference (ACSAC 2007). IEEE, pp 421–430
28. Naik N, Jenkins P, Savage N, Yang L (2019) Cyberthreat hunting. Part 1: Triaging ransomware using fuzzy hashing, import hashing and YARA rules. In: 2019 IEEE international conference on fuzzy systems (FUZZ-IEEE). IEEE, pp 1–6
29. Nguyen AM, Schear N, Jung H, Godiyal A, King ST, Nguyen HD (2009) MAVMM: Lightweight and purpose built VMM for malware analysis. In: 2009 annual computer security applications conference. IEEE, pp 441–450
30. Oleksiuk D (2015) Building reliable SMM backdoor for UEFI based platforms. Web log. http://blog.cr4.sh/2015/07/building-reliable-smm-backdoor-for-uefi.html
31. Or-Meir O, Nissim N, Elovici Y, Rokach L (2019) Dynamic malware analysis in the modern era: A state of the art survey. ACM Comput Surv 52(5):88, 48
32. Payne BD (2012) Simplifying virtual machine introspection using LibVMI. Sandia report SAND2012–7818, Sandia National Laboratories

33. PCILeech. GitHub (2021) https://github.com/ufrisk/pcileech
34. Riley R, Jiang X, Xu D (2009) Multi-aspect profiling of kernel rootkit behavior. In: EuroSys '09: proceedings of the 4th ACM european conference on computer systems. ACM, pp 47–60
35. Rutkowska J, Tereshkin A (2008) Bluepilling the Xen hypervisor. Presentation at Black Hat USA 2008
36. Sharma A, Sahay SK (2014) Evolution and detection of polymorphic and metamorphic malwares: a survey. arXiv:1406.7061
37. Sihwail R, Omar K, Ariffin KAZ (2018) A survey on malware analysis techniques: static, dynamic, hybrid and memory analysis. Int J Adv Sci Eng Inf Technol 8(4–2):1662–1671
38. Smalley S (2002) Configuring the SELinux policy. NAI Labs report #02–007
39. Socała A, Cohen M (2016) Automatic profile generation for live Linux Memory analysis. Dig Invest 16(suppl):S11–S24
40. Song D, Brumley D, Yin H, Caballero J, Jager I, Kang MG, Liang Z, Newsome J, Poosankam P, Saxena P (2008) BitBlaze: a new approach to computer security via binary analysis. In: Information systems security: proceedings of the 4th international conference, ICISS 2008. Springer, Berlin, pp 1–25
41. The Pmem Memory acquisition suite. Rekall Forensics (2015). http://blog.rekall-forensic.com/2015/04/the-pmem-memory-acquisition-suite.html
42. Volatility framework: Volatile memory extraction utility framework. Volatility Foundation (2020). https://github.com/volatilityfoundation/volatility
43. Yu M, Qi Z, Lin Q, Zhong X, Li B, Guan H (2012) Vis: Virtualization enhanced live forensics acquisition for native system. Digit Investig 9(1):22–33

Printed in the United States
by Baker & Taylor Publisher Services